ROYAL IRISH ACADEMY

Ireland's Academy for the sciences and humanities

NATURAL AND CULTURAL LANDSCAPES - THE GEOLOGICAL FOUNDATION

PROCEEDINGS OF A CONFERENCE

9-11 SEPTEMBER 2002
DUBLIN CASTLE, IRELAND

Edited by Matthew Parkes

Conference Sponsors

GSNI Geological Survey of Northern Ireland

GSI Geological Survey of Ireland Suirbhéireacht Gheolaíochta Éireann

RIA

An Chomhairle Oidhreachta The Heritage Council

DEPFA BANK

ProGEO The European Association for the Conservation of the Geological Heritage

UNESCO

IUGS

European Geoparks

ENVIRONMENT AND HERITAGE SERVICE

ROYAL IRISH ACADEMY
19 Dawson Street
Dublin 2

ISBN 1-904890-00-8

British Library Cataloguing-in-Publication Data.
A catalogue record for this book is available
from the British Library.

Recommended citation:
Parkes, M.A. (ed.) 2004. *Natural and Cultural Landscapes - the geological foundation. Proceedings of a Conference 9-11 September 2002, Dublin Castle, Ireland.* Royal Irish Academy. Dublin.

Printed by Leinster Leader, Naas, Co. Kildare, Ireland.

CONTENTS

** indicates the paper was presented as a poster. All others were oral presentations.*

NATURAL AND CULTURAL LANDSCAPES - THE GEOLOGICAL FOUNDATION

PROCEEDINGS OF A CONFERENCE

9-11 SEPTEMBER 2002, DUBLIN CASTLE, IRELAND

Editorial

The Conference arose as an initiative of the National Committee for Geology of the Royal Irish Academy, with the support of other science national committees, and the Academy itself. Its intent was to reach out to a wider audience of people from many different areas of life, with an exploration of the innumerable ways in which geology and geomorphology are the absolute foundation to both natural landscapes and the myriad ways in which humans have occupied and made use of the Earth's resources and thus shaped the landscape. In an ambitious way geologists attempted to communicate with anybody with an interest or a stake in the landscape, about the role of geology in the development or the management and future of the landscape that we live in. To some extent this broke the mould of normal communications about geology and the landscape, which are often only *between* geologists.

The call for papers generated a far greater response than was anticipated, and the original schedule was expanded to three days, but still many contributors could only be included as a poster, rather than an oral presentation. Whilst the focus of the meeting was the Irish Landscape, there was a strong international dimension to the topics explored through both invited speakers from overseas, and the involvement of ProGEO in the meeting, especially in one session and with additional business meetings during the Conference. Many valuable lessons can and have been learnt from the experience of the members of ProGEO throughout Europe (the European Association for the Conservation of Geological Heritage). The application of measures to protect and manage both Irish and foreign landscapes in a sustainable way were presented by participants, but an equally important theme was the participation of communities in their own landscape issues, through such approaches as European Geoparks. Taking ownership of the process of presentation, and communication about particular landscapes was also shown to be important and a variety of methods were demonstrated.

The organisation and running of the Conference could not have taken place without the fundamental input of the two Geological Surveys - the Geological Survey of Ireland and the Geological Survey of Northern Ireland, both of which have shown a strong commitment to explaining the geological foundation of the landscape to new audiences. Without the valuable financial support of the Heritage Council and the DePfa Bank (the main sponsor of the Royal Irish Academy's Conference Programme) and the other sponsors - The Environment and Heritage Service of Northern Ireland, UNESCO, International Union of Geosciences, ProGEO and The European Geoparks Network - the whole Conference could not have taken place and the organisers are especially grateful for all the support. It is also a pleasure to acknowledge the hard work of Rebecca Gageby, Laura Mahoney and Debbie McEntee of the Royal Irish Academy, in the organisation of the Conference and also the hard work of Elaine Roche, Sophie Preteseille (also for final proof reading), Debbie Harte, Aiveen Kemp and Maura Ni Cearbhaill in the running of the event. The Advisory Committee, the Organising Committee, the full National Committee for Geology all played significant roles, and we acknowledge the support provided by the National Committee for Climate Change to bring Professor Turner from the USA for his Keynote Presentation at the Conference.

Lastly, this volume contains written versions of almost all oral and poster papers from the Conference providing a more permanent record which we hope will continue to be useful and of interest to those who could not attend. All contributions have been edited, but not too 'heavily', so that the author's style and message should shine through in each case. There is also a section of colour plates, in which images from various papers are reproduced in colour, which in a few cases is necessary to understand the diagram. A few papers were not followed up by a written version for various reasons, but the original abstracts are included here within the relevant session. Any omissions of abstracts or other errors are the responsibility of the Editor. However, opinions expressed in papers within this volume are the author's own, and should not necessarily be taken to be the views of the Organising Committee, the National Committee for Geology or the Royal Irish Academy.

It should be noted that the enthusiasm and motivation of the audience to continue the mission of the Conference clearly extended beyond the final discussion of the Conference Declaration, and a Steering Group subsequently explored ways of furthering the work. The results of this are presented in the final chapter and will hopefully see you, the reader, taking further steps along the journey this book has started you upon.

OPENING SESSION

In: Parkes, M.A. (Ed.) 2004 *Natural and Cultural Landscapes - The Geological Foundation*, 11-17, Royal Irish Academy, Dublin

BOG BANK, ROCK FACE AND THE FAR FETCH OF POETRY

Seamus Heaney, Nobel Laureate, MRIA
Opening Address 1

Abstract: The address is concerned with poetry and geology as kindred ways of responding to the mystery and makings of the planet we live on, our at-oneness with it and our distinctness from it.

Beginning with observations about the achievment of poets such as Hugh MacDiarmid (who "created a language in which to express the relationship of man to stones and stars"), Heaney goes on to link the poet's task of "giving glory to things just because they are" with the geologist's, whose first concern is with "the here and nowness of the world or its there and thenness...There is a kind of artistic detachment and visionary perspective about geological discourse that match the scope and fetch of poetry itself."

Heaney goes on to explore the ways in which some poets (including himself) have written out of a geological feel for landscape and have thereby added to the cultural and political meaning of the environment. After a discussion of an English poem dealing with the "fern forests" that went to the making of the coal in the mines, he talks about the preoccupations behind his own poems about bogland and the Giant's Causeway and touches briefly on work by Wallace Stevens and MacDiarmid.

He ends with the observation:"what we settle for and are settled into by the givens of our natural and cultural landscapes can be creatively unsettled...in the words of MacDiarmid, 'Let men find the faith that builds mountains/Before they seek the faith that moves them."

What poets and geologists do, at their most heroic, can be expressed in the admiring terms once applied to the work of the Scottish poet Hugh MacDiarmid. MacDiarmid, according to the cultural commentator Neal Ascherson, went fearlessly up to the universe and wrung answers from it. He took on the cosmos single-handed and created a language in which to express the relationship of man to stones and stars. (1) To put it another way, poets and geologists at their most heroic are like W. B. Yeats in his poem, 'The man and the echo': they ask a question of the stone. (2) And I repeat 'at their most heroic' because in their more accomodating moods, they are ready to enter into a less questioning relationship with 'earth's diurnal course', as William Wordwsorth called it, happy to be rolled round in it, to recognize their adaptability to it and to celebrate their planetary at homeness 'With rocks and stones and trees.' (3)

Poets and geologists begin with sensation and move to a state that is somewhat closer to meditation. Robert Frost's account of the the writing process -- 'sight, excite, insight' -- probably applies also to the stages of geological enquiry. The work is in getting the process started, getting from the sight to the insight, from the natural encounter to the cultural accretion, from the delight to the wisdom. And in order to give more substance to what I mean by such a statement, I'll read a poem by the American poet Wallace Stevens. It's called 'The Irish Cliffs of Moher' and even before it begins, the poet's mind has been excited enough to make the move from the base of the cliff to the base of the spirit:

> Who is my father in this world, in this house,
> At the spirit's base?
>
> My father's father, his father's father, his --
> Shadows like winds
>
> Go back to a parent before thought, before speech,
> At the head of the past.
>
> They go back to the cliffs of Moher rising out the mist,
> Above the real,
>
> Rising out of present time and place, above
> The wet green grass.
>
> This is not landscape, full of the somnambulations
> Of poetry
>
> And the sea. This is my father or, maybe,
> It is as he was,
>
> A likeness, one of the race of fathers: earth
> And sea and air. (4)

Whether you are one of the first people gathered round an upright megalith or a lab technician measuring the radioactive component in a lump of pre-Cambrian rock, you are likely to be susceptible to this poem. It does indeed ask a question of the stone: 'Who is my father?' And even if the answer is not given in theological terms, we sense that the question itself is more than biological: the large mystery of origin, of how and why, is looming in the background. It doesn't much matter, from the poet's point of view or from ours, that the poem was inspired by a postcard of the cliffs rather than by a visit to the actual place. What matters is the fact that the postcard got him started on his poet's job, which is to keep facing the phenomena and to find a language and a music that match them and possess their own phenomenal carrying power.

The geologist, of course, needs a language also, but its purpose is much more technical. Its ultimate effect may be the same as the poet's -- a geologist's account of the formation of a cliff or a mountain is bound to inspire some kind of awe and prolong the moment of contemplation -- but even so, the words on the page of the geology book are there first and foremost to record and to inform rather than to promote a reverie. A poet's words, on the other hand, will have come out of deep processes of ear and association. Their very sound is an essential part of their meaning. It is because of the gravity of the utterance in Stevens's poem that we gain a sense of the sheer cliffness of those cliffs. And this musicality, so fundamental to poetry, is an inducement to mind and imagination to begin an excited reverie, to pass through the matter of Moher and into its meaning. To discover a new settlement between Moher and mind, body and being:

> Who is my father in this world, in this house,
> At the spirit's base?
>
> My father's father, his father's father, his --
> Shadows like winds
>
> Go back to a parent before thought, before speech,
> At the head of the past.

This has what I call the far fetch of poetry. This is the familiar world brought out from its familiarity, delivered from the veil of the usual, coming to and coming in upon us as something new and strange. Stevens is here fulfilling one of the poet's callings which has been defined as the job of giving glory to things just because they are, a definition which could surely be applied with equal justice to the job of the geologist. Geologists face the facts before they force the issues. Their first concern is with the here and nowness of the world or its there and then-ness, although this may well lead to questions of its how and whyness. They read the ground rules of earth itself and tell the time by the planet's body clock. In fact, there is a kind of artistic detachment and visionary perspective about geological discourse that match the scope and fetch of poetry itself.

There was, in fact, a strong geological element in one of the first poems that helped to form my own taste in poetry. I didn't read it in any of the prescribed school anthologies nor was I introduced to it by an enthusiastic teacher. It lacked, therefore, the seal of official approval and for that reason it probably came home to me with more persuasive force. I did encounter it in an English class, but in one of those exercises that used to be called "appreciation and analysis". Usually the texts presented on such occasions were extracted from books full of difficult vocabulary and complicated syntax which you were then asked to unravel and to clarify. This time, however, what was on offer was a poem by Wilfred Owen, the English poet who had written and fought and died in the trenches during the First World War. It was called 'Miners' and I believe it will repay our attention here again this morning.

The poem begins with Owen the soldier at home on leave, sitting by his coal fire, enjoying a reverie that carries him far beyond the walls of the room; it ends on a large note of sympathy and solidarity as the reverie leads the poet to realize that his comrades at the front and and the men who mined his coal have much in common. They all labour at the base of the social and economic system, and even as the system keeps overlooking them and grinding them under, their deprivations and sufferings sustain it. Yet the poem is not as politically oriented as I'm making it sound: it's not so much soapbox oratory as a fireside daydream. The sounds of coal burning in a grate, the little soft rifts and shifts of it, lead the poet's mind back into the softness of the coal's prehistory when it was vegetation beginning its slow geological transformation:

> There was whispering in my hearth,
> A sigh of coal,
> Grown wistful of a former earth
> It might recall.
>
> I listened for a tale of leaves
> And smothered ferns,
> Frond-forests, and the low sly lives
> Before the fawns.
>
> My fire might show steam-phantoms simmer
> From Time's old cauldron,
> Before the birds made nests in summer
> Or men had children.
>
> But the coals were whispering of their mine
> And the moans down there
> Of boys who slept wry sleep, and men
> Writhing for air.
>
> And I saw white bones in the cinder-shard,
> Bones without number;
> For many hearts by coal are charred
> And few remember.
>
> I thought of some who worked dark pits
> Of war, and died
> Digging the rock where Death reputes
> Peace lies indeed.
>
> Comforted years will sit soft-chaired
> In rooms of amber;
> The years will stretch their hands, well-cheered
> By our life's ember:
>
> The centuries will burn rich loads
> With which we groaned,
> Whose warmth shall lull their dreaming lids
> While songs are crooned;
> But they will not dream of us poor lads
> Left in the ground.

(5)

Here we have a poem that answers the preoccupations expressed by the title of this conference: these lines are really about the geological foundation of the natural and cultural landscapes of England. Owen may be a disaffected soul, but he is still an English patriot. You sense that his recognition of the injustice of the social system is not going to overhwhelm his attachment to the culture of the fireside. His vision of his island home may not be as radiant as John of Gaunt's in Shakespeare's play, where England is a 'scepter'd isle', a 'precious stone set in a silver sea,' (6) but Owen still loves and celebrates the dark carbon foundation of the country's landscape -- and celebrates it for its own sake, not just because it constitutes the mineral base of the country's economic prosperity. His lines express a deep *tír-ghra* - the Irish for patriotism, literally 'land-love', a tender and tenacious devotion to the ground itself. After all, the smothered ferns down in the geological strata are as important to the artistic life of the poem as the smothered groans of men suffering in the trenches and the mines. Which is to say that 'Miners' is yet another instance of imagination pressing back against the pressures of reality - in this case against the reality of what we would nowadays call the military and industrial complex. And it is worth noting that the source from which that resistant action springs is every bit as geological as it is ideological.

To the geologist, of course, coal is a youngster among the rocks, but peat is the total new-born; bogland is like the soft fontanel of the planet, although I must admit that in an early poem of my own. I thought of it as a specifically Irish rather than a general planetary phenomenon. Again, a natural landscape that had emerged from immemorial geological process was being made over into a cultural landscape. In fact, when I wrote 'Bogland', some time around 1967 or -8, I had Wilfred Owen's poem about coal and coal miners at the back of my mind. Owen's poem drifts and dreams its way back through the frond forests in much the same way as my own poem drifted down through the rush-beds and marshlands of the Irish midlands, down into the lacustrine levels that subsist beneath them. But also at the back of my mind was that need to press back: I wanted to counter the negative associations that clung round the word 'bog', and to reverse its cultural charge, go against the grain of terms like 'bogman' and 'bog-trotter' and 'bog-ignorant'. And so I ended up playing with the word 'pioneer', which originally meant someone who dug to undermine the enemy's defences.

What I wanted to undermine was the demeaning force of those terms which were the imposition of an old imperial hegemony dating back to the English Reformation and the Elizabethan conquest. This dispensation had created a vision of Ireland as a natural and cultural wilderness, and I was out to revise that. It struck me that if modern American identity was constructed on the basis of the outward bound movement of unhistoried settlers (pioneers in a new sense), and industrial England's identity constructed to a large extent on a foundation of coal (mined by pioneers of the original sort), then the quest for a modern Irish identity could be imagined as a downward dig through levels of bog that were at once geological, archaeological and historical:

We have no prairies
To slice a big sun at evening --
Everywhere the eye concedes
To encroaching horizon,

Is wooed into the cyclops' eye
Of a tarn. Our unfenced country
Is bog that keeps crusting
Between the sights of the sun.

They've taken the skeleton
Of the Great Irish Elk
Out of the peat, set it up,
An astounding crate full of air.

Butter sunk under
More than a hundred years
Was recovered, salty and white.
The ground itself is kind, black butter

Melting and opening underfoot,
Missing its last definition
By millions of years.
They'll never dig coal here,

Only the waterlogged trunks
Of great firs, soft as pulp.
Our pioneers keep striking
Inwards and downwards,

Every layer they strip
Seems camped on before.
The bogholes might be Atlantic seepage.
The wet centre is bottomless.

(7)

That last image of the wet centre, by the way, must derive to some extent from the geography lessons taught in Anahorish Public Elementary School in the late 1940s, in the course of which we were told that the shape of Ireland resembled the shape of a saucer: it had a high rim of mountains round the edge and a low-lying centre. And since most of the saucers in our house always had some tea spilled into them, the wetness of the centre was for me a natural association.

What I ask myself now, however, some thirty-five years after the poem was written, is how far it was an unconscious answer to another, very different natural and cultural landscape, one that we also heard much about in our geography classes. This was the basalt-based and battlemented north-east of Ulster, Fortress Fair Head, as we might call it. Louis MacNeice, for example, had written a poem about Belfast where he declared that 'The hard cold fire of the northerner/Frozen into his blood from the fire in his basalt/Glares from behind the mica of his eyes...' and MacNeice went on to associate this northern hardness with a 'country of cowled and haunted faces' over which 'The sun goes down with a banging of Orange drums/While the male kind murders each its woman/To whose prayer for oblivion answers no madonna.' (8) I was very conscious of this poem when I was writing 'Bogland', and of another one, 'Carrickfergus', where MacNeice talks about the pier shining with lumps of crystal salt, and where 'The Scotch Quarter was a line of residential houses/But the Irish Quarter was a slum for the blind and the halt.' (9)

What MacNeice was providing, in effect, was a high cultural overlay for a subcultural understanding which identified the north Antrim landscape with the culture of Ulster protestants. The presbyterianism of the Lowland Scots planters and the difference of their Ulster Scots language set most of the Ulster population east of the Bann apart from those west of it, politically as well as religiously, with results that were variously expected and unexpected. It meant, for example, that a landmark as geologically neutral as the Giant's Causeway could come to be identified in some shadowy way with unionist rather than nationalist Ulster. The promenade and Castle Walk at Portstewart, on the other hand, patronized as they were by a mostly catholic crowd, signified the nationalists' position -- slightly to the side, that is, and slightly to the west.

All of this was and is, as I say, very shadowy, but even if it's a debatable proposition, there's still enough in it to prompt my conjecture that the bogland poem was, among other things, an unconscious answer to that image of north-east Ulster as a basalt bawn set in a Scots-Irish sea, its rock faces stern as the face of a Tudor reformer in a Holbein portrait. My own bog banks and turf stacks, on the other hand, would probably have been regarded by some of the constituents in north Antrim as too romantically nationalist altogether, too reminiscent of the Connemara landscapes of Paul Henry, in so far as Henry's images of the west of Ireland had a cultural pigmentation as brown and fibrous as the robes of any pre-reformation Franciscan friar.

I am caricaturing the situation, but I am doing so for a purpose. Over the past forty years, everyone in Ireland, but in Northern Ireland especially, has become highly aware of these cultural signalling systems. There has been a variety of reasons for this new alertness, some having to do with the general intellectual climate which has been permanently affected by the deconstructing impulse, and some having to do with a political self-scrutiny induced first of all by the troubles and then, more recently, by the peace process. Whatever the reason for it, the awareness itself is a welcome development. Loyalties and local affections, cultural heritage and religious faith are all very well, but perspective is necessary if the very thing we all share, namely our *tír-ghra*, our love of the land itself, is not to be reduced and limited by the influence of party politics. So the far fetch of poetry is again called upon to deliver us from the veil of the usual, to find a new language to link star and stone, to bring us back 'to a parent before thought, before speech,/At the head of the past.'

Obviously, I don't mean to suggest that the local should be leached of its intimacies and cherished characteristics. I mean rather that the light in which it is viewed should be as disinterested and far-sourced as possible. The heaven on earth of the north Antrim coast, for example, can be retrieved just by going there. The natural impact of the basalt columns of the Giant's Causeway are a standing rebuke to the false consciousness that would annex them to any ideology or creed, and in the late 1980s I found another way to salute the mind-clearing and sense-widening reality of the place in a short poem which I would like to read. What happened to me at the time can be expressed, perhaps, in a line by the American poet Robert Lowell, where he says 'I found my exhaustion the light of the world.' (10) Certainly by the late 1980s I found myself, like many another, exhausted by the woeful, wasteful, repetitive violence, by the recriminations and retrenchments, at once understandable and exasperating, that marked life for everybody in Northern Ireland. We were all being dragged down. 'Heaviness of being,' I wrote at the time, 'and poetry/Sluggish in the doldrums of what happens' -- but I then went on to allow myself a hope for 'air to brighten,/Time to be dazzled and the heart to lighten.' (11)

At a personal level, the hope was fulfilled by a series of forty- eight short poems that had more light and air in them than anything I had previously written. Which is not to say that they actually deserted the ground. On the contrary, the physicality of places and things, including the Giant's Causeway, was both their light source and their launching pad. The following is no. xxxix of the series, and it's about the sensation of being enthroned among the causeway rocks in that little seat of power they call 'the wishing chair':

> When you sat, far-eyed and cold, in the basalt throne
> Of 'the wishing chair' in Giant's Causeway,
> The small of your back made very solid sense.
>
> Like a papoose at sap-time strapped to a maple tree,
> You gathered force out of the world-tree's hard ness.
> If you stretched your hand forth, things might turn to
> stone.
>
> But you were only goose-fleshed skin and bone,
> The rocks and wonder of the world were only
> Lava crystallized, salts of the earth
>
> The wishing seat gave a savour to, its kelp
> And ozone freshening your outlook
> Beyond the range you thought you'd settled for
>
> (12)

Education, Robert Frost once remarked, changes the plane of regard. And so does geology and, in the best instances, poetry. I was delighted, therefore, to find one of the above stanzas quoted as an epigraph to that admirable geological handbook, *A Story through time*. In the early stages of the correspondence about to-day's talk, I was sent Dr Patrick McKeever's account of the formation of the scenic landscapes of the twelve northern counties of Ireland, or 'Ireland (north)', (13) as he denominates the area, and once again I was impressed by the art and stealth of the whole enterprise. The book is sponsored jointly by the Geological Survey of Ireland and the Geological Survey of Northern Ireland and is funded, understandably, by the Special Support Programme for Peace and Reconciliation. I say 'understandably' because in this book we don't hear the familiar terminology that speaks of either 'Northern Ireland' or 'the north of Ireland'; what we get instead is the deliberately defamiliarizing term 'Ireland (north)' -- and 'Ireland (north)' relates the familiar landscapes *not* to the usual cultures and contentions that so often condition our responses to them, but to a far deeper background, a geological perspective, a vista of elemental strife -- millions of years of melting ice sheets, drifting continents and erupting volcanoes. This educational text, in other words, provides a thoroughly scientific account that still manages to be as thrilling as any of the old creation myths. Again, the aim is to create a kind of original awe and restore us -- not necessarily reconciled but at least newly alert, psychologically and philosophically -- to who we are and where we are. When we go to the coast or the mountain as sightseers or rock-readers, the sheer factuality of earth and stone will often bring us to our senses and induce this kind of mood -- call it what you like, Edenic or pre-Socratic or simply transfigured. Ideally, we will be moved to say what the apostles said on the mountain after they witnessed the transfiguration of their lord. Ideally we will want to repeat the simple powerful words, 'It is good for us to be here.' Once that emotion is aroused, the individual has a chance of reawakening to himself or herself and to the world. A chance to relate once again to that parent at the head of the past.

There is no greater paean to stones and stoniness as the minders and the measurers of our being on earth and being human than Hugh MacDiarmid's rhapsodic ode entitled "On a raised beach". (14) This is the poem that Neal Ascherson had in mind when he spoke so highly of MacDiarmid's achievement, and I want to let you hear some of it before I conclude by reading of a final short lyric of my own. There are many other poems I could have chosen to read and many passed through my mind as I thought about this address: W. H. Auden's temperate allegorical meditation, 'In Praise of Limestone', for example, or the French poet Eugene Guillevic's *Carnac*, his response to the rocks of his native Brittany, or the Polish poet Zbigniew Herbert's lyric in honour of the pebble as a thing of uncompromised integrity. But because of the element of manifesto it contains, MacDiarmid's work more or less chose itself.

'On a raised beach' was inspired first and foremost by the poet's experience of living on the small island of Whalsay in the Shetlands. He had gone there in the early nineteen thirties, in a state of emotional and physical exhaustion, retreating like a prophet to the desert, another middle-aged writer about to find his exhaustion the light of the world. But he might never have found a way to let that light shine had he not become friendly with a geologist called Thomas Robertson who was doing field work on the island at the same time. From Robertson, MacDiarmid gained access to the raised beaches on the adjacent island of West Linga. His biographer tells (15) how in August 1933, the poet would go across to one such beach called Croo Wick, with local men who were fetching shingle for mixing into concrete. As they rowed backward and forward, MacDiarmid stayed on the spot with his notebook and no doubt set down the first impresssions that would later be worked into the poem. But from the geologist Robertson, the poet also gained access to a technical vocabulary that commingled with and greatly extended two other vocabularies which he already possessed: one derived from the Biblical narrative of creation and redemption and the other from the theories of historical transformation and political liberation offered by Marxism. The result was a poem which should be issued to every student who opts to study geology, and that should be read in memory of every geologist who has, if you'll excuse the expression, buried the hammer.

The poem takes great artistic risks, preaching and piling up words in ways that I at first found more ridiculous than revealing. But it has an epic attack, a rhetorical reach equal to its conceptual grasp, and I therefore propose to indulge in a different kind of pile-up, a series of short, representative quotations from it. The speaker, remember, is to be imagined lying at his ease in sunlight, on one of the natural balconies of planet earth, so his mood is correspondingly elevated. "Nothing has stirred," he says, near the beginning of the poem,

Nothing has stirred
Since I lay down this morning an eternity ago
But one bird. The widest open door is least liable to
intrusion,
Ubiquitous as the sunlight, unfrequented as the sun.
The inward gates of a bird are always open.
It does not know how to shut them.
That is the secret of its song,
But whether any man's is ajar is doubtful.
I look at these stones and know little about them,
But I know their gates are open too,
Always open, far longer open than any bird's can be....
(16)

*

We must be humble. We are so easily baffled by
appearances
And do not realize that these stones are one with the
stars.
It makes no difference to them whether they are high
or low,
Mountain peak or ocean floor, palace or pigsty.
There are plenty of ruined buildings in the world but
no ruined stones....
(17)

*

I must get into this stone world now.
Ratchel, striae, relationships of tesserae,
Innumerable shapes of grey,
Innumerable shapes,
And beneath them all a stupendous unity,

Infinite movement visibly defending itself
Against all the assaults of weather and water,
Simulataneously mobilized at full strength
At every point on the universal front,
Always at the pitch of its powers,
The foundation and end of all life.
Let me try them with the old Norse words - hraun,
Duss, ronis, queedaruns, kollyarum;
They hvarf from me in all directions
Over the hurdifell.... (18)

*

They came so far out of the water and halted forever.
That larking dallier, the sun, has only been able to play
With superficial by-products since;
The moon moves the waters backwards and forwards,
But the stones cannot be lured an inch farther
Either on this side of eternity or the other.
Who thinks God is easier to know than they are?
Trying to reach men, any otherwise, than they are?
These stones will reach us long before we reach them....
(19)

*

Do not argue with me. Argue with these stones.
Truth has no trouble in knowing itself.
This is it. The hard fact. The inoppugnable reality,
Here is something for you to digest.
Eat this and we'll see what appetite you have left
For a world hereafter.
I pledge you in the first and last crusta,
The rocks rattling in the bead-proof seas.
(20)

I don't think there's any need for commentary on this. The experience of reading a poem is fleeting and fragile, but it's as real and/or as hallucinatory as the experience of eyeballing a cliff. You can put whole generations through prescribed poetry courses and find no quantifiable rise in the spiritual or emotional finesse of the population. You can bring people by the busload to Co Clare or by the boatload to Whalsay and all they may have to show for it is a snapshot that doesn't at all measure up to what they experienced. But to admit this is not to deny the reality and potential effects of the geological frisson and the poetic fetch. A piece of petrified wood on a schoolroom shelf can keep a meaning and a mystery for a lifetime: the particular specimen I'm thinking of is th subject of the following poem, which has no ambition to rise to the epic heights scaled by MacDiarmid. It's a lyric called "Relic of memory", about a piece of wood that changed my plane of regard because it had gone through a process of silicification in the waters of Lough Neagh:

The lough waters
Can petrify wood:
Old oars and posts
Over the years
Harden their grain,
Incarcerate ghosts

Of sap and season.
The shallows lap
And give and take:
Constant ablutions,
Such drowning love
Stun a stake

To stalagmite.
Deaf lava,
The cooling star,
Coal and diamond
Or sudden birth
Of burnt meteor

Are too simple,
Without the lure
That relic stored --
A piece of stone
On the shelf at school,
Oatmeal coloured.
(21)

This talk has been based on a hypothesis that might be formulated as follows: what we settle for and what we settle into because of the givens of our natural and cultural landscapes can be creatively unsettled. The given can be reimagined in the light of a world that is geological as well as historical. Even if we don't go the whole distance with the poet John Hewitt who wrote 'I have turned to the landscape because people disappoint me', (22) we can still agree with, and should even celebrate the poet MacDiarmid who wrote 'Let men find the faith that builds mountains/Before they seek the faith that moves them.' (23)

CITED WORKS

1. See Neal Ascherson's sleeve note on *Whaur extremes meet*: Hugh MacDiarmid, a record produced in 1978 by Tuatha Music Ltd, Station House, Lashan, Alton, Hampshire (TU1).

2. W. B. Yeats, *The poems, a new edition*, ed. Richard J. Finneran (London, 1983), 345.

3. "A slumber did my spirit seal," *Wordsworth, poetry & prose* (Oxford, 1921), 72.

4. Wallace Stevens, *The collected poems of Wallace Stevens* (New York, 1982), 501-2.

5. Wilfred Owen, *The poems of Wilfred Owen*, ed. Jon Stallworthy (New York, 1985), 112-3.

6. William Shakespeare, *Richard II*, Act 2, Scene 1, lines 40, 47.

7. Seamus Heaney, *Opened ground, poems 1966-1996* (London, 1998), 41-2.

8. Louis MacNeice, *The collected poems* , ed. E.R. Dodds (London, 1966), 17.

9. MacNeice, *The collected poems*, 69.

10. Robert Lowell, *Day by day* (New York, 1977), 7.

11. Heaney, *Opened ground*, 357.

12. Heaney, *Opened ground*, 383.

13. Patrick J. McKeever, Ph. D., *A Story through time* (Landscapes from Stone, 1999), 1.

14. Hugh MacDiarmid, *The complete poems of Hugh MacDiarmid*, Volume 1 (London, 1978), 422-33.

15. Alan Bold, *MacDiarmid*, (London, 1990), 339.

16. MacDiarmid, *The complete poems*, 423.

17. MacDiramid, *The complete poems*, 425.

18. MacDiarmid, *The complete poems*, 426-7.

19. MacDiarmid, *The complete poems*, 427.

20. MacDiarmid, *The complete poems*, 430.

21. Heaney, *Opened ground*, 27.

22. John Hewitt, *The collected poems of John Hewitt*, ed. Frank Ormsby (Belfast, 1991), 67.

23. Hugh MacDiarmid, *The complete poems*, 427.

In: Parkes, M.A. (Ed.) 2004 *Natural and Cultural Landscapes - The Geological Foundation*, 19-21, Royal Irish Academy, Dublin

LANDSCAPE AND THE SHADOWS OF TIMES PAST AND PLACES BEEN

Pádhraig S. Kennan, Geology Department, University College Dublin, Ireland

Opening Address 2

People of knowledge - geologists, learned in the names of stones, and conversant with stratas and basalts .... observant and particular - they love to be particular and when they excite an exclamation of wonder, they never sympathise with it, but treat it as a matter of course that you should be astonished, and they the astonishers (Mrs. and Mrs. S.C Hall. 1841-2-3. Ireland: Its Scenery, Character, etc. 3 vols., London)

INTRODUCTION

Why is a beautiful landscape universally enjoyed, and all the more so when not familiar? Why do the purple heathers and yellow furze on a Wicklow slope catch every eye? Why is the darkened mystery of a remote granite glen remembered - sometimes with a shiver - or the ruggedness of a rocky coast? Perhaps, the relaxing experience of a landscape - an encounter enjoyed by most, if not all, without quite knowing why - is simply not for analyzing. Whatever the answer, I doubt whether rock plays much part - whether more than a very few are ever conscious of that rock beneath the vegetation and the veneer of soil. I certainly doubt whether any element of the geological structure is other than rarely appreciated even when the rocks are exposed to full view - as a sunlit granite monument rising from the sea near Cruit Island in Donegal for instance, or as layered sandstones in the cliffs of North Mayo? Except for that glorious exception, the Giants Causeway in Antrim, do the rocks themselves remain effectively unseen, and afterwards unremembered, as cliffs become just cliffs and the view nothing but a view - even if enjoyed none the less for that? This teacher thinks so, for he has often experienced the response that even a short geological introduction brings.

Relax for a while on the beach in the small harbour at Loughshinney in north County Dublin on a summers day. Sun glistens on the water. Only a few fishermen unloading their catch, the cries of seagulls and happy children break the silence. Cliffs of layered limestones frame the scene to the west and a harbour wall does so to the east. Out to sea, Lambay Island lies in the near distance. Some, perhaps a little restless, may actually notice large contortions of the limestone layers and wonder. How many, however, would realise that these are the distant ripples of earth movement on a global scale away in Europe and long ago? How many would see in these rocks the fossil remains of animals that lived some 350 million years ago? How many would know that these had been entombed as the limestones were deposited on a sea floor at a time when the climate and the place was Bahama-like no less? Who, looking towards Lambay Island, would see a belching volcano spewing hot lava and ash about and who, not in the know, would believe that those lavas and ashes are still there to be seen on the Dublin coast at Donabate and Portrane 450 million years later? You see, even the casual tourist enjoying an ice cream can enter the world of the geologist with ease - a world of never-ending gradual change and, on occasion, awesome catastrophe. A walk at Loughshinny in the future may never be quite the same again. It will be a walk in the past and, at the best of times, time travel can disconcert.

A GLOBAL ODYSSEY

A history of some 2000 million years duration is archived in the rocks of Ireland. Those rocks can record a rain drop falling, a tide coming in, the sun shining, the footsteps of a animal venturing ashore, perhaps for the first time ever. They can, with no trouble at all, record the rising of mountain chains and the movements of continents. Though the history they preserve is one of boundless detail and, of course, conflicting interpretation, a few key events shape and colour the rest. It is no different from any other history - except that it goes back a bit further.

With regard to the Irish landscape as it is now, the oldest important event was the crunching collision 400 million years ago of Laurentia and Eurasia - two continents then already very old. That collision established the grain of Ireland, the NE-SW trends of, among others, the Derryveigh Mountains in Donegal and the Ox Mountains in Mayo, and is expressed in the very shape of the island. An ocean, the Iapetus Ocean, vanished as these continents met along a line that now runs from Clogher Head near Drogheda to the Shannon Estuary. No way could an ocean simply disappear without volcanoes erupting in great arcs from Lambay to the Chair of Kildare and Arklow and on down to Waterford and, over in the West, from Connemara up to Tyrone - the evidence for all is still there in one form or another. The leading edges of both continents, and any rocks that had accumulated on the ocean floor in between, were severely deformed. Some melted to later recrystallise as the granite masses of Leinster, Galway, Newry and Donegal. Metals that had collected on the Iapetus floor were, by devious geological processes, to become the ores that have made significant contributions to the Irish economy since Bronze Age times. In tandem with these gains, the local and global problems stemming from the use of geological windfalls came to be increasingly appreciated. The industrial landscape of Ireland had beginnings long ago. Hopefully, it will never speak too much of greed and waste.

The collision that married the two parts of Ireland, if they may be so termed, took place some distance south of the equator. Subsequent northward drift brought a tropical climate and a warm shallow sea in which limestones and sandstones of Carboniferous age were deposited. These had followed, and were to be followed by, red desert sandstones. The easily weathered limestones floor the Central Plain of Ireland that extends, beyond a peripheral ring of tougher older rocks, into the bays of Galway, Sligo, Dublin and elsewhere. Where those limestones have no cover, as in north Clare, the outcome is the astonishing rock-desert landscape of the Burren with all its drainage banished underground. This is a truly strange and beautiful place.

As time passed, millions of years of it, the northern margin of this limestone sea moved southwards bringing great deltas of sand to be colonized later by tropical forests. The transition from limestone to sandstone is, perhaps, hardly better seen in Ireland than on a drive southwards from the dry Burren uplands to the more homely, less well-drained landscape of south Clare and Loop Head. The forests, compressed and carbonized to coal by burial and time would, in tandem with the smelting of metal, fuel the Industrial Revolution. Small Irish remnants of that coal were mined till recently at Arigna in Leitrim, and on the Castlecomer Plateau in Kilkenny where exhausted rusting machines for crushing and washing anthracite now stand forlorn. The physical, economic and social landscapes of this windy and misty island intermesh in the rocks of warmer equatorial times.

Carboniferous times saw turmoil too. Croghan Hill, in County Offaly, is a deeply-eroded volcanic centre of this age as are the volcanic rocks that outcrop in County Limerick - some with hexagons that remind of the Giants Causeway. Things were clearly stirring at depth. In fact, across central Europe, another round of deformation was taking place - continental migrations and jostlings simply never stop. The Carboniferous and older rocks of Munster were pushed about and far-flung ripples, you will remember, reached Loughshinney. The E-W grain of Munster on geological and physical maps, the ridges of Old Red Sandstone in Cork and Kerry, the indentations of the Kerry coast, the rivers flowing west to east - all are landscape derivatives.

Weathering and erosion were to rip many pages, some possibly blank, from the Irish rock record of what happened between the end of the Carboniferous and the opening of the North Atlantic between Ireland and Greenland some 50 million years ago. That opening allowed molten basalt lavas to erupt in huge quantities - the precursors of those that flow frequently on Iceland today. The basalt makes the broad Antrim Plateau. There, the relatively young lavas overlie chalk horizons on which dinosaurs might have walked. The white chalk and its basalt cap is beautifully seen on a sunny morning on the Antrim Coast Road or, in the afternoon, across the water on Rathlin Island.

Hand in hand with the basalt eruptions, further melting at depth resulted in the granite of the Mourne Mountains. The Giants Causeway, the Antrim Plateau, the Glens of Antrim sculpted by rivers flowing off the basalt, the Mournes that sweep down to the sea - and Slemish forever associated with Saint Patrick - none of these striking landscape features would exist at all if an ocean had not opened and eruption had not happened. In their absence, the Antrim landscape might now compare with the chalk landscapes of south-eastern England. Is it not altogether remarkable how the geological responses to the movements of continents long ago can make the ups and downs and roundabouts of even a Sunday stroll, not to mind a cross-country ramble? The movement of the sea, and of the clouds overhead, can serve to remind that the rocks underfoot are moving too as continents drift about - slowly as a fingernail grows.

It was that early collision, and the subsequent journey northwards through the tropics, that created the geological framework of Ireland. It was flowing and falling water that carved the physical landscape from the rocks beneath. In Ireland, an Ice Age provided some final decoration. In the mountains of Mayo, Donegal, Wicklow and elsewhere, ice scooped out corries and cirques. Mountain glaciers deepened and straightened old river valleys - Glenmalure, Glendalough and Glenmacnass in Wicklow and Killary fjord in the West. Melting ice dumped armies of drumlins in Clew Bay and many more across Sligo, Monaghan and Down - and a surface veneer of clay and winding esker ridges of sand almost everywhere. The place would be very different place if that ice had not arrived 2 million years ago.

The ice departed about 10,000 years ago. Afterwards, as the rivers returned to work, glaciated valleys gradually lost their ice-sharpened profiles. Changing climates styled new coats of vegetation to clothe everything in turn. Man contributed to the evolving landscape by clearing vegetation, planting fields, domesticating animals, constructing defenses, damming rivers, and building networked towns and cities. Sadly, something absolutely new was to reach into every pastoral and remote place - the noise of accelerated mechanical lives.

THE GEOLOGICAL MAP - OUR UNIQUE PLACE

Every home should have a geological map with its patchwork of colours. Those colours, originally a gift from Goethe, reveal the distribution and the ages of the rocks. On the Irish map, for instance, granites are red and the Carboniferous limestones of the Central Plain are blue. Younger rocks tend to be in greens and yellows, e.g., the Antrim chalk, the New Red Sandstone under Belfast, the small, gypsum-rich, patch hidden away in Cavan. The environments of the times they saw - marine, desert, glacial or whatever - are written in the rocks themselves.

What has this map to do with landscape? A great deal, for the very simple reason that the rocks, differently-coloured on the map, weather and erode in their own different ways in the open. Limestones, soft and easily removed, are the rocks of low-lying plains and indented bays - which explains, as it were in passing, how it might be possible, with a good following wind, to free-wheel across the whole of County Westmeath. It is not chance that sees Errigal in Donegal, the Sugarloaf in Wicklow, the Bens of Connemara or the Old Red Sandstone ridges of Cork and Kerry all stand proud - tough quartzites and sandstones resist. However, as always, there are some complications. Lugnaquillia, the highest point in Wicklow, is roofed by flat-lying schist (an up-market slate). Stood on end, schist and slate, can floor a valley. Still and all, with a geological map in hand, everybody should be able to begin to appreciate why a geological history might be reflected not only in the outline and topography of an entire island, but in the distinctive character of every county, of every townland, of every parish, of every stone wall that bounds every farm field - even, perhaps, and without knowing it, in that special affinity all have with their home place.

The Stone-Age and Bronze-Age inhabitants of Ireland saw different landscapes for landscapes tend always to be remoulded to changing current needs - and by the conquests of armoured knights, bulldozers and multi-wheeled trucks. It is always, of course, a fight against climate. The island, if abandoned, would see every man-made arrangement quickly undone in defeat. But the lie of the land was never changed nor would it change on any human timescale. The Sugarloaf in Wicklow, for instance, would have been a reference for every traveller - the rock clothed differently, certainly, but always the familiar shape. Every sailor seeking shelter would have sought the same familiar places. Though climate and farming methods may have changed, the familiar slopes and hollows were always there for every farmer who worked a field. In their fashioning of familiar surroundings, the rocks of Ireland made a unique enduring heritage shared by all who ever came to live.

The rocks, and the landscapes they support, link more than every Irish person. The island only began to shape when the Atlantic opened. It was anonymous as part of that continent formed 350 million years before, and for perhaps 200 million years before that, two parts at least had been on separate journeys. Any even older history is seen just barely - there are 2000 million-year-old traces in the rocks of Mayo and, amazingly, even some 4000 million-year-old events on other continents have escaped geological reworking and erasing. What remains important, however, in the context of landscape, is that more recent events were unmistakably shared with other nearby places. The zone of collision that closed the Iapetus Ocean, continues into Britain and beyond and, to the west, in Newfoundland and Appalachia.

The geology of every county is a unique part of the geology of Ireland. The history preserved in the rocks of the entire island is, in turn, a unique fragment of a geological history that directly influences the landscapes and, in countless ways that would be easily appreciated and understood, the lives of many others not too far away. Our bit of the record is effectively something that is in our safe keeping. What if, by not wishing to understand and by failing to protect, that fragment were to become the last jig-saw piece that was found lost?

A CONCLUSION OF SORTS

During the making of 'Written in Stone' for the Geological Survey of Ireland some years ago, it proved impossible to film a view that had been part of my geological life for many years. That well-known view westwards over the Upper Lake in Glendalough, Co.Wicklow, of granite, schist, falling water and the aftermath of glacier ice, was not used. A film company had bought an option, and though the circumstances were special and pertained to professional film making, I appreciated for the first time that a popular view that I had thought open and public, and not just the postcard, could be for sale. I began to wonder about futures of increasing limitation, of toll-gates and of landscape view-bites. Impossible, I continually remind myself - but then, as a schoolboy watching international rugby, I could never have imagined future rugby-less days on which, I might, if I so wished, pay real money to play fantasy football.

The rocks store the oldest history of Ireland. For that reason no less than for the resources they deliver, they merit deep researching. However, even a little knowledge can add greatly to everybody's appreciation of the link between the contrasting landscapes of Kerry, Leitrim, Derry, of every other county and parish, and the rock architectures beneath. The landscapes and the rocks are both treasures for everyone to appreciate. For that reason, I hope the future sees a wealth of explanations, of films and web pages made, of travelling lecturers dispatched to every corner, and well-funded, public educational centres opened. And, for certain, I hope never to see any unreasonable qualification on the freedom to enjoy the Irish landscape on one's own two feet - and, especially, in total ignorance of what anyone might claim it all means.

SESSION 1

HOW DID IRELAND'S LANDSCAPES EVOLVE?

Lough Doon, Dingle Peninsula, Co Kerry, where the rock lip of the corrie exhibits spectacular striations and grooves, gouged out by other rocks at the base of the ice glacier as it flowed from the corrie. Photograph: Matthew Parkes

In: Parkes, M.A. (Ed.) 2004 *Natural and Cultural Landscapes - The Geological Foundation*, 25-27, Royal Irish Academy, Dublin

IRELAND'S BEDROCK - A MAGICAL MOSAIC

Paul D. Ryan, Department of Earth and Ocean Sciences, NUI, Galway, Ireland.
Email: Paul.ryan@nuigalway.ie

KEYNOTE ADDRESS

Abstract: The Geology of Ireland evolved over two eons (2,000,000,000 years), some 20% of all of time. It is about one million times older than Irish culture. However, like our culture, this evolution is recorded in printed works, but these are not books, they are maps - the Geological Map of Ireland. This article introduces this map and shows how it records a history of the various environments this small country has experienced during its journey through deep time.

INTRODUCTION

The Irish crust, in spite of its extreme antiquity (circa 2.0Ga, 1Ga is equal to 1,000 Ma or 1,000,000,000 years), is not unusually old for continental crust. Remarkably, however, Ireland has been situated near the margins of a tectonic plate for approximately 55% of this time. Plates are the 120km thick rigid shells, called the lithosphere, that comprise the outer layer of the earth, which move to facilitate the cooling of the hot, viscous interior. Their rigidity means that their interiors are geologically quiet, whilst their margins are sites of major tectonic features such as volcanoes and earthquakes. Where plates move away from each other oceanic lithosphere is generated out of heavy volcanic rock, basalt. Where plates move towards each other, the heavy oceanic lithosphere returns to the earth's interior, melting as it goes, to produce lighter andesite and granite which create continental lithosphere. This is happening in the Andes today. The heavy basalt sinks low to form ocean basins which trap sediment, whilst the lighter continental rocks form the landmasses and the continental shelves. Where plates actually collide the continents are thickened and thrust up into great mountains, such as the Himalayas, with rocks in their interiors becoming metamorphosed by heat and pressure. Ireland, although currently lying safely in a plate interior, has enjoyed a cataclysmic past as it has been in the centre of two supercontinents, Rodinia and Pangea (Figure 1).

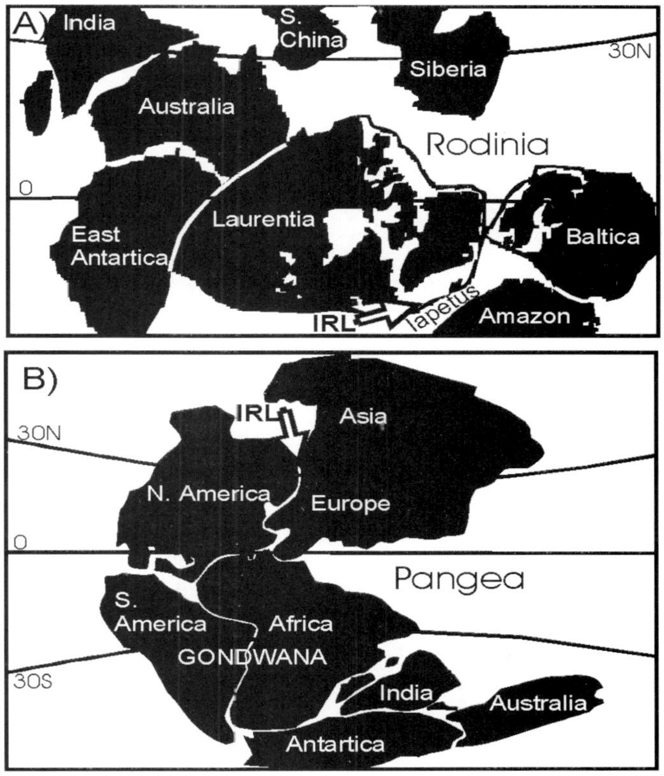

Figure 1a. A palaeogeographic reconstruction of Rodinia at 750 Ma.
Figure 1b. A palaeogeographic reconstruction of Pangea at 200Ma

THE BIRTH OF RODINIA

The oldest rocks lie in Ulster where, approximately 2.0 Ga igneous rocks were created by the under-thrusting of one tectonic plate beneath another. These rocks were part of an ancient continental mass, called Laurentia made up of North America and Greenland. For a period from about 1.3 Ga to 0.9 Ga Laurentia docked with the other continents of the world to form a super continent, Rodinia (Figure 1). The collisions of the plates to form this early super-continent created great mountains during the Grenvillian Orogeny. Traces of this event can still be found in north Mayo at Bellmullet. After the Grenvillian event, northern Ireland lay south of the equator in the Arequipa embayment on the west side of South America. Rodinia experienced two glaciation events, the oldest of which (750 Ma) may have frozen the oceans of the earth, glacial deposits similar to those created by the Pleistocene glaciations are found in Donegal and Connemara although they were then in the tropics! About 600 Ma an ocean opened, the Iapetus Ocean, splitting this super-continent. Figure 1 shows Rodinia breaking up and the Iapetus ocean beginning to form. Many of the rocks of Ulster and northern Connaught were formed during this time (Figure 2 - Neoproterozoic).

Life in the oceans was comprised of multi-celled organisms like jelly-fish, that did not have a hard skeleton. The large amounts of sand that were deposited in the opening ocean basins now provide a source of quartz and aggregate, and contorted limestones have now become the famous Connemara marble. The synchronous volcanism produced considerable deposits of gold.

THE STORY OF IAPETUS

The southern tip of Ireland was part of a micro-continent, Avalonia formed by a 600 Ma Andean style mountain building event. This lay on the other side of the Iapetus ocean at the margin of Gondwana, a continent comprising Africa, South America, Antarctica and Australia (Figure 1b). The Iapetus ocean, which was probably more than 2000km wide, closed (500-420 Ma) with Laurentia colliding with Avalonia. The rocks of Ireland from Clew Bay to Leinster were forged from the volcanoes and sediments that formed during this event which we term the Caledonian orogeny (Figure 2 - Lower Palaeozoic). At the close of this orogeny Ireland became a site of extreme volcanic activity with large granite batholiths being emplaced in north, west and south (Figure 2 - granite). These would have been capped by large cone shaped volcanoes such as those found in the Andes today.

The first hard shelled marine organisms appeared rapidly at the start of the Cambrian (543 Ma). One of nature's great unsolved mysteries. The volcanic rocks, sediments and granites provide a source of base metals, aggregate and architectural stone.

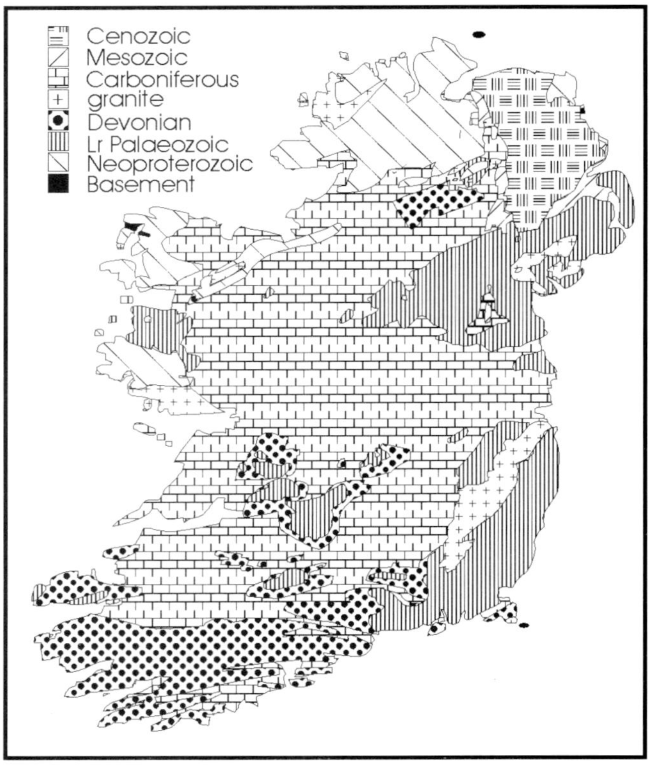

Figure 2. A simplified map of Ireland's bedrock geology.

THE FORMATION OF PANGEA

Ireland then became involved in the formation of a second super-continent, Pangea (Figure 1b). Oceans to the south of Avalonia (in France and Spain) closed to dock Laurentia and Gondwana. This event, the Variscan orogeny (405-280 Ma), involved the formation of deep basins, which contain the red Devonian sandstones of Munster and the ubiquitous grey Carboniferous limestones with their mineral wealth, and the formation of the mountains of Munster as part of Himalayan style ranges in Europe (Figure 2 - Devonian and Carboniferous).

Life began to colonise the land just prior to the Devonian period and by the Carboniferous trees had evolved to cover parts of this forming super continent in great forests which later became fossilised as coal deposits. Perhaps the world's earliest four legged animal footprints are preserved in Valentia Island. The limestones that formed in the shallow seas had much bacterial activity which produced sulphur that reacted with metals in the volcanic Lower Palaeozoic basement to form our considerable lead and zinc deposits. The limestones are an important source of aggregate and when mixed with associated muds can be turned into cement.

THE ATLANTIC IS BORN

The break-up of Pangea took place during the 'Age of the Dinosaurs' (180-64 Ma) led to the formation of the deep Atlantic basins that surround this island. In Greek mythology, the giant Iapetus is the father of Atlantis, and remarkably the new ocean opened down the line of the old, this may be due to weak rocks being thrust deep into the lithosphere during the final collisions associated with the closure of Iapetus. The warm seas that existed in a 'greenhouse earth' at this time teemed with life that is now trapped in them as petroleum deposits. Most of these rocks exist in basins offshore, however some are preserved in Antrim (Figure 2 - Mesozoic). Ireland separated from Canada, an event recorded in the 60 Ma volcanic rocks of the Giants' Causeway formed when the volcanoes of Iceland were near our shores (Figure 2 - Cenozoic, alternatively termed the Tertiary). This event uplifted the western seaboard and rejuvenated our scenery. The most recent geological event is the Pleistocene glaciation (2.0 Ma to 10,000 years) that has sculpted most of our landscape. We now live in a warm inter-glacial period and the ice may return in a few thousand years. We don't know. However, we theorise that the Atlantic will close again and that our map will keep changing.

The Mesozoic-Cenozoic boundary was the time of the extinction of the dinosaurs. Many palaeontologists believe that they did not become completely extinct, but were the ancestors of the birds. The giant land animals of the late Mesozoic did not survive the climate changes associated with the opening of the Atlantic. A meteorite collision in Chicxulub, Gulf of Mexico at exactly this time is widely believed to have contributed significantly to this process. What is perhaps less well known is that a fragment of this meteorite impacted in the North Sea and maybe Ireland was assaulted by the huge resultant tsunamis. Ireland's offshore petroleum deposits formed in the offshore basins that opened at this time.

The rejuvenation and sculpting of Irish topography in geologically recent times, plus the spectacular lava flows of the Giant s Causeway have produced one of Ireland's most valuable resources, our scenery. The growing Geotourism Industry reflects how much this is valued. One of the great challenges that faces society is how we continue to manage the exploitation of our mineral wealth and preserve this island's natural beauty formed over eons.

FURTHER READING:

Figure 2 is based on 1:750,000 National Geology Map published by the Geological Survey of Ireland (www.gsi.ie)

A good description of the process of plate tectonics can be obtained from the United States Geological Survey website at
http://pubs.usgs.gov/publications/text/understanding.html

A clear overview of the geological evolution of the island of Ireland and its scenery is given by:

Williams, M.D. and Harper, D.A.T., 1999. *The making of Ireland : landscapes in geology.* Immel, London, 98pp.

Detailed more academic accounts of the geology of Ireland can be found in:

Holland, C.H., 2001. *The geology of Ireland,* Dunedin Academic, Edinburgh, 531pp.

Woodcock, N.H. and Strachan, R.A., 2000. *Geological history of Britain and Ireland.* Blackwell Science, Malden, MA, 423pp.

In: Parkes, M.A. (Ed.) 2004 *Natural and Cultural Landscapes - The Geological Foundation*, 29-32, Royal Irish Academy, Dublin

THE ICE AGE INHERITANCE OF THE IRISH LANDSCAPE

Jasper Knight, formerly Glacial Research Group, University of Ulster now at Department of Geography, Loughborough University, UK.

Abstract: Upland, lowland and coastal landscapes of Ireland record the effects of glaciation during the late Midlandian period (~ 25,000-13,000 years BP). Upland landscapes (> 200 m OD) have glacially-smoothed and streamlined bedrock surfaces and show such characteristic landforms as corries, cross-valley moraines and erratics. Lowland landscapes (< 200 m OD) are mainly covered by glacial drift and dominated by drumlins and ribbed (Rogen) moraines, which may be overlain by eskers, end moraines and outwash fans formed during ice retreat. Coastal landscapes may contain features characteristic of both upland and lowland glaciation, masked by the effects of later changes in sea-level which have reworked glacially-carried sediment into sand dunes, estuaries, and present-day and raised beaches. The glacial inheritance of these distinctive upland, lowland and coastal landscapes has also given rise to different ecological, archaeological, agricultural and settlement patterns, showing the ways in which human activity has adapted to the physical nature of these landscapes. The close cultural connection between the Irish landscape and its peoples is an evolutionary and symbiotic one, and is responsible for the distinctive cultural 'character' found in upland, lowland and coastal landscapes of Ireland.

INTRODUCTION

As a whole, the present-day landscapes of the island of Ireland are celebrated for their scenic diversity, greenness and long history of human occupation (Schouten 1994; Mitchell and Ryan 1997). The distinctive landscapes of individual regions, such as the Burren, Connemara, Joyce's Country, the Causeway coast and Dingle peninsula, are also linked very closely to the physical characteristics of the underlying rocks and surface sediments which give the landscape its overall shape (morphology) (Orme 1970, Mitchell and Ryan 1997).

Any landscape can be considered an amalgam of three elements (geological, geomorphological, and human) which develop over different spatial and temporal scales (Fig. 1). Geological landscape elements include rock type and rock structures that are developed on and have continuity over entire regions, often forming fault-bounded or geologically-discrete upland blocks such as the Mourne (Co Down), Derryveagh (Co Donegal) and Ox Mountains (Co Sligo). Geomorphological landscape elements are developed upon and in response to the geological basement. In Ireland, typical geomorphological features include drumlins, eskers and other glacial landforms, sand dunes and soil and river patterns. Human landscape elements, including patterns of settlements, roads, agriculture and archaeological features, are developed on a local scale upon and in response to broader landscape patterns imposed by the geomorphology. The time-scale and spatial-scale over which these three landscape elements develop therefore decreases upwards (Fig. 1). The close connection between these three elements - Ireland's Ice Age inheritance - is demonstrated by the glacial geomorphic terms drumlin and esker, both of which are derived from the Irish language (Charlesworth 1957). Ireland can therefore be considered not only the 'home' of glacial geology, but also a landscape in which human presence and development is most closely connected to this glacial inheritance (Aalen 1977; Mitchell and Ryan 1997).

During the last (late Midlandian) Ice Age (~ 25,000-13,000 years BP) glacial ice occupied the northernmost two-thirds of the island of Ireland, spreading radially outwards from dispersal centres in northern and western Ireland towards coastal lowlands and onto adjacent continental shelves (Fig. 2a). As the ice sheet grew and decayed in tandem with hemispheric-scale climate changes, distinctive assemblages of landforms were developed in upland, lowland and coastal regions of Ireland (Fig. 2b). Because of the recent geological age of the late Midlandian glaciation, and its geographical extent across much of Ireland, these landform patterns are still strongly imprinted on the landscape today and are a first-order control on overall landscape morphology (Knight 2000).

This paper aims to briefly describe the range of glacial features found in upland, lowland and coastal landscapes, focusing in particular on examples from northern and western Ireland where these features are often best demonstrated. The development of human activity and cultural 'character' in these landscapes is dependent on the provision and distribution of natural resources (including, soils, vegetation and water) which are strongly influenced patterns of glacial features.

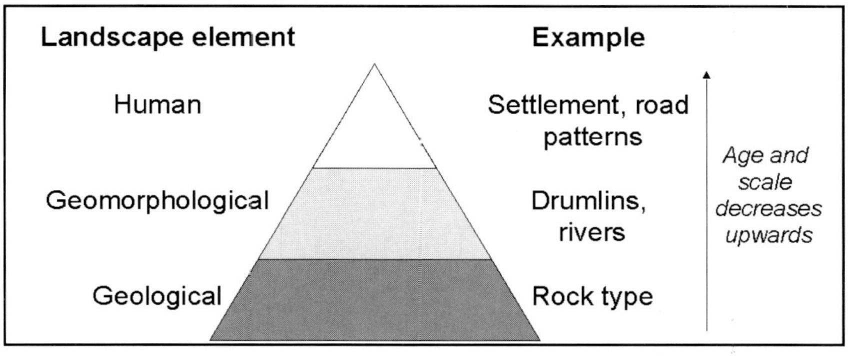

Figure 1. Diagram illustrating the relationship between geological, geomorphological and human landscape elements, with examples of each.

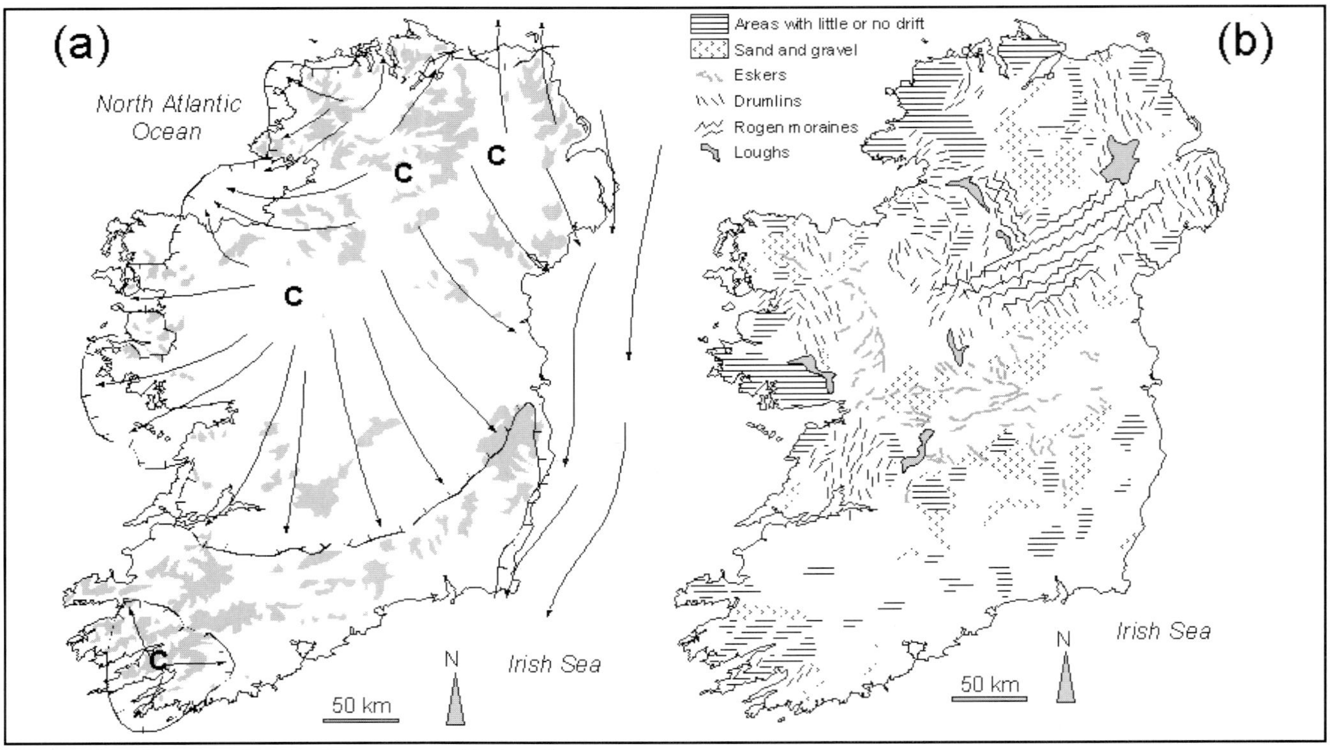

Fig. 2. (a) Map showing the ice centres (C), ice flow vectors (arrows) and maximal ice limits (ticked line) of late Midlandian ice in Ireland (after McCabe, 1985). The shaded area indicates land above 200 m OD. (b) Map of Ireland showing the distribution of glacial landforms as indicated in the key (adapted with additions after McCabe 1985).

UPLAND LANDSCAPES

Upland areas (> 200m OD) of Ireland are underlain mainly by igneous and metamorphic bedrock. These upland areas first experienced the effects of glaciation as ice accumulated in snowbanks and corries. Today, upland areas are characterised by the general absence of glacial drift (except in valleys) and dominance of bare bedrock surfaces, such as in Connemara (Co Galway), Fanad and Lough Finn areas (Co Donegal), and Killarney (Co Kerry). These bedrock surfaces are often smoothed, rounded and streamlined as a result of subglacial abrasion. Large-scale features developed in these landscapes include whaleback-shaped hills, rock drumlins and roches mountonnées, which give a rounded and rolling appearance to the upland profile (Fig. 3). These landforms are oriented in the direction of ice flow, which may be topographically controlled. Upland corries are often associated with U-shaped trough valleys (former valley glaciers) which may contain cross-valley moraines and other ice-retreat features. Corries are particularly well developed at high levels in the Nephin Beg mountains (Co Mayo) and Macgillicuddy's Reeks (Co Kerry). Former valley glaciers occupied many upland and coastal valleys, such as Glenveagh, Poisoned Glen, Lough Swilly (all Co Donegal) and Killary Harbour (Co Mayo). Large erratic boulders, marking ice limits, often litter plateaux such as The Rosses (Co Donegal) and Ox Mountains.

Upland areas located outside or above ice limits may have jagged or broken crests as a result of periglacial weathering. Loose material detached by periglacial activity can potentially accumulate as rock glaciers or screes, which are well developed on the friable quartzite bedrock slopes of Muckish and Errigal (Co Donegal).

As a result of (mainly) glacial erosion processes, upland landscapes today are drift-poor and largely reflect the orientation and disposition of underlying bedrock structures. Upland peat bogs are often developed on these bedrock surfaces and support distinctive types of vegetation and agriculture, and patterns of fields and land-holdings. Associated with the easily-exploited bedrock resources and open vistas of upland areas are found Neolithic to Iron Age archaeological sites of ritual, symbolic and defence importance (Mallory and McNeill 1991).

LOWLAND LANDSCAPES

Lowlands (< 200 m OD) dominate large areas of the western, central and northern midlands of Ireland (Fig. 2a), and are underlain mainly by sedimentary bedrock (Ordovician and Silurian sands and grits and Carboniferous limestones). Surface sediments comprise mainly late Midlandian glacial till and water-sorted sand and gravel. Total drift thickness varies considerably but can be over 50m, which may reflect the easily-eroded bedrock beneath. Glacial till and sand/gravel were deposited mainly during ice advance and ice retreat stages respectively. During ice advance, landforms such as drumlins and Rogen moraines developed beneath the ice (Fig. 4), parallel and perpendicular to ice flow direction respectively. These bedforms, which number many thousands, are a significant landscape component across lowland Ireland (Knight 1999) and are composed mainly of glacial till, although bedrock and sand and gravel are also found within them. The widespread distribution and regular spacing of drumlins and Rogen moraines gives

Figure 3. The upland landscape near Glencolumcille, Co Donegal. Ice flowed from left to right down the centre of the valley, shaping drumlins on the valley floor. Valley sides are developed in bare bedrock with scattered surface erratics (foreground).

a distinct 'pattern' to the Irish lowland landscape (Knight 1999), which is accentuated by the presence of wetter inter-drumlin hollows and loughs (Orme 1970). Drumlins and Rogen moraines are particularly well developed in Counties Down, Armagh, Monaghan, Cavan and Roscommon. Landforms comprising sorted sand and gravel are also found in lowland areas. These landforms accumulated beneath the ice (as eskers), at the ice margin (moraines) and in front of the ice margin (outwash spreads, kames and deltas). Eskers are located particularly across the Irish midlands (Warren and Ashley 1994) (Fig. 2b).

Both glacial till and sand/gravel landforms have a positive relief (i.e. are upstanding in the landscape), and thus determine the topography and river patterns of lowland areas. The presence of lowland accumulations of sediment have also acted to divert the flow direction of rivers, such as the River Bush (Co Antrim). Further, the relief and composition of these landforms also impact strongly on other landscape characteristics such as soil and vegetation type, which are manifested today in distinctive patterns of agriculture. For example, the lush grass found on drumlins in north central Ireland has given rise to the dairying agriculture of that region (Orme 1970). The physical nature of lowland landscapes has also impacted other present-day landscape elements, including road and settlement patterns (Knight 2001). In drumlin areas roads follow the dryer hillslopes and farms are located, often one per drumlin, on hill crests. Eskers also serve as dry, raised routeways through areas of lowland raised bog, and provided ancient east-west communication routes across the Irish midlands (Mitchell and Ryan 1997). The tree-less lowlands and dense patterns of settlements and land-boundaries also attest to the long period of human occupation and agricultural exploitation of the landscape (Mallory and McNeill 1991). Typical archaeological features of lowland areas include raths and crannogs (lake settlements). Raths and similar structures are often located in association with lowland sand and gravel (Aalen 1977).

COASTAL LANDSCAPES

Areas located near present-day coasts can show the glacial features characteristics of both upland and lowland landscapes. Corries, usually found in upland areas, can be found also near present sea-level on Achill Island (Co Mayo), indicating substantial contemporary depression of the snowline (Farrington, 1953). Likewise, drumlins are found as 'drowned' features in Strangford Lough (Co Down) and Clew Bay (Co Mayo). Glacial sediments were deposited and these landforms shaped when relative sea-level was over 100 m lower than today and ice margins extended off present-day coasts. Following ice retreated from these shelf positions, glacial sediments were reworked in a shoreward direction as relative sea-level recovered and transgressed across this barren landscape (Knight 2000). The extensive sandy beaches, sand dunes and estuaries that characterise much of Ireland's northern and western coasts (Fig. 5) were formed as sea-level stabilised from ~ 6000 years BP onwards (Carter et al. 1989). The close association of these landforms suggests they are related to a common forcing mechanism, in this case sediment reworking between the nearshore, onshore and intertidal zones (Knight 2000). Good examples of sandy beaches and sand dunes are located at the mouths of the rivers Bann (Co Londonderry), Erne (Co Sligo) and Gweebarra (Co Donegal). Dunes are also found associated with spits, such as Inch and Rossbeigh (Dingle Bay, Co Kerry) and Mullinasole (Co Donegal).

The physical resources of coastal landscapes include fresh water (from incoming rivers), freshwater and marine fish, shellfish and seaweed. Coastal locations also offer excellent trading, defence and settlement sites. A range of examples from the north of Ireland can illustrate human exploitation of these coastal and intertidal resources. One of the earliest dated sites of the Irish Mesolithic occurs at the tidal limit of the River Bann at Mountsandel (Woodman 1978), located on a promontory where the river narrows. In Strangford Lough, Mesolithic to Bronze Age oyster middens and medieval fish traps record exploitation of the intertidal flats and ebb-tide currents between the drumlin

Figure 4. The lowland landscape near Dungannon, Co Tyrone. Drift-dominated drumlins form a regular pattern in the landscape and are associated with well-drained agricultural land.

islands (McErlean et al. 2002). The dunes around the estuary of the Gweebarra River contain artefacts that suggest the dunes were developing 2000-3000 years BP (Shaw and Carter 1994). Such dynamic activity in the coastal zone demonstrates the combined Ice Age inheritance of abundant glacially-carried shelf sediment, and postglacial changes in relative sea-level.

CONCLUDING REMARKS

The Ice Age inheritance in Ireland is manifested differently in upland, lowland and coastal landscapes. Broadly, upland landscapes are dominantly erosional, lowland landscapes are dominantly depositional, and coastal landscapes show the effects of postglacial reworking and modification of glacial sediments and landforms (Knight 2000). In addition, the surface distribution of different types of glacial sediment (glacial till, sand and gravel) strongly influences the provision of natural resources such as soil type and water availability, and different ecological, archaeological, agricultural and settlement patterns. The Irish landscape can therefore be considered a palimpsest of geological, geomorphological and human landscape elements (Fig. 1), reflecting the processes associated with glaciation and climate change during the late Midlandian and Holocene.

Considering the Ice Age inheritance of the Irish landscape - and some of its impacts on the landscape's present-day characteristics - is useful because it has implications for certain landscape management problems. These may include issues such as the boundaries of administrative and/or political areas, landscape resource conservation and exploitation, regional economic planning, sense of regional identity and distinctiveness and tourism. The nature of the physical landscape can therefore impact on human culture in very subtle ways, also manifested in regional patterns of dialect, art, song, literature and folklore. Collectively, these elements give rise to the distinctive cultural 'character' of discrete upland, lowland and coastal landscapes of Ireland (Knight 2001). Examples may include Inishowen (Co Donegal), Tory Island (offshore Co Donegal), Rathlin Island (offshore Co Antrim), Connemara (Co Wicklow) or the Dingle peninsula (Co Kerry).

Figure 5. The coastal landscape of western Co Donegal. Upland areas (foreground) are drift-free whereas lowland areas are choked with sand in the form of sand dunes and estuaries. This sand was reworked onshore during late Pleistocene and Holocene sea-level rise.

The close cultural connection between the Irish landscape and its peoples is an evolutionary and symbiotic one. Effective management of regionally-distinct upland, lowland and coastal landscapes may be important in both conserving the Ice Age inheritance of Ireland in these key areas, and in maintaining the cultural 'character' of these glacial landscapes.

REFERENCES

Aalen, F.H.A. 1977. *Man and the Landscape in Ireland.* Academic Press, London.

Carter, R.W.G., Devoy, R.J.N. and Shaw, J. 1989. Late Holocene sea levels in Ireland. *Journal of Quaternary Science*, 4, 7-24.

Charlesworth, J.K. 1957. *The Quaternary Era.* Edward Arnold, London.

Farrington, A. 1953. Local Pleistocene glaciation and the level of the snow line of Croaghaun Mountain in Achill Island, Co. Mayo, Ireland. *Journal of Glaciology*, 2, 262-267.

Knight, J. 1999. Problems of Irish drumlins and Late Devensian ice sheet reconstructions. *Proceedings of the Geologists' Association*, 110, 9-16.

Knight, J. 2000. Geology, glaciation and the development of landscape character in Ireland. *Geology Today*, 16, 148-153.

Knight, J. 2001. A geocultural classification of landscapes in Northern Ireland: implications for landscape management and conservation. *Tearmann*, 1, 113-124.

Mallory, J.P and McNeill, T.E. 1991. *The Archaeology of Ulster.* Institute of Irish Studies, Queen's University, Belfast.

McCabe, A.M. 1985. Glacial geomorphology. In: Edwards, K.J. and Warren, W.P. (eds) *The Quaternary History of Ireland.* Academic Press, London, 67-93.

McErlean, T., McConkey, R. and Forsythe, W. 2002. *The Maritime Cultural Landscape of Strangford Lough.* Blackstaff Press, Belfast.

Mitchell, F. and Ryan, M. 1997. *Reading the Irish Landscape.* 3rd ed. Town House Press, Dublin.

Orme, A.R. 1970. *The World's Landscapes. 4. Ireland.* Longman, London.

Schouten, M.G.C. 1994. Nature conservation in Ireland: a critical decade. *Proceedings of the Royal Irish Academy*, 94B, 91-95.

Shaw, J. and Carter, R.W.G. 1994. Coastal peats from northwest Ireland: implications for Late-Holocene relative sea-level change and shoreline evolution. *Boreas*, 23, 74-91.

Warren, W.P. and Ashley, G.M. 1994. Origins of ice-contact stratified ridges (eskers) of Ireland. *Journal of Sedimentary Research*, A64, 433-449.

Woodman, P.C. 1978. *The Mesolithic in Ireland: Hunter-Gatherers in an Insular Environment.* British Archaeology Reports, British Series 58, Oxford.

In: Parkes, M.A. (Ed.) 2004 *Natural and Cultural Landscapes - The Geological Foundation*, 33-35, Royal Irish Academy, Dublin

THE EVOLUTION OF THE BASALTIC LANDSCAPE OF NORTHEAST IRELAND

Paul Lyle, School of the Built Environment, University of Ulster, Jordanstown, County Antrim, Northern Ireland, BT37 0QB.

Abstract: The basalt lavas of the Antrim Plateau cover approximately 2500km^2 in northeast Ireland and are the result of volcanic eruptions in the Tertiary Period about 60 million years ago, as North America separated from Europe and the North Atlantic Ocean opened and widened. The process of sea-floor spreading continues and Iceland today shows many of the volcanological processes that formed the Antrim basalts. This basaltic landscape is unique in Ireland, unlike the other peripheral upland areas of the island, the Antrim Plateau is formed of younger, post-Carboniferous rocks. The stages in the development and evolution of the Antrim Plateau are examined and interpreted. An initial eruptive phase produced the Lower Basalts after which there was a prolonged period of volcanic quiescence before a second cycle of eruption produced the Upper Basalts. In the dormant period between the Lower and Upper Basalts, localized volcanism in north Antrim produced the lavas of the Giant's Causeway area. The unique features of the eruption environment responsible for the distinctive columns at the Giant's Causeway are discussed. A result of the prolonged pause in volcanic activity across northeast Ireland was the formation of residual deposits of bright red soil known as laterite at the top of the Lower Basalts. When volcanism re-commenced in Upper Basalt times, this soft laterite was baked in places by contact with the hot lava to form a hard brittle material called porcellanite. This provided the raw material for the Neolithic axe factories at Tievebulliagh and at Brockley on Rathlin Island. The alteration from fresh basalt to laterite to porcellanite is an excellent example of the progressive creation of geological resources in the crust. Following exploration and discovery, such resources become available for human exploitation, after which there are discernible changes in landscape and lifestyle.

INTRODUCTION

The basalt lavas of northeast Ireland cover approximately 2500km^2 in northeast Ireland and are the result of volcanic eruptions in the Tertiary or Neogene Period, about 60 million years ago, as North America separated from Europe and the North Atlantic Ocean opened and widened. This process is referred to as sea-floor spreading and continues today. The active volcanism that occurs on Iceland results from its position on the Mid-Atlantic Ridge and the continuing expansion of the North Atlantic at the rate of a few cm per year.

Many of the volcanic processes seen in contemporary volcanic areas such as Hawaii or Iceland are similar to those that produced the Antrim Lava Group. The Antrim Plateau consists of a thick succession of lava flows from a series of elongate cracks or fissures, a landscape unique in Ireland, as unlike the other peripheral upland areas of the island, these hills are formed of relatively young, post-Carboniferous rocks.

FORMATION OF ANTRIM LAVA GROUP

The geochemistry and volcanology of the Antrim Lava Group have been discussed by a number of authors, notably Tomkeieff (1940), Patterson (1955), Lyle (1980 and 2000) and Lyle and Preston (1993). The volcanism in Antrim occurred in two cycles (Lyle,1980). In the first cycle of eruption an initial explosive phase was followed by widespread fissure eruptions to produce the Lower Basalt Formation (LBF). The end of this phase was marked by a prolonged period of volcanic quiescence before renewed activity at the beginning of the second cycle produced the Upper Basalt Formation (UBF). See Fig.1, stratigraphy of the Antrim Lavas. A consequence of the pause in volcanic activity at the end of Lower Basalt times was a deep weathering of the Lower Basalt surface under warm wet conditions. This weathering produced a widespread and thick deposit of material called laterite, the Interbasaltic Formation (IBF), a bright red soil formed by the leaching out of many of the basalt components such as silica, to leave a residual deposit rich in iron and aluminium. The development of these laterite deposits was to prove significant in the cultural and industrial development of this part of Ireland.

Formation of the columnar basalts of the Giant's Causeway area

In the dormant period, localized eruptions in north Antrim produced the characteristic lavas of the Giant's Causeway which formed in a different environment of eruption than the Lower and Upper Basalts, thus accounting for their characteristic columnar structure. These flows differed from the preceding Lower Basalts in that they were high volume flows, erupted at high effusion rates to produce flows commonly 30m thick compared with 5-10m thick flows in the Lower Basalts. In addition they show complex jointing patterns as best seen at the Giant's Causeway. Here a typical flow consists of a lower regular set of columns, usually near vertical and often 6-sided as seen at the Causeway itself. This zone passes abruptly up into a zone of irregular narrow, often curved flows that in turn passes up into a crudely columnar zone of widely spaced columns. Tomkeieff (1940) described the multi-tiered columnar jointing of the Causeway flows by adopting the terminology of classical architecture (Fig.2). The regular often vertical columns at the base of the flow he likened to the colonnade while the upper curvi-columnar and irregular zones

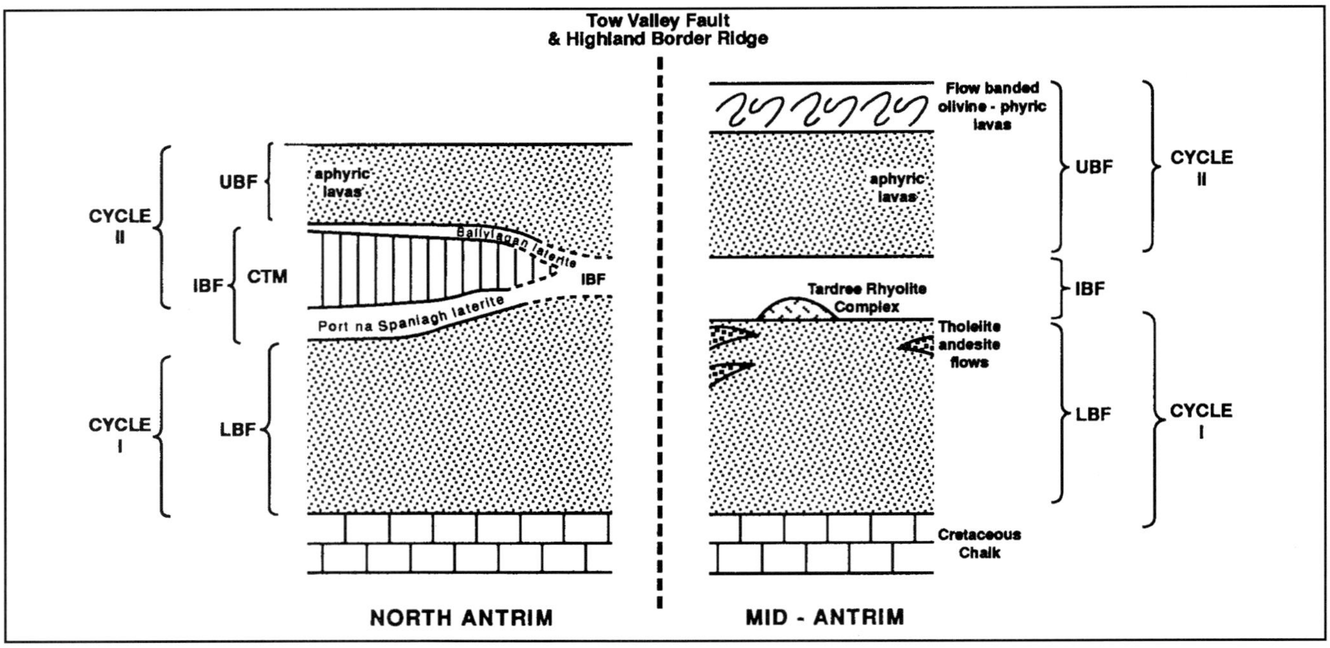

Figure 1. Above. The stratigraphy of the Tertiary Antrim Lava Group (after Lyle and Preston, 1993).
Figure 2. Below. Sub-divisions of columnar basalt flows and their origin in classical architecture (after Lyle, 1996).
Figure 3. Right. The Giant's Organ.

he compared to the entablature of the building. This terminology has been adopted to describe multi-tiered lava flows around the world. The terminology has been modified by later workers to limit the term entablature to the curvi-columnar zone, with the upper irregular zone referred to as the upper colonnade. This is well illustrated at the Giant's Organ, see Fig.3. Here the sharp contact between the lower colonnade (a) and the entablature (b) is clearly seen. The upper colonnade is seen at (c).

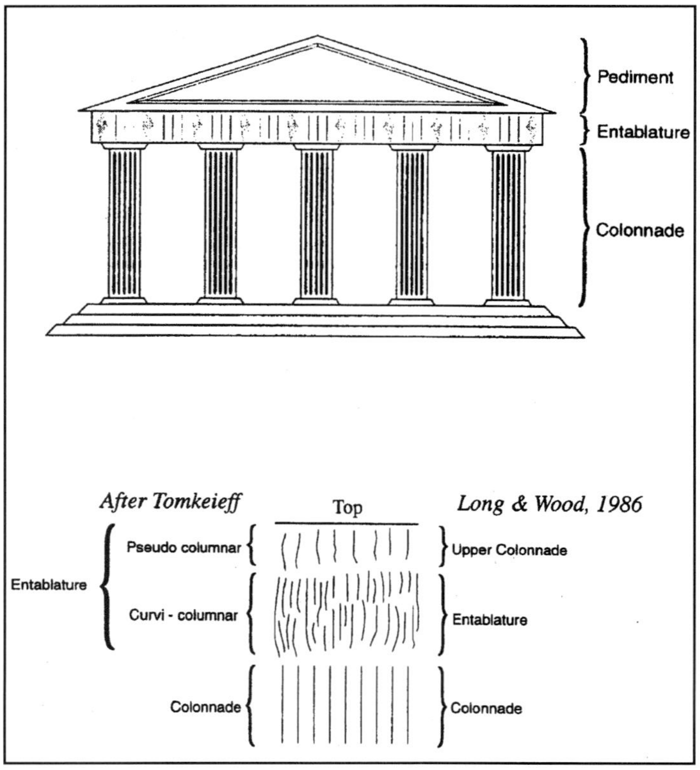

Work by Lyle and Preston (1993) and Lyle (2000) has established the unique characteristics of the environment of eruption in which these distinctive columnar lavas formed. During the Interbasaltic quiescent period, deep river valleys were carved into the Lower Basalt surface, including a valley some 1.5km wide at the Giant's Causeway. (See Fig.4). High volume lava flows filled the river valley, in the case of the valley at the Giant's Causeway the lava pond was about 100m deep at its maximum. The river occupying the valley was displaced and the river water formed a shallow and probably temporary lake on the surface of the hot flow. Columns are caused by the formation of contraction joints as the lava cools.

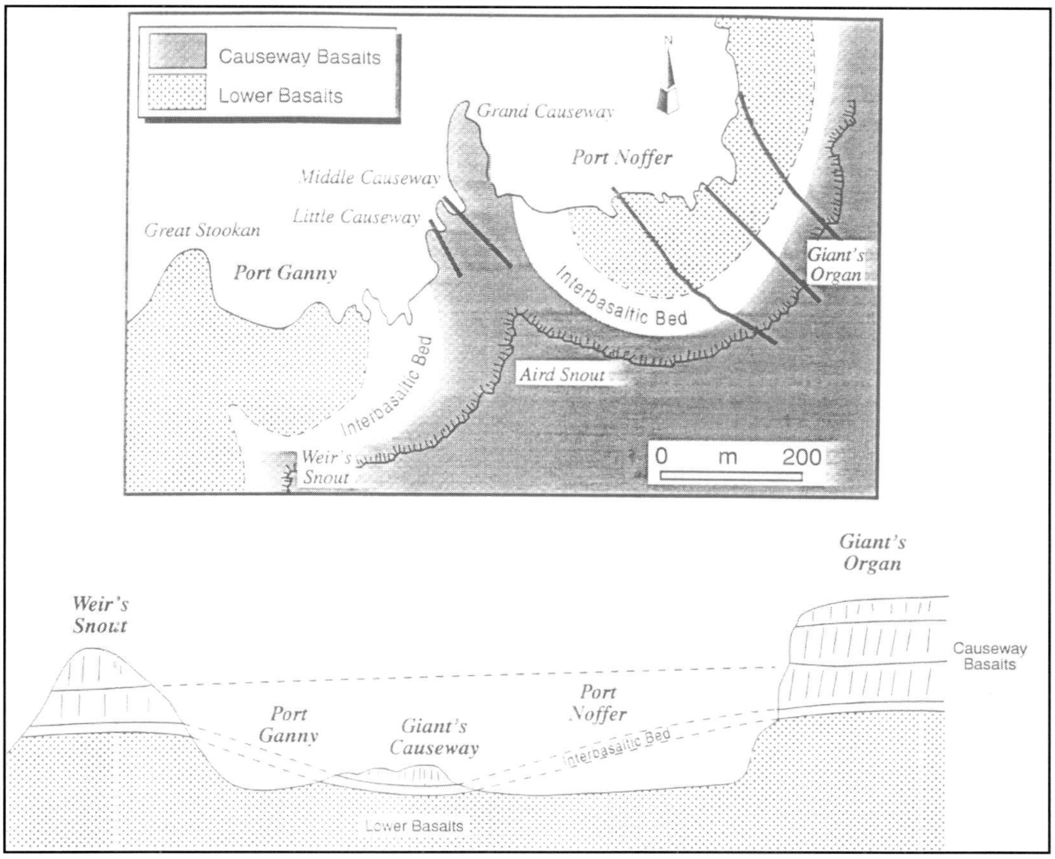

Figure 4. Geological map and cross-section of the Giant's causeway. (After Wilson and Manning, 1978).

Lava flows such as these are erupted at temperatures of around 1100 degrees C and lose heat very rapidly to their surroundings, predominantly through their top and bottom surfaces. Thermal contraction produces sets of roughly parallel joints or fractures at right angles to the cooling surfaces and these joints, moving inwards towards the centre of the flow as it solidifies, takes the form of a cooling front (see Figure 5a). Cracks are started at many points on the surface of this cooling front and three-pronged cracks at angles of about 120 degrees occur at each of these points. As these cracks propagate they intersect to form irregular polygons of 3, 4, 5, 6 or 7 sides. See Figure 5b. With continued cooling these polygonal cracks will move inwards to form polygonal columns, roughly at right angles to the horizontal cooling surfaces. See Figure 5c. Contraction of the columns along their length produces the characteristic "ball and socket" joints that divide the columns horizontally. See Figure 5d.

The Causeway is the colonnade of the first flow of the Causeway basalts that filled the river valley on the Lower Basalt surface.

The "organ pipes" at the Giant's Organ, see Fig.3, are the colonnade columns of the first flow of the Causeway basalts and above this can be seen the very sharp contact with the upper columnar zone, the entablature. There is evidence that this type of complex jointing is the result of modification of the cooling pattern of the hot flow interior by the influx of water. Early-formed fractures, called master joints, on the lava surface allowed this water to pass down into the interior of the flow. The effect is to speed up the cooling in the upper part of the flow and to form the columns of the entablature with narrow, curved columns growing rapidly downwards and the more regular columns of the colonnade growing upwards from the base, at a slower rate. The entablature/colonnade contact seen in the Giant's Organ represents the meeting of the upward moving colonnade cooling front and the downward moving entablature cooling front. See Fig. 6.

Resource exploitation in northeast Ireland

Following the eruption of the Causeway basalts, a further prolonged period of volcanic inactivity led to the formation of a second layer of laterite in north Antrim, the Ballylagan laterite, see Fig.1, and then renewed volcanism produced the Upper Basalt Formation, representing a return to conditions similar to those of Lower Basalt times. In a number of places fragments or blocks of laterite from the Interbasaltic surface were engulfed by the hot magma of the Upper Basalts. Notably this occurred in two places where magma was present in volcanic vents, at Tievebulliagh near Cushendall and at Brockley on Rathlin Island, off the north coast of Antrim. At these localities the soft red laterite was baked by the very high temperatures of the magma to form a hard, dark blue material known as porcellanite, so named for its superficial resemblance to porcelain. This porcellanite provided the raw material for the production of axes in Neolithic times which were then used by Neolithic farmers to clear the thick forest cover and prepare for planting and grazing, leading to substantial landscape modification. Both the porcellanite localities are relatively small outcrops and are in remote areas of the countryside. The fact that they were found and exploited is an indication of the awareness the Neolithic people had of their landscape.

The deposits of laterite present over much of Antrim were exploited as a source of iron ore and played a major part in the industrialization of this part of Ireland in the 19th century. In areas such as around the head of Glenballyemon, near Cushendall are the adits used to extract the iron ore from the hillsides and the disused narrow-gauge railway lines used to transport the ore. Nearby in the village of Newtown-Crommelin are the remains of a peat-fired smelter. Currently fresh basalt is extensively quarried in Antrim as a source of aggregate for road surfacing and as a component in concrete manufacture, these uses all leading to substantial landscape modification as part of modern infrastructure expansion and renewal.

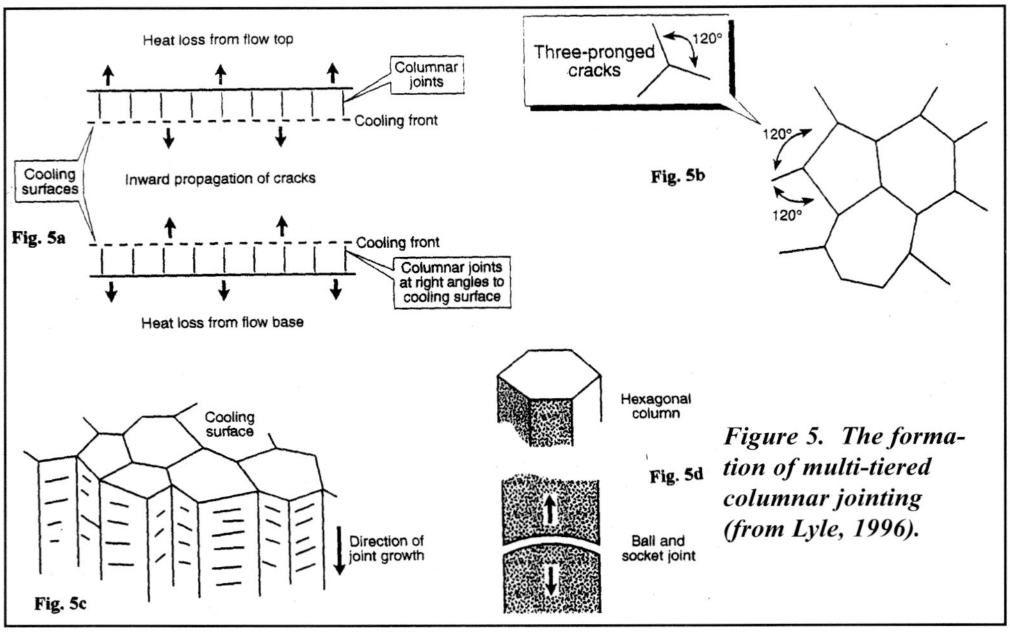

Figure 5. The formation of multi-tiered columnar jointing (from Lyle, 1996).

The alteration of fresh basalt to laterite to porcellanite is an excellent example of the progressive creation of geological resources in the crust. Following exploration and discovery, such resources become available for human exploitation, after which there are discernible changes in landscape and lifestyle. In Antrim the sequence of materials produced by successive geological processes ie basalt-laterite-porcellanite has been exploited in a reverse order. The most altered material was used by the oldest inhabitants of the area during the Neolithic, the weathered basalt was used by the early industrialists in the19th century and the freshest material is currently in use. The common factor however is the exploitation of resources leading to marked changes in lifestyle and landscape.

REFERENCES

Lyle,P. 1980 A petrological and geochemical study of the Tertiary basaltic rocks of northeast Ireland. *Journal of Earth Sciences, Royal Dublin Society*, 2, 137-152.

Lyle, 1996 *A geological excursion guide to the Causeway Coast*. Belfast : Environment and Heritage Service.

Lyle, P. 2000 The eruption environment of multi-tiered columnar basalt lava flows. *Journal of the Geological Society, London*, 157, 715-722.

Lyle, P. and Preston, J. 1993 Geochemistry and volcanology of the Tertiary basalts of the Giant's Causeway area, Northern Ireland. *Journal of the Geological Society, London*, 150, 109-120.

Patterson, E.M. 1955 The Tertiary lava succession in the northern part of the Antrim plateau. *Proceedings of the Royal Irish Academy*, 57B, 79-122.

Tomkeieff, S.I. 1940 The basalt lavas of the Giant's Causeway district, Northern Ireland. *Bulletin Volcanologique*, 2, 6, 89-143.

Wilson, H.E. and Manning, P.I. 1978 *Geology of the Causeway Coast*. Vol. 1 Memoir of the Geological Survey of Northern Ireland.

Figure 6. The formation of multi-tiered lava flows by water influx along master joints (after Lyle, 2000).

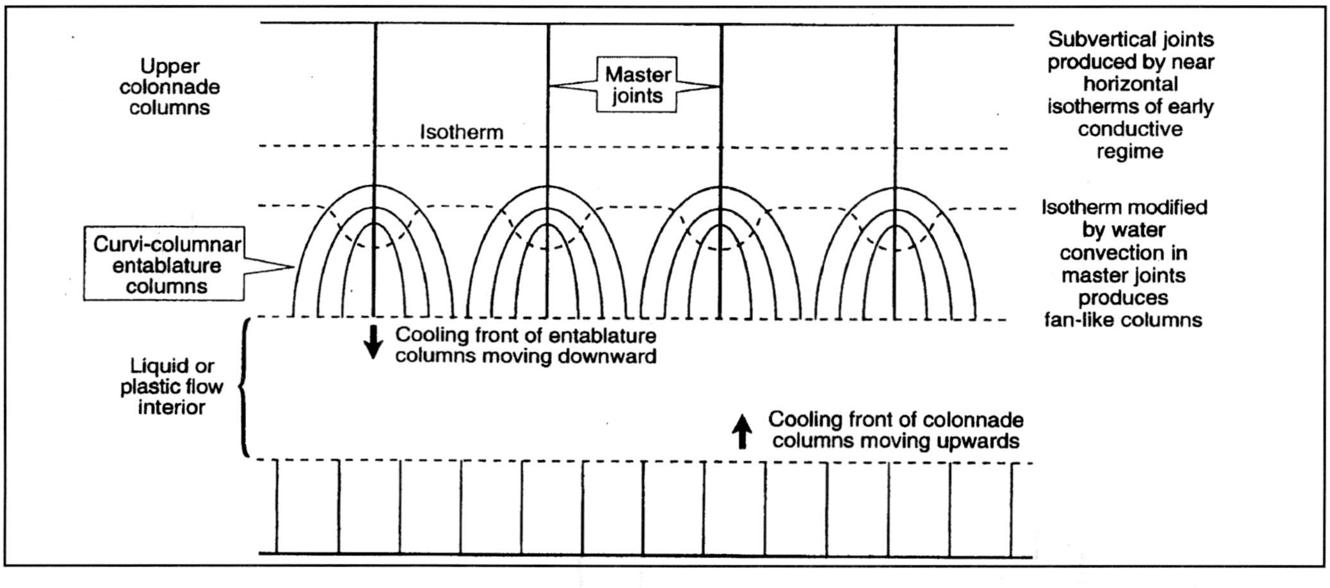

In: Parkes, M.A. (Ed.) 2004 *Natural and Cultural Landscapes - The Geological Foundation*, 37-40, Royal Irish Academy, Dublin

TORTOISES, HARES AND THE EVOLUTION OF THE IRISH LANDSCAPE

Michael J. Simms, Department of Geology, Museums and Galleries of Northern Ireland, Ulster Museum, Botanic Gardens, Belfast BT9 5AB, UK.
Email: michael.simms.um@nics.gov.uk

Abstract: There is a striking topographic contrast across Ireland between predominantly limestone-floored lowlands and uplands developed on other rock types. This arises from the fundamentally different way limestone and other rocks are removed; the former by dissolution, a low energy process enhanced by vegetation, the latter by erosion, a high energy process inhibited by vegetation. In Ireland countless 'soft days' over the last 60 million years have been more effective at removing limestone than other rock types. Limestone uplands survived only where protected by a cover of insoluble rock, such as sandstone or mudstone, which has been stripped away relatively recently. Large-scale removal of considerable thicknesses of limestone across Ireland has increased the relief of non-limestone uplands by isostatic uplift. Denudation across the Irish landscape has led to changing outcrop patterns of limestone and other rocks, resulting in profound long-term changes in topography and drainage patterns.

INTRODUCTION

Geomorphologists have long been divided between the study of karst landscapes, due to limestone dissolution, and landscapes formed, on largely insoluble rocks, by erosion. Indeed, Paul Williams (1978, p. 260) noted "the usual exclusion of karst from discussions of fluvial geomorphology". However, dissolution and erosion are two very different processes. Factors which enhance one may inhibit the other, or vice versa. Consequently the denudation rate, the rate at which rock is removed by dissolution and/or erosion, varies considerably across a geologically diverse landscape such as that of Ireland and must be considered in order to understand how the Irish landscape has reached its present form. A fuller account of the subject of this paper can be found in Simms (in press).

ROCK REMOVAL : EROSION AND DISSOLUTION

Removal of largely insoluble, primarily silicate-based, rocks occurs largely by erosion, the physical transport of rock particles. Dissolution may cause weathering but without erosion the weathered rock (saprolite) may remain in situ for thousands of years and reach tens of metres in thickness. Fluvial erosion, by rivers and streams, is a relatively high energy process and requires a minimum velocity before even loose particles, let alone unweathered bedrock, are moved. This threshold velocity is lower for particles of 0.1-1 mm diameter than for either larger or smaller particles, due to increased mass and molecular cohesion respectively (Fig 1). In contrast limestone denudation in Ireland is largely by dissolution, a relatively low energy process in which even static water can prove effective. Figure 1 shows this difference between erosion and dissolution; one curve compares particle size vs. flow velocity needed to transport them; the other compares the length of scallops (concave hollows formed through the uneven dissolution of limestone by flowing streams) vs. flow velocity. For the former there is no erosion if water flow is less than ~0.1 m/s, while the latter shows that scallops form, and hence limestone is being dissolved, at much slower flow velocities.

Vegetation has profound, though very different, effects on erosion and dissolution rates. Soil and weathered rock is bound together by plant roots while the aerial parts of plants absorb rain energy and form a barrier between flowing water and the soil beneath. In Ireland, countless 'soft days' support abundant vegetation but it seldom rains hard enough to cause much erosion, even on exposed river

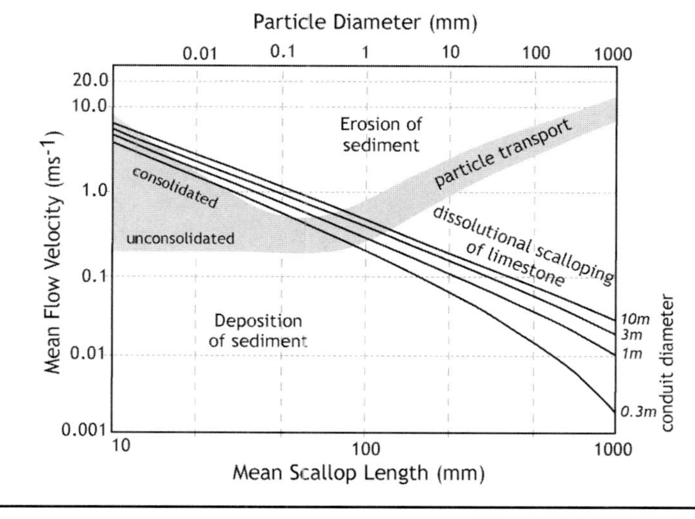

Figure 1. Relationship between water flow velocity, erosion and transport of particles, and dissolutional scalloping of limestone.

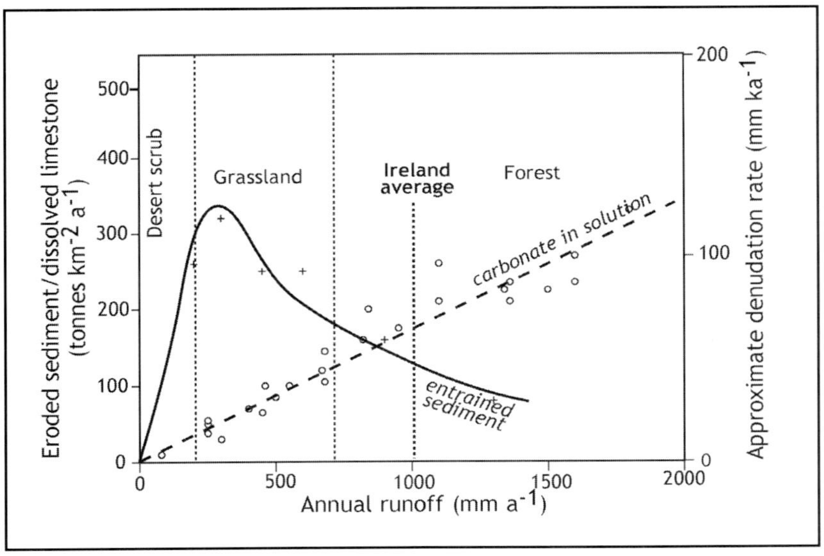

Figure 2. Relationship between annual runoff, sediment eroded from silicate outcrops, and limestone dissolved in soil-covered karst regions.

banks. Furthermore, most of this sediment is actually eroded from stream channels where erosion rates are high, with little erosion on the intervening ground. Although vegetation inhibits erosion, it does not prevent water percolating down through soil to dissolve limestone beneath, and so dissolution affects the entire limestone outcrop rather than just narrow channels. CO2 concentrations are higher in soil-water, due to respiration of plant roots and soil organisms, and so vegetation actually enhances dissolution. Unlike erosion, a direct relationship exists between limestone dissolution and runoff because dissolution is not dependent on flow velocity or the extent of the vegetation cover. Figure 2 shows that in Ireland long-term limestone dissolution rates exceed long-term erosion rates.

This 'Tortoises and Hares' model for landscape evolution (Simms, in press) fits well with what we see in the Irish landscape. Carboniferous limestones underlie more than a third of southern Ireland and mostly form lowlands below 100 m OD (Fig. 3). Adjacent non-limestone outliers and inliers commonly rise tens to hundreds of metres above these limestone lowlands (Fig. 4). A similar pattern is seen in Connemara where outcrops of marble are occupied by ribbon lakes while adjacent quartzite outcrops rise hundreds of metres above them. Much of this relief can be ascribed to the contrasting rates of dissolution and erosion. However, there are limestone hills and plateaux that appear anomalous in light of the previous discussion, e.g. the Burren in Co. Clare and the Rock of Cashel in Co. Tipperary. Dissolution rates would lower the Burren to near sea level within a few million years, as occurred on the Gort lowlands to the east. However, it is clear that in most cases these limestone hills owe their survival to protective covers of siliciclastic rock that were stripped relatively recently. Tertiary weathering produced thick, but mechanically weak, saprolites across many silicate outcrops, though not on limestones. These were then easily removed by glacial and periglacial processes during the Pleistocene. The Burren and other limestone uplands may well have been unroofed in this manner, while much of the low ground of southern Connemara probably represents an etch plain, a weathering front exhumed by glacial erosion.

The implication is that pre-Pleistocene relief between limestone outcrops and other rocks may well have been significantly greater than today.

UNLOADING AND UPLIFT : ISOSTASY

Landscapes do not develop solely through erosion and/or dissolution. Removal of rock causes isostatic uplift in compensation. Were it not for isostasy then long-exposed land surfaces would be reduced to uniformly low peneplains regardless of the geology. Where denudation rates vary across an area according to lithology, as for Ireland, so the isostatic uplift reflects the average denudation rate. If the more rapidly removed lithology, in this case limestone, is a significant proportion of the total outcrop then isostasy will lead to uplift of more slowly eroded siliciclastics. Since the limestone outcrop today accounts for ~40% of Ireland

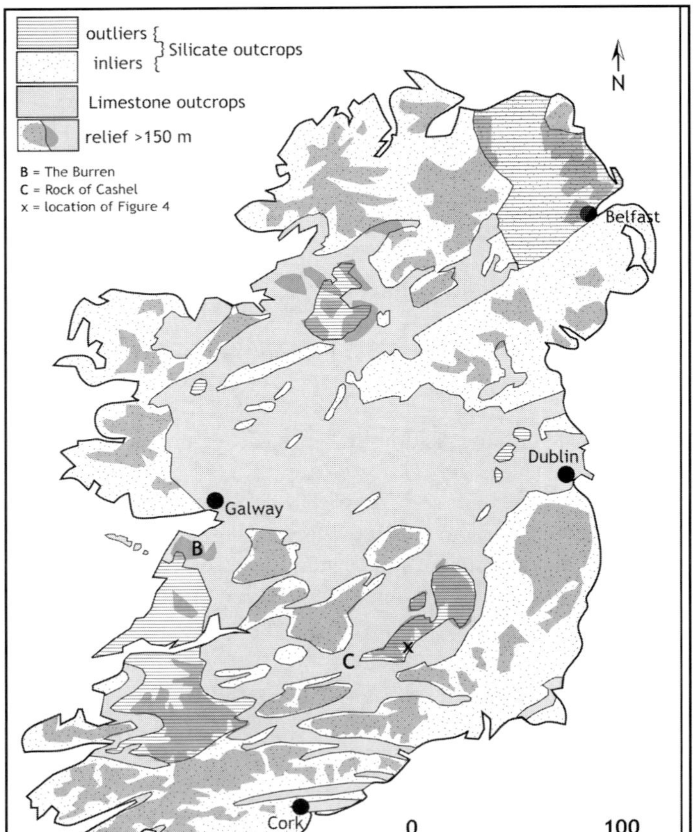

Figure 3. Distribution of lowlands and uplands (>150 m) relative to lithology (limestones vs. silicates) across Ireland.

Figure 4. The Irish landscape epitomised. Looking south from the mudstones and sandstones (Upper Carboniferous) of the Slieveardagh Hills, across the limestone (Lower Carboniferous) lowlands, to the sandstones (Devonian) of Slievenamon, south-east Ireland.

Figure 5. Sequence of stages in the landscape development of a mixed limestone and siliciclastic outcrop pattern, showing the effect of limestone denudation and isostasy in generating relief.

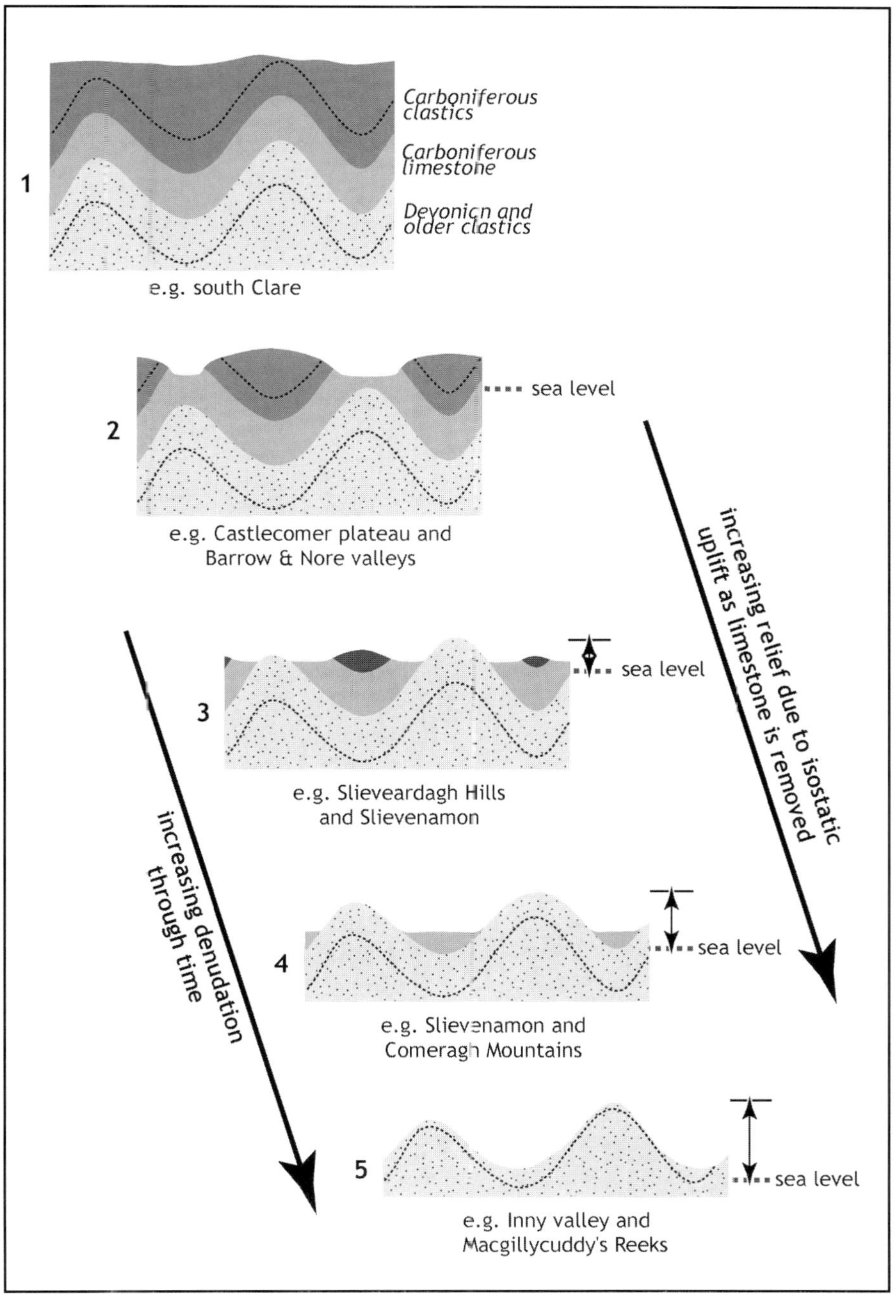

(30,000 km^2) (Fig. 3), with hundreds of metres thickness removed during the Tertiary, this must have been an important factor in landscape evolution.

CHANGING OUTCROP PATTERN AND LANDSCAPE EVOLUTION

If rocks have been folded then surface lowering causes the outcrop pattern to change over time. If certain rock types are then removed more rapidly than others then the long term topographic changes may be dramatic. This is seen across many parts of Ireland, with a gently folded thick limestone sandwiched by thick siliciclastics (Fig 5). The previous discussion, and Fig 2, show that in Ireland limestone is removed by dissolution faster and more evenly than adjacent insoluble rocks, where erosion is focused largely on stream courses. Consequently a valley will develop where limestone is first unroofed on an anticlinal crest and will be flanked by hills on the siliciclastics in adjacent synclines (Fig 5/2). Ultimately, however, siliciclastics beneath the limestone will also be unroofed along the anticlinal crest while the siliciclastics above the limestone will be stripped away in the adjacent synclines. The result is that the valleys shift from anticline to syncline and the hills from syncline to anticline (Fig 5/4). Examples of all of these stages of landscape evolution are present in southern Ireland, particularly in the south-west. Evidence from dated cave fills indicates that the advanced stages had already been reached in parts of Ireland by the mid-Tertiary but any investigation of drainage patterns must consider the full sequence of development.

CONCLUSIONS

The low limestone plains and often striking hills of sandstone, granite, quartzite and other rocks that form the Irish landscape are a consequence of marked differences in the way that limestone and other rocks are removed; the former by slow continuous dissolution, the latter by sporadic erosive flood events. Over tens of millions of years it is the 'Tortoise of Dissolution' that wins the race, with the resulting isostatic uplift actually increasing the differential relief between limestone lowlands and the uplands formed on other rocks. Limestone uplands, such as the Burren, seem less anomalous when it is recognised that the limestone was protected by an insoluble cap rock until relatively recently. Because the valleys and lowlands are located largely on the limestone outcrops, so the changing outcrop pattern of limestones over time must have caused profound changes in the location of hills and valleys as the landscape evolved.

REFERENCE

Simms, M.J. in press. Tortoises and Hares: dissolution, erosion and isostasy in landscape evolution. *Earth Surface Processes and Landforms*.

In: Parkes, M.A. (Ed.) 2004 *Natural and Cultural Landscapes - The Geological Foundation*, 41-44, Royal Irish Academy, Dublin

A COLD-WATER CORAL CARBONATE MOUND POPULATION ALONG THE WESTERN MARGIN OF THE PORCUPINE BANK

P.W. Readman and B.M. O'Reilly, Geophysics Section, Dublin Institute for Advanced Studies, 5 Merrion Square, Dublin 2, Ireland, E-mail: pr@cp.dias.ie; P.M. Shannon, Department of Geology, University College Dublin, Dublin 4, Ireland, E-mail: p.shannon@ucd.ie

Abstract: The detailed structure of a cold-water carbonate mound population has been identified and resolved along the western margin of the Porcupine Bank, west of Ireland using deep-tow sidescan sonar data. The mounds occur in water depths of 800-1000 m and are circular to elliptical in shape and up the 250 m high. They form an impressive seabed landscape. Large-scale sedimentary bedforms, produced by strong NE-flowing contour-parallel currents define a teardrop-shaped pattern around the largest mounds. A model based on the observed size distribution and coral growth rate allows the age structure of the population to be derived using the measured biological growth rates for coral framework builders. A well-defined correlation is found between the growth and colonisation of the mound population with early Holocene/late Pleistocene climate change and related variations in the bottom currents.

Coral reefs are normally associated with warm water tropical regions of the Earth, but during the last decade or so large-scale carbonate build-ups, which resemble tropical reefs in many respects, have been recognized in the north-east Atlantic stretching from Norway to Iberia. These carbonate "mounds" are an important part of the sub-sea landscape on the margins of the Porcupine Seabight and Rockall Trough to the south and west of Ireland (Fig. 1) where they form concentric reef-like structures up to 1 km across and 250m in height. Several mound provinces are defined within these basins where "bioherms" occur either singly or as large clusters in populations, which align themselves along lines of equal water depth. The cold-water coral species *Lophelia pertusa* and *Madrepora occulata*, both first described by Linnaeus (1758), are the main framework constructers of these build-ups and unlike tropical corals flourish far below the photic zone at water temperatures of between 4° and 12° C. The build-ups are populated by a large variety of metazoan and protozoan species that form a biological ecosystem as diverse as those found in tropical coral reefs.

Thomson (1874) first described the occurrence of cold-water corals from the region west of Ireland, but it was not until the next century that the importance of these corals in constructing large reef-like frameworks was recognised. It is only relatively recently that the environmental and oceanographic controls on the formation and growth of these spectacular seabed structures, such as seabed type, nutrient sources and seabed current strength, have been investigated. One school of thought relates the formation of mounds to hydrocarbon seeps from deep reservoirs along tectonic faults (Hovland et al. 1994) or from gas hydrate deposits in response to climatic and sea-level change during glacial to interglacial transitions (Henriet et al. 2001). Carbonate mounds are very likely to contain a geochemical and biological record of deep-sea environmental change induced by palaeoclimatic varia-

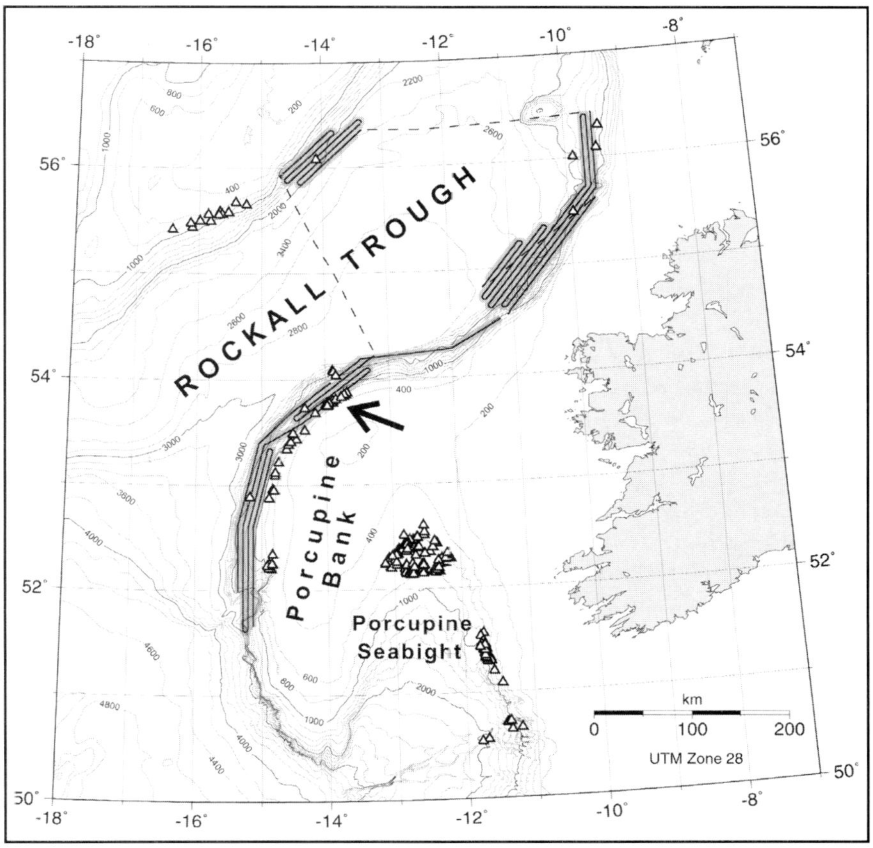

Figure 1. Map of cold water carbonate mound occurrences (triangles) in the region west of Ireland (Croker and O'Loughlin 1999). Areas marked in grey are the regions surveyed by the TOBI sidescan survey system (see Fig. 2). Arrow indicates the region of the continental shelf and slope discussed in this work. Bathymetry contours (GEBCO97) are at 200 m intervals.

Figure 2. The TOBI instrument being deployed from the stern of the Dutch research vessel PELAGIA. The instrument is towed at ca 400 m above the seabed in water depths of up to 5000 m. It can resolve seabed features tens of metres across.

tions over tens of thousands of years. They are also a significant sink for excess carbon dioxide and their growth is potentially important in regulating climate change. Global occurrences of shallower-water carbonate mound bioherms are common in the geological record during the last 500 million years of Earth history. The most famous Irish example is the Waulsortian carbonate mounds that occur within the Lower Carboniferous strata of the Irish Midlands.

The detailed structure of a large cold-water carbonate mound population was resolved along the western margin of the Porcupine Bank, west of Ireland, during a deep-tow TOBI (Towed Ocean Bottom Instrument) sidescan sonar survey, undertaken on behalf of the Irish Petroleum Infrastructure Programme (Fig. 1). The TOBI Instrument (Fig. 2), developed at the Southampton Oceanography Centre, emits very high frequency sound that ensonifies the seabed and is scattered and reflected back to acoustic sensors onboard the instrument to produce detailed images of seabed features (Fig. 3). This deep-towed sidescan sonar technology resolves not only the shape and geometry of the mound structures but also the acoustic properties of the seabed substrate around them.

The mounds imaged along the Porcupine Bank occur in water depths of between 800 and 1000 m. They are circular to elliptical in shape and range in size from 50 to 850 m across and are up to 250m high forming an impressive seabed architecture resembling a strange urban landscape perched high upon the continental shelf (Fig. 4). Large-scale sedimentary bedforms at 700 - 900 m water depth are inferred from acoustic backscatter zonation that defines a "tear drop" pattern around the largest mounds (Fig. 3). These bedforms, which envelop the mound structures, are produced by the strong NE-flowing contour-parallel currents that are present all along the length of the eastern margin of the Rockall Trough (Kenyon 1987). The larger mounds are more elliptical than the smaller ones, which generally are more numerous on the down current side of the larger structures (outside the region shown in Fig. 3). This change to more elliptical shapes as the size of the mounds increases suggests that streamlining effects caused by the strong pole-ward bottom currents strongly influence the geometry of the mound structures.

The frequency distribution of the mound sizes in the entire population (Fig. 5) has some interesting properties that are ultimately believed to reflect the large scale biological and physical processes at work in the oceanic environment that led the population to acquire these properties. First, the smaller mound structures are many times more numerous than the largest structures and second, there is a systematic tendency for them to become gradually more elongate as

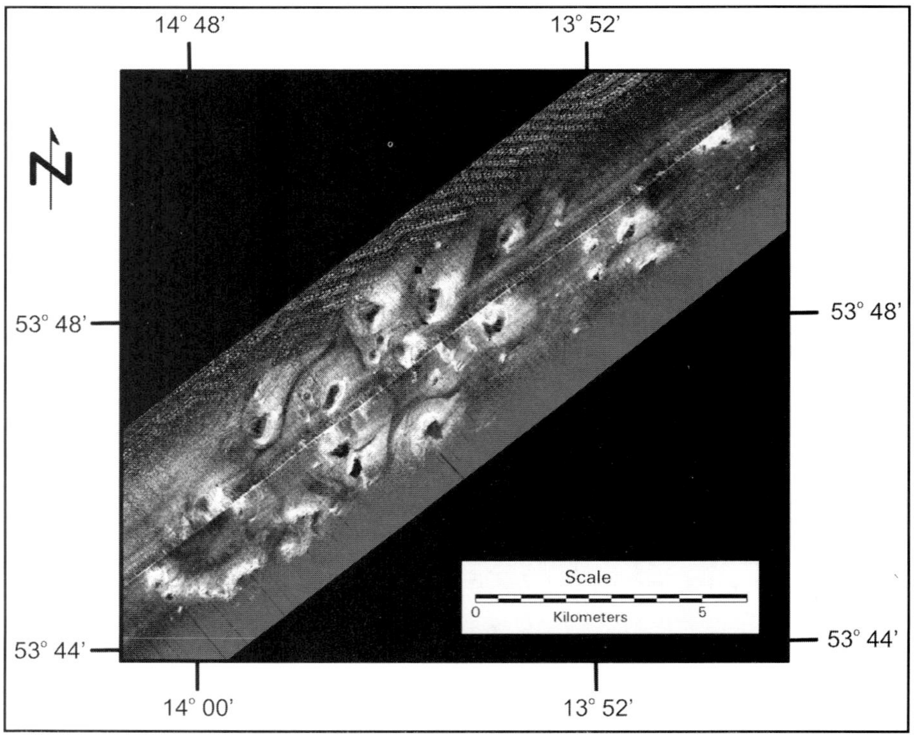

Figure 3. Extract from TOBI sidescan sonar mosaic showing carbonate mounds on the eastern shoulder of the Porcupine Bank. A 3.5 kHz profile across these mounds is shown in Fig. 4.

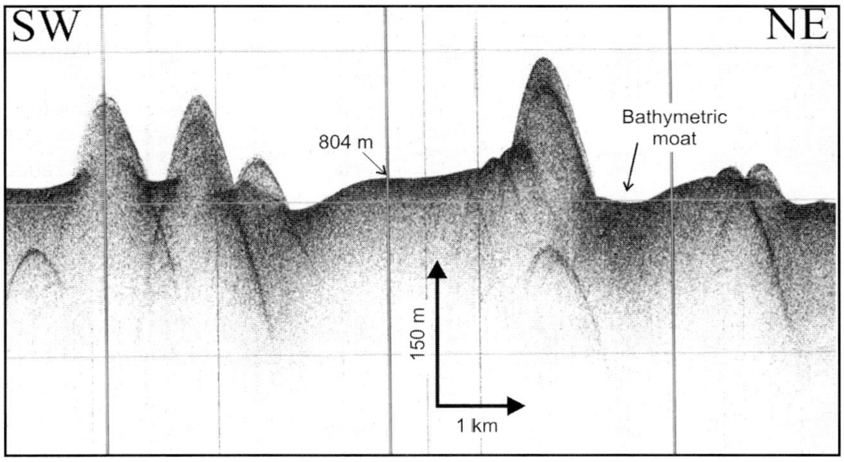

***Figure 4.** A 3.5 kHz echosounder record of carbonate mounds on the eastern shoulder of the Porcupine Bank. The profile shows a cross-section of part of the mound population shown in Fig. 3. Bathymetric moats, produced by the scouring action of bottom currents are well developed on the NE (downstream) side of each mound where "tear drop" shaped bedforms are seen on the sidescan sonar records (see Fig. 3). Sediments (silt and sand) are being transported by the currents to the northeast. From O'Reilly et al. (2003).*

they increase in size (Fig. 5). The form of the population size distribution follows closely a power law distribution, as distinct from the "normal" or Gaussian distribution which is common in many biological populations. Power law distributions have no characteristic statistical average or mean and occur widely in a large variety of both natural and anthropogenic phenomena across disciplines such as economics, ecology and earthquake seismology (Newman 2000, and references therein).

The size distribution of human settlements from villages to the largest towns and cities follows such a power law, known as Zipf's Law (Gabaix 1999). Many interacting and complex socio-economic parameters give rise to the distribution and growth of human settlements. The same is likely to be true for the ecological and oceanographic parameters that have governed the development of the Porcupine Bank mound landscape, so that the analogy with cities may be very relevant in understanding how a mound landscape evolves.

The analogy with human settlements is indeed relevant since the form of the population distribution is ultimately determined by the biological growth rate of the mound framework and the rate at which the coral framework builders colonise the substrate. In terms of this analogy the mound framework growth is equivalent to the growth of individual towns and cities and the colonization rate to the rate at which new centres of growth are established within a developing urban landscape. Using similar ideas, a relatively simple development model can explain the observed statistical properties of the mound population distribution and also the role played by bottom currents.

Initially, these bottom currents are important in establishing a suitable environment for the growth of carbonate mounds and the establishment of new mound settlements or colonies. They transport nutrients into the mound framework and keep it free of the growth-stifling sedimentary debris which is generated during submarine landslides and introduced by rivers from nearby landmasses. However there are also negative feedback effects on coral growth and calcium carbonate production resulting from the drag force which these strong currents generate, particularly across the larger mounds. These feedback effects lead to an upper limit to the size that a mound can ultimately grow. In a similar way there is an upper limit to the size of a great metropolitan city when environmental pressures that grow very rapidly with its increasing size outweigh the advantages of the city's further growth.
The development of a streamlined elliptical form for the larger mounds in the evolving mound landscape minimises the negative growth-inhibiting effects of the drag forces

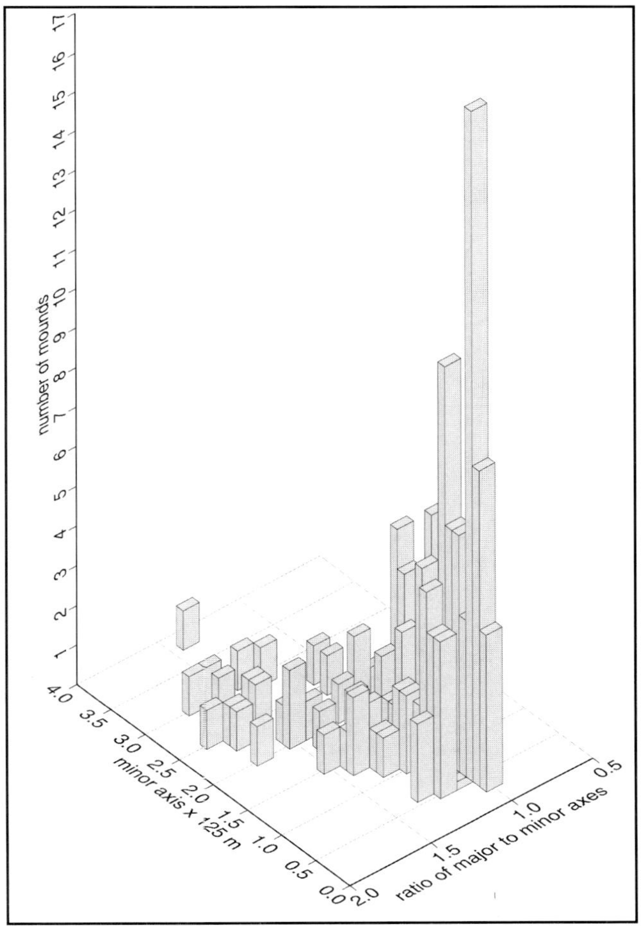

***Figure 5.** Frequency distribution of mound sizes and elongations. Mound elongation (the ratio of the lengths of the major to minor mound axes) increases with mound size (minor axis) so that the fluid drag due to the bottom currents is minimised (see text for discussion). From O'Reilly et al. (2003).*

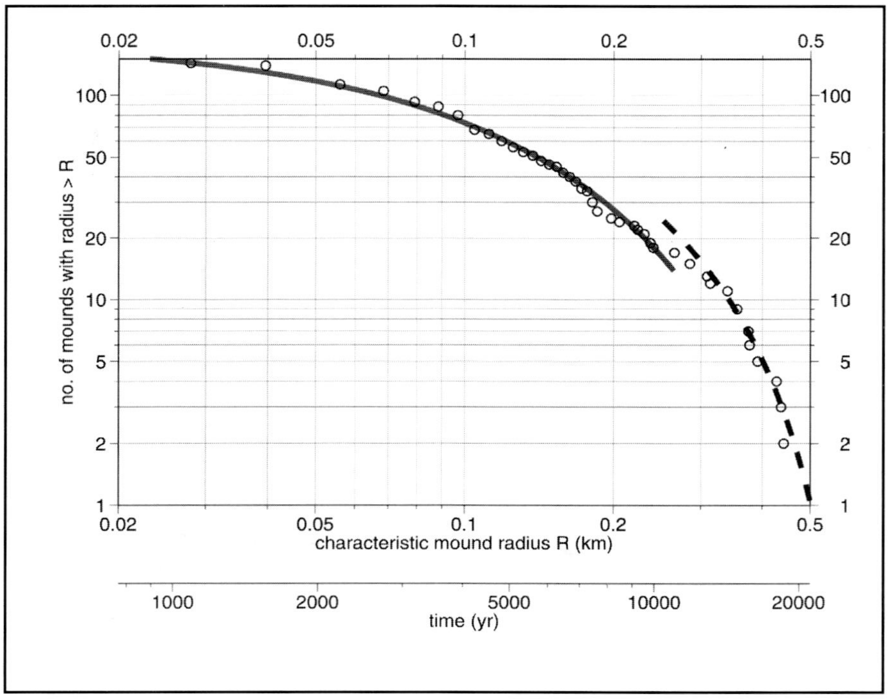

Figure 6. A cumulative frequency curve for the mound size distribution shown in Fig. 5. The number of mounds exceeding a given radius is plotted for the entire range in mound scale. The open circles are measurements. The solid curves are trends predicted by a 'development model' for the growth of the entire population in Fig. 5 (see O'Reilly et al. 2003 for details of this model). This development model allows an age for different mound sizes to be estimated. Note the change in slope of the measured frequency curve at ca 10 ky and the small kink at 8 ky suggesting correlations with known palaeo-climatic variation in the northeast Atlantic during the Holocene and at the last glacial to interglacial transition. From O'Reilly et al. (2003).

generated by the otherwise beneficial bottom currents. A relatively simple model for the development of the mound population that incorporates the effects of bottom currents, allows its age structure to be derived using measured biological growth rates for the coral mound framework builders (see O'Reilly et al. 2003 for details). There is an interesting coincidence between certain properties of the observed distribution of mounds, the age structure of the entire population predicted by the model, and climatic change in the late Pleistocene and Holocene (see Fig. 6). This suggests that there is a subtle but well-defined correlation between the growth and colonization patterns of the biological species, comprising the mound population, with climate change and related variations in the strength (and perhaps direction) of these bottom currents. The climatic and environmental changes, which both modified and sculptured the glaciated landscapes of Ireland (Mitchell 1986), appear to be as important in moulding the submarine landscapes along the extensive western shelf and continental slope regions of Ireland.

ACKNOWLEDGEMENTS

This publication uses data and survey results acquired during a project undertaken on behalf of the Rockall Studies Group (RSG) of the Irish Petroleum Infrastructure Programme Group 2. The RSG comprised: Agip (UK) Ltd, Anadarko Ireland Company, ARCO Ireland Offshore Inc, BG Exploration & Production Ltd, BP Exploration Operating Company Ltd, British-Borneo International Ltd, Elf Petroleum Ireland BV, Enterprise Energy Ireland Ltd, Mobil Oil North Sea Ltd, Murphy Ireland Offshore Ltd, Phillips Petroleum Exploration Ireland, Saga Petroleum Ireland Ltd, Shell EP Ireland B.V., Statoil Exploration (Ireland) Ltd, Total Oil Marine plc, Union Texas Petroleum Ltd and the Petroleum Affairs Division of the Department of the Marine and Natural Resources.

REFERENCES

Croker, P.F. and O'Loughlin, O. 1999. *A catalogue of Irish carbonate mud mounds*. Carbonate mud-mounds and cold-water reefs conference, University of Ghent, Belgium, 7-11 February.

Henriet, J.P., De. Mol., B., Vanneste, V., Huvenne, D., Van Rooij and The Porcupine-Belgica 97, 98 and 99 Shipboard Parties. 2001. Carbonate mounds and slope failures in the Porcupine Basin: a development model involving fluid venting. In: Shannon, P.M., Haughton, P.D.W. and Corcoran, D.V. (eds). *The Petroleum Exploration of Ireland's Offshore Basins*. Geological Society, London, Special Publications, 188, 375-383.

Gabaix, X. 1999. Zipf's Law for Cities: An Explanation. *Quarterly Journal of Economics*, CXIV, August, 739-767.

Hovland, M., Croker, P.F. and Martin, M. 1994. Fault-associated seabed mounds (carbonate knolls?) off western Ireland and north-west Australia. *Marine and Petroleum Geology*, 11, 232-246.

Kenyon, N.H. 1987. Evidence from bedforms for a strong poleward current along the upper continental slope of northwest Europe. *Marine Geology*, 72, 187-198.

Linnaeus, C. 1758. *Systema Naturae*, 10th Edition, Laur Salvius, Stockholm.

Newman, M. 2000. The power of design. *Nature*, 405, 412-413.

Mitchell, F., 1986. *Shell Guide to Reading the Irish Landscape*. Country House, Dublin, 228pp.

O'Reilly, B.M., Readman, P.W., Shannon, P.M. and Jacob, A.W.B. 2003. A model for the development of a carbonate mound population in the Rockall Trough based on deep-towed sidescan sonar data. *Marine Geology*, 198, 55-66.

Thomson, C.W. 1874. *The depths of the sea*. Macmillan, London, 527pp.

SESSION 2

INTEGRATING GEODIVERSITY AND BIODIVERSITY - A HOLISTIC APPROACH

Ben Bulben, Co. Sligo. Photograph: Matthew Parkes

In: Parkes, M.A. (Ed.) 2004 *Natural and Cultural Landscapes - The Geological Foundation*, 47-52, Royal Irish Academy, Dublin

GEODIVERSITY - LINKING PEOPLE, LANDSCAPES AND THEIR CULTURE

Michael Stanley, Geodiversity Consulting, Park House, 8 King St., Ripon, N.Yorkshire, UK, HG4 1PJ. Email: geodiversity@btinternet.com

KEYNOTE ADDRESS

Abstract. Wherever we live we see and use the buildings and the landscape around us. To be able to understand their relationship and interdependence we need a concept to link them together. Geodiversity is the link between people, landscapes and their culture through the interaction of biodiversity, soils, minerals, rocks, fossils, active processes and the built environment; this definition goes further than that of 'Geodiversity in Nordic Nature Conservation' published by the Nordic Council in 2000. There may be disagreement with the need to coin a new term, but it is time for one concept to give an holistic view of the environment around us. In short biodiversity relies on and is a part of Geodiversity. This paper describes the intimate linking within geodiversity looking at animals and plants as opportunists; the imagery of Lewis Carroll in 'Alice in Wonderland', which uses local people, structures and landscape as the basis for his surreal stories, and British World Heritage Sites, all of which are classic examples of geodiversity.

INTRODUCTION

Wherever we live we are surrounded by, observe and enjoy the aesthetics of the buildings and the landscape around us. To be able to understand their relationship and interdependence we need a concept to link them together. Geodiversity is that concept. Geodiversity is the link between people, landscapes and their culture through the interaction of biodiversity, soils, minerals, rocks, fossils, active processes, and the built environment. You may disagree with the need to coin the term, but it is time for one concept to give an holistic view of the environment around us. In short biodiversity relies on and is a part of Geodiversity.

This definition of Geodiversity extends the term as described in "Geodiversity in Nordic Nature Conservation" (Geodiversitet I Nordisk Naturvard) published in 2000 by the Nordic Council, in Swedish with an English summary, edited by Carl Erik Johansson, with contributions from colleagues Veli Suominen, Markus Alapassi, Steen Andersen, Kristjan Geirsson, Annika Jansson and Lars Erikstad who was probably the first Earth scientist to use the word Geodiversity in 1993 at the Malvern International conference (O'Halloran et al 1994). They describe Geodiversity as the variation of the bedrock, deposits, landforms, and processes, which form and characterise the Earth's crust. The publication is peppered with maps of deposits and landforms and their effect on land use; drawings of structures, and fine illustrations of some spectacular features across Scandinavia including new bedrock being eroded on the Reykjanes peninsular in Iceland, the landform of a fossil meteorite impact crater at Soderfjarden, south of the Finnish city of Vasa, beautifully delimited by the fertile 'polder' agriculture of the crater floor, and those moon-like pseudo- or rootless craters, the result of a basalt lava erupting on to shallow water, on the southern shore of Myvatn, the lake of midges, also in Iceland. The book not only illustrates different aspects of geodiversity but also shows how and who should manage and protect it; the full English version is eagerly awaited.

But Geodiversity is even more fundamental than the Nordic definition as it is the result of dynamic interactive processes between landscape, fauna, flora and our culture that have dictated where people have located and developed villages, towns, cities, industries, transport routes and

Moon-like pseudo- or rootless craters, the result of a basalt lava erupting on to shallow water, on the southern shore of Myvatn, Iceland

The landform of a fossil meteorite impact crater at Soderfjarden, south of the Finnish city of Vasa, beautifully delimited by the fertile 'polder' agriculture of the crater floor.

Durham Cathedral on a river meander.

how resources are utilised. Just think where your own town or city is sited. Is it on a hill near a river but out of the flood plain like Ripon in North Yorkshire or is it protected by high cliffs of a meander core like the City of Durham, or is it built along the banks of a main waterway, but with good drainage and foundations for buildings on river terraces like Dublin and London? Proximity to a water supply and a defensible position are the two common denominators in the location of the majority of settlements, and almost every large town and city can trace its development back to one or both of these two factors. Does yours?

GEODIVERSITY BEFORE BIODIVERSITY

Colleagues in Tasmania have a saying "Geodiversity before biodiversity" and that ethos must be promoted in all our dealings with others, as biodiversity is merely a part of Geodiversity. Plants need minerals to grow and many plants, are restricted to soils from particular rock types; animals and plants have exploited those rocks and the resultant habitats and landscapes. Like *Homo sapiens* they are opportunists. The biodiversity of Britain has many examples but a particularly striking one is the Manx Shearwater (*Puffinus puffinus*), a petrel, which exploits the soft bands as nest holes in the ultrabasic layered complex rocks of the peaks of the Cuillins on the Hebridean island of Rum. This is the largest of the 4 islands off the coast of the Isle of Skye and has been owned and managed as a natural laboratory by the Nature Conservancy and its successor Scottish Natural Heritage since 1957. The name Rum is from the Norse Rom oe meaning wide island; it is 13 km by 14.5 km. The shearwater colony is in excess of 100,000 birds making it one of the worlds largest and probably representing 20% of the world population. The other islands off Skye are Eigg, Muck and Canna.

The jagged peaks of the Cuillins, in contrast to the gentler landscape of sandstones, basalts and granites in the north and west of the island, are a fine example of an eroded magma chamber of a volcano that erupted during the Tertiary period 65 million years ago. The remnant is the layered complex, composed of peridotite and allivalite (both olivine rich) bands that weather to a soft sand in which the shearwaters burrow; the birds do not nest in other soils on the island. The other amazing fact about the shearwaters is that every year they return to their nest holes, all above 450m on the southern peaks, from their wintering grounds off Brazil, a distance of over 6,400km.

Rum is also home to several diverse plant communities, which are restricted to particular soils with minerals derived from the underlying rocks. Arctic sandwort, (*Arenaria norvegica* ssp. *Norvegica*), northern rock-cress (*Arabis petraea*), mountain sorrel (*Oxynia digyna*), mountain everlasting (*Antennaria dioica*) and Scottish asphodel (*Tofieldia pusilla*) prefer the cobalt, chromium and magnesium rich soils of the ultrabasic rocks of the Cuillins. Rum is one of the best sites in the United Kingdom for open rocky Calaminarian vegetation characterised by the presence of arctic sandwort, the rare ultramafic species while other uncommon basiphiles include purple saxifrage (*Saxifraga oppositifolia*), mossy cyphel (*Minuartia sedoides*) and moss campion (*Silene acaulis*). The habitat is developed on terraces on the peridotite of Ruinsival. The ultrabasic rocks also have an extensive development of the black bog-rush (*Schoenus nigricans*)-rich form characteristic of northern Atlantic wet heaths and show the development of *Schoenus* in response to base-rich flushing. These rocks also have examples of European dry heaths but are particularly note worthy for the presence of species-rich heath (*Calluna vulgaris - Erica cinerea* and *Thymus praecox - Carex pulicaris*) occurring on the steep south facing slopes. The associated flora includes a number of northern and arctic-alpine species, such as viviparous sheep's-fescue (*Festuca vivipara*), alpine meadow-rue (*Thalictrum alpinum*), alpine bistort (*Persicaria vivipara*) and alpine saw-wort (*Saussurea alpina*); the invertebrate fauna contains large populations of rare and local burnet moths.

The ultrabasic rocks on Rum provide nest holes for opportunist birds

The ultrabasic rocks on Rum provide nest holes for opportunist birds but the limestone pavement of the Burren - the word means a rocky place - is home to a great variety of opportunists including some species found nowhere else. This biodiversity is a result of the geological diversity, as pockets of soil rather than continuous cover and terracing are produced from solution of the different limestone lithologies. The Burren has a typical karst aspect with the famous slab-like clints and grikes, the gaps between the clints; clint is a Scandinavian term for rock and grike is a Yorkshire dialect word for a crack between the stone flags of a floor in a house, surely also Norse in origin. In the Burren they are known prosaically as slabs and scailps.

The clints may break up to form shattered limestone pavement where soil pockets tend to be continuous and solution by cyanobacterium such as Nostoc sp can create solution hollows on the clints. Grikes are home to nationally rare species such as baneberry (*Actaea spicata*) and hoary rock-rose (*Helianthemum oelandicum piloselloides*) the Irish and Pyrenean sub species, which is confined to the Burren. Maidenhair fern (*Adiantum capillus-veneris*) a rare species in the British Isles is a characteristic species of the deeper grikes on the Burren. Hart's tongue fern (*Phyllitis scolpendrium*) is another plant benefiting from the higher humidity in the grikes. Animals are less obviously restricted but the Burren green moth (*Calamites tridens occidentalis*) is found nowhere else.

WORLD HERITAGE SITES

Surprisingly the Burren is not a World Heritage Site, although 40,000ha are designated as a Special Area of Conservation (SAC), but there is no doubt that as Ireland now has about 90% of the world's limestone pavement and 70% of that is the Burren, its importance is immense. However several natural and cultural World Heritage Sites are excellent examples of geodiversity.

St Kilda out beyond the Outer Hebrides, off the west coast of Scotland, is a group of six islands. Hurta, Soay, Dunn, Oiseval, Boreray and Conachair mark the site of a volcano that erupted 65 million years ago when the Giant's Causeway and the island of Staffa, off the coast of Mull in the Inner Hebrides, were also formed. The rocks, which now form this spectacular group of islands, resulted from the slow cooling of molten magma deep beneath the erupting volcano. Magmas of different composition formed the dark grey gabbros, plutonic igneous rocks rich in basic rock-forming minerals, which run from Soay, Hirta and Dun to Boreray, and the lighter pink granite, a more acidic plutonic igneous rock, of Conachair and Oiseval.

After centuries on the islands the population requested that they be moved off and left in 1930.Shortly after the St Kilda house mouse became extinct as no grain or other human food was available. So after nearly 1200 years living on the island with people, it was probably introduced by the Norse colonisation in the 8th century, it disappeared. Another earlier extinction on St Kilda was the Great Auk, once a familiar sight to sailors in the North Atlantic, in 1840, what is believed to have been the last Great Auk recorded in the British Isles, was killed by the St Kildans on Stac an Armin. It is said that they thought it had caused a violent storm, and they suspected it was a witch! The last Great Auks in the world were killed in Iceland in 1844. Fortunately mounted specimens exist in museums with a particularly fine specimen in the Glasgow Museum collections.

Seabirds formed a major part of the St Kildans diet especially puffins, gannets and fulmars. Puffins used to be eaten as a snack, just as we might eat a bag of crisps; records show that in 1867 the local population ate nearly 90,000 puffins.

St Kilda has the world's largest colony of gannets nesting on Boreray, the largest colony of fulmars in the British Isles, nearly 65,000 in 1999 and one of the most extensive groups of vernacular building remains in Britain. The layout of the 19th-century village remains but there are over 1,400 stone-built cleits for storing food and fuel, which are scattered all over the islands, even on the remote sea stacs.

St Kilda is a good example of Geodiversity which links the islands' geology and its outward expression as the rocky landscape, now home to vast colonies of sea birds and, perhaps unusually for a natural World Heritage Site, also links the culture of the people, their history and some of the St Kildans material culture remains.

CULTURAL GEODIVERSITY

Perhaps the most famous World Heritage Site in Britain is the Neolithic to Bronze Age Stonehenge, a cultural site with a diverse past showing several stages of construction, each utilising the local stone, and the not so local stone in the case of the Bluestones (dolerite) from the Prescelli hills in West Wales. Religious sites capitalise on their position in the landscape and the builders of the first stage emphasised this by using bright white chalk dug on site, from the Cretaceous chalk of Salisbury Plain, for the bank. Half a millennium later Bluestones were imported from Wales and erected as a double crescent, and two centuries later still local Sarsen stones, silicified sandstone, each weighing over 25 tonnes, were brought to Stonehenge from the Marlborough Downs, 30km to the north. A processional avenue was added, marked with parallel banks and ditches and leading into the principal entrance. Outside were set the Heel Stone and its partner, marking the alignment with the midsummer sunrise. Over the next 700years Stonehenge changed several times and an altar stone became the focal point. Perhaps one day English Heritage, the custodians of the built heritage in England may be persuaded to re-instate Stonehenge back to this final stage rather than the ruinous apology for a monument we see

Stonehenge

today. However English Heritage announced on 1st August 2002 a £57 million plan to improve the view of the henge by laying the main A303 road in a tunnel for 2km, closing the A344, creating a 32km sweep of grassland between the henge and Avebury and building a new below-ground visitor centre so that the structure will be once again isolated in the landscape.

Many landscapes in Britain have religious structures that use position and materials to emphasise their existence. The Neolithic Thornborough henges are sited across a relatively high fluvio-glacial gravel plateau, which flanks the River Ure in North Yorkshire. Three almost identical and equally-spaced henge monuments which all share the same north-west/south-east alignment are placed approximately 550m apart and the alignment extends for nearly 1.7km. A number of round barrows are scattered across the surrounding landscape.

Such a concentration is clearly unusual, and the design of the henges also sets Thornborough apart from most comparable later Neolithic complexes. Their size is almost identical, each possessing a diameter of around 240m, while each henge is defined by a massive ditch and bank, interrupted by a pair of entrances, and an outer ditch, which is irregular and segmented in outline. Recent excavations have demonstrated that the latter was associated with an outer bank emphasised with white gypsum locally dug. The design is matched only by the three almost identical, but less distinct, monuments a few kilometres south east and further downstream at Nunwick, Hutton Moor and Cana Barn, just north of Ripon and the Big Rings site at Dorchester-on-Thames in Oxfordshire. Sadly the complete lack of directional signage and interpretation at the Yorkshire sites is typical of the poor general level of cultural awareness by public servants for the public in Britain.

The gypsum in Permian age strata continues to have a significant effect on the local landscape of Ripon and its culture. Lying beneath a cover of glacial boulder clay (till) and upper Permian siltstones and mudstones on the west, and till and Triassic sandstones on the east, it is dissolved by groundwater creating caves and cavities, which cause the land above to collapse. The British Geological Survey has mapped these solution hollows and many are very visible and control where development has and can take place in and around Ripon. The solution follows jointing in the gypsum and not surprisingly most of the City of Ripon has developed out from the stable Market Place and Cathedral areas leaving the north west-south east aligned depressions free from any development. Occasionally cavities cause the more competent Triassic sandstones to spectacularly collapse. One of these known as the Big Hole of 1834 by Ripon railway station must have often been seen by Lewis Carroll, an inveterate traveller and user of the new railways and of course one of Britain's most famous authors, as his father was chaplain and later a Canon to the first Bishop of Ripon Rev. Charles Longley, later Archbishop of Canterbury, and his father's old tutor at Christ Church, Oxford. The young Charles Lutwidge Dodgson, latinisation and reversing of his second and first names gives Lewis Carroll, stayed every year in Ripon for a number of weeks while his father was examining students for confirmation, before he himself became a student at Christ Church in 1851. Dodgson also began his long interest in photography while in Ripon and became one of the most famous Victorian portrait photographers with sitters ranging from the Pre Raphaelites -Rossetti, Holman Hunt and Millais, to Ruskin, Lord Tennyson, politicians and the actress Ellen Terry.

The Big Hole of 1834 by Ripon railway station

Dodgson's Alice in Wonderland, Through the Looking Glass, Hunting the Snark and his other literary works are the fruit of his classical education, his observations of life and the local landscape, the rapidly changing society, and his own experiences imbued with an analytical, cruel and satirical sense of humour and imagination. A very visually astute cleric who was a man of his time despite an apparent life of seclusion in an Oxford College. He would have known the 16th century misericords in Ripon Cathedral especially the Mayor's seat with the misericord of a griffin chasing rabbits down a hole, the Saxon crypt with its sloping tunnel like entrance, and other misericords of animated creatures, pig's with bagpipes (Hunting the snark), griffins and the blemyas, figures with heads on legs and also in Alice, on the Bishop's misericord all only a stone's throw from the canon's residence, where the young Dodgson stayed when in Ripon. It is not impossible to see where the idea, for a white rabbit followed by a young girl who fall down a big hole and meet other curious animals and have surreal activities, comes from.

Dodgson would have been equally familiar with Durham Castle and Cathedral sited on the incised meander of the River Wear. The late 11th century Romanesque Cathedral begun in 1093 and finished within 40 years with the transepts having the earliest high level ribbed vaults in Europe. Massive vaults, massive columns, flying buttresses, geometrical decoration, all in sandstone from the adjacent outcrop. So this cultural World Heritage Site sits on a geomorphological feature controlled by its geology, built by people with an understanding of natural materials - the local Coal Measure (Westphalian) sandstones and located as an impressive, aesthetically pleasing suite of structures that seem to grow out of the landscape, but easily defended because of their location.- classic geodiversity.

King Edward I of England's dominating ring of immense fortresses in North Wales represents Europe's most ambitious and concentrated medieval building project, designed to prevent the recurrence of two incredibly expensive military campaigns. After Edward's first campaign in 1277 against the Welsh, he was able to pin down his adversary Llewelyn ("The Last") in Snowdonia and on Anglesey. This gave him time to build the now largely ruined castles at Flint, Rhuddlan, Builth Wells and Aberystwyth, and upgrade existing Welsh castles. Edward's first attempt to subsume the Welsh was Llewelyn's second uprising in 1282 where he was unsuccessful, and determined not to have to fight for the same land three times he extended his ring of fortifications in an extremely expensive display of English power and organisation. Local stone for building and lime for mortar was preferred, but certainly Rhuddlan used a purple sandstone from St Asaph, the local Carboniferous Limestone and red Triassic sandstone from Chester for dressings. The castles at Harlech, Caernarfon and Conwy, though nearly contemporary, display a unique progression towards the later and more highly evolved concentric design of Beaumaris; Ruthin and Holt were enclosure castles with towers. The vast programme cost £80,000 in the years 1277-1304 and another £15,000 from 1304 to 1330, probably in excess of £50million in today's terms. It involved an extensive exercise in recruitment and impressments from all over England. Craftsmen and labourers were sent to Wales through Chester and Bristol, with the vast majority via Chester. In 1282-3 around 8000 woodcutters for clearing, 1000 diggers, 400 carpenters and 150 masons were used; 3500 men were working on Beaumaris in the summer of 1285 roughly equal to about 15% of the total number of men working in London. Soldiers were present to guard workers as they travelled to the sites. Bonuses were paid for quick work, in kind, and wages docked for absenteeism.

Conwy Castle

The ten castles including the town walls of Caernarfon and Conwy, were the work of the master military architect/engineer of his age, Master James of St George, whose work was well regarded by Edward who showered him with cash and honours. We too recognise his skills, as this Edwardian ring of stone represented by Harlech, Conwy, Caernarfon and Beaumaris, is now a cultural World Heritage Site. Each of the castles was integrated with a bastide town, an idea borrowed from Gascony in southwest France, where Edward I was duke - the town and castle mutually reliant on each other for protection and trade. The bastides were always populated with English settlers, the Welsh permitted to enter the town during the day but not to trade and or carry arms. It wasn't until the 18th century that the Welsh would have towns they could truly call their own. Impressive structures in an impressive landscape moulded to provide a ring of stone to impose a culture and subsume another; a complete takeover and a prime example of Geodiversity - people, landscape and their culture.

Giant's Causeway, one of Ireland's natural World Heritage Sites, natural but like St Kilda, with a cultural dimension. The succession of lava flows and interbasaltic beds, the regular columnar jointing of the Tholeitic basalts, the Causeway of 40,000 columns and the cliff exposures of columnar and massive basalt, and the spectacle they present are the substance of the World Heritage Site. The site is, however, not currently fully accessible as the lower coastal path was closed on safety grounds in 1995 and remains closed. It is not beyond imagination to construct an engineering solution to the problem of the path being removed by landslips.

The Causeway, on the Antrim coast, is so called because of the legend of Finn Mc Coul (Fionn Mac Chumhaill). The story says that Benendoner, a Scottish giant, came over to Ulster by building the causeway. Finn McCoul's wife suggested he dress up as a baby in order to frighten Benendoner. The idea worked as the Scottish giant thought that if this is the size of the bairn (baby) what size must the father be and he fled back to Scotland destroying the causeway, which can now only be seen at both its ends in County Antrim and on Staffa. Legend also notes that Finn was responsible for creating Lough Neagh in the middle of Ulster by picking up a clod of earth and throwing it at another giant in Scotland. He mis-aimed and it fell into the sea and became the Isle of Man halfway between Ireland and England. These are myths, told to explain the then unexplainable, but the truth is that Finn was a folk hero of the Iron Age of about 400-300 BC and leader of the Fianna warriors raised by the High King Fiachadh of Leinster.

These examples show the importance of Geodiversity to the public understanding of the world around them. Scientists have tended to compartmentalise our world to fit our own specialisms, but the public see the world as a whole and this requires interpretation to make it more easily understood. The link between the landscape, people and their culture needs to be emphasised through raising awareness, understanding and increased knowledge to the public and decision makers.

One way is to produce information for travellers in trains so that the passing landscape is not just a blur of hills, trees, buildings and other structures, but a succinct interpretation in jargon free language of the view. Informative guides and leaflets are frequently single subject specific, but more are beginning to explain other interests as writers, publishers and organisations realise that people see the whole picture rather one small part.

Creating new trails designed specifically to explain the geodiversity is one obvious way of promotion. The Chalk Way based on the extensive chalk wolds of England, wolds that have been used for millennia as travel routes for trade, is proposed to begin in East Yorkshire. Excavations and existing structures reveal that Rudston on the Yorkshire Wolds was a centre for the trade of stone axes as specimens from almost every axe factory in Britain have been found there. It also has religious significance as the largest standing stone in Britain at 8m high stands adjacent to the much later church, and an array of cursii stretch out into the wolds around. Trade may have developed as a result of pilgrimage/ homage to the site as it always does where people congregate. Rudston is one of many possible stops with interpretation on the route for walkers, cyclists and drivers.

Archaeologists have for many years fired the public's enthusiasm by high profile excavations such as York's Coppergate that revealed Jorvik -Viking York. Biologists hooked the public many years earlier with a string of television presenters who made the life of animals and plants exciting -usually their sex life figured largely. Historians have made countless forays into peoples lives through book and film, but few in Britain have looked at the whole picture except perhaps WG Hoskins in his seminal work "The making of the English Landscape" published in 1955 which gives an historical perspective of landscape development. A more holistic view of the natural and cultural heritage could use cities, towns and villages, sitting in their landscapes as teaching aids to this approach. 'Adopt a monument' schemes like that in Edinburgh and new art/sculpture in town and country need encouragement as do architects, sculptors and engineers to label stone used - what it is and where it is from. Local materials should be used when possible in sculptures, for buildings, for bridges and town and city trail guides such as that written by Patrick Wyse Jackson for Dublin encourages the viewer to see those buildings, bridges and sculptures in a different light.

But one area that needs further development, well certainly in Britain, are site management plans for all designated sites. What is the point in designation for public benefit and posterity if the site is not managed for posterity and public benefit? Handrails along the cliff on Staffa, when there are perfectly flat and safe basalt columns to walk on to visit Fingal's Cave seems to be Health and Safety gone mad as these structures ruin the integrity of the site; there are also warning signs that "waves can be dangerous" on the Giant's Causeway, but in the litigious society that we live in it can be appreciated that there are potential cases for negligence that could be brought by visitors against the National Trust for Scotland and the National Trust respectively who own

Giants Causeway warning signs

the sites. Surely a policy of invoking a disclaimer for all visitors could be introduced at the main information point - *This is a natural site and we will not compromise its integrity by introducing physical structures alien to the site even in the interests of public safety. You walk here at your own risk. We will however take every care that is possible to ensure your safety* -

Many natural sites, such as Brimham Rocks in Yorkshire, have no physical barriers to prevent roaming over the high sculptured rocks. It is difficult to see why Staffa and the Giant's Causeway should.

But the greatest threat to landscapes in Britain and elsewhere is the threat of "development". Whether it is removal of limestone pavements for garden and municipal rockeries, archaeological despoliation through removal of field boundaries and deep ploughing that removes cultural features and monuments or industrial landscape emaciation, because people think that it is ugly or unsightly. For example when coal spoil heaps, often the only tangible evidence of former mining in the area, are flattened the heritage disappears in the name of planning tidiness. We can make a difference by giving help and advice, by surveying areas to ensure decision makers are aware of the unique local materials that are used in the local area, by giving public lectures on local geodiversity. In some areas the local plans only allow for traditional materials to be used as in Hovingham, North Yorkshire where warm yellow Upper Jurassic limestone walls are used with red pantile roofs, a tradition probably in use since the Roman takeover almost 2000 years ago

Just as there are Biodiversity Action Plans in place in every county in England there should be Geodiversity Action Plans (GAP) implemented as suggested in the Earth Science Strategy published by the Royal Society for Nature Conservation in May 2000. There should be a national plan and local GAPs designed to be proactive and complimentary. National Plans would integrate with national and regional planning to strive for a simplified designation system rather than the plethora of terms currently in use and a requirement for all designated areas and sites to have management plans by 2005. But as implored by our colleagues in Tasmania we must put Geodiversity before biodiversity.

In: Parkes, M.A. (Ed.) 2004 *Natural and Cultural Landscapes - The Geological Foundation*, 53-56, Royal Irish Academy, Dublin

THE PATTERN OF GEO- AND BIODIVERSITY CONSERVATION IN POLAND

Zofia Alexandrowicz, Anna Dyduch-Falniowska and Wojciech Mróz,
Polish Academy of Sciences, Institute of Nature Conservation, al. A.Mickiewicza 33, 31-120 Krakow, Poland, Emails: alexandrowicz@iop.krakow.pl, mroz@iop.krakow.pl

Abstract: Poland has a very differentiated geological background. Geological formations, surface morphology and soil cover are the chief determinants of habitats differentiation. To estimate relations between the various geological and biological strands two digital maps have been joined using GIS technology. The first of them shows the geodiversity of Poland and the second one, created in the frame of "CORINE Biotopes" project, comprises Polish natural sites. The interpretation of these maps indicates the considerable integration of geo- and biodiversity, particularly in southern Poland. The case study of Silesia-Cracow Upland is presented in the article as the example of a coherent nature conservation network.

INTRODUCTION

Poland - the country situated in central part of Europe is spread over three main geotectonic structures of the continent. There are East-European Precambrian Platform bordered with the Tornquist - Teisseyre Zone from the West-European Palaeozoic Platform while the Alpine orogen cover the southern segment of the area (Fig. 1). In the northern and central part of Poland structural units of the bedrock are covered with glacial and fluvioglacial sediments of Scandinavian ice sheets. Quaternary deposits exceed 100-200 m here and landforms typical of glacial areas are dominant. On the contrary in southern Poland Quaternary deposits occur as a thin discontinuous mantle and structural forms of the relief are widespread. Different geological formations of Palaeozoic, Mesozoic, Tertiary age or even older ones crop out. This spatial duality is the distinctive feature of the Polish relief. Physico-geographical features following this pattern is defined mainly by nature diversity, namely geo- and biodiversity.

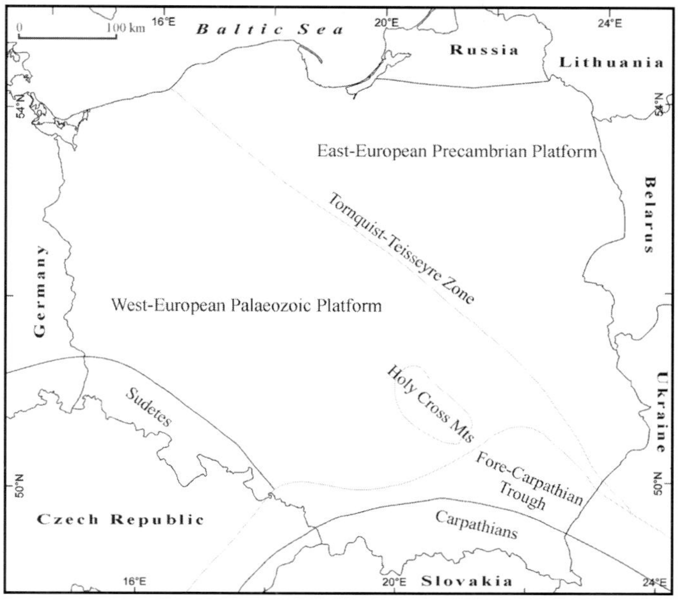

Figure 1. Main geotectonic structures of Poland.

INTEGRATION OF GEO- AND BIODIVERSITY

Geological formations, surface morphology, soil cover and climate are chief determinants of habitats differentiation. To estimate relations between various geological and biological strands two related maps have been joined using the GIS technology (Fig. 2). On the first of them the geodiversity of Poland is presented in five-class scale (Alexandrowicz and Kozlowski 1999). Two highest classes (1 and 2) represent about 12% territory of Poland. Two next classes (3 and 4) spread over 37% of the country while the remaining area has relatively lowest differentiation. Natural values and their state preservation, degree of anthropogenic transformation and usefulness for sustainable regional and local socio-economic development are common criteria the evaluation of geodiversity components. The distribution of protected geosites in particular regions is more or less adequate to distinguished classes of the geodiversity. The network of the geoconservation is gradually supplemented with a special regard to the important geosites, representative for particular regions and the European framework included as a part of GEOSITES/IUGS Programme (Alexandrowicz 1998, 1999-ed., Wimbledon 1998-ed., Alexandrowicz and Poprawa 2000, Gawlikowska 2000, Wroblewski 2000).

Biological patterns comprising large natural sites exceeding 100ha, specified as key habitats are offered as a second component of the joined map (Fig. 2). It represents nearly 20% of the country being a potential part of coherent conservation network in Europe (Dyduch-Falniowska et al. 1999).

The interpretation of two mentioned overlapping maps indicates the considerable integration of geo- and biodiversity. A mosaic pattern of differentiated classes of geodiversity well corresponds with the distribution of natural sites in mountain and upland areas of the southern Poland. Areas of an average geodiversity form belts in North Poland. They are connected mainly with ranges of terminal moraines forming long ridges. The important natural sites are distributed along them. Middle Poland has a lowest geodiversity. This area is considerably transformed by human agriculture activity. Interesting habitats of plants and animal concentrate mainly along river valleys and lakes. The presented method of overlapping mentioned

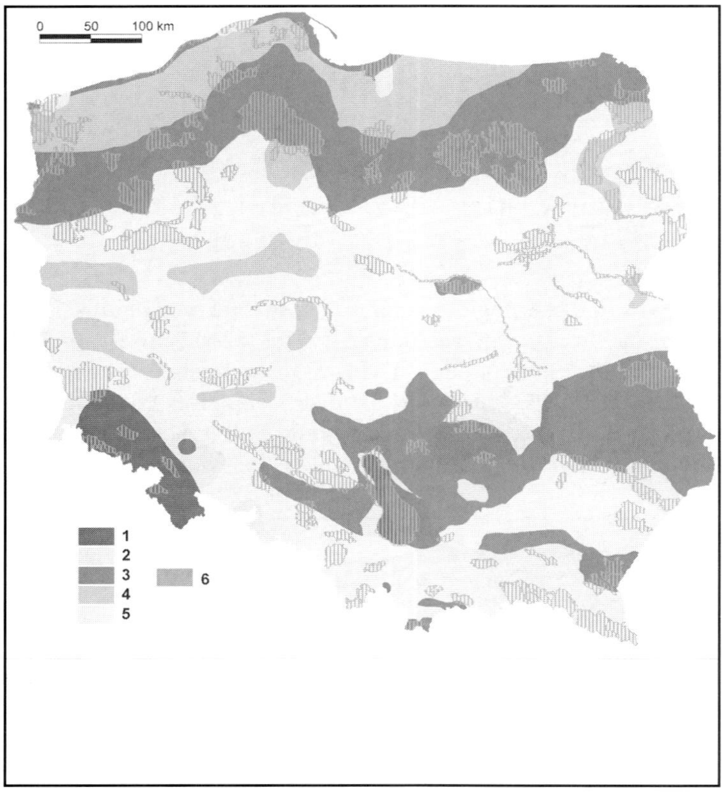

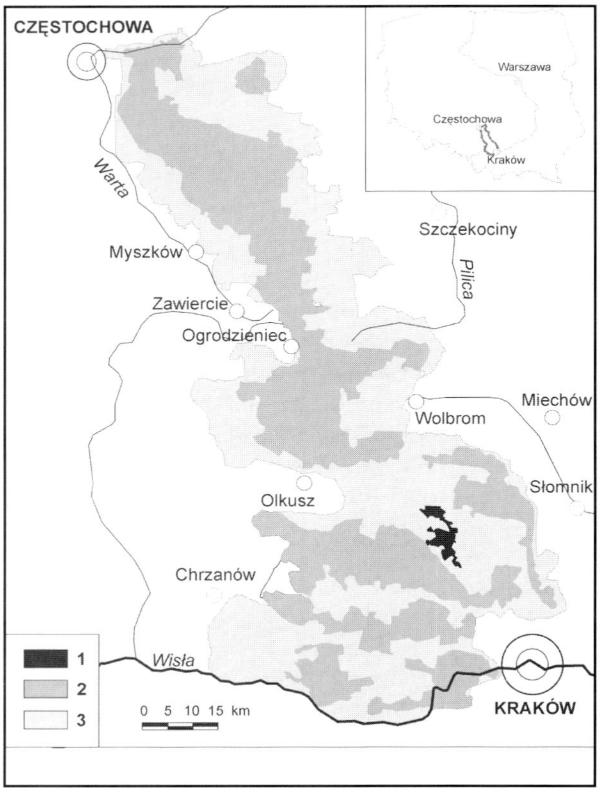

Figure 2. above left. Joined pattern of geodiversity and biological habitats of Poland. 1 - 5 classes of areas according to geodiversity from highest (1) downward (5); 6 - large natural sites (biosites) exceeding 100 ha as key habitats.

Figure 3. above right. Polish Jura Chain - Krakow-Czestochowa Upland 1 - Ojcow National Park, 2 - Complex of Jurassic Landscape Parks, 3 - buffer zone.

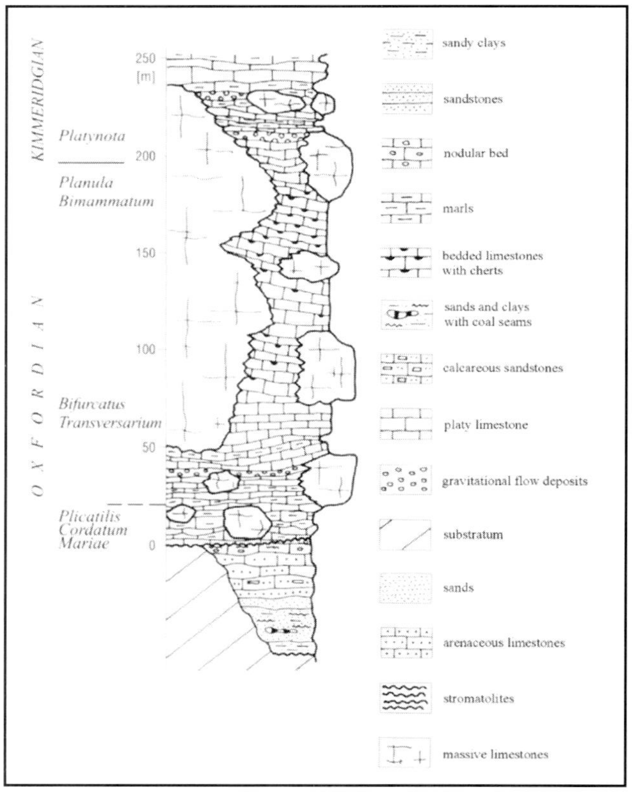

Figure 4. Lithostratigraphic profile of Upper Jurassic deposits in Cracow region (after J. Matyszkiewicz 2001, simplified). Explanations on the figure.

maps should be indicated as a way of the development of nature conservation in a holistic manner.

COHERENT NATURE CONSERVATION NETWORK - A CASE OF SILESIA-CRACOW UPLAND

Many areas representing the key values for both geological heritage and natural habitats are known in Poland. One of them is the Silesia-Cracow Upland of southern Poland. It is very instructive for cross-disciplinary investigations in geo- and biodiversity applied to nature conservation. This geological region encloses the Variscan bedrock covered with the Epivariscan Platform as well as the northern tectonic border of the Fore-Carpathian Trough. Geological formations representing the time span since Middle Devonian up to Holocene crop out in several localities. A part of them has been selected as candidates on the European List of Geosites (Alexandrowicz and Alexandrowicz 1999). The supplemented list of upland 20 geosites encloses, now.

The most interesting part of the Silesia-Cracow Upland called the Polish Jura Chain is situated between Krakow and Czestochowa (Fig. 3). It is almost wholly protected as Ojcow National Park (21 km^2) and Complex of Jurassic Landscape Parks (about 1200 km^2) joined together by the buffer zone. In the total area there are 27 nature reserves and several hundred of nature monuments and other categories of small sites. Geological conditions and specific geomorphology, particularly karst phenomena determine the individual character of the region. High geological and geomorphologic valorisation of Polish Jura Chain promotes this area to the proposed international network of Geoparks UNESCO (Alexandrowicz and Alexandrowicz 2000). Thick complex of Upper Jurassic (Oxfordian and lowermost Kimmeridgian) massive and bedded limestones is the main geological component of the region well reflected in its relief (Fig. 4). The massive limestones are typical carbonate cyanobacterial-sponge buildups comparable with similar ones in the Swabian Alb. They form hardrocks with several groups of limestone tors rising on the large Tertiary planation surface (Figs 5 and 6). Limestone tors are generated by both joint and karst, modified by caves and rocky-shelters with traces of prehistoric cultures. The plateau gently sloping to the north-east is incised by numerous rocky valleys and gorges (Fig. 7). The ruins of medieval castles crowning rocky hills are enriching the natural landscape.

Figure 5. View from the Zborow Hill over the Tertiary planation surface of the upland near Czestochowa, partly covered with the pine-oak forest and crowned by limestone tors. On karstic Jurassic rocks the xerothermic vegetation developed. Photo by Z. Alexandrowicz.

Diversified relief and differentiated microclimatic conditions result in great differentiation of habitats and the same, in a great natural richness of area. The phenomenon of Jura chain lies in that it has preserved this richness despite the many hundred years of man's impact. Moreover the fragmentation of habitats, which in part is the effect of man's husbandry, contributed much to natural richness of this region.

The total number of vascular plant species is about 1600, i.e. more then 50% of Polish flora and the number of mosses reaches 400 (Babczynska-Sendek et al. 1992, Jedrzejko et al. 1992). Among the vascular plants there are two endemic species, known only from this area, i.e. *Galium cracoviense* and *Cochlearia polonica*. It is also the place of the occurrence of rare birch *Betula oycowiensis*. The specific feature of the flora is high percentage of xerothermic and mountain plants.

Approximately 100 plant communities were described from the area. About 70% of the area is occupied by synantropic communities, encountered in fields, gardens, parks, uncultivated grounds etc. Forest communities cover 20% of the region. Previously, large areas of plateau were

Figure 6. Rocky group of Jurassic massive limestone separated by enlarged fissures on the plateau near Krakow. Photo Z. Alexandrowicz.

overgrown with pine-oak forests. Most of these forests are now greatly changed. In dry and sandy places there occur poor pine forests. Mixed oak-lime-hornbeam forests are found on rich limestone substratum. Northern and eastern, shady valley slopes are overgrown with forests of mountain character, such as Carpathian beech wood or fertile Sudetic beech wood; their sites in the Jura are of a relic southern character. Sunny opposite slopes are covered by thermophilous beech forests. Among non forest communities, particularly worthy of notice in xerothermic and calcareous vegetation, represented in the Jura by natural and semi-natural grasslands developed on rocks and their surroundings (Fig. 8). Patches of natural grasslands belonging to the association *Festucetum pallescentis* are particularly worthy of mention.

Figure 7. Fragment of rocky dry valley incised in large Tertiary planation surface within the cultivated landscape. Photo Z. Alexandrowicz.

The fauna of Silesia-Krakow upland is also extraordinary rich. It is estimated that the described region is inhabited by more than half the animal species recorded in the Polish fauna (Szeptycki, Warchalowska-Sliwa 1992). Characteristic features of the invertebrate fauna of this region are the presence of xerothermic elements and post-glacial relics, and high percentage of Carpathian species.

Figure 8. Dry meadows of Festuco-Brometea class (typical of Southern Poland uplands) develop in specific abiotic conditions e.g. in feet of calcareous rocks. They distinguish themselves with very high plant and animal diversity. Photo J. Perzanowska.

Among xerothermic elements, comprising taxa associated with open and warm habitats are species confined to grassland as well species living on rock and sand. In respect of zoogeography these are Sub-Pontic and Sub-Pannonic species. Specific to the mammal fauna of the region is a very rich group of insectivore species with almost all species of bats known from the country. Among them the most interesting species are: the seriously threatened lesser horseshoe bat *Rhinolophus hipposideros*, the greater horseshoe bat *Rhinolophus ferrumequinum*, sporadically noted in this area and the extremely threatened Geoffroy's bat *Myotis emarginatus*, all connected with rocky habitats.

Taking into consideration all different natural values the region has been selected as a site of European importance within the framework CORINE biotopes programme. Selection of these sites was based on different criteria, from zoological and botanical, through habitat, to geomorphologic ones. The whole area has been identified as a so-called complex-site, and particularly valuable 29 objects situated in this area have been selected as its subsites. Their selection was based on the presence of 10 CORINE plant species (*Cochlearia polonica, Cypripedium calceolus, Pulsatilla patens, Microstylis monophyllos, Betula oycoviensis, Botrychium simplex, Epipogium aphyllum, Dianthus gratianopolitanus, Galium cracoviense, Larix polonica*) and 4 animal species (*Rhinolophus hipposideros, Myotis myotis, Myotis bechsteinii, Vertigo angustior*). Other, very important criterion of sites selection in CORINE programme and also in actually projected NATURA 2000 network is the occurrence of natural habitats of European importance listed in Annex I of Habitat Directive (Mroz W., Perzanowska J. 2001, Habitat Directive 1992). The identification of habitats is based mainly on vegetation characteristics but in some definitions abiotic features are also included. More than 10 types of these habitats were identified in the area. Between them there are also priority ones - semi-natural dry grasslands (code 6210) and alluvial forests (91E0). The rich pattern of natural habitats of Polish Jura Chain fit together with the structural landscape in a harmonious way.

REFERENCES

Alexandrowicz Z. 1998. Representative geosites of Poland and their status of conservation. *Geol. Balcanica* 28, 3-4, 37-42.

Alexandrowicz Z. (ed.) 1999. *Representative geosites of Central Europe*. Pol. Geol. Inst. Special Papers 2.

Alexandrowicz S.W. and Alexandrowicz Z. 1999. Selected geosites of the Cracow Upland. In: *Representative geosites of Central Europe* (Z.Alexandrowicz - ed.). Pol. Geol. Inst. Special Papers 2, 53-60.

Alexandrowicz Z. and Alexandrowicz S.W. 2000. Draft project of Jurassic Geopark in the Krakow-Czestochowa Upland (southern Poland). Ann. Meeting ProGEO in Prague. Abstracts: 6-7.

Alexandrowicz Z. and Kozlowski S. 1999. From selected geosites to geodiversity conservation - Polish example of modern framework. In: *Towards the balanced management and conservation of the geological heritage in the New Millenium* (D.Barettino, M.Vallejo, E.Gallego - eds), 40-44.

Alexandrowicz Z. and Poprawa D. (eds) 2000. *Geodiversity conservation of the Polish Carpathians* (English summary). Pol. Geol. Inst., Warszawa, pp.142.

Babczynska-Sendek B., Kimsa T. Wika S., 1992. Vegetation of the Krakow-Czestochowa Upland in circumstances of anthropopressure. (English summary). *Pradnik. Prace Muz. Szafera* 5, 47-63.

Council Directive 92/43/EEC of 21 May 1992 on the conservation of natural habitats and of wild fauna and flora. Official Journal of the European Communities. Brussels.

Dyduch-Falniowska A., Kazmierczakowa R., Makomaska-Juchiewicz M., Perzanowska J., Zajac K., 1999. *Natural sites in Poland.* Institute of Nature Conservation PAS, Krakow.

Gawlikowska E. 2000. *Geodiversity conservation of the Lower Silesia* (English summary). Pol. Geol. Inst., Warszawa, pp. 72.

Jedrzejko K., Klama H. Zarnowiec J., 1992. Bryophytes of the Krakow-Czestochowa Upland. (English summary). *Pradnik. Prace Muz. Szafera* 5, 47-63.

Matyszkiewicz J. 2001. A role of the Cracovian region in the Late Jurassic sedimentation of the Cracow-Czestochowa Upland (southern Poland) (English summary). *Przegl. Geol.* 49, 8, 724-727.

Mroz W., Perzanowska J., 2001. Habitat Directive - natural habitats of European importance in Poland. (English summary). *Chronmy Przyrode Ojczysta* 57, 5, 55-73.

Szeptycki A., Warchalowska-Sliwa E., 1992. Characteristics of the fauna of the Krakow-Czestochowa Upland. (English summary). *Pradnik. Prace Muz. Szafera* 5, 149-159.

Wimbledon W. (ed.) 1998. A first attempt at a geosites framework for Europe - an IUGS initiative to support recognition of world heritage and European geodiversity. *Geol. Balcanica* 28, 3-4, 5-32.

Wroblewski T. 2000. *Geodiversity conservation of the Gory Swietokrzyskie region* (English summary). Pol. Geol. Inst., Warszawa, pp.88.

In: Parkes, M.A. (Ed.) 2004 *Natural and Cultural Landscapes - The Geological Foundation*, 57-60, Royal Irish Academy, Dublin

THE ROLE OF GEOLOGICAL HERITAGE IN THE NATURA 2000 NETWORK (HABITATS DIRECTIVE, 92/43/EEC): A LOCAL STUDY IN NORTHERN ITALY (LIGURIA REGION)

Gerardo Brancucci, Dipartimento POLIS, Università di Genova, Stradone S. Agostino 37, 16123 Genova, Italia. Email: brancucci@arch.unige.it
Maurizio Burlando, Parco naturale regionale del Beigua, Corso Italia 3, Savona, Italia, E-mail: dirbeigua@tin.it
Valentina Marin, Dipartimento POLIS, Università di Genova. Email: alentinamarin@inwind.it
Guido Paliaga, Dipartimento POLIS, Università di Genova. Email: gpaliaga@arch.unige.it

Abstract: In 1992 the European Union issued the Habitats Directive (92/43/EEC) which concerns the preservation of natural and seminatural habitats and the wild flora and fauna of communitarian interest. The purpose was to promote the preservation of biodiversity in the Union with the adoption of measures for preservation and restoration of the areas called Sites of Community Importance (SCIs). The guardianship was given to Natura 2000 Network, an ecological network which also concerns the Special Protection Areas (SPA) identified by the Birds Directive (79/409/EEC).

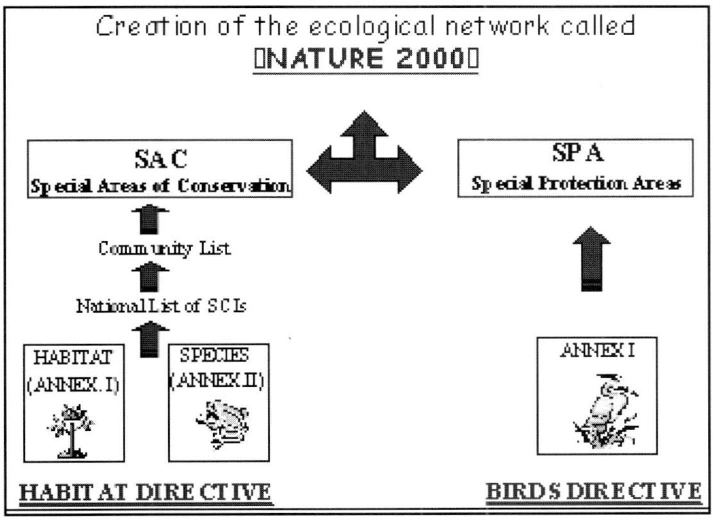

Figure 1. above. Relations between Natura 2000 network, habitat directive and Birds directive.

The EU Habitats Directive provides for protection and improvement actions of the natural habitats and the main purpose is the preservation of animal and plant species which live there. Nevertheless it is clear that geology, together with geomorphology and in general other physical features, are the basic elements of an ecosystem. Because of this connection between physical environment and life, it is clear that it is necessary to preserve geodiversity as it is an essential element for biodiversity. An habitat often appears significant or exclusive because of the presence of animal or plant species which live there thanks to the geological-geomorphological characteristics. So geodiversity - though it is not so much considered by Habitats Directive, even if some habitats can be defined as sites of geological interest - has a fundamental role when we have to consider the necessary measures for the preservation and the management of a natural habitat.

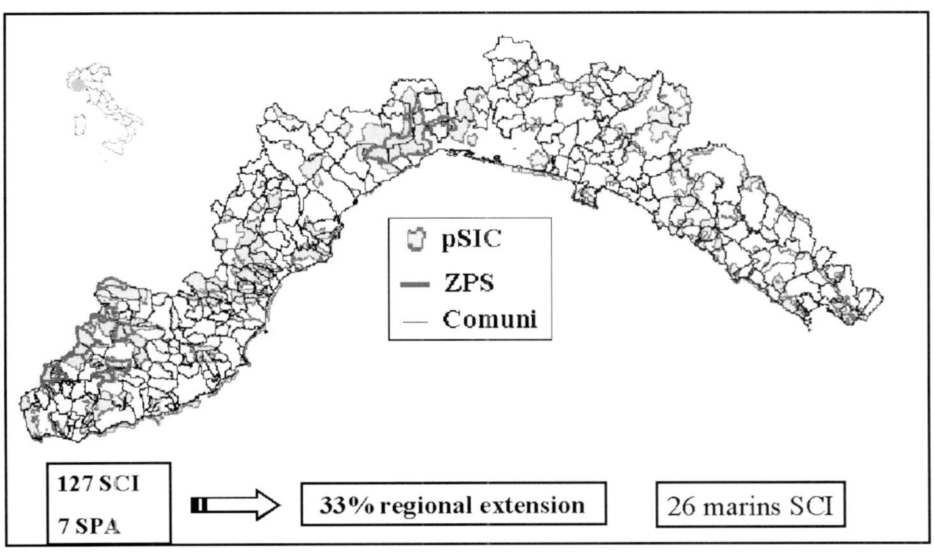

Figure 2. left. SCIs and SPA extension in Liguria region (Italy).

In 1995 the Italian Government started "BIOITALY" Program which, following the provisions of Habitats Directive, identified some areas which could become Sites of Community Importance (SCIs) and in 1997 Habitats Directive became a National rule. The areas chosen by "BIOITALY" Program were proposed to the examination of the European Commission in June 1997 and these SCIs are 2.425 (in the European Community they are

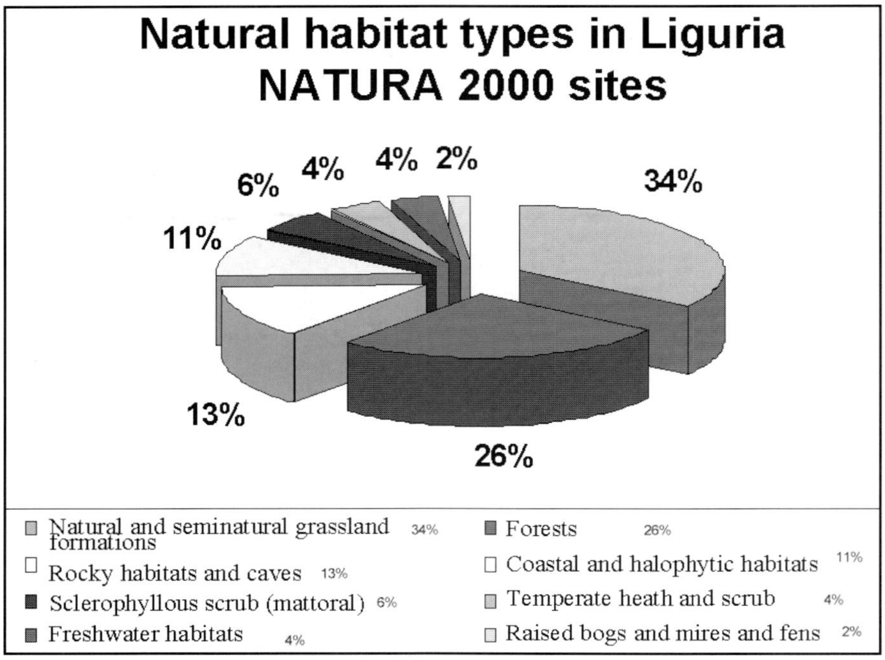

Figure 3. Natural habitats types in Ligurian SCIs.

in total 14.912). In the Liguria Region, in the Northern Italy, the research has identified 127 SCIs (26 of them are marine sites).

In this context we examined the SCI proposed for Liguria and we put in evidence the geological-geomorphological characteristics of them. For the study we used two points of view:

1. analysis of sites where geological-geomorphological features were significant elements but always connected to biological elements ;
2. analysis of sites where geological-geomorphological features were more important than biological elements

1. For the first point of view, the analysis of habitat types of Annex I points out the presence of either habitats with specific geological-geomorphological features or other types though always related to abiotic issues. Within Ligurian habitats we found in 100 SCIs 30 different habitat types that could be considered "of geological interest", showing one of the following features:

- habitats of species with strong dependence from the substratum (calcareus, silicious, etc...);
- important from a geomorphological point of view (cliffs, rocky habitats, etc...);
- governed by geological and hydro-geological processes (rivers, estuaries, bogs, etc...).

2. For the second point of view we studied the presence of singular geological sites inside local Natura 2000 areas and we found 94 geosites in 15 areas.

Some examples in Liguria region show the fundamental role of geological heritage in the preservation and management of many natural habitats, where it is impossible to separate geodiversity from biodiversity.

THE SITE OF COMMUNITY IMPORTANCE (SCIS) CALLED "FINALESE - CAPO NOLI"

In this site geological-geomorphological contents are really important but connected with main biological elements. A wonderful sight in a karst area with different geomorphological attributes: particularly, significant elements of epigeal and hypogeal karstic phenomena, with remarkable scientific value about distribution and morphogenetic mechanisms are signalised. The Finalese - Capo Noli area also shows further biological and environmental peculiarities, connected with geological-geomorphological contents (flora with strong connection with calcareous substratum) and with scenic, didactic and pleasant attributes such as to justify a detailed work of inventorying.

In the area there are 21 habitats of community interest (5 priority habitats); 100 species (81 birds, one priority species of animal *Callimorpha quadripunctata* and one prioritary species of flora *Campanula sabatia* de not); 27 sites of geological interest.

THE SITE OF COMMUNITY IMPORTANCE (SCIS) CALLED "PIANA CRIXIA"

In this site geological-geomorphological contents are prevailing over biological aspects. The sweet slopes cultivated of the Langhe di Piana Crixia in the small homonymous regional park, are at sectors roughly interrupted by gullies, interesting erosive forms really exclusive in Liguria, almost sand waves and clay where the pioneer vegetation seem to try to float. The park's emblem is for the spectacular and gigantic mushroom shaped stone .The natural monument is made up of a large mass of ophiolitic rock held by a column of detritus modelled by the erosion of the flowing of superficial water. In the area there are only 4 habitats of community interest, none of them priority.

THE SITE OF COMMUNITY IMPORTANCE (SCIS) CALLED "BEIGUA" PARK

In this site geological-geomorphological contents help to increase extraordinary biological value of the site. The Park is commonly identified with the massif of Mount Beigua, an imposing rock rampart dominating the coastal area.

Summits reach almost 1,300m rising a few kilometres from the sea. The main ridge of the mountain group - often char-

Figure 4. Rocca di Corno, Finalese - Capo Noli SCIs. (ph. G. Paliaga)

Figure 6. **Cistus Albidus,** *Finalese - Capo Noli SCIs. (ph. G. Paliaga)*

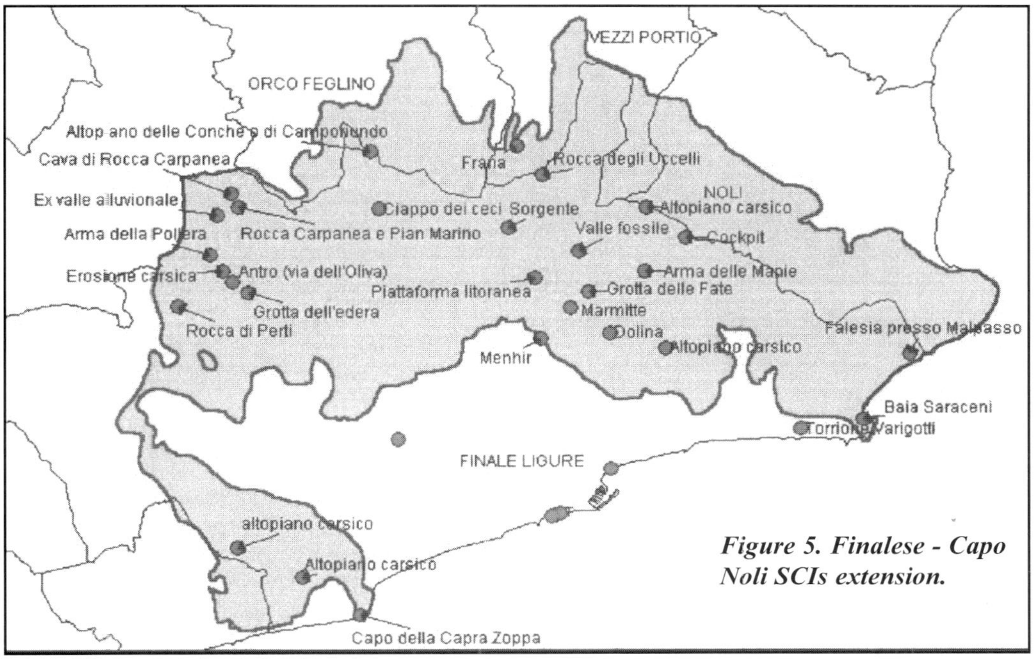

Figure 5. Finalese - Capo Noli SCIs extension.

acterized by plateau - runs almost parallel to the coast and comprehends mountain tops about 1,000 - 1,3000 meters high. The southern slopes - extremely steep - are deeply cut by short streams that flow rapidly towards the sea. The northern sides gently decline towards the Padana plane crossed by long, winding streams. The large quantity of springs and surface waters is due to the presence of the sea close to the mountain ridge. The proximity to the sea is the causes of high quantities of fallen rain or snow, of frequent 'orographic fogs' and abundant snowfalls in the winter. A tormented geological history forged the heart of the park mainly built up of metamorphic rocks - 'ophiolites' in particular - locally covered by sedimentary sequences.

Among the many species whose interest is not only aesthetic but also geobotanical there are 'glacial relics' and the 'plants of the serpentine. The former group comprehends arctic and alpine species that descended from the Alps during the ice ages. The substratum of ophiolite rocks - toxic for most plants - can host only some peculiar species. Some of them - 'relics growing on serpentine' - can grow also elsewhere but the concurrence of stronger plants overcame them. Other species are exclusively or mainly love growing on serpentine soils and need the presence - though in various quantities - of serpentine rocks to survive.

In the area there are 25 habitats of community interest (5 priority habitats); 109 species (97 birds, one priority species of animal *Callimorpha quadripunctata*); 4 sites of geological interest (but not all the area has already been analysed from this point of view).

The last examples and the whole research show the fundamental role of geological heritage in the preservation and management of many natural habitats in Liguria Region, where it is often impossible to separate geodiversity from biodiversity.

Anyway this research put in evidence that geological-geomorphological characteristics - sometimes very significant - have never been rightly appreciated.

So finally two hypothesis can be formulated for further studies:

- a comparison between Liguria and the other similar Mediterranean areas (firstly in France, Spain and Greece).
- a proposal which may include in Natura 2000 Network a proper directive concerning the careful consideration of geological heritage as a whole and as an element of biodiversity.

REFERENCES

Nicosia E. (2001), *La rete NATURA 2000 in Regione Liguria", Congress Proceedings "SIC e ZPS: valorizzazione e salvaguardia della biodiversità in Liguria,* Genoa, Italy, 15th February 2001.

Marin V. (2001), *Valutazione analitica dei SIC (Siti di Importanza Comunitaria) proposti dalla Regione Liguria: comparazione tra valenze biotiche ed abiotiche*, tesi di Laurea, Genoa, Italy.

Brancucci G., Burlando M. (2001), *La salvaguardia del*

patrimonio geologico: scelta strategica per il territorio. L'esperienza della Liguria, Franco Angeli Ed., Milano, 96 pp.

Geological Survey of Sweden (2000), *The importance of a basic geological perspective in Society*, Uppsala, Sweden, 8 pp.

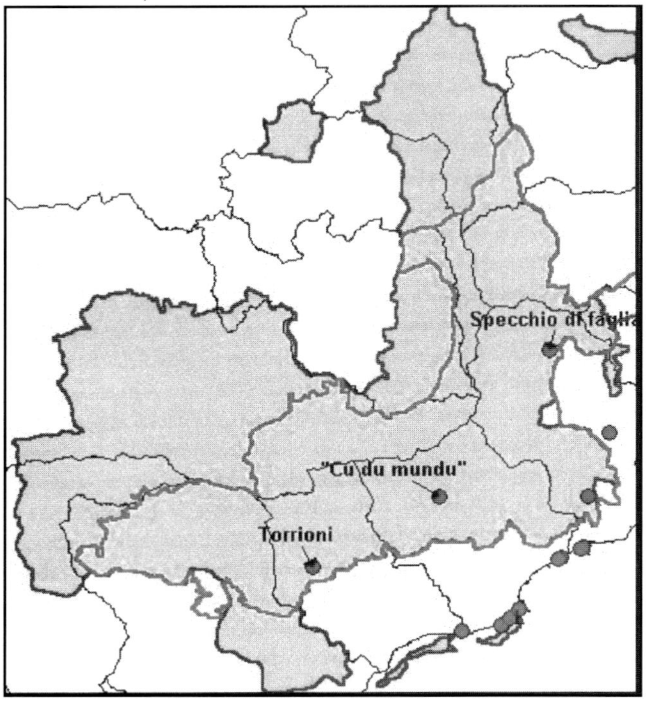

Figure 7. middle bottom picture. The Piana Crixia fungus, an erosion form in Piana Crixia SCIs. (ph. G. Brancucci)
Figure 8. above: Beigua Park SCIs extension.
Figure 9 bottom left. Gargassa valley, Beigua Park SCIs. (ph. G. Paliaga)
Figure 10. top left picture. Lajone torbiera, Beigua Park SCIs. (ph. G. Paliaga)
Figure 11. left. Beigua Park SCIs. (ph. G. Paliaga)
Figure 12 bottom right.* Violetta Odorata, *Beigua Park SCIs. (ph. G. Paliaga)

In: Parkes, M.A. (Ed.) 2004 *Natural and Cultural Landscapes - The Geological Foundation*, 61-64, Royal Irish Academy, Dublin

GEODIVERSITY AS AN EXPLANATION FOR BIODIVERSITY IN THE BURREN NATIONAL PARK

N.R. Moles [1] and R.T. Moles [2] [1] School of the Environment, University of Brighton, Cockcroft Building, Brighton BN2 4GJ, U.K. Email: n.moles@brighton.ac.uk [2] Department of Chemical and Environmental Sciences, University of Limerick, Ireland. Email: richard.moles@ul.ie

Abstract: A holistic understanding of landscape evolution and biodiversity requires scientists to cross-link between the disciplines of geology, soil science and ecology and to study landscapes at scales ranging from square kilometers to square metres. This paper explores the intricate links between landforms, solid geology, surficial sediments, soils and vegetation in the southeastern part of the Burren, a glaciated karstic landscape in western Ireland. At a landscape scale (measured in square km), large-scale patterns in the distribution of vegetation and species are linked to bedrock geology and the distribution of glacial sediments. At a landscape element scale (measured in hectares), weathering and mass movement processes have created a pattern of alkaline soils in exposed limestone till on drumlin flanks, and more acid soils in colluvium on lower slopes. At a patch scale (tens of square metres), geological influences on vegetation include the occurrence of patches of clay-rich soil derived from glacially transported shale-rich boulders. At a site scale (1-10 m), boulder-strewn surfaces inhibit access by grazing animals thus resulting in more successful seed production than in exposed grassland. Effective conservation of biodiversity in the Burren requires management action to preserve soil and maintain plant heterogeneity.

INTRODUCTION

The Burren is a c.750 square km area of glaciated and karstified limestone uplands in northern County Clare. The region has maritime climate with a high rainfall of c. 1200-2000 mm/year (Met Éireann 2002). The Burren National Park, managed by National Parks and Wildlife Service (formerly Dúchas), was established in 1991 with an initial area of 1.150 hectares. Situated on the south-eastern edge of the Burren (Fig. 1), the Park includes examples of habitats typical of the Burren landscape namely limestone pavement, grassland, hazel scrub, deciduous woodland, ephemeral lakes (turloughs) and fenland. The Park area is centred on the hills of Mullach Mór (191m), Sliabh Rua (188m) and Knockanes (207m) (Fig. 2).

Despite Ireland's relatively depauperate flora, grass and heathland vegetation in the Burren is rich in species. Although it comprises only 1% of the land-mass of Ireland, the area hosts over 700 species of flowering plants and ferns, and some three-quarters of Irish native species (Webb and Scannell 1983). Values of 40 to 50 species per square metre are often recorded (O'Donovan 1987), which is as many species per unit area as grasslands elsewhere in Europe that are considered to be of high conservation status. In the Burren, some plants from northern ('Arctic') regions reach their southern limits, some plants from southern Europe attain their northern limits, and some plants from high mountains ('alpines') grow close to sea-level. In addition, calcicole and calcifuge species are found in close proximity, despite the predominantly calcareous bedrock of Carboniferous limestone. The Burren is also rich in fauna, notably birds and butterflies. 46 species of birds are known to breed in the National Park area (Dúchas 1991).

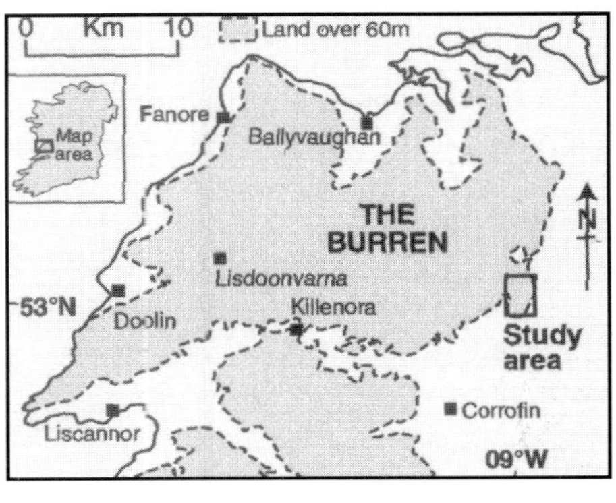

Figure 1. The location of the study area in northern Co. Clare, and Ireland.

Many factors combine to create conditions for such high levels of biodiversity. These include climate, topography, and past and present human management of the landscape. Of particular importance however is the soil composition, since soil is the interface between the substrate and the vegetation. Mapping and analytical studies of soils in the Park area reveal a high degree of spatial heterogeneity that is directly related to geological processes (Moles and Moles 2002).

GEOLOGY, GLACIATION AND SOIL COMPOSITION

The study area (Fig. 2) is dominated by an outcrop of generally gently-dipping limestones of Visean (Lower Carboniferous) age, known as the Burren Limestone Formation (Drew 2001). The lower eastern slopes of Mullach Mór and Sliabh Rua are formed of massive limestone of the Maumcaha Member, whereas the stepped profile of the summit areas and western slopes comprise the overlying, well-bedded limestones of the Aillwee Member. Dark grey chert forms isolated nodules and irregular, semi-continuous beds. Otherwise the non-carbonate content of the limestones is low, and clay 'wayboards' between limestone beds were not observed. However, on the northwest flank of Sliabh Rua and on Knockanes there are outcrops of a breccia comprising limestone fragments in a shale matrix with some black shale clasts up to 0.5m in diameter. The restricted outcrop distribution indicates that this unit forms a tongue-shaped body with a lens-shaped transverse section up to 5m thick and 250m across. It is interpreted as

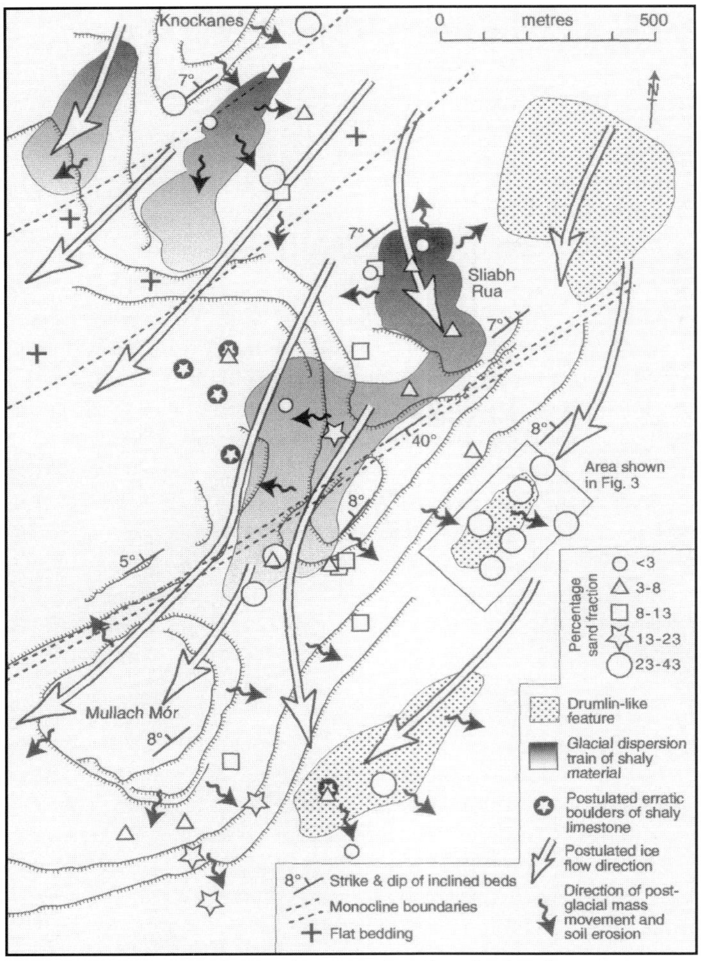

Figure 2. Map of the study area showing bedrock structure and the scarps of prominent limestone beds, and the % sand fraction in soil samples from 38 sites. Also shown are drumlin-like landforms comprising limestone-rich till; glacial dispersion trains of shaly material and shale-bearing erratics; generalised ice-flow directions; and directions of Holocene mass movement and soil erosion that explain the observed distribution of clay- and silt-rich shale-bearing soil.

an intraformational breccia, produced by slumping of a semi-lithified mixture of shale and limestone down a palaeoslope, between periods of limestone formation.

The Burren was at the northern margin of the region affected by the Late Carboniferous Variscan Orogeny. In the study area, beds dip northwest at 5-8° and are folded into asymmetrical monoclines with localised dips in the range 0-40° southeast. These folds have a pronounced affect on the morphology of hills such as Mullach Mór and Sliabh Rua (Drew 2001). Two major joint sets trending north-south and east-west occur throughout the Burren. Some north-south joints acted as conduits for mineralising fluids which deposited quartz and calcite, and silicified the limestone wallrocks.

Quaternary glacial erosion (Farrington 1965; Warren 1991) has accentuated aspects of the bedrock geology, especially the scarp features in the Aillwee Member limestones. Ice-flow was predominantly to the southwest, as indicated by glacial striae on limestone pavements, the northeast-southwest alignment of hillforms, and southwest-facing cliff features formed by ice-plucking of large blocks of limestone on the lee side of hilltops. Ice-moulded deposits of glacial till form drumlin-like landforms on hill flanks. In the study area the largest such feature is on the eastern flank of Mullach Mór (Fig. 2). It reaches a maximum length of c. 500m and width of c. 200m, and the till is estimated to be c. 25m in depth. A similarly shaped but smaller drumlin-like landform, 250m long and 8-10m deep, occurs on the south-east flank of Sliabh Rua (Fig. 3). Unweathered till, comprising predominantly limestone but with 5-10% by volume granite and metamorphic rocks, is preserved in the cores of these landforms. Elsewhere, the shallow glacial till inferred to have mantled the bedrock immediately after the last glaciation (Finch and Synge 1966) has subsequently weathered to a sandy soil that is prone to erosion. A distinctive, reddish and clay-rich soil is associated with the outcrops of shale-limestone breccia. Shale-rich material derived from this breccia was glacially transported to form a diffuse dispersion train extending southwards from the outcrops. Isolated patches of shale-rich soil occurring up to 1.2km from these outcrops may have formed by weathering of ice-transported boulders of shaly breccia.

Mass movement and soil erosion occur under contemporary conditions and have been much more active in the past (Drew 1983), indeed the paucity of soil cover may result largely from soil erosion following Bronze Age deforestation (Jelicic and O'Connell 1992; Moles et al. 1999). Extending downslope to the east and south-east of the Mullach Mór 'drumlin' is a depositional apron of sediment generally <4m in depth. A smaller and more patchy depositional apron occurs downslope of the Sliabh Rua 'drumlin' (Fig. 3). The 'aprons' are attributed to rainfall-promoted decalcification of the glacial till and erosion of resultant sandy soils on the steep flanks of the landforms, with re-deposition on the less steep limestone surfaces downslope. The presence of often relatively deep (1-2m) soil cover on shale-rich parent materials is attributed to Holocene weathering to form a clay-rich soil, which may have been more resistant to erosion than more sandy soils.

To aid explanation of a complex set of relationships between geology, soils and vegetation, examples are selected at four spatial scales.

LANDSCAPE-SCALE ASSOCIATIONS BETWEEN GEOLOGY AND VEGETATION

At landscape scale (measured in square km), there is often a clear pattern in the distribution of vegetation that is linked to solid geology and the distribution of glacial sediments. Chert bands in the limestone and silicified wallrock of hydrothermal veins are more resistant to erosion and form raised linear features, with associated acid peaty soils supporting *Calluna vulgaris* and *Pedicularis sylvatica*. The shale-rich breccia on the hill-flanks of Sliabh Rua and Knockanes has weathered to produce a clay-rich acid soil. Until recent times this was largely covered by blanket peat

Figure 3. (a) A topographic map of the Sliabh Rua drumlin-like landform (located in Fig. 2) showing areas of grassland (shaded), limestone pavement (white) and cliff (heavy dashed lines), and locations of 73 sample sites. (b) and (c) show data obtained from these sites: (a) soil pH values in six class intervals; (b) the occurrence (black circles) of three representative plant species that illustrate the edaphic zones developed by pedogenesis and soil erosion (see text).

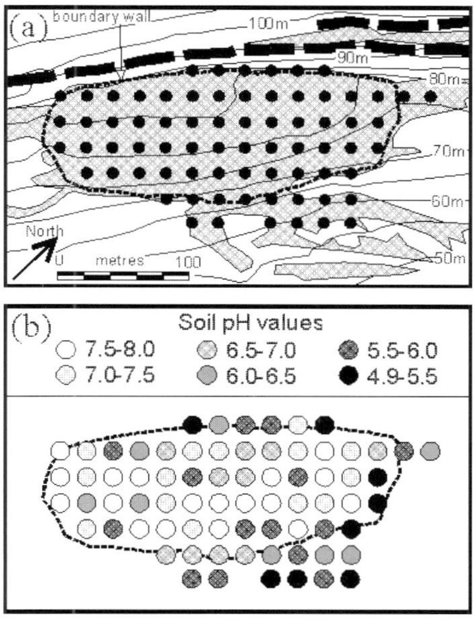

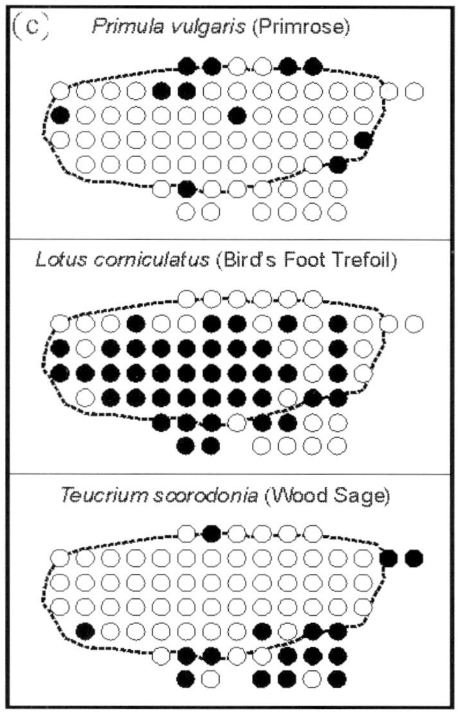

which supported a rich heathland community dominated by *Calluna vulgaris* and *Erica cinerea*. Since 1980, the peat and associated heathland plants have disappeared from the patch on Knockanes, but survive intact on Sliabh Rua.

Shallow, sandy soil is widespread under patches of grassland (Fig. 2). This soil is interpreted as the product of long term decalcification of a previously limestone-rich glacial till. In some places where this soil is sufficiently thick (more than c. 10cm), an acid, peaty soil surface horizon has developed, supporting heathland species such as *Calluna vulgaris* and *Potentilla erecta*. But in general the peaty horizons are shallow and discontinuous, interspersed with limestone gravels supporting for example *Dryas octopetala*, *Antennaria dioica*, *Epipactis atrorubens*, *Listera ovata* and *Senecio jacobaea*, and calcareous rendzinas supporting *Linum catharticum* and *Seslaria caerulea*. The outcome is a complex mosaic of vegetation formed of both calcifuge and calcicole species.

Soils on the drumlin-like landforms are somewhat decalcified but are in general neutral in pH, and peaty horizons are absent. These landforms support patches of distinctive grassland vegetation in which gentians (*Gentiana verna* and *Gentianella amarelle*) and orchids (*Orchis mascula*, *Dactylohiza fuchsii*, *Gymnadenia conopsea*) are often abundant. Some are dominated by *Knautia arvensis*, which provides a very distinctive show of flowers in late summer.

LANDSCAPE-ELEMENT SCALE ASSOCIATIONS BETWEEN GEOLOGY AND VEGETATION

At landscape-element scale (measured in hectares), a more intricate pattern is evident in the relationship between vegetation, glacial deposits and soils that formed from them. To illustrate this we describe the drumlin-like feature lying between approximately 65-90m OD on the eastern flank of Sliabh Rua, surrounded by limestone outcrop (Fig. 3). It is in the form of a humpback ridge with flank slopes typically in the range 18-25°. Weathering has decalcified the uppermost till to produce a sandy soil that contains fragments of granite. Soil erosion under past and contemporary conditions has removed this soil downslope. In combination, these two processes have produced a distinctive edaphic spatial pattern. At the centre of the patch, below the ridge peak, erosion of the decalcified sandy topsoil has exposed the underlying, more calcareous subsoil. Strongly decalcified colluvium has accumulated on the more gentle slopes at the patch edges. On gentle slopes near the summit of the 'drumlin' occur soils of intermediate characteristics, often with shallow decalcified surface horizons, creating three edaphic zones.

The vegetation response is equally distinct (Fig. 3, Moles et al. 2003). Near the ridge peak, shallow, discontinuous soils support for example *Rosa pimpinellifolia, Senecio jacobaea*, *Antennaria dioica*, *Linum catharticum*, *Geranium robertianum*, *Epipactis atrorubens*, *Hieracium pilosella*, *Sanguisorba minor*, *Hypochoeris radicata* and *Seslaria caerulea*, though with a recent invasion by *Pteridium aquilinum*. Downslope in the intermediate zone, a greater number of grass species is present, and forbs such as *Knautia arvensis*, *Gentianella amarelle*, *Orchis mascula*, *Dactylohiza fuchsii*, *Trifolium pratense*, *Primula vulgaris*, *Lotus corniculatus*, *Fragaria vesca*, *Plantago lanceolata*, *Taxacum officinalis*, *Achillea millefoliium* and *Campanula rotundifolia*. The distribution of *Gentianella amarelle* is especially clearly restricted to a narrow altitude band on this patch. Near to patch edges in the third edaphic zone the vegetation merges from grassland to grass-heath, with dominant *Calluna vulgaris*, and *Geranium sanguineum*, *Potentilla erecta*, *Centaurea nigra*, *Teucrium scorodonia* and *Polygala vulgaris*.

PATCH- AND SITE-SCALE ASSOCIATIONS BETWEEN GEOLOGY AND VEGETATION

At patch scale (measured in tens of square metres), clear linkages between vegetation and geology are also evident. Areas of limestone pavement support species such as *Rosa pimpinellifolia*, *Hieracium anglicum*, *Senecio jacobaea*, *Epipactis atrorubens* and *Seslaria caerulea*. In contract, patches of clay-rich soil derived from shale-bearing erratics support a distinctly different community, including grasses such as *Dactylis glomerata* and *Festuca rubra*, and forbs, especially *Potentilla erecta*, *Pedicularis sylvatica*, *Geranium sanguineum*, *Galium verum*, *Hypericum pulchrum*, and *Lathyrus montanus*.

At site scale (variation over distances of 1-10 metres), geology may impact on grassland biodiversity in a crucial manner. Grassland on surfaces without stones and boulders is generally heavily grazed throughout the year, which prevents flowering and seed formation. Thus while this grassland may be formed of many species, very few are able to reproduce. It may be the case that they contribute little to longer-term maintenance of biodiversity. Where grassland occurs between larger stones and boulders, protection from grazing results in much more successful seed production. Stone and boulder-strewn surface therefore supply the seed rain which maintains biodiversity in more exposed grassland. A similar effect has been described in Mediterranean grassland subject to grazing pressure (Pagnotta et al. 1997).

Shallow, sandy soils are easily eroded, so that limestone pavement has encroached on sward and species-rich habitat is lost. Erosion may be exacerbated along tracks used by grazing animals where stocking levels are too high, or the seaonality of grazing is wrongly set (Moles 1992; Moles 1995). Dunging by cattle and goats, especially along tracks and in preferred grazing grounds, locally alters acidity and nutrient status thereby allowing calcifuges and calcicoles to co-exist even in small patches.

CONCLUSIONS

A complex and dynamic mosaic of soils and surface conditions has developed in the Burren National Park. This can be attributed to the spatial heterogeneity in soil parent materials created by localised occurrences of shale-bearing rock, multiple episodes of glacial transport of siliceous material into the limestone region, and patterns created by the subsequent weathering, erosion and mass movement of these materials. The presence of various glacial deposits, erratic boulders and stony surfaces has a profound impact on site-scale biodiversity. Heterogeneity in soils and surface conditions creates habitat heterogeneity: habitats support differing assemblages of plant species. Therefore, the present rich biodiversity of the Burren is strongly related to geology and landscape evolution.

ACKNOWLEDGEMENTS

Financial assistance was received from Dúchas. We thank Kate Moles for field assistance, Paddy Gaffikin, John McAlister and Yoma Megarry at Queen's University, Belfast for undertaking laboratory analyses, and Maura Pringle for drafting the maps.

REFERENCES

Drew, D.P. 1983 Accelerated soil erosion in a karst area: The Burren, western Ireland. *Journal of Hydrology* 61, 113-124.

Drew, D.P. 2001 *Classic landforms of the Burren karst*. Geographical Association. Sheffield, U.K. 52pp.

Dúchas 1991 *Burren National Park* (colour brochure).

Farrington, A. 1965 The last glaciation in the Burren, Co. Clare. *Proceedings of the Royal Irish Academy* 64B, 33-39.

Finch, T.F., Synge, F.M. 1966 The drifts and soils of West Clare and the adjoining parts of counties Kerry and Limerick. *Irish Geography* 5, 161-172.

Jelicic, L. and O'Connell, M. 1992 History of vegetation and land use from 3200 B.P. to the present in the north-west Burren, a karstic region of western Ireland. *Journal of Vegetation History and Archaeobotany* 1, 119-140.

Met Éireann 2002 1961-1990 Mean Annual Rainfall map of Ireland. http://www.met.ie/climate/rainfall.asp accessed 2 September 2002.

Moles, N.R. and Moles, R.T. 2002 Influence of geology, glacial processes and land-use on soil composition and Quaternary landscape evolution in The Burren National Park, Ireland. *Catena* 47, 291-321.

Moles, R.T. 1992 Trampling damage to vegetation and soil cover at paths within the Burren National Park, Mullach Mór, Co. Clare. *Irish Geography* 25, 129-137.

Moles, R.T. 1995 Change at the edges of Burren grassland patches, 1979-92. In Jeffrey, D., Jones, M. and McAdam, J. (eds) *Irish grasslands: Their biology and management*. Royal Irish Academy. Dublin. pp. 246-252.

Moles, R.T., Hayes, K., O'Regan, B. and Moles, N.R. 2003. The impact of environmental factors on the distribution of plant species in a Burren grassland patch: implications for conservation. *Biology and Environment: Proceedings of the Royal Irish Academy*, 103B, 139-145.

Moles, R.T. and Travers, I. 1981 *Habitat map of Mullach Mór, Co. Clare, Ireland* (scale 1:7500). OPW, Dublin.

Moles, R.T., Moles, N.R. and Leahy, J.J. 1999 Radiocarbon dated episode of Bronze Age slope instability in the south-eastern Burren, County Clare. *Irish Geography* 32, 52-57.

O'Donovan, G. 1987 An ecosystem study of grassland in the Burren National Park, Co. Clare. Ph.D. thesis, University of Dublin (Trinity College), Dublin.

Pagnotta, M.A., Snaydon, R.W. and Cocks, P.S. 1997 The effects of environmental factors on components and attributes of Mediterranean grassland. *Journal of Applied Ecology* 34, 29-42.

Warren, W.P. 1991 Fenitian (Midlandian) glacial deposits and glaciation in Ireland and adjacent offshore regions. In J. Ehlers, P.L. Gibbard and J. Rose (eds) *Glacial deposits in Great Britain and Ireland*. A.A. Balkema, Rotterdam. pp. 79-88.

Webb, D.A. and Scannell, M.J.P. 1983 *Flora of Connemara and the Burren*. Royal Dublin Society and Cambridge University Press. Cambridge, U.K.

In: Parkes, M.A. (Ed.) 2004 *Natural and Cultural Landscapes - The Geological Foundation*, 65-68, Royal Irish Academy, Dublin

LANDSCAPE AS A SUPPORT FOR BIODIVERSITY: THE *ARRIBAS DO DOURO* CASE STUDY (INTERNATIONAL DOURO NATURAL PARK, NE PORTUGAL)

M. I. C. Alves*, A. Monteiro, N. Ferreira***, G. Dias*, J. Brilha* & D. I. Pereira***
***Universidade do Minho, Campus de Gualtar, 4710 - 057 Braga, Portugal.**
**** Parque Natural do Douro Internacional, 5200 Mogadouro, Portugal.**
***** Instituto Geológico e Mineiro, Apartado 1089, 4466-956 S. Mamede de Infesta, Portugal.**
E-mail (1st author): icaetano@dct.uminho.pt

Abstract: The fluvial canyon of the international Douro and Águeda rivers (NE Portugal-Spain border) supports a relevant natural and cultural heritage, which led to its recent integration in the portuguese conservation system - the International Douro Natural Park (IDNP). The spectacular steep cliffs, regionally named *Arribas do Douro*, are the major landscape feature of the region. They were developed during the incision of the Douro River and its tributaries on the Iberian Meseta, crosscutting granitic and metasedimentary terranes under the control of the fracture network. The landscape reflects the stream capture of the Tertiary Douro Basin towards the Atlantic by an active Quaternary down cutting. This geological setting provides one of the most important nesting habitat in Europe for vultures, eagles and black storks. The floristic heritage found on the rock slopes and ancient floodplains is also relevant. Evidence of the close relationship with the geological characteristics is also given by the millenary human occupation of the abrupt slopes, profiting from the microclimatic conditions for the production of wine, olive and almond. In conclusion, the *Arribas do Douro* landscapes have a relevant geological heritage and provide the support for the IDNP biodiversity. Therefore, the nature conservation policies must take this strong linkage into account.

INTRODUCTION

In 1998, Portugal assigned the status of Natural Park to the deep valleys of the international Douro and Águeda rivers - the International Douro Natural Park (IDNP) - in the Portuguese-Spanish border, north-eastern Portugal (Fig. 1.a). The 851 km^2 of the park include 130 km of these international rivers and a narrow contiguous area of the old peneplain landscape known as Iberian Meseta. The establishment of this park was due to the high relevance of the geomorphology, wildlife, natural flora and cultural heritage.

The biotic richness of the area and its association with important geological features are the two main natural aspects for IDNP management. In spite of that, the IDNP geological heritage has not yet been object of a global inventory and characterisation. Since 2001, a three-year research project is under development, aiming to improve the IDNP geological knowledge and to create scientific instruments as a support for the sustainable management of the resources (Dias and Brilha, 2002).

The holistic approach to conceive Earth heritage conservation integrating several contexts such as landscapes, habitat type and traditional land use was already discussed by several authors (for instance, Larwood and Prosser, 1998; Erikstad, 1999). The IDNP is an example of how sustainable management of protected areas could be better attained if all aspects of nature conservation are considered, as Schelhas (2001) states. This type of management also increases the protected areas relevance in environmental education (Biderman and Bosak, 1997).

The main aim of this paper is to show how a particular geological setting influenced biodiversity in International Douro Natural Park and to reinforce that nature conservation policy must be implemented having this strong linkage in consideration.

GEOLOGICAL SETTING

The IDNP is located in one of the most complex geological areas of the Central Iberian Zone (CIZ), having a diversified and tectonically imbricate geology (Carvalhosa, 1959; Silva and Ribeiro, 1994; Pereira, 2000). It is dominated by a sequence of autochthonous units largely covering the geological scale, from the Neoproterozoic to the Cenozoic, and a sequence of allochthonous units that belong to the nappes front of the Morais Massif. These successions include the Precambrian Massif of Miranda do Douro, one of the rare evidences of the pre-Variscan basement in CIZ, as well as a great diversity of Hercynian granitic rocks.

LANDSCAPE EVOLUTION

The Iberian Meseta is a large geomorphological unity, developed in granitic and metasedimentary rocks of the Hesperian Massif. The Meseta is an old and extensive plain, one of the largest in Europe, located about 700 m above sea level.

The Douro River, in the IDNP area, carved an entrenched valley cutting the high plateau (Fig. 1.b). The channel orientation and a great number of incised meanders are clearly controlled by the fracture network (Fig. 1.c). The major fractures are late hercynian and some of them were reacti-

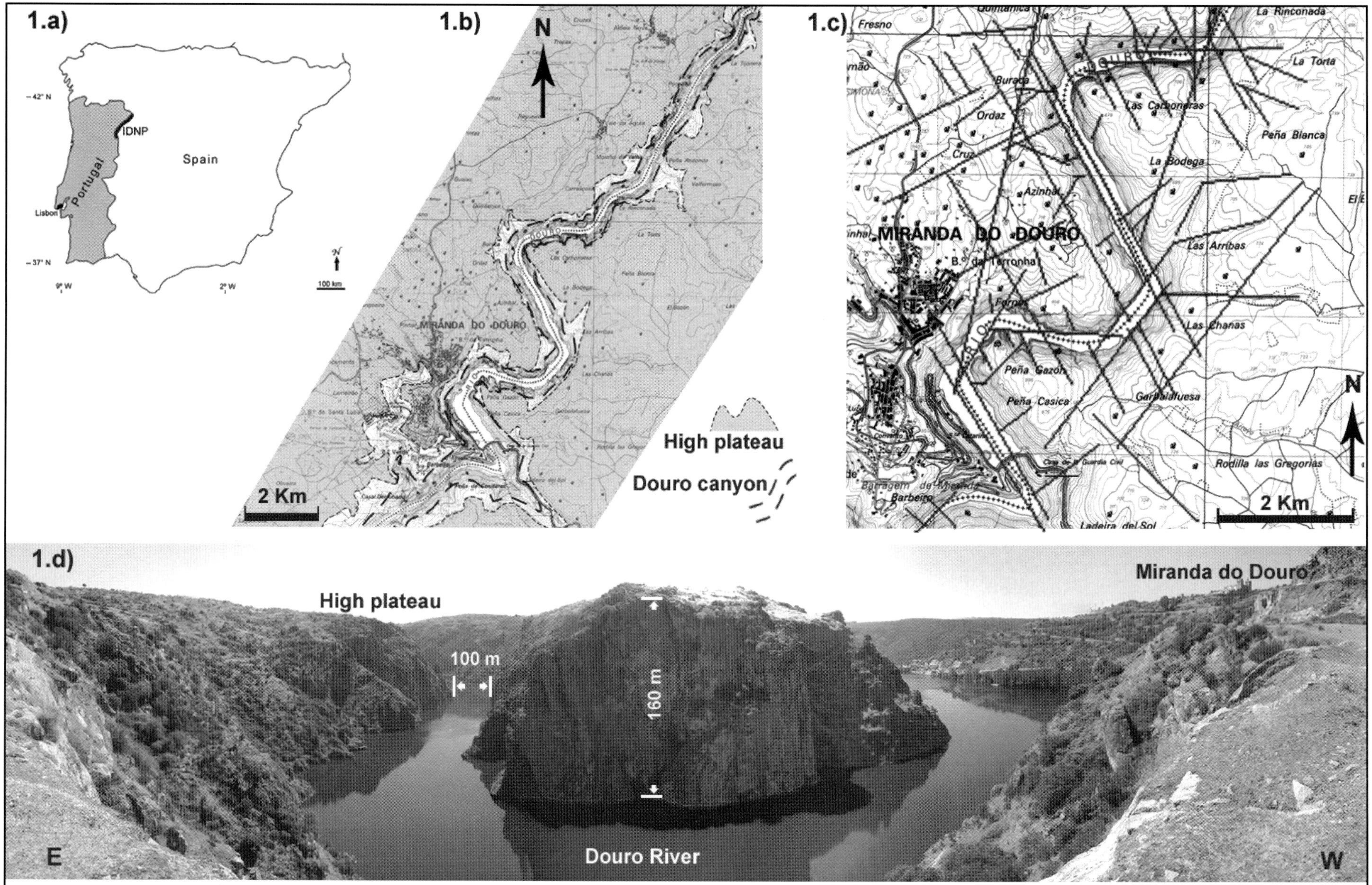

Figure 1. Location of the International Douro Natural Park (IDNP, NE Portugal) (1.a). Geomorphological features (1.b,d) and fracture pattern (1.c) at the Miranda do Douro region (northern IDNP).

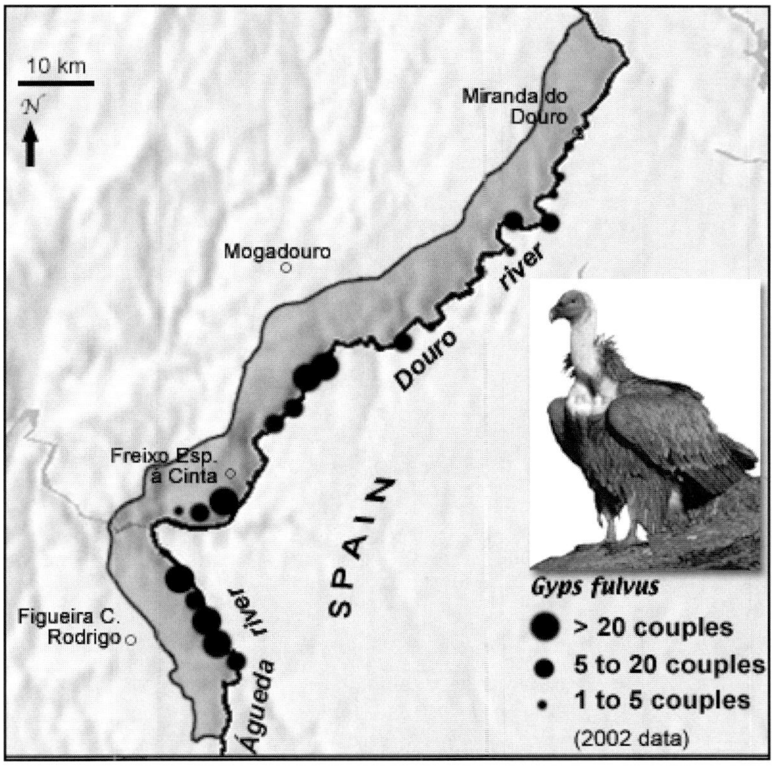

Figure 2. Distribution of the Griffon Vulture (**Gyps fulvus*****) in the International Douro Natural Park (dark shaded area).***

Table I - Vertebrates found in the IDNP area.	Total species Numbers	Total species number in Portugal (%)
Mammals	29	42%
Birds	150	37%
Reptiles	17	59%
Amphibians	11	65%
Fishes	14	33%
Total	221	50%

vated during the Alpine orogeny. The main orientation of the international Douro course is influenced by the NNE-SSW to NE-SW fractures. Locally, the valley is under the control of NW-SE, N-S, and E-W fractures, which promote the occurrence of incised meanders.

The narrow Douro valley is canyon-type. The spectacular canyon escarpments are regionally known as *Arribas do Douro* (Fig. 1.d). The river flows 200-300 m below the plain surface. The longitudinal profile of the Douro River is very abrupt, descending from 550 m to 130 m in only 90 km inside the IDNP. These geomorphological features allowed the building of six hydroelectric dams.

The present day Douro River represents a snapshot of an ancient fluvial evolution. The former Douro basin has been considered an intracontinental basin of cratonic type. Its western boundary is a relatively flat hercynian border that remained essentially passive during the Cenozoic (Santisban et al. 1996). Probably until Zanclean, fluvial drainage in NE Portugal was endorheic, towards the Douro basin. The major hydrographical reorganization in the area occurred during Miocene-Pliocene times. In the Late Pliocene, an important fluvial network was already captured by the fluvial pattern that drained towards the Atlantic. Therefore the endorheic Tertiary Douro Basin turned into a generally exorheic basin by the end of the Pliocene (Pereira et al. 2000). The *Arribas do Douro* landscape reflects an active Quaternary downcutting of the basement, associated with the incision of the Douro River and its tributaries. This process is controlled by the Atlantic base level and by the fracture network.

CLIMATE AND LANDSCAPE

The present day geomorphological features induce different climatic characteristics between the high plain and the valley. Climate in the valley has Mediterranean characteristics, being more prominent to the south, at Freixo de Espada à Cinta - Barca de Alva region, and is more humid and softer than in the plateau. The mean annual temperature in the valley is 15°C, with few days below 0°C in the winter.

These particular climate conditions allowed the establishment along the Douro and Águeda rivers of fauna and flora typical of southern latitudes. Sun radiation, air temperature and moisture conditions inside the canyon change as function of the side slopes height, their orientation and declivity, thus supporting an enormous variety of floristic communities. Another evidence of the close relationship between geological, geomorphological, and climatic characteristics is given by the millenary human occupation of the slopes taking advantage of a particular microclimate for the production of wine, olive, almond and orange.

LANDFORMS AND BIODIVERSITY

Inside park boundaries, there are rich communities of fauna and flora, with important populations of endangered species such as cliff nesting birds. There are 221 species of vertebrates including Mammals, Birds, Reptiles, Amphibians and Fishes, representing 50% of the total number of species in Portugal and an important number of Iberian species (Table I).

As already shown, two complementary geomorphological units characterize the IDNP: the high plateau and the canyons. A large diversity of landforms occurs on the canyon sidewalls, such as vertical cliffs, steep dipping slopes, some with inverse dip, convex surfaces, domes, turrets, castellated blocks, angular blocks, split blocks, foliation slabs, wedges, rock caves, cliff-foot caves, clefts and rocky steps. This diversity of morphological features, as well as the nearby high plain, offers a diverse flora and fauna habitat, which are important for the ecological balance of this park.

The plateau is an ecological mosaic of woods, groves,

corn-lands, and other plantations. This diversity allows the existence of insect species and wild mammals, as the roe deer, the wild boar, the wolf, the otter (in the main rivers), bats, and also reptiles and amphibians. There are also autochthonous species of cattle and sheep, which contribute for the nourishment of the eagles and vultures communities. The plateau around the main valleys is dominated by pastureland, sclerophyllous scrub, *Quercus* sp. and *Juniperus oxicedrus* woodland.

Several species of animals and plants have their main habitat inside the canyon. The geological and ecological features are a benefit for cliff nesting species, in particular for the eighteen bird species, which are the IDNP emblematic group, that build their nests in colonies on rocky clefts, caves and steps (Monteiro, 1998). Only considering birds, there are 300 couples of Griffon Vulture (*Gyps fulvus*) (Fig. 2), 70 couples of Egyptian Vulture (*Neophron percnopterus*, elected for the park symbol), 18 couples of Golden Eagle (*Aquila chrysaetus*), 14 couples of Bonelli's Eagle (*Hieraaetus fasciatus*), 10 couples of Peregrine Falcon (*Falco peregrinus*), 15 couples of Eagle Owl (*Bubo bubo*), 15 couples of Black Stork (*Ciconia nigra*) and more than 100 couples of Chough (*Pyrrhocorax pyrrhocorax*). *Arribas do Douro* is one of the most important areas in Europe for this group of species. The bare rocks are also inhabited by floristic species growing in fractures and clefts, some of them millenary. In the canyon several species are frequently found, such as the alder, the arbutus-berry, the ash-tree, the juniper, the lotus-tree, the oak, the oak evergreen and also plantations of wine, olive, almond, chestnut and orange in the southern area of the park.

This rich biodiversity was responsible for the classification of this region as Site of Community Importance (SCI) and Special Protection Area (SPA).

CONCLUSIONS

The International Douro Natural Park (IDNP) presents a richness of natural features constituting a reference in the portuguese natural park system. In this natural park the most significant geomorphological feature is the fluvial canyon of the Douro River and its tributary, the Águeda river, dissecting the high plateau of the Iberian Meseta and crossing various geological terranes where a large diversity of landforms are exposed. This relevant geosite, locally known as *Arribas do Douro*, the IDNP *ex-libris*, conditioned the establishment of many ecosystems. This is an ecological region of great value, due to a large number of species, particularly rupicolous birds, sedentary and migratory, which search this area for nesting. The valley/plateau landscape features provide the support for the food chain equilibrium of the various ecosystems.

Due to the strong links between IDNP geodiversity and biodiversity, conservation policies must be undertaken together. This holistic approach is very important for park management and, consequently, for environmental education.

ACKNOWLEDGEMENTS

This work has been developed in the scope of the project PNAT/1999/CTE/15008, sponsored by FCT (Portuguese Science and Technology Foundation) and ICN (Portuguese Nature Conservation Institute).

REFERENCES

Biderman A. and Bosak W. (1997). Environmental education in protected areas as a contribution to heritage conservation, tourism and sustainable development. In: *National Parks and Protected Areas: Keystones to Conservation and Sustainable Development*, Nelson J.G. and Serafin R. (eds.), vol. 40, NATO ASI Series, Ser. G, Ecological Sciences, Springer, Berlin, pp. 93-102.

Carvalhosa A. (1959). Geológica de Portugal na escala 1:50 000. Notícia explicativa da folha 15-D (Figueira de Castelo Rodrigo). Serv. Geol. Portugal, Lisboa, 17 pp.

Dias G. and Brilha J. (2002). Raising public awareness of geological heritage: a set of initiatives. This volume.

Erikstad L. (1999). A holistic approach to secure geoconservation in local physical planning. In: *Towards the Balanced Management and Conservation of the Geological Heritage in the new Millenium*, Barettino D., Vallejo M. and Gallego E. (eds.), Sociedad Geológica de España, Madrid, pp. 69-72.

Larwood J. and Prosser C. (1998). Geotourism, conservation and society. *Geologica Balcania*, 28(3-4), 97-100.

Monteiro A. (1998). Fauna do Douro Internacional. In: *Entre Duas Margens. Douro Internacional*, Azevedo J. (ed.), Mirandela, pp. 35-62.

Pereira D.I., Alves M.I.C., Araújo M.A. and Cunha P.P. (2000). Estratigrafia e interpretação paleogeográfica do Cenozóico continental do norte de Portugal. *Ciências da Terra* (UNL), Lisboa, 73-82.

Pereira E. (coord.) (2000). Carta Geológica de Portugal à escala 1:200 000, folha 2. Inst. Geol. Min., Lisboa.

Santisban J.I., Martín-Serrano A., Mediavilla R. and Dabrio C.J. (1996). South-western Duero and Ciudad Rodrigo basins: infill and dissection of a Tertiary basin. In: *Tertiary basins of Spain: the stratigraphic record of crustal kinematics*, Friend P.F. and Cristino J.D. (eds.), University Press, Cambridge, pp. 196-202.

Schelhas J. (2001). The USA national parks in international perspective: have we learned the wrong lesson? *Environmental Conservation*, 28(4), 300-304.

Silva A.F. and Ribeiro M.L. (1994). Carta Geológica de Portugal na escala 1:50 000. Notícia explicativa da folha 15-B (Freixo de Espada à Cinta). Inst. Geol. Min., Lisboa, 48 pp.

In: Parkes, M.A. (Ed.) 2004 *Natural and Cultural Landscapes - The Geological Foundation*, 69-72, Royal Irish Academy, Dublin

GEODIVERSITY AND BIODIVERSITY OF KARST LANDSCAPES: THE EXAMPLE OF SARDINIA

Jo De Waele, Department of Earth Sciences, Cagliari University, Via Trentino, 51- 09127 Cagliari (Italy). Email: geoam@unica.it

Giuseppe Grafitti, Gruppo Speleologico Sassarese - Società Speleologica Italiana, Via dei Navigatori 7, 07100 Sassari. Email: ggfranc@tiscalinet.it

Abstract: Landscapes are usually considered as complex systems of superficial natural landforms covered by a more ore less developed and diversified land-cover (vegetation, agricultural and human elements). These territories are usually sorted in more or less homogeneous landscape units reflecting different intrinsic (natural and geological) characters and distinct ways of utilisation of the soils.

Among all landscapes, the karstic ones are probably the most interesting and fascinating, also because the "landscape" continues underneath the topographic surface. In fact, where in other lithologies human exploration stops at the surface - except in mining regions where artificial voids have been made by Man - in karst areas extremely complex cave systems have sometimes developed, forming a unique subterranean landscape with its own characteristics and peculiarities. Furthermore, this type of landscape is among the most conservative ones, and inside the caves many evidences (sediments, archaeology, palaeontology, cave dwelling fauna, etc.) that enable palaeogeographic, palaeoenvironmental and palaeoclimatic reconstruction can be found.

The example of the Island Sardinia (Italy) is illustrated: here almost 8 % of the territory is constituted of karstic areas where carbonate rocks with ages ranging from Cambrian to Miocene crop out. More than 2500 natural karstic caves have been surveyed for a total length of over 320 km; 40 % of these are located in the Mesozoic limestones of Central-east Sardinia and 30 % in the Cambrian Iglesiente Sulcis area in Southwest Sardinia. Other interesting karst landscapes are situated in the Northwest of the Island (Capo Caccia-Punta Giglio) and in the central-eastern part (Tacchi area). Geodiversity and biodiversity of these karst complexes give interesting information on local and regional palaeogeography, palaeoenvironment and palaeoclimate, and enables classification of some of these areas as extremely important geo-ecosystems of international interest.

INTRODUCTION

During the Upper Oligocene-Lower Miocene the Sardinian-Corsican micro-plate moved in an anti-clockwise rotation and translation to its actual position in the middle of the Tyrrhenian Sea. This particular origin of the Island of Sardinia makes it more related to the Southern part of the European continent instead of Italy.

From a geological point of view Sardinia is the only Italian region in which Early Palaeozoic sedimentary sequences diffusely crop out (*Iglesiente, Nurra, Gerrei-Sarrabus* etc.), giving rise to unique peculiar landscapes and morphologies.

In the past 10 million years the isolation, probably interrupted only briefly during the most pronounced Late-Quaternary Sea level low stands, has also determined an independent biological evolution making Sardinia one of the most interesting biogeographical regions of Europe (AA.VV. 1983).

KARST IN SARDINIA

Sardinia is characterised by a Palaeozoic basement mainly composed of intrusive and metamorphic rocks upon which have been deposited various sedimentary and volcanic units ageing from Mesozoic to recent times (Carmignani et al. 2001). Only little less than 8 % of all these outcropping rocks are carbonates and all these areas can generally be defined as karst terranes. The carbonate rocks of Sardinia are present in the four geological Era; the most important karstic complexes are the Sulcis-Iglesiente, composed of limestones, metalimestones and dolomitic limestones of Cambrian age (South-West Sardinia), the calcareous-dolomitic outcrops of Central-East Sardinia (Supramonte s.l., comprising Supramonte s.s., Gulf of Orosei, Monte Albo, Monte Tuttavista, the Island of Tavolara, Golfo Aranci and the Tacchi or Tonneri) and of the Nurra (North-West Sardinia), the Eocene nummulithic limestones of the Salto di Quirra (South-East Sardinia) and the Miocene limestones and marly limestones particularly common in the Sassari province (North Sardinia).

More than 2500 caves are known on the island; the longest subterranean system is situated in the Codula Ilune canyon and reaches a total development of more than 40km (De Waele 1999). Furthermore 8 caves are opened to the public, among which the Nettuno cave (North-West Sardinia), with about 200,000 visitors every year. Aside the underground karst, Sardinia is rich in different geomorphological karst landforms among which several sinkholes, canyons, polje, grikes and various microforms are to be mentioned (De Waele et al. 1999).

KARST LANDSCAPE AND GEODIVERSITY

The Sardinian karst areas can be subdivided according to geochronology, geography, lithology and/or morphology. The karstic phenomena have developed differently in relation to geology, palaeogeography and palaeoclimatology of the area, and most karst areas are identified by typical morphologies and landscapes.

In South-West Sardinia, in the region called Iglesiente-Sulcis, karst is developed in dolomitic limestones of the Lower-Middle Cambrian, representing one of the oldest karstified areas of the world. At least five karst cycles, from Upper Cambrian to now, have left their sometimes extremely complex signs in the landscape, both on surface and underground. Structurally these rocks have been subjected to three important orogenies (Caledonian, Hercynian and Alpine) with the consequent deformation of the sedimentary sequences that make geological interpretation quite complex. Surface landforms are not well developed, due to today's sub-arid climate, that does not allow the rapid formation of surficial karst forms, to lithology, prevalently dolomitic and recrystallised and thus less soluble, and to intense fracturing of the rocks that enhances rapid infiltration and less surface solution. On the other hand, subterranean karst is extremely well developed and the area has been defined one of the most "karstified" of Italy. Another very important aspect that makes this karst area unique in Europe is the presence of Pb-Zn-Ag-Ba mineralisations, sometimes of karstic origin, that have been exploited since more than 2,000 years and which are now abandoned. Some of these mine sites constitute important places of natural, cultural and scientific interest, included in the Geomining, Historical and Environmental Park of Sardinia, supported by UNESCO and officially instituted in the year 2001.

Other smaller Palaeozoic carbonate outcrops, for a total surface of 60 square km, occur in several areas of Sardinia and are mostly referable to Silurian and Devonian periods, but the small dimensions of these limestone limbs does generally not allow the development of interesting landforms or cave systems.

The three Mesozoic karst areas, Supramonte s.l., Tacchi and Nurra, beside their geographical differences, can be easily distinguished from a structural and morphological point of view.

Supramonte s.l., that comprises the central homonymous massif, the Gulf of Orosei, Monte s'Ospile, Monte Tuttavista, Monte Albo, the island of Tavolara and Capo Figari is a collection of Mesozoic carbonate outcrops characterised by alpine tectonics with slide-slip faults, wide folds, syncline and monocline structures. The dolostones and limestones cover a lapse of time between Middle Jurassic and Upper Cretaceous and are underlain by Palaeozoic granites and metamorphic rocks. The total thickness of the carbonate sequences reaches 800 meters and these karst areas cover a surface of more than 450 square km (almost 2% of the Island's surface). The karst landforms are extremely well developed and particularly in the largest outcrops (Supramonte, Gulf of Orosei) great cave systems, beautiful rock arches, impressive sinkholes, polje and uvala, lapiez and long canyons have formed. The same characteristics, although less well developed, can be found on Monte Albo, the Island of Tavolara, Capo Figari, Monte Tuttavista and Monte s'Ospile.

More to the South, many other Middle-Upper Jurassic dolostones and limestones crop out, separated by deep tectonic valleys in alone-standing mountains named Tacchi or Tonneri. These table like mountains are characterised by almost horizontal decimetric and metric carbonate beds for a total thickness that reaches 300 meters, covering a total of almost 260 square km and easily distinguishable form Supramonte s.l. by structure and morphology. In fact, morphology is controlled by the almost horizontal stratification and vertical fracturing, creating a typical ruin-like landscape locally enriched by sinkholes, polje and minor karst landforms.

In the extreme NW of the Island, in the Nurra region, other important Mesozoic carbonate outcrops form several karstic areas that cover a total surface of 150 square km, corresponding to 0.6% of the entire Island's surface. These karst massifs, composed of a discontinuous sedimentary succession dated from Lower Trias up to Upper Cretaceous and separated by tectonic dislocations, cover a metamorphic basement and are surmounted by Miocene and Pliocene sedimentary and volcanic rocks. The thickness of this succession ranges from 300m in the Capo Caccia area to less than 10 in the northernmost area and the layers are disposed in monoclinal cuesta-like morphologies constituting the flank of wide folds, mostly faulted in a concentric way. Superficial karst landforms are developed only on the largest outcrops, with sinkholes, poljes and many microforms (lapiez, solution flutes etc.). Underground karst phenomena are very well developed especially in the pure limestones of Upper Jurassic and Cretaceous of the coastal area of Capo Caccia-Punta Giglio.

Other minor and scarcely karstified Mesozoic outcrops occur in the Gulf of Palmas (Monte Sari North of Porto Pino and Maladroxia on the Island of Sant'Antioco) for a total surface area of 20 square km.

The first Tertiary carbonatic rocks are the nummulitic limestones of the Salto di Quirra in South-East Sardinia, with a thickness of the sedimentary sequence that can reach 300 meters. This carbonate plateau, which is reminiscent of the Mesozoic Tacchi typical of the area just to the North, covers a surface of 40 square km and is intensely karstified with numerous sinkholes and a very well developed subterranean drainage pattern.

The most extensive carbonate cover in Sardinia is the one formed by the Miocene marly limestones that occupy almost 500 square km (2%), mainly concentrated in the North of the Island, around the city of Sassari, but present also in other areas. These limestones are sometimes very

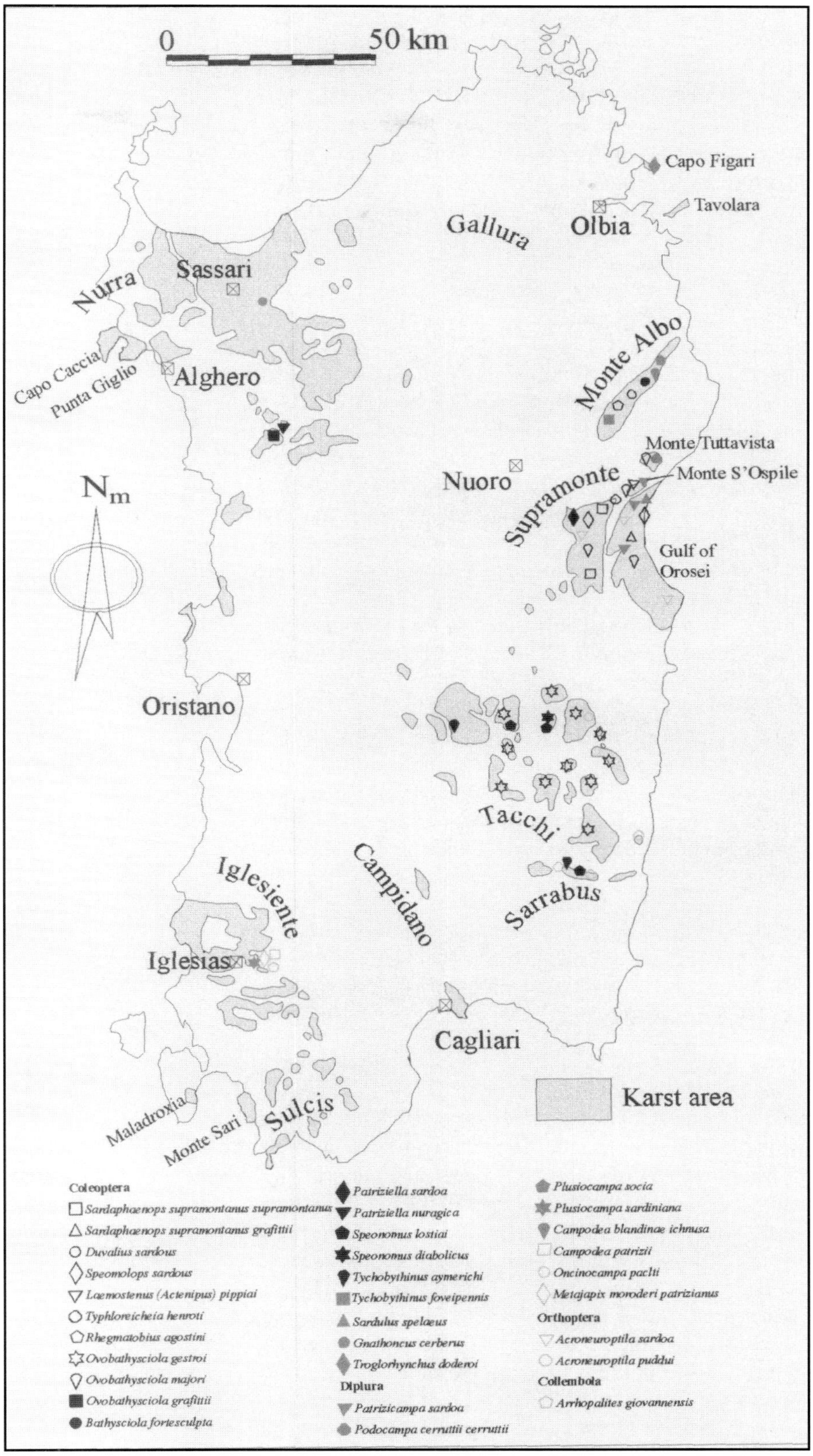

rich in fossils, have an almost horizontal stratification locally disturbed by neotectonic events. In the largest Miocene outcrops of North Sardinia karst phenomena are sometimes very evident with beautiful landforms (sinkholes, polje, karren etc.) and well developed cave systems. Most of the time these karst landforms and horizontal cave passages are imposed on marly surfaces or volcanic layers that form local base levels of the underground water flow. In Cagliari province, where the outcrops are much smaller, karst microforms can be abundant on the purest microcrystalline limestones.

KARST AND BIODIVERSITY

The cave dwelling fauna of Sardinia comprises at present more than 500 species that represent a great heritage of biodiversity (Grafitti et al. 1999; Grafitti 2001). Many species are troglobe and troglofile endemisms with interesting biogeographical stories, especially Insects (Coleoptera, Orthoptera, Diplura and Collembola), Spiders (Pseudoscorpiones, Aranea, Palpigrada and Opiliones), Crustaceans (Isopoda, Copepoda and Amphipoda), Diplopods (Polydesmida, Calipoda, Blaniulida and Iulida), Chilopods (Litobiomorpha) and, in conclusion, Mollusca (Gasteropoda) (Cassola 1982; Grafitti 2001).

In this short paper only the most interesting species of the Insect group are considered, giving brief comments on their biogeographical significance, generally related to the above mentioned geological history of the Island.

Insecta-Coleoptera: *Sardaphaenops supramontanus supramontanus* Cerr. & Henr., *S. supramontanus grafittii* Cas. & Giach.; *Duvalius sardous* (Dod.); *Speomolops sardous* Patr.; *Laemosterus* (*Actenipus*) *pippiai* Fiori; *Typhloreicheia henroti* Jeann., *Ty. elegans* Jeann.; *Rhegmatobius agostini* Jeann.; *Ovobathysciola gestroi* (Fairm.); *Ov. majori* (Reitt.); *Ov. grafittii* Ramp. & Sbord.; *Ovobathysciola* n. sp.; *Bathysciola fortesculpta* Jeann.; *Patriziella sardoa* Jeann.; *P. nuragica* Ramp. & Zoia; *P. muceddai* n. sp.; *Speonomus lostiai* (Dod.); *Sp. diabolicus* Jeann.; *Tychobythinus aymerichi* (Dod.); *Ty. foveipennis* (Dod.); *Sardulus spelaeus* Patr.; *Gnathoncus cerberus* Auzat; *Troglorhynchus doderoi* Sol. & Sol. (Casale and Vigna Taglianti 1996; Caccone and Sbordoni 2001; Casale 2001; Grafitti 2001). From a biogeographical point of view only some Carabidae and Cholevidae are to be mentioned. First of all *Sardaphaenops*, a genus exclusive for Sardinia, having close relationship with the genus *Aphaenops* of the Pyreneean caves and species of the genus *Apoduvalius* of Cantabrian caves (Spain); *Speomolops,* also an endemic genus of Sardinia, related to the two genus *Zariquieya* and *Molopidius* of the Pyrenees and Catalonia and to the genus *Henrotius* of the Balearics. *Duvalius sardous* (Dod.) instead is related to species of the same genus in Provence (France), in Catalonia and on the Balearics. Among the Cholevidae the genus *Ovobathysciola*, exclusive for Sardinia, with four species, has ancestral parents such as *Anillochlamys* and *Spelaeochlamys* genus of caves in East Spain; *Patriziella*, another endemism of Sardinia, would derive from the same genetic line of *Ovobathysciola*; finally, *Speonomus*, genus known also with troglobian species from caves in France and Spain.

Diplura: *Patrizicampa sardoa* Condé; *Podocampa cerrutii cerrutii* Condé; *Plusiocampa socia* Condé; *Pl. sardiniana* Condé; *Campodea blandinae ichnusa* Condé; *C. patrizii* Condé; C. (*Monocampa*) *emery spelaea* Condé; *Oncinocampa paclti* (Condé & Sendra); *Metajapix moroderi patrizianus* Pagés (Ramellini 1995). Diplura have a great biogeographical interest; the very well evolved genus *Patrizicampa*, Sardinian endemism, doesn't have any connection with other European genus; *Campodea patrizii* Condé, *Plusiocampa socia* Condé and *Pl. sardiniana* Condé instead present affinity with species of France, of Catalonia and of the Balearics.

Orthoptera: *Acroneuroptila sardoa* Bacc., *Acroneuroptila puddui* Cad. The genus *Acroneuroptila* belongs to a group of genus of paleomediterranean origin, the Petaloptilae, which at present have a discontinuous Mediterranean distribution going from the Middle East, over Greece, Italy and Spain up to Morocco.

Collembola: *Arrhopalites giovannensis* Cassag. and Delam. Deboutt., have affinities with species of Spain and France.

CONCLUSIONS

Even though the size of the article does not allow more than a broad overview of the geological and biological interest of the Sardinian karst landscapes - a more detailed contribution will be presented in an international scientific review - even this short outline marks the geological and biological importance of the Island in the framework of the European region. Sardinia represents one of the most remarkable geological schools of Europe, and the variety of rocks of different ages has caused an extremely rich and remarkable karst heritage. Furthermore, thanks to its Tertiary and Quaternary geological history and the relative isolation from the European mainland since millions of years the fauna and flora have evolved differently from the continental ones, composing a very rich endemic biological heritage. This fact is even more evident in and on karst areas where relics of ancient entomologic fauna and calcophyle flora have been preserved, confirming the post-Mesozoic geological history of the Island.

ACKNOWLEDGEMENTS

This research has been performed in the framework of the National Project COFIN 2001-2003 entitled "Geosites in the Italian Landscape", regional co-ordinator Prof. Antonio Ulzega.

REFERENCES

AA.VV. (1983) - Il popolamento animale e vegetale della Sardegna. *Lav. Soc. ital. Biogeogr.*, N. S., 8, (1980): 874 pp.

Caccone A. & Sbordoni V. (2001) - Molecular biogeography of cave life: a study using mitochondrial DNA from Bathysciine beetles. *Evolution*, 55 (1): 122-130.

Carmignani L., Oggiano G., Barca S., Conti P., Salvadori I., Eltrudis A., Funedda A., Pasci S. (2001) - Geologia della Sardegna. Note illustrative della Carta Geologica della Sardegna a scala 1:200.000. *Memorie Descrittive della Carta Geol. D'Italia* 60: pp. 1-283.

Casale A. (2001) - Biospeleologia e fauna cavernicola. Atti Conv. *Biospeleologia dei sistemi carsici della Sardegna*, Cagliari, 10 giu. 2000, G. Piras and F. Randaccio (eds), 7-12.

Casale A. & Vigna Taglianti A. (1996) - Coleotteri Carabidi di Sardegna e delle piccole isole circumsarde, e loro significato biogeografico (Coleoptera, Carabidae). *Biogeographia*, 18, (1995): 391-427.

Cassola F. (1982) - Il popolamento cavernicolo della Sardegna. *Lav. Soc. ital. Biogeogr.*, N. S., 7, (1978): 615-755.

De Waele J. (1999) - Il Catasto delle grotte della Sardegna: strumento per la gestione e la tutela dei territori carsici. Atti della III Conferenza Nazionale ASITA, Napoli 9-12 novembre 1999, pp. 695-700.

De Waele J., Di Gregorio F., Piras G. (1999) - The twenty most important karstic geosites of Sardinia. In "*Towards the Balanced Management and Conservation of the Geological Heritage in the New Millenium*", Barettino D., Vallejo M. & Gallego E. (Eds.), pp. 155-161, Madrid.

Grafitti G. (1999) - La fauna cavernicola della Sardegna. Un patrimonio da salvare. Anthèo, *Boll. Gruppo speleo-archeol. "G. Spano", Cagliari,* 3: 33-39.

Grafitti G. (2001) - Osservazioni sulla fauna cavernicola della Sardegna. Atti Conv. *Biospeleologia dei sistemi carsici della Sardegna*, Cagliari, 10 giu. 2000, G. Piras and F. Randaccio (eds), 13-33.

Grafitti G., Casale A., Delitala G. & Manca I. (1999) - Contributions to the knowledge of subterranean fauna of Sardinia: a work in progress. XIV Intern. Symp. of Biospeleology (Makarska, 19-26.IX.1999): 90.

Ramellini P. (1995) - Materiali per un catalogo topografico dei Dipluri italiani (Diplura). *Fragmenta entomol.*, 27 (1): 15-50.

In: Parkes, M.A. (Ed.) 2004 *Natural and Cultural Landscapes - The Geological Foundation*, 73-76, Royal Irish Academy, Dublin

GEOLOGICAL BACKGROUND FOR BIODIVERSITY IN THE EASTERN FENNOSCANDIA, ESTONIA AND LATVIA

Y.J. Systra, Mining Institute, Tallinn Technical University, Kopli 82, 10412, Tallinn, Estonia, Email: systra@staff.ttu.ee

Abstract: The biodiversity of the Eastern Fennoscandia, Estonia and Latvia is strongly influenced by geological factors such as bedrock composition and geochemistry, thickness of Quaternary deposits, variations in elevation and slope orientation, the occurrence of faults and fractured bedrock zones as pathways for groundwater, colour of bedrock and others. The most important feature having effects on biodiversity and the plant growth is the composition of the bedrock as an accessible source of macro- and trace elements. Geological background may be used for biodiversity study and selection of new protected nature territories.

DIFFERENCES IN THE CLIMATE

The ancient Fennoscandian Shield and its southern-western marginal zone, where Precambrian rocks are buried under Paleozoic platform cover, are represented by igneous, metamorphic and sedimentary rock of different composition (Fig. 1). The territory extends from tundra birch forests only in river valleys in the Kilpisjärvi, North Finland (69° N) to broad-leaved forest in Central Latvia (57° N) and the most of it is flat or hilly, only in North-western Finland do the highest tops reach 1-1.3 km. Although microclimatic conditions and solar exposition influence the vegetation: isolated mountain tundras on the hill tops higher than 450 m a.s.l. exist in the Paanajärvi National Park during last 9.5 thousand post Ice-age years with special tundra species *Loiseleuria procumbens*, *Phyllodoce caerulea*, *Diphasiastrum alpinum*, *Dryas octopetula*, *Juncus trifidus* and etc., while southern species as *Convallaria majalis*, *Fragaria vesca* can been found in deep valleys, for many places the most important geological feature seems to be the geochemistry of bedrocks.

BEDROCK GEOCHEMISTRY

The role of the carbonaceous rocks to vegetation was noted in the beginning of last century in Kilpisjärvi, where the first Malla protected territory was created in 1916 (now Malla Strict Reserve), and in Oulanka River valley (Pesola 1928). More serious investigations of plant and animals geochemistry and the influence of the different elements to biota were began in 60-70-es of the last century (Thornton 1983a). They showed that healthy plant growth and development is possible when soils contain enough necessary elements as nutrients. Plants have specific requirements of different major (C, O, N, H, Ca, P, S, K, Mg, Na) and trace (Fe, Cl, Cu, Zn, B, Si, Mo, Co, Mn, I, Se, Ni, Sn, V, As, F) elements. The ultimate source of these nutrients are in that

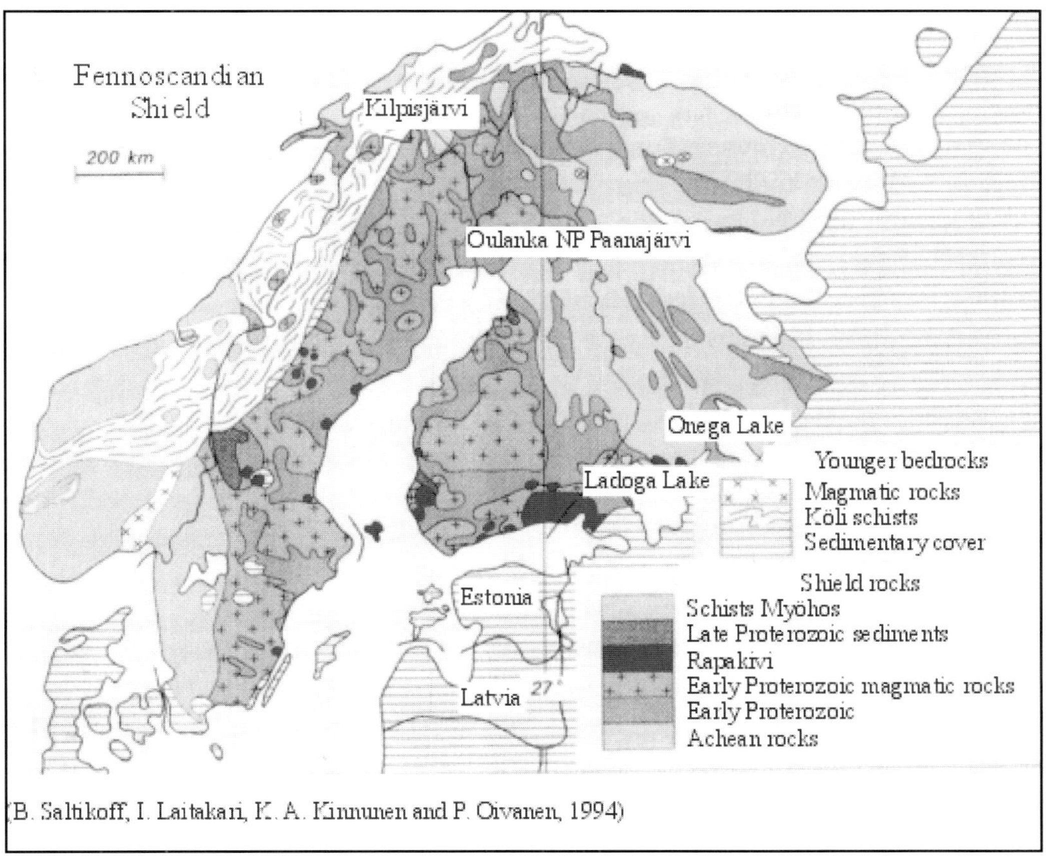

Figure 1. Geological map of Fennoscandia (by Saltikoff, B., Laitakari, I., Kinnunen, K.A. and Oivanen, P., 1994, translated into English). Places of study are shown.

Figure 2. Fault scarp Ruskeakallio (Brown Cliff) on the north shore Lake Paanajärvi is one of the most famous botanical sites of the northern Fennoscandia. Here are listed some tenth rare plants.

part of Earth crust accessible to biota, mainly soil and groundwater. Some of the trace elements such as As, B, Cd, Cu, F, Pb, Hg, Mo, Ni, Se and Zn at high concentrations may be potentially toxic (Thornton 1983b; Trofimov 2000).

SOME EXAMPLES OF THE GEOLOGICAL FEATURES' INFLUENCE ON BIODIVERSITY

The ancient core of the Fennoscandian Shield in Russian Karelia and in Eastern Finland consists of granites and gneisses, more than 2.65 Ga old, that form the Archean basement of the Karelian Craton. These rocks contain 65-70 % of SiO_2 and insignificant amounts of nutrients required by biota (Mg, Fe, B, Mn, Cu, I, Zn, Cl, Co, Mo, Se, Ni, Sn, W, As, F). Typical taiga vegetation grows on acidic soils, whereas all rare plant species are restricted to areas underlain by Paleoproterozoic volcanic and sedimentary sequence, which have been preserved from erosion only within deep syncline cores. Soils formed from mafic volcanics, carbonaceous rocks, black schist have neutral pH and contain all required nutrients.

Vendian-Paleozoic sequence in the studied territory consists of outcrops in the North Estonia Vendian and Cambrian clay and sandstone in the lower part, Ordovician and Silurian limestone and dolomite in the middle part, Devonian sandstone and dolomite in upper part, which crops out in the South and South-Eastern Estonia and in the northern part of Latvia.

PAANAJÄRVI NATIONAL PARK (RUSSIA) AND NEIGHBORING OULANKA NATIONAL PARK (FINLAND)

An unusually rich diversity of vegetation characterizes Paanajärvi-Oulanka area near the Arctic Circle (66°10'-66°30' N). About 600 vascular plants, and more than 400 species of mosses and lichens are recorded here, of which 10 % are rare or endangered, which is twice the corresponding figure for surrounding terrain. This difference can be primarily attributed to bedrock geochemistry, as the Oulanka and Paanajärvi National Park (Fig. 2) is underlain by Paleoproterozoic volcanics and sediments - including carbonates, while the bedrock in the surrounding region consists of diorites, granites and gneisses of the Karelian Craton (Systra 2000). Altitude also has an important influence on the distribution of natural vegetation types; at an altitude of 450 m the treeless rocky mountain-tundra zone begins with tundra species. Slope aspect and exposure also has a marked effect on microclimate. Thus, southern species inhabit south-facing slopes and deep valleys with greenhouse effect whereas northern species persist under favourable conditions on cold north-facing slopes and in deep shady valleys. One cliff - Ruskeakallio (Brown Cliff) on the northern shore of a lake, composed by albitite dyke with numerous carbonate veins, is about 60 m high and about 200 m long. This vertical sunny wall with hanging gardens is unique with its botanical rarities like *Gypsophila fastigiata, Asplenium ruta-muraria, Draba cinerea, Draba hirta, Potentilla nivea, Androsace septentrionalis, Hackelia diflexa* and more than another 20 rare plants are listed here. During the last 150 years many generations of famous Scandinavian, and after the last war, Russian botanists have visited it. Now it is one of the most picturesque sights of the Paanajärvi NP, but is closed for free visiting to save the unique vegetation. In this park is also the last place, where the lichen *Usnea longissima* is found during the last 50 years in Eastern Fennoscandia (Kotiranta et al. 1998).

An additional effect is the colour of the bedrock, where well exposed, as dark rocks accumulate more solar energy. The deep valley of the River Oulanka - Lake Paanajärvi - River Oulankajoki, which is incised through the Maanselkä uplands, the highest points of which rise up to 400 - 600m above sea level forms a special migration corridor. This river system begins at 200m a.s.l. on the western side of ridge and flows eastwards via Lake Paanajärvi (136m) into Lake Pääjärvi (109m) which discharges into the White Sea. There are major differences between the plant, bird, fish

Figure 3. Folded dolomite layer in the Paleozoic thrust on the slope of the Pikku-Malla Mountain is the place of the richest biodiversity of the Malla Strict Reserve. Kilpisjärvi, Finland.

and animal populations of the Atlantic western and Arctic eastern slopes of the Maanselkä ridge. Over the last decade a number of new species of birds and plants have been found both sides of the ridge.

KILPISJÄRVI AREA, NW FINLAND (69°N)

Geologically the territory belongs to the Caledonide nappes eastern marginal zone. The oldest volcanic and sedimentary rocks belong to Archean basement, cratonised 2.7-2.8 Ga ago. On this metamorphosed and folded basement lie Early Cambrian basal conglomerate, silt-rich and quartz-rich layers, still in an autochthonous position (Lehtovaara 1995). A dolomite is typical of the Jerta Nappe. It forms layers 1-40 m thick. The most of rare plant finds: *Erigeron acer*, *E. uniflorus*, *Silene uralensis*, *Woodsia alpina*, *W. glabella*, *Veronica alpinum* etc. A narrow zone along the folded dolomite layers on the slope, during the flowering of *Dryas octopetala* seems to be nature's flowerbed (Fig. 3). The influence of dolomites continues down along the slope and in the spring water. Numerous small springs discharge in the lower part of slopes and the territories surrounding have a very rich composition of vegetation. About 470 vascular plant species are listed in the surroundings of Kilpisjärvi (Eurola and Virtanen 1991), but most of the rare species can be found only on dolomites.

In this area two botanical protected territories exist: Malla Strict Reserve (under protection since 1916, as strict reserve since 1938) and Saana Protected Territory (from 1988). They include all big outcrops of dolomites.

ZAONEZHJE PENINSULA, ONEGA LAKE

Large, at about 10 thousand sq. km, the Onega synclinorium is composed of Early Proterozoic sediments and volcanics, including dolomite, shungite-bearing black schist, mafic lava and sills. About 800 vascular plant species are listed here and many rare plants from Red data book of Karelia (Kotiranta et al. 1998).

WESTERN AND NORTHWESTERN LADOGA LAKE AREA

The marginal zone of the Svecofennian Terrane is known as a territory of mantled granite gneiss domes. Narrow synclines between domes are presented by the same Early Proterozoic volcanics and sedimentary rocks, which are overlain by younger Kalevian rhythmical sediments. The central part of Lake Ladoga occupies Riphean Lagoga graben (aulacogene), cropping as mafic volcanics on the Valaam archipelago islands. The large water-body as the biggest lake of Europe - Ladoga, shows strong influence to microclimatic conditions. For agriculture and vegetation the shore zone without early frosts is much more suitable than higher territories 100-150m above lake level, where frosts begin in the autumn 2 weeks earlier. The Precambrian basement is divided into blocks by numerous fault zones, what in relief reflect as deep narrow valleys. In these deep valleys some northern species of mosses and lichens are found. Total number of vascular plants listed reaches to 800, while in territory of the Karelian Craton to northeast it is half as many.

NATIONAL PARKS OF ESTONIA

There are 4 national parks in Estonia: Lahemaa NP in the North Estonia, on the southern shore of Finnish gulf, Vilsandi NP - on the small island to west from Saaremaa Island, Karula NP on Karula Rise in Southern Estonia and Soomaa NP in the Southwestern Estonia in the basin of Pärnu River. Lahemaa NP (created 1971) is located mostly on the Ordovician carbonaceous rocks and has a rich biodiversity, with more than 1200 vascular plants. Deep river valleys are formed by river water. It worked out riverbeds in soft sedimentary rocks. Specific marine vegetation characterizes seashore and small islands. Vilsandi NP (under protection from 1912, NP from 1994) is located on small rocky islands and islets in the Baltic sea. Although they are composed of Silurian limestone and dolomites, the park has importance as a place rich in different birds. Karula NP was created in 1995 to save specific landscapes, where low hills are partly occupied by agriculture lands, partly by forests. Hills were covered during the Ice-age by thick moraine cover. The moraine, coming from north Ordovician and Silurian dolomite and limestone area, contains enough Ca and Mg, which reflects in rich vegetation. Soomaa NP was created in 1995 to save large mire areas in natural conditions , that were formed here formed during the last 5 thousand years. The reason for forming these mires is that NW part of Estonia is uplifting quicker (3mm yearly) than SE part (near 0mm). The river mouth is uplifting and leads to formation of large mires.

NATIONAL PARKS OF LATVIA

The most of the territory of the Kemeri NP (under protection 1977, NP from 1997) not far from Riga occupies the Kemeri Great Moorland - a typical high bog. The sulphur-springs complex is a unique nature monument of Europe and used in a resort built in 1930s. These waters are result of interaction between Devonian dolomite rock and the wide massif of the bog. The NE part of the park includes the coastal water area of the Gulf of Riga and the sandy beach and dunes, dry medows and dry old pine dune forests. Park is rich in rare plants (Pilats, 2000). Gauja NP (established in 1973, under protection since 1922) protects a unique ancient valley of the Gaauja River with the numerous valleys of tributaries and side ravines. The valley was formed as result of geological processes of faulting and weathering in Devonian rocks. Along the up to 85 m deep valley more than hundred cliffs, caves and other interesting objects can be met. The territory has a rich cultural history, only listed are 221 cultural monuments, total number of historical sites is more than 500. Slitere NP (established 2000, from 1921-Slitere Nature Reserve) has a very rich biological diversity. There are both broad -leaved forests and elements of taiga and tundra together on a small territory. The territory of park reflects the development history of the Baltic Sea with a high cliff on the Baltic Ice Lake ancient coast.

CONCLUSIONS

Many geological attributes and processes, including geochemistry, physical-mechanical properties and porosity of bedrock, denudation, differences in relief, and landscape morphology, absolute elevation and special migration corridors, exert a strong effect on biodiversity. Different stability of rocks, fracturing and block movements are important relief-forming factors. Fractured rocks of fault zones also form pathways for groundwater. Mineralized groundwater supplies numerous springs on the slopes, giving life to small mires with very rich vegetation.

It should be appreciated that in the all studied territory the richest biodiversity occurs only in the places with rich geological diversity. All nature protected territories: reserves, national parks and nature parks usually have both rich geological and biological diversity, but the importance of the geological background is not always understood.

ACKNOWLEDGEMENTS

Field work in the Kilpisjärvi area was done with financial support of the LAPBIAT grant HPRI-CT-2001-00132.

REFERENCES

Eurola, S., Virtanen, R.. 1991. *Key to the vegetation of the northern Fennoscandian fjelds*. Kilpisjärvi Biological Station, University of Helsinki. 28 pp.

Kotiranta, H., Uotila, P., Sulkava, S., Peltonen, S.-L.(eds.) 1998. *Red Data Book of East Fennoscandia*. Ministry of the Environment, Finnish Environment Institute and Botanical Museum; Finnish Museum of Natural History. Helsinki. 351 pp.

Lehtovaara, J.J. 1995. Kilpisjärven and Haltin kartta-alueiden kallioperä. Summary: Pre-Quaternary rocks of the Kilpisjärvi and Halti map-sheet areas. Geol.Survey of Finl., Espoo. 64pp.

Pesola, V.A. 1928. Kaltsiumkarbonaatti kasvimaantieteellisenä tekijänä Suomessa (Summary: Calcium Carbonate as a Factor in the Distribution of Plants in Finland). - *Ann. Soc. "Vanamo"* 9(1). 246 pp

Pilats, V. 2000. Text. In: Esmeralds of Latvia. Ministry of Environmental Protection and Regional Development. 16 pp.

Saltikoff, B., Laitakari, I., Kinnunen, K.A. & Oivanen, P. 1994. Helsingin seudun vanhat kaivokset ja louhokset. Geologian tutkimuskeskus, Opas 35. 64 s.

Systra, Y.J. 2000. Geological Heritage of Northwestern Russia: experience in conservation. *Mem. Descr. Carta Geol.d'It.* LIV (1999), Roma, pp.361-366.

Thornton I. (ed.). 1983a. Applied Environmental Geochemistry. *Acad. Geol. Ser. London*. 501 pp.

Thornton I. 1983b. Geochemistry Applied to Agriculture. In: Thornton I. (ed.) 1983a. Applied Environmental Geochemistry. Acad. Geol. Ser. London. 501 pp.

Trofimov, V.T.(ed.). 2000. *Ecological functions of the lithosphere*. Moscow University Press. 432 pp.

In: Parkes, M.A. (Ed.) 2004 *Natural and Cultural Landscapes - The Geological Foundation*, 77, Royal Irish Academy, Dublin

LANDSCAPE CHARACTER AND EARTH HERITAGE: DEVELOPING A MORE HOLISTIC VIEW OF THE LANDSCAPES OF NORTH WEST SCOTLAND

John E. Gordon, Rebecca Hughes, Ken Monteath and Colin C.J. MacFadyen,
Scottish Natural Heritage, 2 Anderson Place, Edinburgh EH6 5NP, Scotland.
Email: john.gordon@snh.gov.uk

Abstract: From the agricultural heartlands of Aberdeenshire to the remote mountains of Wester Ross, Scotland has a great variety of landscapes. This variety results from the interaction over different timescales of both natural and human influences on the land. In this study, we use GIS to examine the links between geology, geomorphology, landscape character and land cover way in North West Scotland.

At a macro-scale, the character of the landscape reflects the underlying geology, perhaps more strikingly than in any other part of Scotland. This is clearly seen in the present landscape expression of the main geological elements - Lewisian plateau surfaces, Torridonian Sandstone mountains and the Moine Thrust. The action of geological and geomorphological processes over long timescales, involving large-scale earth movements, weathering and erosion has emphasised the lines of geological weakness, giving a strong 'grain' to the landscape, seen for example in the orientation of the main sea lochs and the fjord-like coast. Weathering and erosion have also acted on the bedrock to shape the distinctive character of individual mountains. Successive Quaternary glaciations and post-glacial processes, including the formation of extensive peat bogs, have left an equally distinctive imprint. All these patterns are well-illustrated on a geomorphological map of the area which shows a great diversity of landform types.

There are also many close links between geology and biodiversity, expressed for example in the range of habitats present (a complex of wet heath, blanket mire and open water bodies on the ice-scoured Lewisian plateaux, grass and moss heaths on the drier Torridonian mountains and calcareous plant communities on the Durness Limestone). The human element in the landscape here is a relatively recent one, predominantly in the last 5000 years, but the impact on landscape character has been significant, especially in terms of patterns of landuse and settlement. This is evident when examining, for instance, the successive extent and patterns of past woodland clearance, plantation forestry and native woodland regeneration. North West Scotland therefore illustrates many fundamental interdependencies, both direct and indirect, between geodiversity, human activity and the landscape of today.

In: Parkes, M.A. (Ed.) 2004 *Natural and Cultural Landscapes - The Geological Foundation*, 78, Dublin, Royal Irish Academy.

TERRAIN DIVERSITY AS INDICATOR OF GEODIVERSITY AND BIODIVERSITY IN DIFFERENT SCALES

Lars Erikstad and Vegar Bakkestuen, NINA, BOX 736 Sentrum, N-0105 Oslo

Abstract: Geomorphological diversity is an integrated part of geodiversity, expressed in the clear links between geology and landforms. Biodiversity spans over the scales from diversity of genes, species and ecosystems. Within ecosystem diversity, geodiversity plays an important role. Different ecosystems (or natural areas as a practical unit) have their characteristics from the rocks, the deposits and the landforms controlling nutrient supply, hydrological conditions, disturbance, slope and exposition. It should therefore be possible to establish links between terrain diversity and the overall geodiversity and biodiversity.

As a source for terrain information, we used digital elevation models (DEM) in different scales. On an overall European scale we used the HYDRO 1k model from USGS EROS data centre giving elevation data in a regular grid of 1kmx1km resolution. On an overall national scale we used the Norwegian national 100x100m DEM. On a detailed scale we computed DEMs with a resolution of 10x10m based on maps with 5 m contour interval. As input for geology we used geological maps in different scales and for biodiversity we used different species databases, vegetation maps and other sources containing information on biodiversity in natural areas.

The terrain variability had a significant potential in indicating ecosystem diversity in all scales. The degree of the relationship between terrain, geology and biodiversity was also clear, but dependent on scale and how much detailed data that was available and with variations between different systems and species.

SESSION 3

GEOSITES: THE CENTRAL AND EASTERN EUROPEAN EXPERIENCE

In: Parkes, M.A. (Ed.) 2004 *Natural and Cultural Landscapes - The Geological Foundation*, 81-89, Royal Irish Academy, Dublin.

IUGS GEOSITES PROJECT PROGRESS - A FIRST ATTEMPT AT A COMMON FRAMEWORK LIST FOR SOUTH EASTERN EUROPEAN COUNTRIES

I. Theodossiou-Drandaki[4a], R. Nakov[3], W.A.P. Wimbledon[b], A. Serjani[1], A. Neziraj[1], H. Hallaci[1], G. Sijaric[2], P. Begovic[2], Sv. Petrussenko[3], Pl. Tchoumatchenco[3], T. Todorov[3], I. Zagorchev[3], M. Antonov[3], D. Sinnyovski[3], A. Diakantoni[4], Ch. Fassoulas[4], G. Fermeli, D. Galanakis[4], A. Koutsouveli[4], A. Livaditi[4], K. Papadopoulou[4], P. Paschos[4], A. Rassiou[4], N. Skarpelis[4], N. Zouros[4], D. Grigorescu[5] Al. Andrasanu[5], Br. Hlad[6], U. Herlec[6], N. Kazanci[7], F. Saroglu[7], A. Dogan[7], H. Inaner[7], M. Dimitrijevic[8], D. Gavrilovic[8], B. Krstic[8], D. Mijovic[8], [1]: Albania (Al, aserjani@yahoo.com), [2]: Bosnia and Herzegovina (Ba, gsijaric@bih.net.ba) including Republic of Srpska (RS, begovic_p@hotmail.com), [3]: Bulgaria (Bg, radnac@router.geology.bas.bg), [4]: Greece (Gr, ren@igme.gr), [5]: Romania (Ro, aalex@geo.edu.ro), [6]: Slovenia (Si, Branka.Hlad@gov.si), [7]: Turkey (Tr, nkazanci@penta.gyte.edu.tr), [8]: Yugoslavia (Yu, miyo@eunet.yu), a Leader of ProGEO- WG1; b. Leader of IUGS Geosites project, UK (B.Wimbledon@ccw.gov.uk)

Abstract: This contribution refers to the progress of the Geosites project in the South Eastern European Countries. The Geosites project is an International Union of Geological Sciences (IUGS) initiative for the inventory and compilation of the global Geosites list. The first common European product of a Geosites framework for Europe appeared in *Geologica Balcanica* special issue "Geological Heritage of Europe" . South Eastern European Countries have contributed greatly in this common effort, and decisive steps for the progress of Geosites project have been achieved with the workshops organized in Athens in 2001 and in Ankara in 2002 by ProGEO-WG1, the working group of ProGEO for SE Europe. The methodology of the inventory, as well as the S. Eastern European countries action plan and a first attempt for a common Geosites framework as an appendix, follow in this paper.

The initiative for the systematic inventory of the most important sites was undertaken by the Global Geosites Working Group, set up by the International Union of Geological Sciences (IUGS) with the support of UNESCO, and active since 1996. This inventory focuses mainly on the needs of the geological sciences research and education; there can be no science and education without site conservation, as it is well known. The project aims to involve the whole geological community in geoconservation, as well as to support any national or international initiatives to protect geodiversity, a resource that is underestimated and completely unrenewable.

This ambitious project started in 1996 as a consequence of discussions at an international level, and it intended to identify geological sites to be included in global site database and to aid protection projects, including a small number of sites to be included in the World Heritage list of UNESCO (Wimbledon, 1996).

In Europe, most of the pilot work for the Geosites project has been carried out by ProGEO, supported by its regional working groups and its national representatives. ProGEO's (the European Association for Conservation of the Geological Heritage) members represent most of Europe's nations.

As a starting point for the project, a SE Europe regional workshop of ProGEO, sponsored by UNESCO, was held in Sofia in 1995, where a strategy for standardisation of the processes of the inventory and the compilation of the global Geosites list was discussed. IUGS's President wrote to all National Committees in 1996 announcing the new initiative.

Some significant milestones with reference to the Geosites project can be enumerated since then. The first common European product of a Geosites framework for Europe appeared in *Geologica Balcanica* special issue "Geological Heritage of Europe", in 1998, after the Belogradchik, Bulgaria ProGEO conference and Geosite IUGS-UNESCO World Heritage workshop (Ishchenko et al, 1998).

The phases of the project described in Belogradchik paper and already stated in Roma during a ProGEO international meeting, are the following:

1) first devise a country framework
2) open a discussion in the country, publish the first attempt of the framework
3) more or less finalize your country framework
4) start to select sites to represent your matured framework
5) discuss frameworks with neighbouring countries, and correlate frameworks between countries
6) finalize a common geosite list for all countries (in your regional working group)

to represent the final regional framework, that is - final representation of
geodiversity of the region

7) place final documentation (standard format) in central database of IUGS.

-During all these phases publish and consult with all, at as many stages as possible-

Specifically for South Eastern European Countries and after a long period with a sinister situation in the region, decisive steps for the progress of Geosites project have been achieved via workshops organized in Athens in 2001 and in Ankara in 2002 by the ProGEO-WG1 (the network of the ProGEO group for SE Europe) with the valuable assistance of IUGS Geosites WG chair. The creation of the first Working Group of ProGEO goes back to 1995 in Bulgaria, during the first sub regional meeting "Conservation of the Geological Heritage in South-Eastern Europe". Today ProGEO WG - 1 members represent: Albania, Bosnia and Herzegovina including Republic of Srpska, Bulgaria, Croatia, FYROM, Greece, Romania, Slovenia, Turkey, and Yugoslavia.

This paper constitutes an attempt of 5th phase of the project, while at the same time phases 1,2,3,and 4 continue regularly in each country of the region (Theodossiou-Drandaki, 2001 and Theodossiou-Drandaki et al, 2001). A South Eastern Geosites frameworks list is presented as an appendix to this paper. This list is the product of fruitful cooperation between numerous geoscientists who constitute the network of ProGEO-WG1. It is a dynamic product, which will develop gradually and systematically, with the involvement and cooperation of all the more specialists, becoming more complete, with a representivity equivalent to the geodiversity of the region. This cooperation has been achieved electronically and with regular meetings and workshops organized in several countries of the region in order to discuss and correlate frameworks with neighbours. During the workshop in Athens, September 2001, an ambitious action plan for the continuation, progress and achievement of SE Europe Geosites was discussed and agreed as following:

1) In the country

Each country will complete its frameworks ("contextual frameworks") list. So, revised versions of Belogradchik lists or new lists will be compiled. Each country coordinator to undertake description of each framework to the format given below.

2) Between countries - join, eliminate or confirm frameworks. Comparison of frameworks between countries will start as soon as possible, but be completed by May.

3) By end May-Mid June (Turkey meeting) a regional framework list will be compiled - to be discussed and finalised at the meeting.

4) Final framework to be presented as a joint WG 1 paper at Dublin conference in September.

5) Countries, country coordinators, must edit the country lists, organise consultation with experts and discuss shared frameworks with the WG members in the countries next door. We aim to combine shared frameworks, and to eliminate those, which are unnecessary. To aid discussion between countries, we agreed to prepare a sheet (format) for each framework in the country list, as follows:

1) Geosite framework

2) Concise description, a paragraph to a page in length, to make it possible to precisely compare frameworks between the countries.

3) Value of framework - is it of country/ Balkan/ Europe/global significance?

4) Suggested occurrence of framework in adjacent countries

What it is important is that with this action plan as a guide, most of the goals have been fulfilled and this joint WG1 paper was presented in Dublin. What remains to be fulfilled for the final framework list of South Eastern European countries, is the continuation of consultation with experts, the valuation of the framework so far partially completed, and the thorough comparison after precise description for each framework. These will be some of the issues of next ProGEO WG1 meeting in Romania next year.

On the basis of the geosites project progress shown in this paper, it is expected that in a medium term the next steps of the Geosites (phases 6 and 7 of accepted methodology) will be achieved and completed and that the richness of regional geosites will be represented in international lists. We look forward to this end.

REFERENCES

Wimbledon, W.A.P., (1996). National site selection, a stop on the way to a European Geosite list. In: Proceedings of the Special Symposium " Geological Heritage in South East Europe", May 1995, *Geologica Balcanica*, 26. 1, 15-27.

Ishchenco et al, (1998): a first attempt at a Geosites framework for Europe- an IUGS initiative to support recognition of World Heritage and European Geodiversity. In: Special Issue "Geological heritage of Europe" (I. Zagorchev and R. Nakov, Sp. Eds), *Geologica Balcanica*, 28. 3-4, pp 5-47.

Theodossiou-Drandaki, I., (2001): framework for the Geosites selection according to the international union of geological sciences (IUGS), and the European association for the conservation of geological-geomorphological heritage (ProGEO); first phase: an attempt for a country framework. Proceeding 9th International Congress of the Greek Geological Society, vol XXXIV/2, pp 795-802.

Theodossiou-Drandaki, I., Papadopoulou-Vrynioti, K., Markopoulou-Diakantoni, A., (2001): framework for the Geosites selection according to the international union of geological sciences (IUGS), and the European Association for the conservation of geological-geomorphological heritage (ProGEO), 2nd phase: opening of a discussion in the country, publishing and more or less finalizing the country framework. Proceeding 9th International Congress of the Greek Geological Society, vol XXXIV/2, 803-810.

Framework	Location (established and suggested)
Stratigraphic	
Phanerozoic	
Holocene	In all the countries
Pleistocene (e.g. marine and continental terraces)	Gr, Tr, ...
Quaternary remnants of alpine glacial and interglacial sediments	In all the countries
Karst cave sediments, their fossils and stone-age tools	Bg, Gr, Si, Tr, Yu,.
Sedimentary sequences of Mio-Pliocene continental grabens from the Aegean Extensional region	Bg, Gr, M, Tr,....
Quaternary and Neogene – Pliocene marine and brackish basins and sequences (e.g. Eastern Para-Tethys sequences of the Fore-Carpathian, Euxinian, Pannonian Basins; continental Pliocene deposits, marine and continental molassic sediments, complete marine cycles-Tr; sedimentary sequences of the grabens from the Aegean extensional region; Full section of Miocene molasse (e.g. Morava, Korcha region, Al; Velenje Pliocene lignite sequence-Si; Lignitiferous basins of Pleistocene, Pliocene , Miocene age (e.g. Macedonia, Thrace, Peloponnese-Gr)	Al, Bg, Gr, Ro, Si, Tr, Yu
Sections in Pre-Adriatic Depression (e.g. Erzeni section, Tirana region-Al)	Al, ?Gr, ?M
Eocene sections with Fe-Ni and Ni-silicate deposits (e.g. Bitincka Erzeni section)	Al, ...
Upper Cretaceous sediments with fish fossils of the Dinaric platform	Si, ...
Mezosoic section of the Tethys and Peri-Tethys (e.g. Mali I Gjere classical section of carbonate siliceous Mesozoic rocks in Ionian zone-Al; Gr)	Al, Bg, Gr, Ro, Si, Tr, Yu
Permian to Oligocene stratigraphic sections, with rocks' inversion and "high pressure" fossils (e.g. Tallaia Mts., Crete)	Gr
Palaeozoic sections of Gondwana and Peri-Gondwana origin	Bg, Gr Tr, Ro, ..
Sedimentary and biological characteristics of time-boundaries (e.g. Pliocene/Pleistocene boundary, Miocene/Pliocene boundary-Gr; Cretaceous/Paleogene boundary, East Balkan and West Fore Balkan-Bg; Jurassic/Cretaceous boundary, West Fore-Balkan and West Balkan-Bg; Permian/Triassic at Slovats-Yu and NW Dinarides-Si; Devonian/Carboniferous boundary at Milivoyevtsi, Druzetich, Valievo-Si; Silurian/Devonian, Varniche-Si, etc.)	Al, Bg, Gr, Ro, Tr, Si, Yu, ...
Phanerozoic stratigraphic and sedimentological phenomena (sequence stratigraphy, unconformities, sedimentary sequences, lithology, intertidal and supratidal carbonate platform sedimentation, climatic cycles, cyclic sedimentation-climatic Milankovitch cycles, condensed sections, "ammonitico rosso", hard-grounds, beach-lagoon sites, storm deposits, rudist biostromes, reefs, neptunian dykes, manganese nodule horizons, Middle-Upper Cretaceous transgression, etc.)	In all the countries
Stratigraphic stratotype sections of Balkan and European values (e.g. Triassic, Zlatibor-Yu; Triassic, Paleogene-Tr; Jurassic-Cretaceous transition with the clay-layers and facies at Bolietin River, Dolni Milanovats-Yu; Tortonian Molasse-Al; etc)	Al, Tr, Yu, ...
Proterozoic	
Precambrian sequences (e.g. Precambrian Pohorje metamorphic Group-Si; Rhodopes-Bg; Gr; Tufanbeyli authochthon-Tr ; Vlasina Complex-Yu)	Bg, Gr, M, Si, Tr, Yu
Archaean	
Archean (??) sequences of Pre-Rhodopian Supergroup (Rhodopes-Bg;Gr)	Bg, Gr

Palaeoenvironmental	
Holocene glacial forms (e.g. Jezerca, Gashi and Korabi highlands-Al; Rila and Pirin Mountains-Bg; remnants of Alpine glacial forms-Si)	Al, Ba, Bg, Si, Yu
High mountain Holocene lakes with glacio-lacustrine sediments (e.g. Rhodope, Rila and Pirin Mountains-Bg; Montenegro-Yu; Sylbica and Jezerca lakes-Al; Socha Valley, Srpenica-Si)	Al, Bg, Gr, Si, Tr, Yu, ...
Loess bearing records of the last 450.000 years (with remnants of large Pleistocene mammals, Serbia, Voivoidina-Yu). Loess/paleosoils alternations (e.g.Pydna-N. Greece)	Bg, Gr, Ro, Si, Yu
Quaternary and Neogene paleosols	Gr, Tr
Quaternary lakes and terraces (along the Black Sea coast and the Danube River-Bg; Ro; along the Black Sea coast-Tr; Bg; Ro; along the Aegean Sea-Gr; Tr; along the Adriatic-Al; Gr)	Al, Bg, Gr, Ro, Tr, Yu
Calcretes for Pleistocene climate	Gr, Tr, ...
Sandstone concretions with peculiar shape and specific conditions (e.g. Upper Miocene, Lower Meotian, Southern Carpathians)	Ro
Messinian salt crisis, Neogene evaporate basins	Gr, Tr, Yu, ...
Bioherms (e.g. in the Paleogene Dikilitash Formation-Pobitite Kamuni Rock Columns-NE Bulgaria)	Bg, Si, Tr
Sequence stratigraphic phenomena of palustrine carbonates (e.g. Upper Cretaceous/Paleocene palustrine carbonates in NW Dinarides)	Si, ...
Palaeokarst and karstification processes (e.g. Benca, Bogazi and Noraj, M. Jurassic break in sedimentation with paleokarst into Liassic massive limestones-Al)	Al, Bg, Gr Tr, Si, Yu
Various marine deep-facies zones in tectonically displaced and imbricated sections (e.g. Jurassic with radiolarites in SW Bulgaria; Jurassic marine with radiolarites, deep-facies zones in aborted tectonic rift-Si)	Bg, Gr, Tr, Si, Yu
Diagenetic features in carbonate rocks (e.g. in Triassic of W. Bulgaria; W. Slovenia)	Bg, ?Ro, Si, ?Yu
Continental red beds of Permian (Bg; Si; Yu, Curraj Eperm-Albania, etc.) and Early Triassic age. Marine red beds of Late Triassic age (e.g. Bg, Yu)	Al, Bg, ?Ro, Si, Tr, Yu,
Reefs of different ages (e. g. Neogene Kalemegdan, L. Permian Petkovitsa Monastery-Yu; Devonian coral reefs-Stegovnik-Si; Oligocene fringing reefs of "Gornji grad" formation-Si; etc)	Bg, Gr, Ro, Si, Tr, Yu
Complex systems as palaeoenvironmental indicators-fluvial, shallow and deep water, etc. (e.g. L. Jurassic fluvio-deltaic palaeoenvironment-fossil soils, channel fills, crevasse splays, with terrestrial plant remains-ferns, horsetails, pteridosperms, ginkgoales, bennettitales and conifers remains with *in situ* spores and pollen and in growth positions (Ponor Quarry-Ro)	Bg, Gr, Ro, Tr, Yu, ...
Euxinic environments of Early Silurian	Tr, ...
Palaeo-ecosystems of continental organisms, marine organisms or mixed	Gr
Palaeontological	
Footprints, fossil leaves (e.g. Santorini island, Gr) in volcanic rocks	Gr, Tr
Quaternary mammal sites (e.g. mammouth sites Nevlje-Si; Belgrade- Yu; Indigenous dwarf elephants in Tilos island, Crete island, Gr)	Gr, Ro, Si, Tr, Yu, ...
Fossilized palm-trees of Plio-Pleistocene age mixed with marine organisms (e.g. SW Peloponnese)	Gr
Banks of Cirripedia of M. Pliocene age permitting the tracing of Pliocene coastline in Mediterranean sea.	Gr
Neogene mammalian sites (including large mammals, e.g. Axios (Vardar) basin with anthropoid Uranopithecus macedoniensis of U. Miocene age-Macedonia,Gr, Pikermi fauna-Attica,Gr, Samos island, Chios island-Gr)	Bg, Gr, Si, Tr , Yu
Neogene mollusc sites (e.g. Miocene-Central N. Bulgaria; reef at Kalemegdan, Belgrade, Karagach Valley, Grgeteg-Yu; Miocene bivalves in Pannonian basin-Si; etc)	Bg, Gr, ?Ro, RS, Si, Tr, Yu,
Upper Pontian (Pannonian G-E) very rich plant association-cineritic beds very rich in leaf impressions, today explored in eight fossiliferous outcrops (Chiuzbaia, N. Romania)	Ro
Petrified forests of various ages (e.g. Miocene-Lesvos island, Gr; Paleogene-Rhodopes, Bg)	Bg, Gr, M, Tr, Yu
Lower Miocene (Burdigalian)-sand bed surfaces with various fossil traces and trails (e.g. Southern part of East Carpathians, Priseaca Valea Sarii)	Ro
Fish and leaf fossils in marsh deposits of Late Miocene	Tr

Lower Oligocene (Rupelian) typical near-shore neritic fossil assemblage (e. g. Southern Carpathians, Suslanesti)	Ro
Oligocene-rich benthic fish association, (e.g. East Carpathians, Pietricica, Cozla Cernegura)	Ro
Paleogene large foraminifera in bioherms (e.g. Eocene Dikilitash Formation-Pobitite Kamuni Rock Columns, East Bulgaria, East Moesian Plate, and East Fore-Balkan; Nummulites & Alveolina assemblage of the Eocene in the North eastern Dinarides-Si)	Bg, Gr, Tr, Si
Late Mesozoic large foraminifera (e.g. Cenomanian in the East Balkan, RS; Lower Cretaceous (Aptian) in the West Fore-Balkan, Bg)	Bg, RS, ...
Mesozoic molluscs (e. g. ammonites and brachiopods of Stara Planina and the Fore Balkan-Bg; Kchira site of Triassic ammonites-Al; Jurassic ammonites, Peloponnese-Gr; corals-Si; etc.)	Al, Bg, Gr, M, Ro, Si, Tr, Yu
Limestone with stromatolites (e.g.Grabon-Leqet e Hotit, Al; Parnassos-Gr)	Al, Gr, ...
Continental deposits with dinosaur remains and nests (e. g. Upper Maastrichtian continental red beds, Hateg and Tustea, SW Romania)	Ro
Fossiliferous Permian sediments (e. g. Upper Permian limestone with brachiopods- "Zazar Formation", Middle Permian Neoschwagerina limestone and Lower Permian limestones with brachiopodes of Dovzanova soteska-Si; Bulog Limestone with ammonites, Sirogojno-Yu)	Si, Yu, ...
Late Carboniferous trilobites, brachiopods, bivalves, crinoids etc. (e.g. S. Karavanke Mts.-Si)	Gr, Si, ...
Late Carboniferous plants (e.g. Svoge and Dobroudzha coal basins-Bg)	Bg, Ro, ...
Namurian goniatites and conodonts (e.g. Devonian/Carboniferous boundary, with nine conodont zones, Milivojevici, Druzetic, Valjevo)	Yu
Devonian fish fossils	Tr, ...
Silurian graptolite faunas (e.g. NW Bulgaria; Tr; Chios island-Gr)	Bg, Gr, Tr
Trace fossils of different environments and age (e.g. Pcinja basin-Yu)	Bg, Si, Yu, ...
Igneous, metamorphic and sedimentary petrology textures and structures, events and provinces	
Paleogene, Neogene extensional basins of the Southern Balkans	Bg, Gr, M, Tr, Yu
Olistostromes, turbidite sequences (e.g. of Late Cretaceous-Paleogene flysch-type basin East Balkan and E. Srednogorie Zones-Bg; E. Pontides-Tr; Paleogene basin of W. Slovenia; Devonian-Bg; Yu; RS; Jurassic-Cretaceous-Bg; Yu)	Bg, Gr, Si, RS, Tr, Yu
Flysch basins of different ages (Nish-Troyan, L. Jurassic-U. Cretaceous-Bg; Yu, L. Cretaceous-Bg; Ro; Tr; Yu, L. Devonian-Yu; Bg, Carboniferous- Belava-Yu, Paleogene-SW Slovenia)	Bg, Gr, Ro, Si, Tr, Yu
Permian eolian deposits (e.g. Central Fore Balkan, Bg)	Bg, Yu, ...
Hercynian molasse	Bg, Ro, Tr, Yu
Marine and continental deposits of Carboniferous age (e. g. Upper Carboniferous fossiliferous cyclic carbonate/clastic sediments of the Southern Karavanke Mts.-Si)	Bg, Gr, Tr, Si
Cambrian sedimentary sequences	Bg, Tr, Yu
Lower Palaeozoic succession of Northern Gondwana	Bg, Tr
Collision volcanism of Quaternary age	Gr, Tr, ...
Quaternary subduction-related volcanism (Volcanic Arc of Southern Aegean Sea, e.g. Santorini island, Milos island, Nisyros island)	Gr
Transform-fault volcanism	Gr, Tr
Extension volcanism of Plio-Quaternary age	Gr, Tr
Basaltic rocks with polygonal columns (e.g. Upper Miocene, West Carpathians and East Carpathians-Ro);	Bg, Gr, Ro, Tr, ...
Intraplate Miocene basalts (e. g. from the Moesian Plate and S. Bulgaria)	Bg, M, Ro, Yu, ...
Paleogene-Miocene volcanic arcs (volcanoes and volcano-sedimentary sequences)	Bg, Gr, Si, Tr, Yu
Upper Cretaceous volcanic and intrusive sequences (e.g. Banat-Timok-Srednogorie-Pontides magmatic Belt)	Bg, Ro, Tr, Yu,
Pillow-lava complexes (e.g. U. Cretaceous Bulgarovo paleo volcano, spectacular outcrops of "Bulgarite" -unique petrographic variety, Eastern Srednogorie-Bg; Dinaridic Belt-Yu)	Al, Ba, Bg, Gr, M, Tr, Yu
Late Triassic rift volcanisms related to the opening of Neo-Tethys Ocean	Ba, Bg, Tr, Ro,Yu
Ladinian volcanism within aborted rifting	Si
Mud volcanoes	Bg, Ro, ...

Sheeted dykes (e.g. Kachinar-Munella-Domgjon, Al; NW Bulgaria)	Al, Bg, ...
Volcanic structures and landforms recent and ancient (basalt lakes, columnar basalts, maars, tuff rings, etc.- Tr; Ro)	Bg, Gr, M, Ro, Tr
Polymetamorphic rocks(e.g.Serbo-Macedonian,Rhodopes-Bg;Gr, Menderes Metamorphic Provinces-Tr; Picari-Al; Pohorje,Kobansko Mts-Si)	In all the countries
Sanidinite facies contact metamorphism	Tr
Amphibolite facies, Triassic with fossils (e. g. Sakar Mountain-Bg)	Bg, Gr
Greenschist facies Mezosoic rocks (e. g. Eastern Rhodopes-Bg; Gr)	Bg, Gr, Tr, Yu
Eclogite and blue-schist high-pressure metamorphism facies (e. g. Derveni-Al; Tr; Syros island, Sifnos island, Tinos island-Gr)	Al, Gr, Tr, M, Yu, .
Extremely high pressure metamorphism (newly found micro diamonds in the Rhodopes of N. Greece)	Gr, ?Bg,
Rift deposits related to the opening of Neo-Tethys Ocean	Tr
Triassic high-pressure metamorphism of Palaeo-Tethys	Tr
Oceanic crust of the Palaeo-Tethys subduction zone	Ba, Tr, Yu
Core complex	Bg, Gr, M, Tr
Ophiolite sections of different ages (e.g. Gjegjan, Kalimash, Runa-Al; Dinaridic Ophiolite Belt-Yu; Complete ophiolitic sequence of Late Proterozoic-Early Cambrian age NW Bulgaria; Yu; Tr, Mesozoic ophiolite sequences Pindos Mts, etc-Gr)	Al, Bg, Gr, M, Ro, RS, Tr, Yu
Precambrian (and Cambrian) ophiolites and island arcs	Bg, Tr, Yu
Mineralogical	
Smaragds in desilicified pegmatites (e. g. Rila Mountain-Bg)	Bg, ?Gr, ?M, ?Tr,
Occurences of large twins (up to 10-15cm) of sanidine in Paleogene trachyandesite (e.g. Mesta Valley-Bg)	Bg, ?Gr
Boudings in marbles, containing scapolite, diopside, amphiboles, titanite, clinozoisite, phlogopite, graphite (Rhodopes-Bg)	Bg, Gr, ?M
Polymorphic modifications of Al_2SiO_3 (sillimanite - kyanite - andalusite) from the Rhodopes	Bg, Gr, Tr, ?M,
Sulphides, Cu-arsenates & Cu-carbonates (e.g. Vratsa Ore district)	Bg
Precious stones and gem minerals	Bg, Gr, Tr, Yu, ...
Zeolite mineralization (+L. Cretaceous magmatism: E. Srednogorie –Bg; ?E. Pontides-Tr; +Pg volcanism, E. Rhodopes – Bg; Gr)	Ba, Bg, Gr, ?M, Tr
Base metal druses and pyrite crystals in Pg deposits (Rhodopes-Bg)	Bg, Gr
Skarn mineralization from the metasomatic zones developed at the expenses of marbles, pegmatites and intrusions (e.g. Kopaonik-Yu, with complete develop zone of metasomatic alteration; Rhodopes – Bg; Gr, Strandja Mountains-Bg; Tr, Appuseni-Ro, Serifos island-Gr)	Bg, Gr, Ro, Tr, Yu, ...
Late Cretaceous tourmaline-bearing pegmatites (e.g. W. Srednogorie - Bg; Dravograd area with the brown magnesium tourmalines, dravite etc., - Si)	Bg, Si, ?Yu
Macrocrystalline gypsum mineralization (e. g. Qafa e Kochiut-Al)	Al, Gr, ?Ro
Typical paragenetic assoc. of PGE Fe-Ni-Cu-Cr (e.g. Bregu i Bibes)	Al
Fe-Cu-S and Fe-Cu-S-Zn-Pb-As-Au-Ba association (e.g. in Mirdita Cu deposits-Al)	Al, ?Bg, Gr, ?Yu
Gold-silver telluride minerals and associated sulphides (e.g. in Upper Neogene-Sacarimb mine in the West Carpathians)	Ro
Neogene evaporite mineral beds (e.g. trona, borates, celestite)	Gr, Tr
Type localities of minerals (e.g. Konyaite, bursaite,etc., Tr; Strashimirite, kostovite, Bg; Moissanite in Vardar Zone and Serb-Mac. masse-Yu; Wulfenite classical locality etc., secondary Pb-Zn mineralisation-Si; Baryte and septarian concretions in Miocene sediments, Shtrihovec-Si; Pseudobrookite in the Uroi neck, South Carpathians-Ro, Nagyagite,petzite, muthmannite, stuetzite, krenerite, krautite, etc in Sacarimb mine, West Carpathians-Ro; Laureonite, kamarizite etc, Laurium-Attica, glaucophane, Syros island-Gr)	Bg, Gr, Ro, Tr, Si, Yu, ...
Sepiolite formations in lacustrine basins	Tr
Thermal spring carbonates	Gr, Tr, ...
Direct laterization of spilite-ceratophyre basement (e.g. Triassic)	RS, ...
Altered rocks-minerals (contact-metasomatic, skarns, epithermal type, etc)	In all the countries

/Albania are in the UNESCO WH List; Divjaka-Karavasta marsh-Al; Tuz Gul-Turkey, etc)	
Quaternary marine coastal deposits (ooids, beachrocks, terraces, sand bars) and erosional forms (e.g. uplifted paleo coast lines-Gr)	Al, Bg, Gr, Ro, Tr, Yu
Recent eolian sand dunes (Black Sea, Aegean Sea, Adriatic Sea)	Al, Bg, Gr, Ro, Tr, Yu
Holocene glacial landscape (e.g. Rila and Pirin Mountains (the last in the UNESCO WH List); Albanian Mountains: Lura, Jezerca-Valbona, Bjeshket e Nemuna; Duuboki Do Hole - Serbia; etc)	Al, Ba, Bg, Gr, ?M, Ro, Si, Yu
Glacial landforms and deposits	Al, Ba, Bg, Tr, Yu
Karst-superficial forms and caves (e. g. Classical karst-superficial forms and caves-Si; Ljubljanica karst river and its tributaries-Si; Durmitor Mountain and canyon of Tara, registered in the World Heritage list-Yu; Samaria gorge, Crete island-Gr, Karstic erosional forms (e.g. Katharos plateau and Nida plateau, Crete island-Gr, Paleorelief of various ages, Parnassos mountain, Elikonas mt, Mainalon mt, Meteoara mt, Cyclades islands-Gr, recent karst-Gr)	In all the countries
Evaporite karst	Tr
Impressive landscape of erosional rock columns, formed in Paleogene sediments (e. g. Pobitite Kamuni Rock Columns (Dikili-tash), a multi valuable category-NE Bulgaria)	Bg, ...
Rock pinnacles and rock bridges in carbonates and marbles	In all the countries
Erosional forms and landscapes formed in clastic sediments (e.g. Melnik Pyramides, Neogene, SW Bg; Belogradchik Rocks in Lower Triassic red beds-Bg, Meteora, Oligocene, Northen Gr; Cappadocia, Paleogene, Tr)	Bg, Gr, M, Tr, ...
Erosional forms in clastic rocks, in volcanics, in intrusive rocks (e.g. sheroidal erosion forms of granite, Tinos island-Gr), in quarries	In all the countries
Canyons and valleys and theirs specific landscape	In all the countries
Peat bogs (e. g. Kopren, Stara Planina-Serbia)	Yu, ...
Landscapes (erosional, karst, volcanic e.g. domes and columnar jointing in Lesvos island, Milos island-Gr, tectonic e.g.hanging valley of Ha, Crete island-Gr, fluvial, aeolian, glacial, in salt massifs and diapirs with dissolution-deposition phenomena, etc.)	Bg, Gr, M, Ro, Tr, Yu
Geotopes resulting from catastrofic phenomena (eg. Phalasarna uplifted harbour, Crete island-Gr, traces-ruptures of recent earthquakes in various greek areas)	Gr
Hydrogeology-Hydrology features	
Hydrological features and aquifer systems, springs, thermal springs, etc (e.g. Lesvos island-Gr)	In all the countries
Astroblemes	
Spherolites (e.g. Kopliku)	Al
Continental or oceanic-scale geological features, relationships of tectonic plates and terrains	
Ophiolites (Late Precambrian-E. Cambrian Ophiolites, West Stara Planina – Bg; ?T; Mezozoic (?) Eastern Rhodopes-Bg, Gr; Mesozoic-Al, Ba, Yu, M, Gr, Tr, RS)	Al, Ba, Bg, Gr, M, Ro, RS Tr, Yu
Suture zones and melanges of different ages (e.g. Llogara-Al; Ankara-Izmir-Tr ; Vardar zone)	Al, Bg, Gr, M, Tr, Yu
Tethys terranes	Bg, Gr, Ro, Si, Tr, Yu
Early Paleozoic Peri-Gondwana and Gondwana terranes	Bg, Gr, Tr, ...
Submarine	
Flooded river valleys and coast lines (Black Sea coast of Bg, Tr; Gr)	Bg, Gr, Tr
Submarine hot springs and sink holes (Izola)	Si
Karstic submarine springs (e.g. Northen Peloponese)	Gr
Historic	
Ancient mines: e.g. Ai Bounar, 4000-5000 B.C (Cu)-Bg, Rudna Glava, (Cu) -Serbia; Rubiku (Cu, Au)-Al; Laurium, Kassandra, etc (Au, Ag)-Gr; Brdo, Ljubija, Prijedor (Fe)-RS; Opaline deposits, Starevo culture, earlier Neolithic (6th millennium B. C.), Divlje Polje, Kragujevac-Serbia; Second largest world Hg mine, Idrija-Si; Sacarimb gold mine, etc	Al, Bg, Gr, Ro, RS, Tr, Yu

Economic	
Oligocene sediment-hosted Mn deposits, NE Bulgaria	Bg
Paleogene epithermal deposits (base metals-Au) (e.g. E. Rhodopes, Serbo-Macedonian massif)	Bg, Gr, M, Tr, Yu
Cu-porphyry, VHMS, vein and epithermal deposits (e. g. L. Cretaceous, Chelopech, Bg high sulfidation type, largest gold deposit in Europe; Bor, Maidanpek, Serbia; Gjegjiani, Al; etc.; Upper Carboniferous polymetalic vein mineralization of Litija-Si; Banat-Timok-Srednogorie-Pontides, Tr)	Al, Bg, M, Ro, Tr, Yu
Strata-bound base-metal deposits (e.g. Vratsa-Izdremets Ore district-Bg; Upper Triassic Pb-Zn Mississippi Valley Type mineralisation of Mežica-Bleiberg Range-Si; Anisian early diagenetic Zn-Pb mineralization in Topla, N. Karavanke Mts.-Si)	Bg, Si, Yu, ...
Magmatic and sedimentary uranium deposits (e.g. Middle Permian "red beds" uranium and copper deposits-Si)	Bg, Ro, Si,...
Emery ores, Naxos island	Gr
Skarn mineralization of Palaeozoic, L. Cretaceous and Paleogene age (e. g. Trepcha in Kopaonik-Yu, Stara Planina (Pz), Srednogorie-Strandja (L. Cretaceous), Rhodopes-Bg, Tinos island-Gr)	Bg, Gr, M, Ro, Tr, Yu
Titanium-Magnetite deposits (e.g. Kashnjeti)	Al, ...
Chromite Deposits (e. g. Bulqiza unique folded deposit and Shkalla-Al)	Al, Gr, ?M, Tr
Middle Triassic Hg epigenetic and syngenetic mineralisation (Idrija)	Si
Asphalt deposits (e.g. Selenica deposits in Tortonian sandstones)	Al, ...
Massive phosphate deposit ±U (e.g. Gusmari-Al)	Al, Gr,
Fe-Ni and Ni-Silicate deposit (e.g. Bitincka-Al)	Al, Gr, M
Metamorphic and sedimentary bauxite (e.g. Triassic and Cretaceous-RS; Amorgos island-Gr)	?Al, Gr, RS, Tr, ?Yu
Non-metallic magmatic and hydrothermal deposits (e. g. native sulphur, fluorite, etc)	Al, Bg, Gr, Tr, ...
Evaporite gypsum-anhydrite deposits (e.g. U. Permian-L. Triassic- RS; Neogene-Bg; Gr; Ro; Tr)	Bg, Gr, Ro, RS, Tr, Yu
Evaporite salt deposits (e.g. Dhrovjani and Kavaja-Al)	Al, Bg, Gr, Ro, Tr
Coal deposits (e. g. Velenje Pliocene sequences- Si; Yu)	Bg, Si, Tr, Yu
Devonian Mercury-Cu-Ba deposits (e. g. Otomalji)	RS, ...
Geothermal fields	Gr, Tr, ...
Structural	
Neotectonic features, active tectonics, extensional and recent relief-building, tectonic windows (North Anatolian Fault Zone-Tr; Idrija fault area -Si; Olympos tectonic window, Tinos island tectonic window-Gr, etc.)	Al, Ba, Bg, Gr, Si, Tr, Yu
Neogene extensional structures from the south Balkans and Turkey (Aegean Extensional Region)	Al, Bg, Gr, Tr
Paleogene extensional structures	?Al, Bg, M, Tr
Alpine Fold-Belt structures (Carpathians-Ro, Balkan Fold-Thrust Belt-Bg; Yu, Hellenides e.g. Vossakos and Aghios Pavlos folds, Crete island-Gr, Albanides, Dinarides-Al; RS; Gr; M; Si; Yu; Pontides, Taurides -Tr)	Al, Ba, Bg, Gr, M, Ro, RS, Tr, Yu
Thrusts series of Crete island	Gr
Triassic tectonics (e. g. Late Triassic-Old Cimmerian structures of W. Stara Planina-Bg; Triassic tectonic graben of Idrija and Middle Triassic aborted rift-Si)	Bg, Si, ?Yu
Permo-Triassic salt domes and tectonics (e.g. Mali I Bardhe dome-Al)	Al, Bg, ...
Upper Permian to Upper Cretaceous platform sedimentation	Ba, Si, Yu
Polyphase deformed metamorphic units (Rhodopes, Sakar and Strandja Mts., Menderes, Serbo-Macedonian Mass)	Bg, Gr, M, Ro, Si, Tr, Yu
Geophysical anomalies	Cr, Yu, ...
Geomorphological features erosional and depositional processes, landforms and landscapes	
Capes (Black Sea, Aegean Sea, Adriatic Sea)	Al, Bg, Gr, Ro, Tr, Yu
Modern marine coastal landforms (spits, bars, beaches, lagoons, deltas, capes)	In all the countries
Modern lakes, wetlands, rivers and rivers' mouths (e.g. the mouths of the rivers drained into the Black Sea-the Delta of the Danube is in the UNESCO WH List; Sreburna wetland-Bg and Ohrid Lake – FYROM	In all the countries

Ancient marble quarries (e. g. Ancient marble quarry,Thassos island-Gr ; Roman marble quarry- Pohorje, Si)	Gr, Si, Tr
Old oil deposits (e.g. Kuchova-Al)	Al, ?Ro
Remains of gold placers washed in ancient times (e.g. Vitosha Mountain-Bg; Northen Greece)	Bg, Gr, Ro, RS, Y,
Ankara melange (the site where the term "melange" was first defined)	Gr, Tr
Geo-history, Geo-archaeology, Geo-mythology	
Sites related to History (eg uplipted roman harbour- Phalasarna, Crete island-Gr), Archaeology (e.g. Kokkinopilos geo-archaeological site, Western Greece, Corinth Canal-Diolkos site, various ancient mines, quarries-Gr, Tr), to Mythology (e.g. Lerni Springs-Gr, Idaion Cave, Crete island-Gr, Girl Tower-Tr)	Gr, Tr, ...
Sites related to archeological discoveries (e.g. Oldest gold objects in the world, 4000 - 5000 years B.C, Varna Necropolis- Bg)	Bg, Gr, ?Ro, ?Tr, .

In: Parkes, M.A. (Ed.) 2004 *Natural and Cultural Landscapes - The Geological Foundation*, 91-92, Royal Irish Academy, Dublin

PROBLEM OF PROTECTION OF GEODIVERSITY IN THE CENTRAL EURASIA COUNTRIES

Il'ja Fishman, Academy of Mineral Resources of the Republic of Kazakhstan, Kunaev street, 105, Almaty 91, Kazakhstan. Email: bekzhanov@nursat.kz

Abstract: Protection of geodiversity in Central Eurasian countries suffers from the economic transformations of recent years, with geodiversity having a low priority. Yet many of the sites illustrated in this paper may be of world importance. The role of ProGEO and how it may advance geoconservation in Central Eurasia is discussed. This includes the need to hold a major meeting in Central Eurasia, as a boost to establishing practical action.

International cooperation considerably assists the organization of Geological Heritage protection work on all the Continents. The ProGEO role is very great in this matter. At first it included the West European countries, and henceforth became spread to the East European and Asia states.

Joining ProGEO is especially important for new independent states (NIS). Many of them are positioned in Central Eurasia. This region includes 8 New Independent States (NIS): Armenia, Azerbaijan, Georgia, Kazakhstan, Kyrgyzstan, Tajikistan and Turkmenia which were the republics of the former USSR. This huge region has nearly 5 million km^2 (including adjacent aquatories in the Caspian Sea).

The common issues for the Central Eurasia Countries are: the large economic transformations; unsustainable development; tourism (especially geotourism) is not developed; no balanced approach to the protection of bio- and geodiversity; an absence of systematic work on the geosite listing; difficulty with international contacts and experience; an absence of participation in ProGEO activity; a geodiversity information deficit; real danger of geosite damage and destruction and the presence of numerous geosites, sometimes of world importance.

The economic transformations throws back this sphere of the ecological work. Many government functionaries are not interested in spreading awareness of natural phenomena, because as the result they will have additional nature safety tasks, as well as the restriction on the giving of land plots. Therefore, the present system, which limits open dialogue and dissemination of information, is more suitable for them.

As the result the numerous natural phenomena, for example in Kazakhstan, are subjected to real threat of damage and even destruction. The beautiful Chernorechensk cave was destroyed by an explosion. Irreversible changes have taken place in the Kyzyltas terrain. This picturesque landscape was disfigured. The Kyzyltas gabbro-essexites have an unusual quality: they sing after a slight blow. These singing stones are being chaotically exploited together with the numerous petroglyphs just as standard building material. In some places the beautiful minerals as well as the rare fossils are being plundered or even destroyed, etc.

The same situation exists in many Central Eurasia countries. Most of them don't take part in ProGEO actions and have not started any systematic work of inventory and protection of their Geoheritage. It is important that many of the geosites in these countries should be included into the World Database.

The modern tectonic appearances, picturesque rocks (Fig. 1) and caves (Fig. 2) of Kyrgyzstan, stone forest (Fig. 3) and other fossils of Uzbekistan, the Kugitan gypsum caves of Turkmenia are among them. In the Caucasus, caves of Gorge, palaeovolcanoes and lakes of Armenia, mud volcanoes of Azerbaijan all attract attention. 12 geosites and

Figure 1. Geosites of Kirgyzstan. "Aeolian Castles" in Kongorchek stow. Outcrops of redcolour conglomerates of Paleogene-Neogene as bizarre vertical walls of "castles" and "bastions ". The data presented by V. Rogalsky.

Figure 2. Cave Chil-Ustun in Visean limestone. It is one of 50 caves of karst region near of Osch city. The data presented by V. Rogalsky.

Figures from top left to right, and l to r on each row.
Figure 3. Geosites of Uzbekistan. Dzharkuduk Geosite "Stone Forest" Upper Cretaceous. The data presented by I. Pjanovskaja & G. Pjanovsky
Figure 4. Canyon Charyn river (North Tien-Shan) in Miocene-Pliocene is well known among tourists as "Valley of Castles". Photo I. Fishman
Figure 5. Stone Glacier "Morainic" in Zailiysky Alatau mountain. Photo I. Fishman
Figure 6. The Cretaceous and Paleogene sediments are in the background. The famous iridium horizon was revealed at the boundary between them as an evidence of the cosmic disaster which led to the dinosaurs death. Photo O. Belyalov.
Figure 7. Monument of Nature "Stone town". Unique landscape complex : granite monadnocks, Alpine lakes and a swamp of the tundra type. The picturesque cliffs are composed of biotite, muscovite, binary and leucocratic granites of the Beloubinsk complex (Permian). Photo V. Izraztsov.

geoareas of Kazakhstan were listed for including into the World Heritage. Some of them are shown here (figs. 4-7). The efficient protection of the geological heritage may be achieved as a result of the integrated approach to the whole natural diversity. This is particularly important for the NIS where the principles of the sustainable development are not yet used.

As one of the ways to protect geological heritage, we suggest to IUGS and ProGEO, to create and present a draft of the Convention on the geodiversity, similar to the existing one on biodiversity. The essence of the Convention is the determination of the minimum necessary level of the geoconservation work by the countries-participants. To draw attention to the problem of the geological heritage we have published the picturesque album "*Geosites of Kazakhstan*".

Another way will be the seminar for the States of the Central Eurasia. The necessity for this became obvious some years ago. The Programme of the seminar should include such subjects as spreading of the UNESCO, IUGS and ProGEO experience, the integrated approach to the bio- and geodiversity, geoparks and geoecotourism along the Silk Road etc. This proposed seminar would be the start of the new working group action. As a result the possibility of professional contacts would exist for the geoconservator of the Central Eurasian countries. The ProGEO specialists, as experts and guides could take part in these meetings. Participation in ProGEO meetings in Europe is otherwise too expensive and impossible for the geologists of Central Eurasia.

One more way of saving of the geological heritage is the development of geoecological tourism. The essence of this task consists of the wide extension of the economic mechanism of ecotourism into the geological heritage. As a result an impetus for the preservation and popularisation of the geological objects will appear.

In: Parkes, M.A. (Ed.) 2004 *Natural and Cultural Landscapes - The Geological Foundation*, 93-96, Dublin, Royal Irish Academy.

GEOLOGICAL HERITAGE OF A SOUTHERN PART OF UKRAINE

V.V. Manuk, Dnepropetrovsk National University, Marks, 36, Dnepropetrovsk, Ukraine, 49044 Email: manuk-geo@mail.ru

Abstract: This paper reviews the main features of geological structure and geomorphology of a Southern part of the Ukrainian crystal shield and parts, connected to it, the Black Sea and Dnipro-Donetsk depressions. The complex combination of these structures causes an abundance and variety of minerals in this region and, an abundance of natural geological monuments (geosites) of various types, age and origin. The various aspects of study and use of natural geological monuments in the country between the rivers Southern Bug and Vovcha are considered.

The region represents an area where three large tectonic elements join: the Ukrainian crystalline shield combined with dislocated metamorphic complexes of Precambrian, Black Sea and Dnipro-Donetsk depressions filled with a cover of Phanerozoic sedimentary rocks from the Devonian up to modern formations. The larger part of the area is occupied by the Middle-Dniper megablock of the Ukrainian shield, tectonically combined in the east, on a zone of the Kriviy Rig-Kremenchug fault (suture), with Ingul-Ingületz, and in the west, on a powerful zone of Orehovo-Pavlogradsky fault, with the Azov megablock of the Ukrainian shield.

The insignificant depth of deposit, connected with the long tendency of the shield to rise and the modern level of erosive cutting-in cause a significant amount of natural outputs of Precambrian formations on a surface. Last, in valleys of such large and average rivers as the Dnieper, Southern Bug, Wovcha, Sura, Inguletz, Kamenka, which cut them through, magnificent picturesque landscapes are created which are valuable not only in aesthetic appeal, but also having large scientific-cognitive importance.

The complex structural-tectonic situation of the region where large structures of different age and origin are joined, and their long existence with periodic reactivation of tectonic movements, has led to various kinds of mineral raw materials. This is connected to a growing interest in natural geological monuments (geosites) described here, and the large variety of types and value of objects, which they demonstrate. To demonstrate this relationship between mineral resources and the importance of geological monuments, it is necessary to examine some examples in the region.

In the zone of the KriviyRig-Kremenchug fault is one of best iron ore deposits in the world, in quantity and quality. The Krivorizke deposit was opened in 1871 by V. Zuyev. From the earliest development in 1875, due to significant efforts of the naturalist and industrialist Al. Paul, and till now deposits of Krivorizky basin have provided the basic part of needs not only of the Russian state, but also were the basic suppliers in the majority of the countries in Central and East Europe. Today it is known not only for iron ores, but also commercial reserves of talc-magnesites and dolomites, suitable for manufacturing high-quality fire-resistant products, appreciable reserves of a wonderful marble, integumentary materials, mineral paints (ochre, mummy, minium). Amphibolites of the Krivorizka suite are quality raw material for the manufacture of a basalt fibre, stone moulding, decorative rubble, and filter materials. In ferrous-siliceous rocks are ore-occurrences of gold and silver.

In the metaconglomerates and metasandstones of the Skelevatska suite of early Proterozoic and carbonate-terrigenous graphite-containing rocks of Gdancevska suite of a Jovtorechensk deposit in northern part of Krivorizki basin are a variety of notable abyssal and metamorphic rocks. Among more than 200 minerals known in rocks of different horizons for Krivorizka series is a treasury of semiprecious and ornamental stone. Most famous, widely known even outside Ukraine is the decorative silicific version of crokidolyt or cummingtonite asbestos, known as "tiger's", "falcon's" or "cat's" eye depending on colours in this silky, iridescent mineral. Finds of various versions chalcedony, amethyst, nephrite, rock crystal, wandering units of goethite, rauh-topaz here are usual. The jesspilites and ferruginous quartzite with microfoliation and microtectonics, egyrin metasomatites, serpentinites, ophyotiffes, and jasper formations are used as ornamental stones.

A wonderful feature of the Krivorozki deposit is its easy availability not only in opencast mines, but also in natural outcrops. This circumstance, and also the scientific value of outcrops on a surface of ancient Precambrian formations containing large deposits of iron ores, explains the assignment of 12 objects of geological heritage from 17 known in the region. The most significant of them are:

- the outcrop of rocks of Krivorizka series of early Proterozoic age in a gorge in Northern Chervona, where the most ferrous horizons are consistently traced
- the Skelevatske outcrop, describing a complete section of Krivorizka series as numerous rocky cliffs decorating a landscape of the left bank of the Inguletz river
- the Inguletz river cliffs
- the Kirov historical and geological monument located on the right bank of river Saksagan with Precambrian flaky slates having not only geological, but also historical value -the most ancient mountains in Ukraine
- Moprovski geosite etc.

The Nikopol manganic-ore basin is one of largest in the world with late Oligocene sedimentary-diagenetic oxide and carbonate ore, characterized by significant stocks of manganese. The deposit was opened in 1883 by the geologist, V.A. Domger, in an outcrop on Solona River, but now natural exposure of manganous ores are no longer exposed. However, a perfect opportunity to study a section not only of ore thickness, but also of all stratigraphic divisions from marine sedimentary formations of Oligocene age up to Quaternary deposits, is given by numerous opencast mines producing manganous ores.

Figure 1. Outcrop Miocene of rocks in Gruschevskian open pit of Nikopolian deposit of manganous ores.

The Grushevski opencast mine of Nikopolski deposit, includes exposed palaeontologically characterized marine sediments of the Palaeogene and Neogene, providing a basic section for the Northern part of East Paratethis (Fig.1). The problems of preserving fulfilled opencast mines or their fragments containing the valuable geological information, are demonstrated here (Gritsenko et al., 1995).

The third deposit comparable in importance with Krivorizki and Nilopolski, is the Samotkansi deposit of Pridnieprovski titanic province, placer titanium-zirconium ore which have been developed by open cast working since 1953. The Miocene Novopetrovska suite of ore-bearing sands, has fragments of section in coastal ledges of the rivers Wovcha, Samara, Nijnya Tersa etc., where the geosites of local importance are protected, but only in Samotkanski open-cast mines for is a complete section accessible for research.

Well known outside Ukraine are the Tokovke and Kudashivske deposits of facing granites, with which many buildings, structures, monuments and underground stations in Moscow, Kiev, St.-Petersburg and many other cities of Europe and Asia are faced. Pink, red and grey porphyryblastic granites of Tokovsky massif are cut through by the steppe rivers Kamenka, Basavluk and their tributaries, with picturesque canyon valley, thresholds and cascades, the best of which are brought in to the Register of the Nature-Reserved fund of Ukraine as geosites of petrological, hydrological and mineralogical types .

Within the limits of distribution of granitoids of Tokovsky massif, at village Sholohovo, in 1932 the unique in Ukraine deposit of pink quartz, suitable for manufacturing of jewellery art products of high quality was opened in 1932.

The prospects of so-called greenstone structures of the Ukrainian shield are concentrated mainly in the limits of the Middle-Dnieper megablock. Most significant of them are Verhivsevska, Surska, Chortomlyk-Solonovska and others, with deposits and shows of ore of gold, molybdenum, copper, tungsten, rare metals etc. A fragment of one of green-stone structures opened by erosion in the Upper Basavluk is a perfect natural geological monument, one of the most picturesque places of Ukraine.

Hypergenesis played a significant role in the formation of mineral richness of the region, and dominated the most part of territory of the Ukrainian shield during practically all of its post Riphean geological history. Complex formation of areal and linear weathering crusts are connected to the numerous deposits of kaolin and porcelain clay. In the southwest part of the Dnepropetrovsk area a deposit of bauxite is located, one of few in Ukraine, connected with lateritic crust of Archaean metabasic rocks.

Figure 2. Rapids in place where the river Kamenka cross over Tokovske granite massif.

The large scale Suhohutorske deposit of silicate nickel ores is connected with a linear crust of weathering of hyperbasic rock of Verhivcevska greenstone structure. The small fragments of this structure were exposed by erosion of the picturesque river Samotkan, in which valleys they form rare rocky outcrops. During 2000-2001, scientists, members of public organization "Orlan" and employees of Management of Ecological Safety, with the financial support of the foundation ISAR carried out a complex investigation of the valley of the river Samotkan with the purpose of creating a landscape park and preservation for posterity of a site of rare beauty with original rare and disappearing vegetative associations, fauna, ethnographic and cultural values of the local population.

In the northeastern area the slopes of the Ukrainian shield are affected by a system of faults and joints of the Dnipro-

Donetsk depression, with axial deposition up to 15 km depth. High quality coal and prospects of petroleum and gas deposits are found in the valley of the River Orel.

The numerous deposits of the Dnieper Brown coal basin are connected with a lagoon-continental Carboniferous phase of the Buchatska suite. All deposits are dated to palaeo-depressions, dismembering a surface of the Precambrian base and accommodating also deposits of secondary kaolin, suitable for use as raw material for porcelain, faience and fire-resistant products.

In the area also are:

- industrial deposits of marl and lime clay, used as astringent materials for manufacture of portland cement, air and hydraulic lime
- the quartz-glauconitic aleurites and sands of the Palaeogene Mejigorska suite
- glauconite with high adsorptive property used in preparation of mineral paints, as an effective fertilizer fine dispersed Palaeogene and Neogene's clay, suitable for manufacturing of claydite gravel
- the quartz sand, which are used in manufacture of building and silicate brick, different kinds of concrete astringent solutions and asphalt mixture etc.

This list of deposits of Middle Pridneprovie, testifies to the paramount importance of the region in the maintenance of Ukraine by the major supply of mineral raw materials and, accordingly, about real prospects of growth of economic potential of the state and well-being of its population.

From this study of unique geological objects, which characterize the Dnepropetrovsk area, as geological monuments, it is necessary to consider 34 objects, of which only 14 are included in "the Register of Nature-Reserve Fund of Ukraine" (Manyuk, 1999). The majority of them have the status of monuments of local importance and only 2 are monuments of state importance. They are characterized by a significant variety and according to Ukraine's classification, include stratigraphic, geochronological, paleontological, mineralogical, petrologic, tectonic, geomorphologic, landscape, hydrological-hydro-geological, glaciological, museum-collection and complex types. Many from them are already mentioned above, some are briefly characterized but for their scientific, cognitive and practical importance the following geosites deserve attention.

- Moprovski historical-geological monument located in the central part of Kriviy Rig and representing picturesque rocks on River Inguletz with alternation of Precambrian ferruginous quartzite and slates. It has stratigraphic importance, also tectonic, as a fragment of Large Tarapakovsko-Lihmanovska anticline.
- Novokievski key section of Pontic, Maeothic, Sarmatic and Konkski stages of Neogene system at southern coast of Kahovski reservoir. In steep breakages of a coast ledge of height till 20-35m the often alternation motley on colouring and structure clays, marls, sands, limestones and dolomites with various fossilized micro- and macrofauna of good safety is observed (Figs 3, 4).

Figure 3. The sediments of sea Neocene in coastal steep of Kachovskian reservoir (Kachovskian geosite).

Figure 4. Neotectonics in Sarmatian clay-carbonatic sediments.

- Bioherm construction as large onkoides of zonal-concentric, shell-shaped structure, of late-sarmatic age in coastal ledge of river Dnieper in area of village Vishetarasovka.
- Bila Skela (White rock). A large batch of outcrops of rocks of Central-Azovska series of late Archaean with the large variety of petrographic types and powerful bodies of white barren quartzites, decorating the banks of the Wovcha river in east of the Dnepropetrovsk area (Fig. 5).

Figure 5. Geosite " A White rock ".

Figure 6. The nature geological monument " Ribalske open-cast mine ".

- Mountain Kalitva. Represents large finitely-starved hill of Dnieper (Riss) glacier combined with boulder loams, fluvio- and limnoglacial sediments complicated powerful glacial disposition.
- Voloski rocks representing outcrops of granites of the Dnepropetrovsk Archaean complex, continuing on a coast flooded by waters of Dnieper one of largest and menacing, at one time for navigation, the Lohanivski rapid.

The production of different kinds of mineral raw material, with which so the region is rich deserves consideration. On the one hand are wounds on the surface of the Earth, and damage of complete landscape systems. On the other hand, they open secrets of entrails, giving a perfect opportunity in large spaces and significant depths to look after those features of geology, tectonics and stratigraphy, which cannot be established without the extraction industries.

One of most well known opencast mines in Ukraine is Rybalsky, on the outskirts of Dnepropetrovsk. It represents a complex natural geological monument deserving preservation from the palaeontological aspect, and also from the stratigraphy and petrology. Today this is a unique place in Europe, where there remain on the surface so-called Mandricovski layers of upper Eocene with unique fossilized fauna, submitted with various molluscs, nummulites, foraminifera, ostracoda, corals, pearlworts, shark teeth and needles of sea-urchins, which inhabited warm coastal waters of a subtropical Eocene basin 35 million years ago (Fig. 6). Also in the mine is one of the most complete Ukrainian sections of Quaternary subaerial and subaqueous sediments submitted by alternation of fossilized soils and loess from Eopleistocene up to the Holocene.

Thus, the Middle Pridneprovie has a powerful fund of natural geological memorials, which demands an inventory, and preservation for posterity. The reliable system of protection of geosites cannot, obviously be created only on the basis of administrative, economic and legal measures. Considerably more effective can be actions directed on popularisation of unique natural objects, on their reasonable use for the scientific - cognitive and educational-tourist purposes, on attraction of youth not indifferent to the nature of their native territory, to all-round scientific research of its geological heritage. We develop a system of the measures necessary for creation of the scientifically proved register of natural geological memorials with subsequent reasonable use for development of recreation-tourist business.

The detailed consideration leaves frameworks of the given article and, besides is a part of the program on preservation of a geological heritage in Southern Ukraine, which introduction requires the further development, specifications and additions. It would be desirable especially to emphasize, that most effective utilization of natural memorials is possible only under condition of wide use of the information concerning other monuments located within the limits of distribution of objects of a geological heritage or nearby from them.

It is necessary to consider first of all, monuments and memorable places connected to the history of Zaporozhye Cossacks, archaeological, ethnic, historical-cultural and architectural monuments, routes of a religious orientation. In structure of complex natural memorials it is possible to involve also valuable geobotanical objects with rare, relic and disappearing kinds or associations of plants, landscape parks, reserves and other natural complexes.

In summary, it would be desirable to begin efforts directed at preservation of a natural environment, following from the education of consciousness of the irreversibility of processes of intervention in an environment and impossibility of restoration of lost inanimate objects created by nature many millions of years ago.

REFERENCES

Gritsenko V.P., Ischenco A.A., Ruosko Yu.A. & Shevchenko V.I. 1995. *Geological natural monuments of Ukraine: study, reservation and rational use*: National Academy of Science of Ukraine, Central Museum of Natural History., Kyiv, pp.61.

Ivchenko, A.S.1998. *Geological monuments of Ukraine*. Mineral resources of Ukraine. I. p.38.

Wimbledon W.A.P., Gerasimenko N.P., Ischenko A.A., Lisichenco G.V., Lisichenko K.V.1999: *Problems of the protection of the geological Heritage of Ukraine*.-Kyiv: NASV.-1-129.

Manyuk V.V. 1998. The natural geological monuments of Dnepropetrovsk region. *Boristhen*, 2 (80), 1-3.

In: Parkes, M.A. (Ed.) 2004 *Natural and Cultural Landscapes - The Geological Foundation*, 97-100, Royal Irish Academy, Dublin

JURASSIC GEOSITES FRAMEWORKS IN BULGARIA

Platon Tchoumatchenco, Geological Institute Bulgarian Academy of Sciences, Acad. G. Bonchev Str. Bl. 24, 1113 Sofia, Bulgaria. Email: Platon@lark.tu-sofia.bg

Abstract: The geosites frameworks for the outcrop of Jurassic rocks in Bulgaria are part of large palaeotectonic-palaeogeographic units in which the sedimentation, distribution of the fauna, depositional sequences, synsedimentary tectonics, and the others elements of the rocks were determined by palaeogeographic conditions. Four large palaeotectonic units existed in: the Moesian Platform Terrane, the Balkan Terrane, theThracian Massif Terrane, and the Nysh -Troyan Foreland Basin. The first three developed from terranes docked to the East European Plate during the Late Carboniferous. In the outcropped Moesian Platform Terrane there exists the Vidin Palaeohorst, the Mihaylovgrad Palaeograben, and the Vratsa Palaeohorst. In the Balkan Terrane the Izdremets Palaeograben and the Matorides Rifted Basin are defined. In the Thracian Massif Terrane there are the Tundzha Diagonal Tilted Basin, Getic-Dragoman Unit, and Supra-Getic - Jurassic Kraishtids Unit. The Nysh-Troyan Foreland Basin was imposed on the Moesian Platform and part of the Thracian Massif during the Callovian and the Late-Kimmeridgian.These large units are regarded as geosites frameworks with the same names.

INTRODUCTION

In recent years, the meaning of geosite and geosites framework have been discussed in the Working Group 1 (WG1) of ProGEO for SE Europe. My opinion is that the geosites frameworks for sedimentary rocks are part of a basin, limited in the time, in which the sedimentation, distribution of fauna, depositional sequences, etc. are predetermined by palaeotectonics and the palaeogeography. The geosites frameworks are made up of a number of single geosites,displaying one or more characteristic features of genetically-related outcropped sediments.The geosites may be, among other things, palaeontological (with an occurrence of numerous fossils), stratigraphical (e.g., the type sections of one stage/substage, or of some lithostratigraphical units, etc.), historical (e.g., the locality where, one stage/ substage, was introduced for the first time in the world or in the country etc,), geomorphological, etc. These single geosites are related between them genetically in one or other geosites framework. The geopark is another classification, a more commercial idea, different from the geosite and the geosites framework. "The geopark is a territory comprising a number of geological heritage sites of particular value for education, science, culture and socio-economic development (Patzak, Eder, 1998).

JURASSIC GEOSITES FRAMEWORKS

During the Jurassic in Bulgaria there existed a few large paleotectonic elements: Moesian Platform Terrane (Peri-Tethys) (in which crop out Vidin Paleohorst, Mihaylovgrad Paleograben and Vratsa Paleohorst), Balkan Terrane (Northern branch of Tethys) (with Izdremets Paleograben and Matorides Rifted Basin), and Thracian Massif Terrane (Intra-Tethys Platform) (with Getic-Dragoman Unit, Supra-Getic-Jurassic Kraishtids Unit, and Tundzha Diagonal Tilted Basin); on the Moesian Platform Terrane and on the western part of the Thracian Massif Terrane was imposed the Nish-Troyan Foreland Basin (Trough). These palaeotectonic units of second order have the value of geosites frameworks (Figure 1).

MOESIAN PLATFORM TERRANE (PERI-TETHYS)

The Jurassic sediments (Figure 2) crop out only in its north-western parts: (a) on the Vidin Palaeohorst - the sedimentation began with Middle Jurassic shallow marine sandstones, followed by coarse-grained limestones, and continues with Callovian-Kimmeridgian lithoclastic limestones and

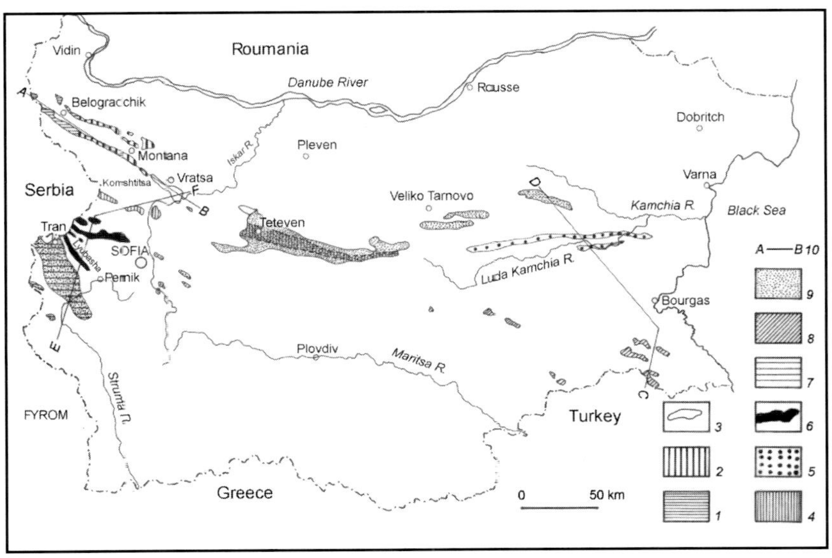

Figure 1. Distribution of the Jurassic Geosites Frameworks in Bulgaria Moesian Platform Terrane: 1 - Vidin Palaeohorst Geosites Framework; 2 - Mihaylovgrad Palaeograben Geosites Framework; 3- Vratsa Palaeohorst Geosites Framework ; Balkan Terrane: 4 - Izdremets Palaeograben Geosites Framework; 5 - Matorides Rifted Basin Geosites Framework; Thracian Massif Terrane: 6 - Getic-Dragoman Unit Geosites Framework; 7 - Supra-Getic-Jurassic Kraishtids Unit Geosites Framework; 8 - Tundzha Diagonal Tilted Basin Geosites Framework; 9 - Nish-Troyan Foreland Basin (Trough) Geosites Framework; 10 - line of Weeler cross-section.

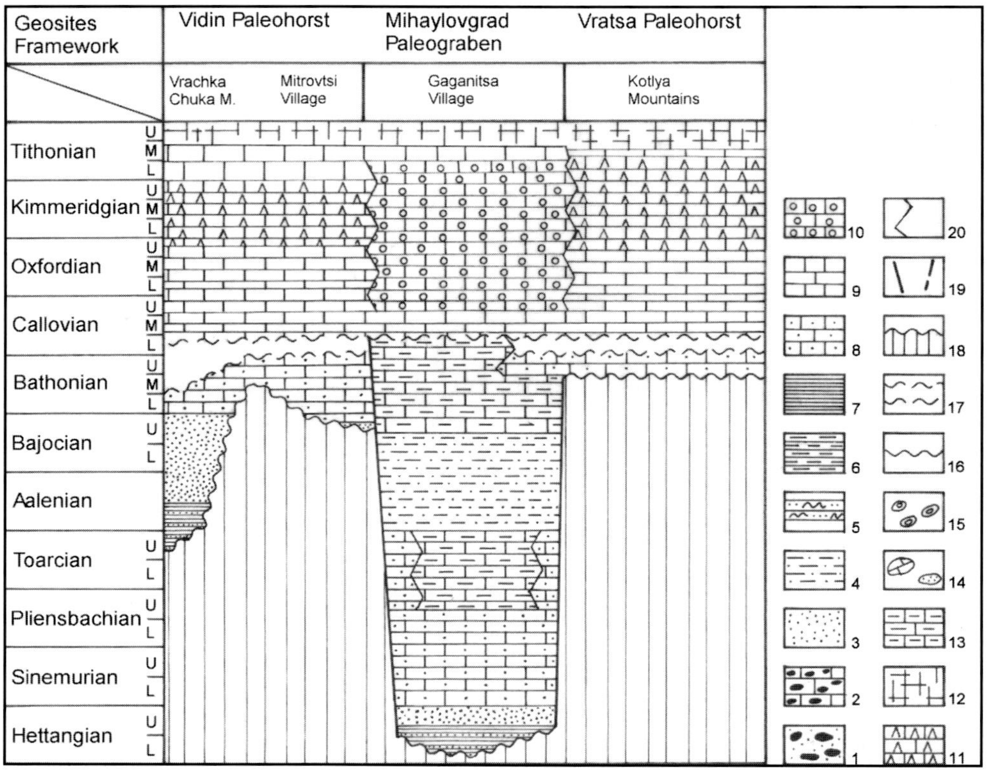

Figure 2. Weeler diagram cross-section (A-B) across the Jurassic of the Moesian Platform Terrane (north-western Bulgaria) (no horizontal, nor vertical scale). Explanation of the lithological symbols of figs 2-4: 1 - conglomerates with sandy matrix; 2 - conglomerates with limy fragments and limy matrix; 3 - sandstones; 4 - aleuritic argillites; 5 - convoluted sandstones; 6 - argillites; 7 - clays; 8 - biodetritic and sandy limestones; 9 - micritic limestones; 10 - nodular limestones ("ammonitico rosso facies"); 11 - lithoclastic limestones; 12 - thick bedded and massive limestones ("Stramberg Facies"); 13 - micritic clayey limestones; 14 - olistoliths; 15 - sideritic concretions; 16 - transgressive surface; 17- surface of submarine break in the sedimentation; 18 - hiatus (Jurassic dry lands); 19 - intra-Jurassic faults; 20 - horizontal passage between rocks.

massive, subreef Tithonian limestones ("Stramberg" type); (b) Vratsa Palaeohorst - the sedimentation started with shallow water Lower, partly Middle and Upper Jurassic marine sediments; (c) Mihaylovgrad Palaeograben - between them, the sedimentation started during the Early Jurassic, by thick continental and continental-marine sediments, followed by relatively deep water sediments- bioclastic limestones, and clayey limestones and/or marls, and Middle Jurassic "black shales with *Bositra alpina*" and marls and clayey limestones during the Bathonian; during the Callovian and the Late Jurassic nodular and sub-nodular limestones of "ammonitico rosso" type, micritic limestones were sedimented, ending with carbonate platform limestones.

BALKAN TERRANE (NORTH BRANCH OF TETHYS)

This terrane is made up of two units: (a) Matorides Rifted Basin (Figure 3) (eastern Stara planina Mountains), fulfilled by Lower Jurassic siliciclastic turbidites (connected with progressive transition with Upper Triassic calciturbidites) and Middle Jurassic "wild flysch" type sediments with large (up to several thousand metres square) Triassic and Jurassic olistoliths, included in black shales; this basin was closed during the Late Bathonian; and (b) Izdremets Palaeograben (Figure 4) (western Stara Planina) with non turbidity, relatively deep water sedimentation (Lower Jurassic continental sandstones and clays and marine bioclastic limestones and marls, Middle Jurassic "black shales with Bositra alpina" and Upper Jurassic "ammonitico rosso" type sediments) .

THRACIAN MASSIF TERRANE (INTRA-TETHYAN PLATFORM)

During the Jurassic only a portion of this terrane was covered by the sea and forming some palaeotectonic-palaeogeographic units: (a) Tundzha Diagonal Tilted Basin (Figure 3) (Strandzha Mountain) (part of the Sevlievo-Tundzha Diagonal Unit) with Lower Jurassic shallow marine sedimentation and Middle Jurassic sediments of the type "black shales with *Bositra alpina"* and Bathonian Zabernovo Exotic Nappe - during the formation of the latter, the Tundzha Basin and the Matorides Rifted Basin , were closed; (b) Getic-Dragoman Unit (Figure 4) (western Bulgaria) - the sedimentation started predominantly by shallow marine limestones during the Middle Jurassic, and continues with Callovian - Upper Jurassic carbonate platform sediments (type "Stramberg"); (c) Supra-Getic-

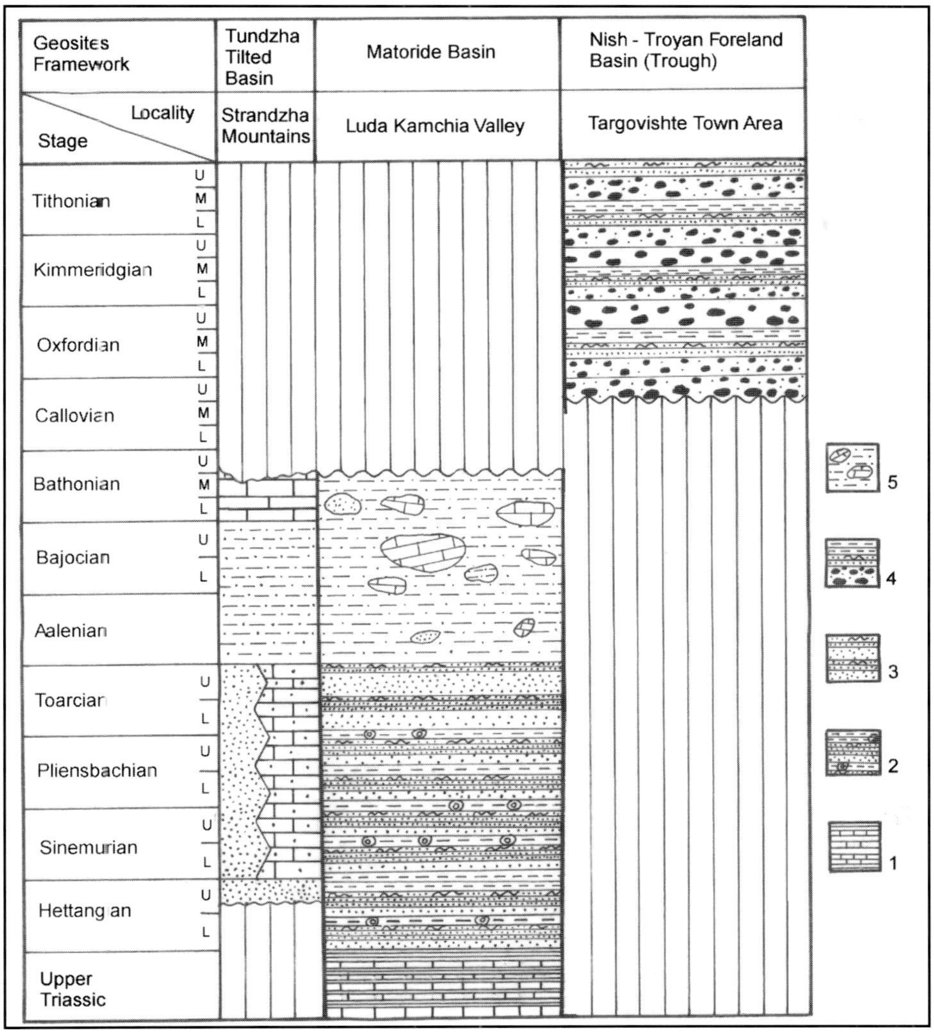

Figure 3. Weeler diagram cross-section (C-D) across the Jurassic in east Bulgaria (no horizontal, nor vertical scale) Types of turbidites: 1- calciturbidites; 2 - distal siliciturbidites; 3 - mesodistal siliciturbidites; 4 - proximal siliciturbidites; 5 - "wild flysch" - black shales with olistoliths.

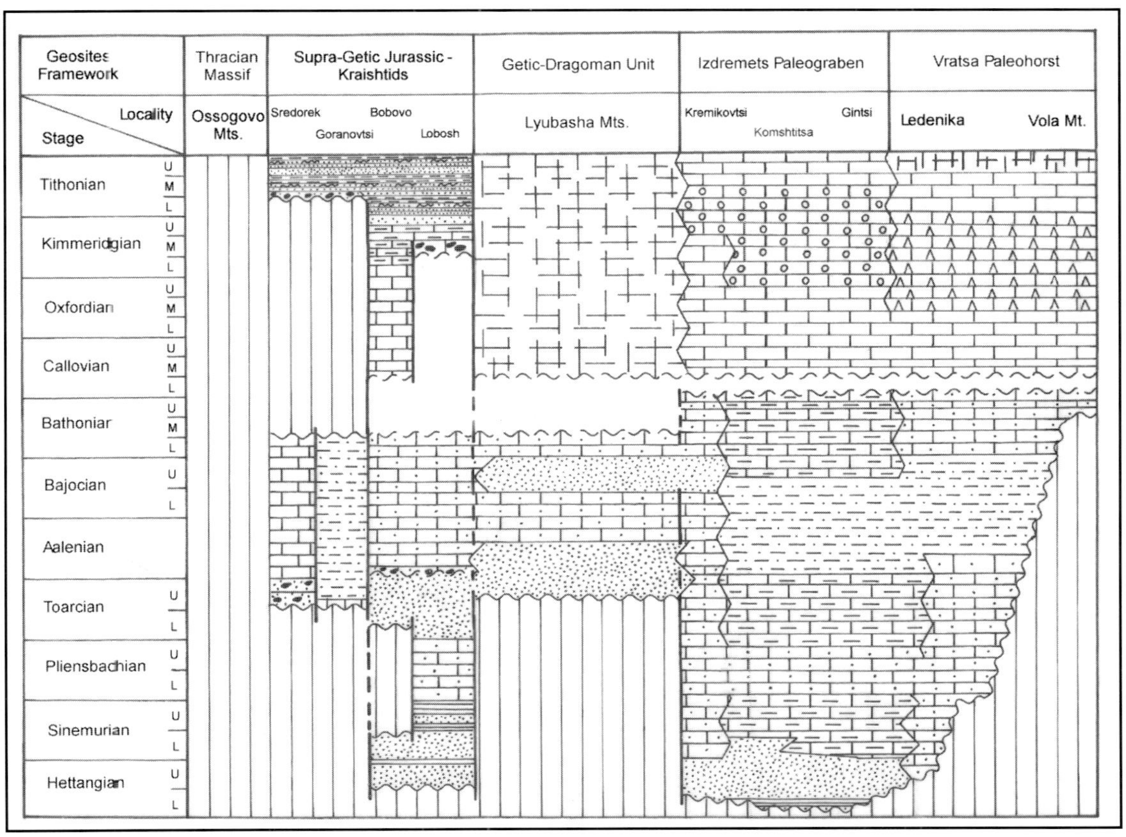

Figure 4. Weeler diagram cross-section (E-F) across the Jurassic in western Bulgaria (no horizontal, nor vertical scale).

Jurassic Kraishtids Unit - series of grabens with Lower Jurassic shallow water sediments (only in the Svetlya Graben) and Middle Jurassic sediments - black shales in some grabens and bioclastic limestones - in others.

NISH-TROYAN FORELAND BASIN (TROUGH)

This is a newly-opened structure after the closure of the Tundzha and Matorides Rifted Basins, which started during the Callovian in eastern Bulgaria (Figure 3) and continued with Upper Kimmeridgian -Tithonian - Lower Cretaceous siliciclastic turbidites (Figure 4), imposed on the Moesian Platform Terrane (in the eastern part) and on the Thracian Massif Terrane, and on its Getic-Dragoman and Supra-Getic-Jurassic Kraishtids Units (on the western part).

CONCLUSIONS

The next work to be done, will be to choose the type area for each geosites framework, which can be used for transborder correlations and comparisons of the geosites frameworks.

REFERENCES

Dodekova, L., Sapunov, I. and Tchoumatchenco, P. 1984. Stratigraphy of the Aalenian, Bajocian and Bathonian sediments in parts of Sud-West Bulgaria. *Geologica Balcanica* 14(2), 3-55.

Patzak, M. and Eder, W. 1998. "UNESCO Geopark". A new programme - a new UNESCO label. *Geologica Balcanica* 28(3-4), 33-35.

Sapunov, I., Cernjavska, S., Tchoumatchenco, P. and Shopov, V. 1983. Stratigraphy of the Lower Jurassic sediments in the region of Kraishte (South-Western Bulgaria). *Geologica Balcanica* 13(4), 3-30.

Sapunov, I., Tchoumatchenco, P., Dodekova, L. and Bakalova, D.1985. Stratigraphy of the Callovian and Upper Jurassic deposits of South-Western Bulgaria. *Geologica Balcanica* 15(2), 3-61.

Sapunov, I., Tchoumatchenco, P. and Mitov, P. 1988. Jurassic development of Northwest Bulgaria. *Geologica Balcanica* 18(1), 3-82.

Sapunov, I., Tchoumatchenco, P., Atanasov, A. and Marinkov, A. 1991. Central North Bulgaria during the Jurassic. *Geologica Balcanica* 21(5), 3-68.

Tchoumatchenco, P. 2002. *Jurassic tectonics of Bulgaria and adjacent areas*. IGCP 430 Workshop II: Ha LongBay, Vietnam, April 1- 5, 2002; 61-62.

Tchoumatchenco P. and Cernjavska S. 1989-90. The Jurassic system in East Stara planina. *Geologica Balcanica* 19(4), 33-65; 20(3), 17-58.

Tchoumatchenco, P. and Sapunov, I. 1998. The Jurassic geological sites in Northwest Bulgaria. *Geologica Balcanica* 28(3-4), 137-142.

Tchoumatchenco, P. and Sapunov, I. 1994. Intraplate tectonics in Bulgarian part of the Moesian Platform during the Jurassic. *Geologica Balcanica* 24(3), 3-12.

Wimbledon, W. et al. 1998. A first attempt at geosites framework for Europe - an IUGS initiative to support recognition of world heritage and European geodiversity. *Geologica Balcanica* 28(3-4), 5-32.

In: Parkes, M.A. (Ed.) 2004 *Natural and Cultural Landscapes - The Geological Foundation*, 101-102, Royal Irish Academy, Dublin

STUDY OF LOCAL LORE IN RUSSIA

Olga A. Mironenko, Central Scientific-Researching Geological Prospecting Museum Academician F.N.Chernyshev (CNIGR Museum), V.O., St. Petersburg, 199106, Russia.
Email: cnigr@mail.rcom.ru

Abstract: A description of the study of local lore in Russia is given. As an example of its importance for geological and landscape conservation the region of Altai is examined.

In Russia with its vast territories and diverse landscapes studies of local lore is very important for awakening people's interest and enjoyment of Nature. The study of local lore -"craevedeniye"- consists of comprehensive study of nature, history, culture and economy of some part of the country, administrative district and/or natural area mainly by efforts of the residential population. Regional museums serve as guidance centres of studying local lore. They are enlightening institutions engaged in the study, exhibition, and popularisation of collections made by students of local lore. This is the prevailing type of a combined museum in Russia. These museums appeared in Russia in the late eighteenth-early nineteenth centuries, when such museums were founded at educational institutions, chambers of State property, and the Geographic Society of Russia. Early in the 20th century there were 60 museums of local lore.

After 1917 such museums were set up in the remotest depths of the provinces for the enlightenment of the nation. By 1920 there were over 250 such museums in Russia and by 1975 over 500. Every museum has a nature department, where the geological structure of the area is shown in expositions; samples of minerals, flora and fauna collected by students of local lore are exhibited. Mass purposeful walking-tours were arranged in 1930s, which were aimed at studying natural resources of the area useful in the national economy. During the Great Patriotic War the students of local lore revealed local resources for defence. In recent decades the students of local lore have been mainly engaged in the revealing, description and registration of natural monuments, including geosites.

Visiting regional museums is a part of the educational programme on Geography, History, and Natural History. In many museums there are circles for children for studying local lore, frequently on a voluntary basis. Walking-tours are arranged during summer vacations for studying the native land.

For example there is a brief information on students of local lore role in the study and conservation of natural landscapes in two Russia Regions: Altai and Altai Republic. In geological respect this territory is a part of the Urals-Siberian Paleozoic folding zone. It includes the southern end of the West-Siberian plate composed of the horizontally bedded Mesozoic -Cenozoic rocks and Mountain-Altai Caledonian structures. It has complicated geological construction. The base complex is composed of Vendian - Lower Cambrian metamorphic shists. These are overlain by Cambrian carbonate - terrigenous deposits with green schists and flint interlayers. The terrigenous and carbonaceous rocks prevail within an Ordovician sequence. Various limestones, terrigenous and rare volcanic rocks represent Silurian deposits. A Devonian section is mainly composed of a thick sequence of effusive-sedimentary rocks. Upper Proterozoic granitoids and Middle Devonian granitoids and subvolcanic porphyrites prevail among igneous rocks. From the end of Late Paleozoic the territory became a land. The Altai relief was formed during the Cenozoic as a result of the youngest tectonic movements: dome uplift and conjugate tectonic settling of the Culundinskaya depression in the North, and also lateral movements along the faults. The Northern tectonic Altai border is expressed in the modern relief as sharp "North Altai face of fault". In the Quaternary the relief was strongly subjected to glacial activity.

The complicated geological structure of the Altai and its disposition to various climate-forming factors caused a wide range of landscapes there. In the northwest at an altitude of 100 to 200m there occurs a lacustrine-alluvial lowland with a loess-like loam cover. Numerous closed lakes are characterized by complex geochemical composition; salt concentration in them varies from almost fresh to brines. Salt-marshes and aeolian landforms make the landscape even more diverse.

To the east an elevated plateau up to 285 m high composed of loess-like loams on Paleogene sandy-argillaceous rocks rises as a terrace from 50 to 100m high. Birch and aspen woods are combined with tilled steppe-like meadows. Gullies of ancient run-offs up to 20km wide are typical. Their floors are composed of aeolian sands, thickets of unique band pineries. On the opposite side of the Ob River, the largest river of the region, hills and ridges of the piedmont from 400 to 500m high rise on Palaeozoic structures.

To the south, the Altai Mountains rise beyond a belt of undulating plains. This is a complex system of ranges, which rise as giant steps from the periphery to the centre of an enormous vault to the height up to 4506m. Remains of old planation surfaces have been preserved on the steps. Landscapes of the Gorny Altai display great diversity. The steppe belt gives way to the mountain woods. There are woods consisting of fir, aspen, rowan, combined with herbs, coniferous woods consisting of cedar, spruce, and fir, and larch park woods. Alpine meadows are widespread in the forest belt. The upper boundary of the forests in some places is situated as high as 2500 m. Fragments of the steppe belt occur in the highlands on flat bottoms of river valleys. In the south-east at a height of 2000 m there are peculiar mountain tundra-steppes. Peaks, horns, corries, and trough-shaped valleys with multitudinous lakes formed in axial areas of the ridges under the action of paleoglaciers. Moraine hills, ranges, traces of landslides, mudflows, and avalanches are numerous. The Alps of the Altai Mountains are covered with glaciers. There are 1330 glaciers; their area is 890 km2. The thickest glaciers occur on the Mt. Belukha. The glaciers feed a lot of mountain rivers. These are mountain streams with waterfalls and valleys with numerous terraces and precipitous limestone banks. Carbonate massifs in the Altai are strongly karstic. Over 400 caves are known there, which contain various forms of dripstones, and archeological findings. Numerous lakes were formed as a result of damming rivers. The Russia's Altai is a fabulous land of majestic mountains, clear lakes, rapid rivers, mountain woods, Alpine meadows, and fertile steppes. It is natural that this territory is rich in people, who love their native land and who did everything in their power for studying and developing this region.

One of the first prominent students of the Altai Territory was P. Shangin (1748-1816). He was a physician, but when he was forty-year-old, he transferred to the Mining and Geological Survey. He spent many years in the mountains searching for ores and industrial stones, at the same time studying botany. He wrote a very interesting paper about results of his work and in 1795 he was elected a member-correspondent of the Academy of Sciences.

G. Spassky, an employee of the Mining Office, made an important contribution to the study of the Altai. He studied history and geography, described a lot of mineral deposits, collected a vast body of information about the animal kingdom. He was engaged in aerography on his own initiative, carried out archeological works, and wrote several historical books. In 1810 he was also elected a member-correspondent of the Academy of Sciences. He compiled and published a three-volume mining dictionary.

A. Bunge, P. Slovtsov, A. Kulibin, F. Gebler, V. Radlov, S. Gulyaev and other local specialists also studied the nature of the Altai. F. Gebler (1781-1850), who was a physician by trade, first opened the glaciers in the Altai. S. Gulyaev (1806-1888), a chief of gold mines of the Altai, made an important contribution to the study of local lore. He had an excellent collection of minerals and fossils, described mineral springs, and studied possibilities for industrial development of natural resources. He wrote many articles and was a very good explorer and propagandist of the Altai. D. Zverev (1862-1924) was the initiator of establishing the Society of Amateurs of Altai Studying (1891), which later became a branch of the Geographical Society. Z. Rostovsky carried out several expeditions aimed at soil study. An outstanding student of local lore A. Velizhanin did everything in his power to convert the Barnaul Mining Museum into the public Museum of Local Lore.

In the museums of local lore along with nature departments there are a lot of expositions devoted to the history of development of local trades. In the Altai the stone-cutting trade was one of the most important. In the Altai Mountains there are a lot of nice fancy jaspers, porphyries, marbles, and granites. A centre for production of large highly artistic goods made of stone has been situated in the area of Kolyvan from the middle of the 18th century. Many goods of masters from Kolyvan are world-famed masterpieces, which are kept in the largest museums of Russia.

V. Vereshchagin (1871-1956), a history teacher, devoted all his life to studying the nature of the Altai. He was an initiator of cognitive school tourism in the Altai. At present three large museums of local lore and plenty of small museums at schools are centres for studying local nature and history. Students of local lore made a great contribution, but the most important was that they infected young people with their enthusiasm and inculcated in them love for their country and aspiration for protecting and reasonable usage of its natural resources.

With the development of the agro-industrial complex the problem of nature protection became very important. In 1932 the Altai Reserve was established for studying the mountain taiga. Over 30 partial reserves were organized for the preservation of wild animals (marals). From mid-70s, inventory and protection of nature monuments became the main task of the nature protection. Museums of local lore and the regional branch of the Russian Society for Nature Protection headed this work. By the early 1980s over 350 nature objects were inventoried: more than 80 lakes, 17 waterfalls, 28 picturesque cliffs, rocks, and karst arcs, 41 caves, mineral springs, and areas of landscapes. Among the most valuable geological monuments of the Altai are the Tigiretsky beryl deposit, Lake Teletskoe, Demidov mines, granite outcrops, terraces of the Katun River, radon springs of the Belokurikha health resort, Korbu Waterfall, Lake Kolyvanskoe, archeological objects. Recently two national parks have been projected there.

The best means of nature protection is the inculcation in people of a love for ambient landscapes and understanding the necessity of their preservation. Love comes by way of recognition and therefore the organized teaching and educational tourism is of prime importance. Wealth of the museums of local lore, accumulated by many generations of students of local lore, presents vast material for developing interesting tourist programmes. Tourist agencies place high emphasis on organizational and educational work on nature protection. Numerous tourist itineraries have been developed in the Altai. Each of them traverses the most valuable and interesting nature objects.

During the last decade our country has undergone hard times. The financial situation of the population is very difficult. But nevertheless, the study of local lore in Russia is still alive. Many enthusiasts, particularly school teachers, continue the best traditions of students of local lore and help schoolchildren to study their native land. A Conference of Young Geologists was held in August, 2002 in St. Petersburg. 450 schoolchildren and their teachers came from many regions of Russia to take part in the Conference. Among them there was a group of schoolchildren from a Gorno-Altaic village with their teacher. They presented several rare mineralogical samples to our museum. This gives hope that nature protection in Russia will further improve.

In: Parkes, M.A. (Ed.) 2004 *Natural and Cultural Landscapes - The Geological Foundation*, 103-106, Royal Irish Academy, Dublin

THE DATA BASE OF ITALIAN GEOSITES INVENTORY

Myriam D'Andrea, APAT, Dipartimento Stato dell'Ambiente e Metrologia Ambientale (Environment Dept...), Servizio Sviluppo Sostenibile (Sustainable Development Serv.) Via Cesare Pavese 305 - 00144 Roma. Email myriam.dandrea@apat.it
Stefano Colacchi, Spectrum Graphics s.p.a., Via dei Cardi 1, 00040 Ariccia (Roma),
Gianfranco Gramaccini, Angelo Lisi and Nicola Lugeri
APAT, Agenzia per la Protezione dell'Ambiente e per i Servizi Tecnici
Via Vitaliano Brancati, 48, 00144 Roma

Abstract: The "Land protection Law" (Legislation 183/89) and following laws and acts assign to the Geological Survey of Italy (SGN) the responsibility for geological knowledge of the Italian territory. According to this, the Geological Survey started, in the year 2000, a project named "Protection of the Italian Geological Heritage" which aims at: 1) drawing up a computerized Data Base (DB) of italian geosites data (coming from previous and new geosites recognitions); 2) setting the guide-lines for the geosites inventory; 3) building a GIS of the italian geosites to the small or medium scale.

This paper offers a technical contribution, from the survey (based on a standard form for the geosites inventory), to the data input into a DB of the survey forms, and finally to the Geographic Information System (GIS) of the Geosites. The goal is to introduce and to clarify the steps for building an unique italian data bank, structuring the data on the geosites to give a complete information on the geological processes and events which have affected and still affect the Italian peninsula.

INTRODUCTION

The proposal of a geosites inventory in the framework of a geoconservation policy, came out in Italy in a conflictual phase regarding protection politics and sustainable development strategies. For any such proposal to survive, it needs to be accepted by the cultural fabric in that "historical moment"; this is even more difficult for the issue of geological heritage, which is difficult to perceive. This is the reason why biodiversity is a well-known concept, while geodiversity is still cloudy in its definition and not as successful. Therefore the geosites inventory (past and ongoing) at local (regional and provincial) and national scale (D'Andrea and Di Leginio 2002), while giving a great contribution to the geological knowledge, need to be accepted and "included in the toolbox" to really become a point of reference for policy and decision making in territorial planning (Burlando 2000; Gisotti and Massoli Novelli 1997; Zarlenga 1996).

One of the main goals of the inventory is the definition of some classes of objects of geological interest, represented whithin the data base by a significant group of sites, describing all of the aspects of the geological history of the italian peninsula, providing a complete information about geological processes and events and their products, as the landforms and the rocks.

The holistic approach to the landscape teaches us that it is always more complex than the sum of the different single aspects; what we see is the result of a process where the human being is only a temporary actor and witness. It is therefore necessary to define and to select the aspects to be considered, using rational criteria suitable to be applied in a sound conservation and planning policy.

The conservation of the geological heritage needs a strategic plan that should take care of:

- the definition of the objects worthy of interest (geosites recognition) and the production of a unified list of geosites for the whole italian territory;
- the acknowledgement by the political and administrative world that a geosites conservation policy is actually needed;
- the inclusion of geosites protection activities in the laws dealing with territory (this is already occurring in Italy, also at regional level);
- the education of the community to the geoconservation and their involvement in the heritage management.

DATA COLLECTION: FROM THE "SURVEY FORM" TO THE DATABASE

A national committee, composed of Italian ProGEO members, has selected a set of information necessary for the survey of a geological heritage site. This set has been structured into the fields of a standard survey form, which has been designed to be readily implemented in a digital version, open to further evolution. After a wide discussion in the academic world, the "test" survey form has eventually been adopted and published (Brancucci et al. 1999).

A first test (with an older version than today) of the survey form has been performed during the "Geosite inventory in the EU-Objective 2 areas of the Liguria Region), by the Dept. POLIS of the University of Genoa (Brancucci & Burlando, 1999). Afterwards, this work has been acquired by the SGN in an agreement with the Dept. POLIS; the survey form has been developed and optimised for use also to input the data coming from previous inventories. The official version of the form is now published (as a downloadable document to be printed) on the dedicated pages of the

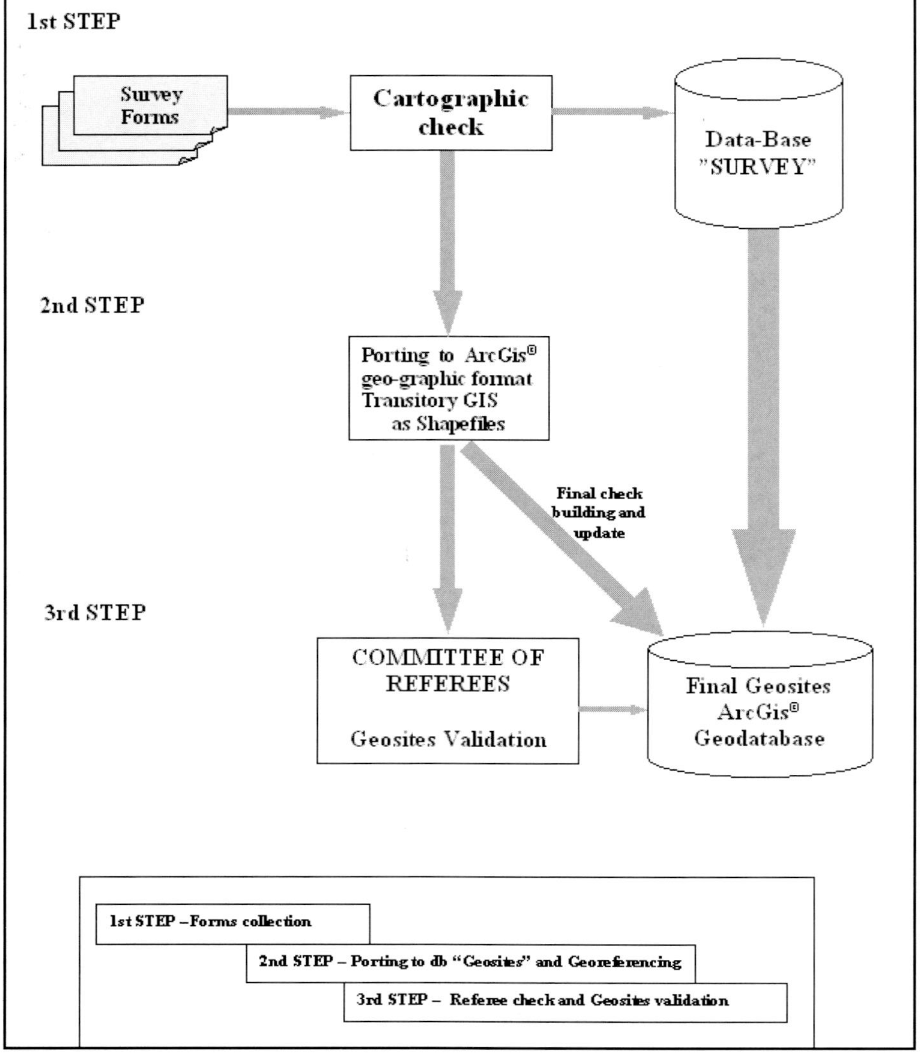

Fig. 1
Process to create and update the Geosites Geodatabase *via* the database of the Survey forms

SGN website, www.dstn.it/sgn.u_ricerca/conservazione .htm, as well as on those of the POLIS Dept., http://www.arch.unige.it/sla/geotopi/index.htm. The digital version of the form will probably be under the form of a web-based application.

The work on the database and GIS design, and on the management of the data collection and transfer into the database is carried out by a working group internal to the SGN, in cooperation with the above mentioned Dept. POLIS in Genoa, and with external technical experts.

The survey form is also tested in several projects involving the SGN and various academic and public institutions (Universities of Rome, Modena and Reggio Emilia, Parks, Italian Geological and Speleological Societies). It is also expected to receive information and proposals of geosites from the ongoing works to prepare the new geological map of Italy at the 1:50,000 scale, a work coordinated by the "CARG" office of the SGN.

THE GEOSITE DATABASE AND GIS

The data from the above described survey forms, after a suitable cartographic check, will fill the database in; it will be storing the information on the geosites, ready to be linked to a GIS, for a geographic based information retrieval and territorial analysis.

The workflow has been structured in three phases, possibly overlapping (Figure 1):

1. collection of the survey forms, cartographic check, insertion of the cartographic coordinates (merely as an info field) in the database "Survey". Here, the central and univocal subject is the survey form itself;
2. porting of the DB "Survey" into the temporary DB "Geosites" and actual geo-referencing of the geosite;
3. validation by a national committee of referees of the sites proposed as geosites and their insertion in the "official" version of the GIS of the italian geosites. Here the centrality and univocity is eventually assigned to the geosite, while possible multiple survey forms dealing with the same geosite are managed through a link to the existing DB "Survey".

PHASE 1: THE STRUCTURE OF THE DB "SURVEY"

As widely known, modern Databases operate through tables. The DB "Survey", consists of several tables, Graphical User Interfaces designed to manage data input and retrieval, and the procedures for printout of the retrieved information.

The design was focused to obtain user-friendly interaction with the DB, both for data input and query. A guidance to the user is given by the so-called "combo-boxes", where pull-down menus provide the user with pre-determined values to choose from, when filling a field in. Also table and internal procedures naming was driven by the need for an easy and well defined DB structure, open to maintenance and evolution interventions.

The core of the DB "Survey" consists of a main table, containing the univocal data identifying the surveyed geosite and its location, and and 9 ancillary tables, designed to store further, more structured, details. Pre-filled tables (Data tables) contain the reference data used to populate the menus mentioned above. Finally, two additional tables store the details of the person and/or institution responsible for the survey.

Multimedia files are also expected to be received from the

surveyors, so they will be managed by the DB as external links. The 9 ancillary tables account for the following informations:

- Administrative details on the geosite location
- Geological characteristics of the geosite, in terms of lithology, chronostratigraphical age, etc.
- Scientific interest and value assessment of the geosite, in different scientific fields
- Accessibility to the geosite
- Soil and/or sea-floor characteristics
- Law constraints already existing or proposed for geosite protection
- Existence of neighboring service infrastructures
- Management of external links to documents on the geosite

PHASES 2-3: THE "GEOSITE" GIS

The most recent development of GIS technology, especially the new commercial standard of ESRI (Environmental Systems Research Institute) products, which have been used by the Geological Survey of Italy for about 15 years now, allows a direct use of common technologies, with special attention devoted to the Microsoft standard.

Therefore, past problems relating to import-export of geo-coded data and to their storage in relational databases (especially concerning the need for fast search on huge databases, that previously required middleware products to improve performance), are now overcome. The conceptual point of view is now turned over. If before, features and geo-graphic elements were central (even in storage and distribution of table indexes) now they are "unpacked" in relational database tables just like their attributes or external tables, and with the same hierarchical level.

Such new solutions bring about several improvements:

- Common technologies: direct access and management to relational database with a standard user and versioning administration
- Fast indexing: as the geographic data is indexed in tables, their accessibility is optimized
- Universal Markup Language (UML) modelling: Geographic objects (including their attributes and constraints) and DB structure can be easily designed through CASE (Computer Aided Software Engineering) tools; afterwards the UML model and the geodatabase can easily be generated by standard procedures.
- Automatic metadata compilation: the data describing the geographic dataset are directly generated within the geodatabase and formatted in an open standard (Extended Markup Language - XML)
- Easy Windows-like data management: the catalogue of the geographic datasets, per-se or within a geodatabase, is managed through a straightforward user interface with drag-and-drop capabilities, and display of the datasets in terms of icons, graphic previews and metadata.

Since a geodatabase can be created without necessarily designing the UML model, the build-up of the geosites database has started from the existing relational DB, which was designed in MS Access to manage the first version of the survey form. We are planning to migrate to SQL Server in the near future.

In terms of design, the geodatabase "Geosites" is not much different from the database "Survey", it only includes a few additional tables, containing both geographic data and additional alpha-numeric information. Figure 2 shows the logical structure of the geodatabase, which includes the tables described in the previous paragraph.

The first additional table refers to possible Synonyms of the same site; such table will keep track of multiple survey forms from different compilers. Another table with detailed and explicit geographic informations, called Infogeo, contains, among others, administrative data, altimetry and bathymetry.

In addition to this, the geodatabase data format, brings about several tables, carrying information (both geographic and not), but its peculiarity is the structure of the datasets into several hierarchical class of objects: tables, feature

Figure 2. The logical structure of the geodatabase.

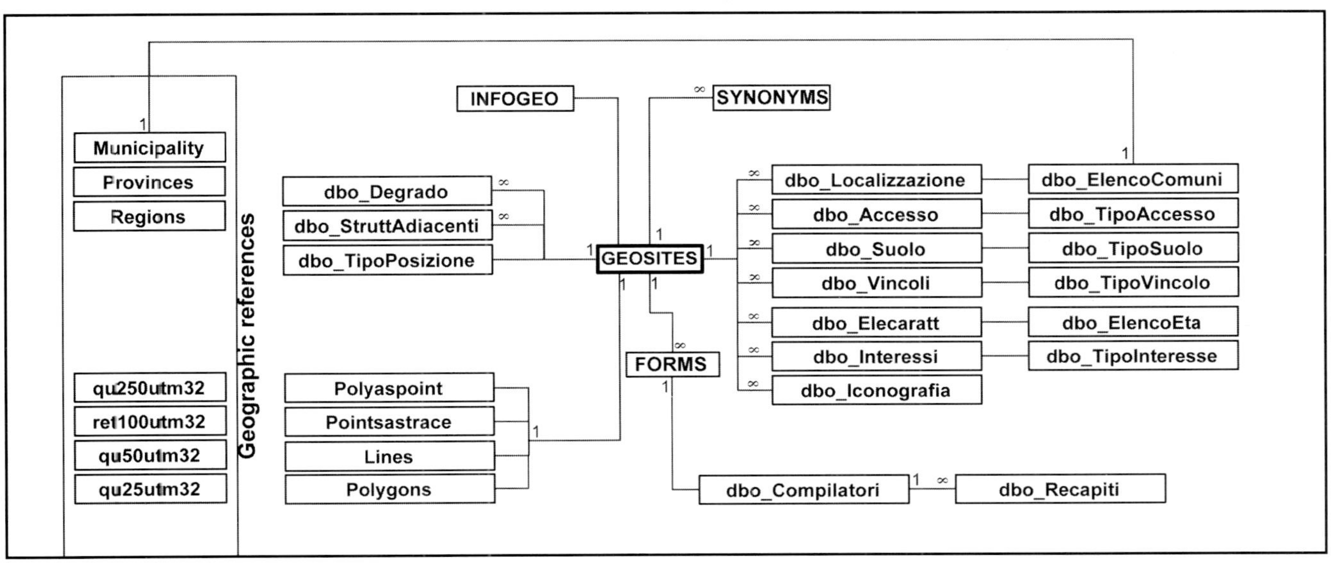

class, feature dataset, relationship class. The Geosite geodatabase consists of two main datasets: the first one contains the actual geosites, organised in different shapes: polygons (if the area is over 5 hectares), polyaspoints (if the area is less than 5 hectares), lines, and, finally as pointsastrace if the site is recognised as distributed over many points instead of a single one. The second dataset, consisting of geographic reference data, will provide the basis for geographic search and indexing of tables. This one will contain the cartographic tiles of the topographic maps officially released by the national mapping agency IGM (Istituto Geografico Militare).

Italian territory extends from 35°49' to 47°09' in latitude, from 6°63' to 18°52' in longitude and its area is about 300.000 square kilometers. Such a physical and geographic shape, a thin and long peninsula along 1200 km from NW to SE, suggested the cartographers to subdivide projections in two zones in order to keep cartographic errors as low as possible. Two zones have been defined both in the old ED40 Reference System (Gauss-Boaga projection; false easting 1.500.000m in the western zone and 2.250.000m in the eastern one) and in the more recent ED50 Reference System (UTM projection; false easting 500.000 m in both zones 32 and 33).

Before the diffusion of GPS, which uses the WGS '84 standard, many central public administrations, in Italy, have been using the ED50 standard with a UTM32 projection extended to 33 zone. Whatever will be the unique standard that is going to be decided in the ongoing discussion phase between national and regional institutions dealing with territorial data, the transformation from a system to another is possible, in terms of geographic positioning, by automatic procedures in GIS sofware. Therefore, for data acquisition in this project, we decided to keep using the above mentioned "extended" UTM 32 projection, as the Geological Survey of Italy has several geographic databases already projected in this standard.

It is expected that the most widely used procedure of geographic data acquisition will be digitizing from video display; this will be based on the IGM topography 1:25.000V, available in 3618 compressed raster TIFF files, geo-coded and clipped on the contour of the cartographic tile of the corresponding paper sheets in order to obtain a seamless mosaic of the map. This will allow a detailed check with an accuracy corresponding to the chosen scale of data acquisition. Other cartographic reference tiles will be the 1:250.000 JOG, the 1:100.000 IGM and the 1:50.000 IGM (new issue).

CONCLUSIONS

In Italy, the recognition of the classes of geological objects deserving attention and conservation policies is now a priority. This would eventually match the guidelines contained in "Action theme 4" of the Pan-European strategy on bio- and landscape diversity (signed in Sophia in 1955 by 55 Ministers of Environment), as well as the principles of the more recent "Rights of Land's memory" (Digne 1991) and "European Convention on Landscape" (Florence 2000).

The setting-up of a system of data and metadata on geosites organised in a Database would be a basic tool toward such conservation goals, also in terms of revaluation and promotion of the territory to which the geosite belongs.

The partial inventory performed in the past by local administrations will provide the first set of data to be included in the geosites database, after their standardisation according to the survey form. The database will then collect new data coming from the works carried out by academic institutions, private firms and non-governmental societies of experts, in cooperation with and sponsored by, the Geological Survey of Italy, as well as from proposals stemming by voluntary actions of geologists or non-professional experts who wish to join the project.

An indispensable tool for an effective result will be the organisation of the database in GIS format, and links to other sources of territorial information. From this point of view, the integration with other geographical databases managed by the SGN, like the Geological Map database, the national landslide inventory, ecological cartography, as well as digital aerial and satellite imagery and other reference thematic maps, will allow a comprehensive approach to territorial analysis to provide a support to political decisions and sound territorial policies.

ACKNOWLEDGEMENTS

The authors wish to thank Dr. Francesco Zarlenga, President Elect of PROGEO, for critical review of the text

REFERENCES

Brancucci G. and Burlando M. 1999. *Geosites inventory in Liguria (Northern Italy) activation phases and its possible future developments*. III Int. Symposium ProGEO Madrid

Brancucci G. Carton M. & Pavia G. 1999. Scheda inventario geositi, *Geoitalia*, 4, 43-49

Burlando M. 2000. Geositi ed aree protette. In: *I Geositi: definizione, gestione, tutela*. Corso di aggiornamento professionale SIGEA, Roma ottobre 2000

D'Andrea M. and Di Leginio M. 2002. Progetto SGN:"Conservazione del patrimonio geologico italiano" I censimenti sui siti di interesse geologico in Italia. *Geologia dell'Ambiente*, anno X, 2, 9-13.

Gisotti G. and Massoli-Novelli R. 1997. *I Geotopi nella pianificazione territoriale*. Atti IX Congr.Ord.Naz.Geol. Roma, aprile 1997, 307-311.

Zarlenga F. 1996. I geotopi, dalla ricerca scientifica alla pianificazione, controllo e gestione. *Geologia dell' ambiente*, 4, 2, 3-6

In: Parkes, M.A. (Ed.) 2004 *Natural and Cultural Landscapes - The Geological Foundation*, 107-110, Royal Irish Academy, Dublin

THE MAP OF GEOLOGICAL SITES OF ALBANIA

Afat Serjani, Adil Neziraj
Institute of Geological Research, Blloku "Vasil Shanto, Tirana, Albania
Email: aserjani@yahoo.com; aneziraj@yahoo.com

Abstract: The first map of geological sites of Albania is the result of research on geological heritage conservation supported by the Geological Survey of Albania and the new contribution for ProGEO strategy in our country. This map has complex geological-geomorphologic, didactical-education, tourist-geotourist values. We are of the opinion that this map will be good help for development of the geotourist industry, for exploration of glacial, karst, erosional and many other rare geological phenomena linked with Mirdita oceanic spreading ophiolites, with sedimentary platforms and basins, and with neotectonic movement.

INTRODUCTION

The first map of geological sites of Albania (Figure 1) was finished on March 2001. The project was undertaken from 1999, after finishing the first inventory of geological sites of Albania. The compilation of this map is a continuation of studies on geological heritage conservation in Albania and is a new contribution for realization of ProGEO strategy in our country. This project was financed by Albanian Geological Survey and realized in the Institute of Geological Research by the authors. The final digital version was realized by M. Luli and A. Avxhiu in "Microstation program"

The map of geological sites of Albania represents the consolidation of the authors' efforts for Geological Heritage Conservation in Albania. Since 1995, in collaboration with geologists of our Institute and from regional geological centres, the authors have done some studies, papers and presentations in national and international symposiums and congresses, especially in ProGEO meeting.

As a base for compilation of this map, served the study on geological heritage conservation in Albania, accompanied by the first inventory of geological sites of Albania and some general studies as: geological (1982) and metallogenical (1984) maps of Albania at the scale of 1:200000; Physical Geography of Albania (1991; 1992) etc. The map is accompanied by short explanatory text and by the list of geological sites with their coordinates. Amongst the geosites, national and local parks and geoparks, woodlands and hunting lands, tourist and climate-curative spots are shown as well. Thus, the map of geological sites of Albania represents a study of complex geological-geomorphologic, didactical-education, tourist-geotourist values. The use of this map will concretely influence a new conception for evaluation of natural beauties in our country, for definition of geotopes and spots which must be protected by human activity and for correlation of activity for environment protection and natural heritage conservation.

METHOD OF WORK

The previous experience in compilation of some capital works and maps in scale 1:200000 was used for composition of this map. Publications in the field of geography and geoenvironment were used as well. Combination of geological, geomorphologic, geographical and topographical data in 1:200000 and larger scales as a complex method helped us to put in geosites in correlation with their surrounding environment. Places of geosites and their coordinates were defined using topographic sheets at the scale of 1:25000.

Geological method consisted in correlation of geological place of geosites according to the tectonical-structural zones and genetic conditions of formation. Stratigraphical sites are shown by conventional sign according to the length of section and the age of formations. Geographical aspect was used during fixation of geomorphologic and especially erosional, glacial, karstic and river-erosional geosites. The place of each geosite was verified and correlated with geological position and geographical relief. "Microstation" applied by M. Luli was chosen as an available program for composition of the map.

The final version at a scale of 1:50000 was chosen for the following reasons: Firstly, in this scale geosites could be shown even in very dense quantity such as in the Albanian Alps region and, secondly, because this scale can be published in one separate sheet suitable for use.

LEGEND

It was not easy for us to present with separate signs of about 400 geosites, geoparks, landscapes, national forests, separate trees etc. and it was very difficult to take into consideration the formation of different kinds of geosites, their colors, forms, real sizes and their presentation conventionally in the map.

Stratigraphic sections are shown by a line, commonly across the strike of sedimentary formations. Magmatic, metamorphic and sedimentary complexes are symbolized by crystal, showing magmatic crystallization process. Metallogenic-economic deposits are presented as a round used usually in metallogenic maps. Tectonic-structural sites are composed in a folding form, but this does not correspond to deep regional faults.

For the largest group of geomorphologic sites, are composed of different shapes, according to the processes of

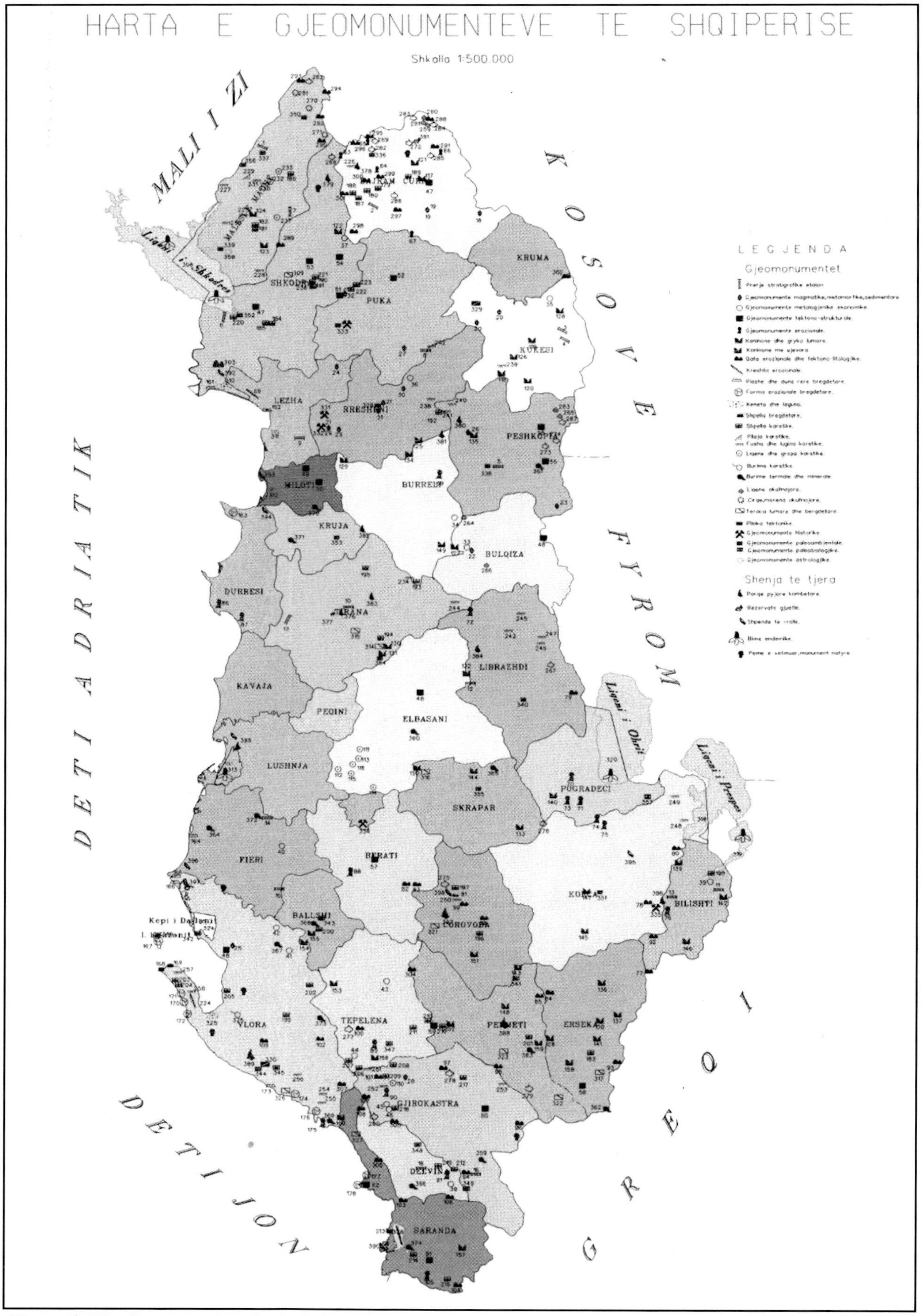

Figure 1. The first map of geological sites of Albania.

their formation. Continental-oceanic scale geosites are shown in the form of contacting plates, while historical sites are put by traditional sign of mines: two crossed hammers. Paleoenvironmental sites, in most cases accompanied by macrofossils, are symbolized by an ammonite in limestone rock. Spherulites are presented by very small rounds. For woodlands a separate conifer plant was chosen, while for hunting lands we have preferred to show a rabbit, or a bird in case of lagoons. With separate signs, are shown old trees, endemic plants, fishes and birds. The legend was composed by A. Serjani and was applied and improved by M. Luli in the computer. Concerning the colours, we have done the correlation of geosites with conditions of formation. In most cases we have applied the colours in analogy to colours used in geological and geographical maps. After many proofs, the size of each site symbol and their numbers in the map was defined as 3-4 x 5-6 mm.

CLASSIFICATION OF GEOLOGICAL SITES

Compilation of this map was based on our study about geological conservation in Albania and especially on second volume: The first inventory of geological sites of Albania (Serjani et al. 1999). For presentation of the full ensemble of natural values of Albania together with geosites, thermal waters, national and local parks, hunting lands and protected forests are included as well.

Geosites are divided and classified in ten main groups. This classification is in accordance to our previous publications and presentation (Serjani et al. 1997; 1998; 2000; Wimbledon et al. 1998).

In the first group are classified stratigraphic sites-sections of Paleozoic rocks in Korabi zone, Mesozoic carbonate sections in Albanian Alps, Krasta-Cukali and Ionian zones, full sections of molasses in inner depressions (Morava section) and in Pre-Adriatic Depression (Patosi section).

In the second group, are included geosites of magmatic, metamorphic and sedimentary rocky complexes, where the most interesting phenomena are stratification and foldings of ophiolite massifs.

Third is the group of economic deposits, where are separated some unique deposits concerning their genetical type, morphology, sizes such as: Bulqiza chromite deposit, Selenica asphalt deposit, some phosphate and oil deposits. Tectonical-structural phenomena are classified in fourth group, including here folding of rocks and some large anticlines and synclines of regular shape, gigantic white salt domes in Peshkopia region and main regional faults.

The widest group of geological sites is that of geomorphologic sites. In this group, are presented the following genetical kinds of geosites:

- Erosional sites: pyramids and towers, ridges and passes, erosional spheres in clays etc.;
- River-erosional sites: canyons, river gorges, waterfalls, seaside erosional landscapes etc.;
- Karstic sites: caves, fields, slopes, valleys, holes. Very interesting, in this group, are Belshi karst lakes on the surface of Dumrea salt diapirs and large karstic fields on the surface of carbonate rocks in Albanian Alps and in Ionian zone (southern mountainous unit);
- Glacial landscapes: during Quaternary, in mountains there are formed a lot of glacial lakes, moraines and cirques. Here, we can note aesthetic lakes in Lura geopark and in Albanian Alps and glacial fields in Korabi highland.
- Erosional lithological sites where are included, mainly, erosional passes formed due to differences of rocks against erosion;
- Neotectonical sites: Shkodra, Ohrid, Prespa lakes, Karavasta and Butrinti lagoons, some river and seaside terraces.

Table 1 shows the quantity of geomorphologic sites and their numbers according to the map.

Table 1

Subgroup of geomorph-ologic site	Quantity according to first inventory	Quantity presented in themap	Numbers on the map
Erosional (surface)	40	41	63-103
River erosional	41	44	11-154
Karst lakes	13	13	104-116
Sea erosion	13	18	155-172
Karst caves	44	45	173-217
Karst fields, valleys	35	35	218-252
Glacial	23	27	253-281
Erosional-lithological	20	20	282-301
Neotectonical	19	21	302-322
Total	248	260	63-322

In the sixth group, are defined Gziqi-Shpali-Shemri ophiolite section, testifying about Mirdita oceanic spreading and subduction of African Plate (Adria Microplate) underneath Orogen (Euro-Asian Plate) in Llogara.

As historical sites are in a separate group, there are classified some old mines such as: Rubiku copper mine, Mborje-Drenove coal mine, Kuçova oil deposit, the finding of Triassic ammonites since 1911 in Kçira section and Rubiku outcrop of clastic-clays pack full of controversies.

Paleoenvironmental and paleobotanical sites are presented in separate groups. In paleoenvironmental sites are classified "storm deposits" full of fossils, transgressions and breaks in sedimentations "hard ground" type, Zverneci outcrops of molasses full of sedimentary forms and prints.

In paleobotanical sites, are included Leqet e Hotit stromatolite section and petrified plants of Pogradeci and Korça coal beds. At least, for the first time, we are evidenced, in Kolpik, spherulites which testify about cosmic activity. Table 2 shows the quantity of different groups of geosites and their corresponding numbers in the map.

Table 2

Groups of geological sites	Number of groups	Quantity according to the inventory	Quantity presented in the map	Corresponding numbers
Stratigraphical	I	16	17	1-17
Magmatic, metamorphic and sedimentary complexes	II	14	15	18-32
Metallogenic-economical	III	14	14	33-46
Tectonical-structural	IV	15	16	47-62
Geomorphologic	V	248	260	63-328
Continental-oceanic and tectonic plates	VI	2	2	329-330
Historic	VII	5	5	331-335
Paleoenvironmental	VII	19	20	336-355
Paleobotanical	IX	2	2	356-3.57
Astroproblems	X	1	1	358

Thermal and mineral water springs are placed according to hydrogeological map of Albania, with respective numbers from 353 to 369. According geographical, tourist and environmental maps, are placed national and local parks and national forests with respective numbers from 370 to 384. At least, with numbers from 385 to 391, are presented hunting lands. As mentioned above, for the completing of natural values of Albania and for giving the more attractive character to the Map of Geological Sites, thermal and mineral watersprings, national parks, natural forest parks, hunting lands and woodland are presented as well.

PRACTICAL IMPORTANCE OF THE MAP

Compilation for the first time of a Map of geological Sites of Albania testifies about a new strategy for our country concerning the evaluation of natural beauties and natural heritage conservation. At the same time, it is a confirmation of a ProGEO strategy in Albania, where the questions of geological heritage conservation began to be studied during the last ten years, after transition period.

Geological outcrops presented in this map, in most cases, are of specific scientific research interest, commonly linked with geological-tectonical development of the Eastern Alpine Mediterranean Chain and Dinarides-Albanides-Hellenides belt. Mirdita oceanic spreading, sedimentary basins and platforms, glacial and karst phenomena are of high scientific and education importance and for exploration by amateurs

Geomorphologic aesthetic landscapes all over Albania, geoparks, natural forests, cold water springs and climate-curative-sporting spots in mountainous regions of Albania are a good perspective for development of tourist industry in the nearest future. The management by foreign companies of some geoparks, tourist and sporting spots will be help for attracting foreign investments and for good management and use of natural beauties of our country.

REFERENCES

Geological Map of Albania in scale 1:200000 (Harta Gjeologjike e Shqiperise) 1982. I. K. Gjeologjike, Tirane.

Gjeografia fizike e Shqiperise Vol. I 1981, Tirane

Gjeografia fizike e Shqiperise Vol. II 1981, Tirane

Metallogenic map of Albania in scale 1:200000 (1982) I. K. Gjeologjike, Tirane.

Serjani A. and Cara F. 1996. List of geological sites of geological importance which should be proposed for protection by the Albanian state. *Geologica Balcanica* 26, 1, 57-60.

Serjani A., Neziraj A. and Jozja N. 1997. Methods and criteria used for classification and selection of geological sites of Albania. ProGEO'97 Meeting, Tallinn, Estonia, 2-4 June 1997. Proceedings of the scientific Conference, pp. 58-67

Serjani A., Neziraj A. and Jozja N. 1998. Preliminary classification of geological sites of Albania. 8th International Congress of the Geological Society of Greece. Patra, May 27-29, 1998

Serjani A. and Neziraj A. 2000. The first inventory of Geological Sites of Albania. Annual Meeting ProGEO'2000 Prague.

Serjani A. and Neziraj A. 2000. Perpilimi i hartes se gjeomonumenteve te Shqiperise shkalla 1:500000. I. K. Gjeologjike, Tirana.

Wimbledon W., et al. 1998. A first attempt at a geosites framework for Europe-An IUGS initiative to support recognition of world heritage and European geodiversity. *Geologica Balcanica* 28, 3-4, 2-23.

In: Parkes, M.A. (Ed.) 2004 *Natural and Cultural Landscapes - The Geological Foundation*, 111-112, Royal Irish Academy, Dublin

THE KANIV MOUNTAINS - A MYSTERIOUS LANDSCAPE ON THE EAST SLOPE OF THE UKRAINIAN SHIELD

Gritsenko V.P. and Taranjuk O.V. (Kyiv)

Abstract: The geological heritage interest of the Kaniv Mountains, near Kiev, Ukraine is briefly described. Extensive ravines in the uplands expose a Jurassic to Quaternary sequence with dynamic geomorphological processes also acting. There is great potential for promotion of the geological interest of the area.

INTRODUCTION

The Kaniv Mountains are a pearl in the geologic heritage necklace of Ukraine. The tops and the slopes are covered by natural forests and cultural plantings. There are meadows, deciduous, coniferous and blended woods. The slopes of mountains are cut up by deep and extended ravines, whose walls expose the geologic pattern transparently. The section consists of Bathonian (Jurassic) marine deposits up to Quaternary. In the Quaternary sequences Ukrainian archaeologists have found cultural strata from the Palaeolithic up to Zrubintsy. In 2001 a unique burial place of a Sarmatian female was found.

GEOMORPHOLOGY

The Kaniv Mountains are situated on the right bank of Dniepro River, 90 km from Kyiv. They occupy a comparatively narrow band up to 35 km wide, and 70 km long from the village of Trakhtemyriv to the village of Moshny. The mountains consist of three main heights - Trakhtemiriv-Buchak (in the north), Kaniv (at the centre) and Moshnogorije (in the south). The altitude is about 250m a.s.l. or 160m above the level of Dniepro River. The upland arc consists of seven ridges, parted by valleys; each ridge consist chains of tops with domed slopes, frequently crossed by deep V-section ravines. The density of the ravine network is the largest in Ukraine. There are many landslides and little mudflows on their slopes. The mountains gradually pass westward into steppe plains, with some agriculture. The broad Dniepro River Valley lies eastward of hills. There are patches of cultural landscapes input in the natural one of Kaniv Mountains. The state reserve of Taras Shevchenko Kiev National University was organized at the centre of the Kaniv Mountains. [Near to the reserve on Tarasova Gora there is also the grave of the great Ukrainian poet and artist Taras Shevchenko].

STRATIGRAPHY

The succession includes (from oldest to youngest):
Jurassic (Bajocian) sands with lignite (25m)
Jurassic (Bathonian) fossiliferous clays (80-100m)
Unconformable Cretaceous (Albian) sands with kaolin (25m)
Unconformable Cretaceous Cenomanian deposits (30m)
Upper Cretaceous glauconitic sandstones with brachiopods, bivalves, gastropods, ammonites and plants
Early Eocene Kaniv Series glauconitic sands (<50m)
Middle Eocene fossiliferous sands (30-45m)
Quaternary deposits: ice (glacial), loess and alluvial

Figure 1. Outcrop of loess on the wall of Ravine Malanchin Potik.

The deposits of glacial origin are widespread and are 1 up to 12m thick. There are introduced sediments of fluvial-glacial, limno-glacial, drift clay, and alluvial-glacial origin. The loess formation is most abundant. During the building of a reservoir dam, in the foundation pit there was one extremely (60m) deep alluvial channel of the Dniepro River.

STUDY HISTORY

The beginning of geological analysis was by Dubois de Montpereux (1832). He noted deformations of sedimentary strata as overfaults. The Jurassic and Cretaceous age was established during the mid 19th century. Since that time there has been a very extensive list of authors who have published on the geology and geomorphology of the area, largely in Russian, of course, with fieldtrip guides such as Baloukhovsky (1958) a good place to find the full record.

Figure 2. Outcrop of flat lying thrust in the middle part of Ravine Malanchin Potik. Callovian clay lies below the Glyadov suite glauconitic sands.

The dislocations (faults and thrusts) of the Kaniv Mountains are the main peculiarity of the geological structure. The genesis of dislocations has been variously connected with tectonic, ice stream (glaciological), gravitational phenomena and clay diapir folding.

TECTONICS

The tectonic pattern is of joint blocks of the Ukrainian shield, with three horsts and two grabens, which separate them. Faulting in the Kaniv Mountains is exhibited as serial thrusts at different scales, due to tectonic and neotectonic forces. Sometimes in good exposures it is possible to observe three and more scales simultaneously (e.g. Kostyanetz ravine). In addition to these patterns in ravines, often landslides and clay diapers folds are observed. In the Dniepro River Basin, after the construction of a series of water reservoirs, the base of erosion was lifted up. Also the active growth of ravines was stopped; their depth is now reduced step-by-step, and outcrops are also lost.

ARCHAEOLOGY

Professor Biliashivsky first dug the mountain top Skiphyan gorodysche (town). He was the reserve founder. As investigation shows this was a place of very strong human activity. The territory of the reserve is rich in archaeological finds. From annalistic sources it is known about the existence of an ancient city Roden' (by the tenth - eleventh centuries) close to Kaniv city. Archaeologists continue to excavate, and ceramic objects of that time are displayed in the reserve museum. A find of 2001 was the burial of the Sarmatian girl, including a cup, bronze mirror and beads.

CONCLUSIONS

The territory of the reserve has significant interest for representatives of many sciences: geologists, archeologists,

Figure 3. Students study geological processes and rock sections in the ravines.

geomorphologists, biologists, historians and ethnographers. There are Natural and Cultural landscapes having a very complex geological basis. The landscapes are varying directly under operating tectonic, erosive, gravitational processes and man's activity. Dynamic changes can still be studied as ravines regularly change, and qualified staff of the reserve can act as guides. A reserve museum can provide the required geological exposition. This reserve could be the sponsor of the first Geopark in the Ukraine.

REFERENCES

Duboia de Montpereux F. 1832. *Geognostische Verhaltnisse in Ost-Galicien und in der Ukraine. - Arch. Miner., Geognosie, Berghau und Huttenkunde.* HRSG von Dr.1. I.B.Karsten, Bd.V, H.2

Baloukhovsky N.F. 1958. *The guide to excursion of meeting on Kaniv. Geological description of district. - International geological congress of Carpathian-Balkan associating.* - Kyiv: P.H. of AS UkrSSR. (in Russian)

See Page 31. Figure 4. The lowland landscape near Dungannon, Co. Tyrone. Drift-dominated drumlins form a regular pattern in the landscape and are associated with well-drained agricultural land.

See page 32. Figure 5. The coastal landscape of western Co. Donegal. Upland areas (foreground) are drift-free whereas lowland areas are choked with sand in the form of sand dunes and estuaries. This sand was reworked onshore during late Pleistocene and Holocene sea-level rise.

See Page 47. Moon-like pseudo- or rootless craters, the result of a basalt lava erupting on to shallow water, on the southern shore of Myvatn, Iceland.

See Page 47. The landform of a fossil meteorite impact crater at Soderfjarden, south of the Finnish city of Vasa, beautifully delimited by the fertile 'polder' agriculture of the crater floor.

See Page 48. Durham Cathedral on a river meander.

See Page 48. The ultrabasic rocks on Rum provide nest holes for opportunist birds.

See Page 49. (Above) Stonehenge.
See Page 51. (Right) Conwy Castle.

See Page 52. Giants Causeway warning signs.

See Page 55. Figure 5. View from the Zborow Hill over the Tertiary planation surface of the upland near Czestochowa, partly covered with the pine-oak forest and crowned by limestone tors. On karstic Jurassic rocks the xerothermic vegetation developed.

See Page 55. Figure 7. Fragment of rocky dry valley incised in large Tertiary planation surface within the cultivated landscape.

See Page 55. Figure 6. Rocky group of Jurassic massive limestone separated by enlarged fissures on the plateau near Krakow.

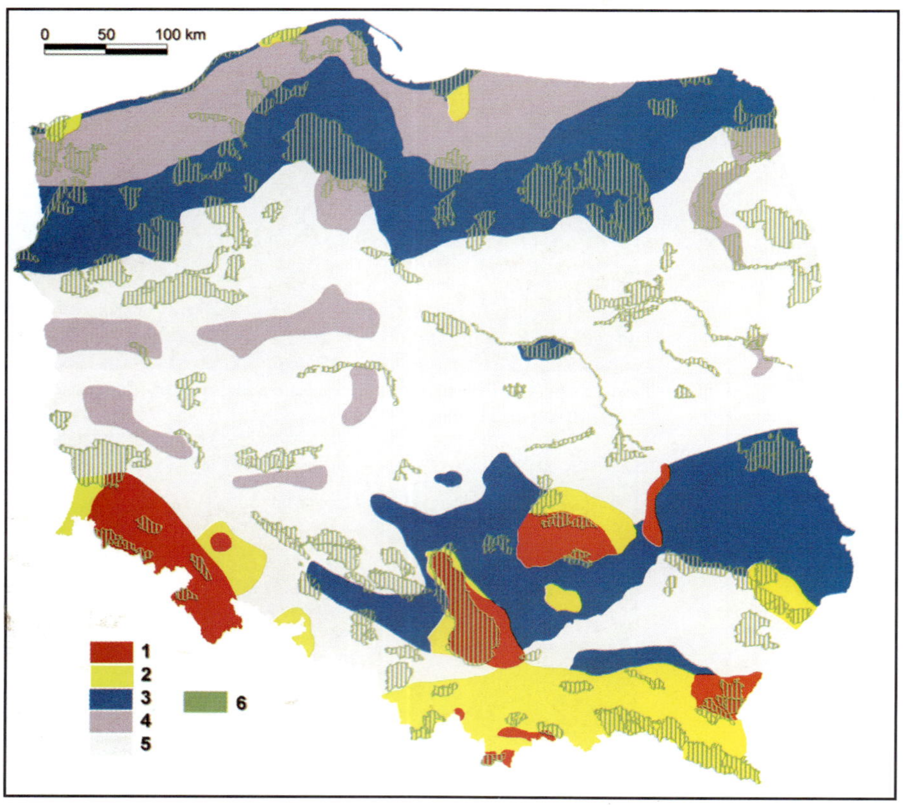

See Page 56. Figure 8. Dry meadows of Festuco-Brometea class (typical of Southern Poland uplands) develop in specific abiotic conditions e.g. in feet of calcareous rocks. They distinguish themselves with very high plant and animal diversity.

See Page 54. Figure 2. Joined pattern of geodiversity and biological habitats of Poland. 1 - 5 classes of areas according to geodiversity from highest (1) downward (5); 6 - large natural sites (biosites) exceeding 100 ha as key habitats.

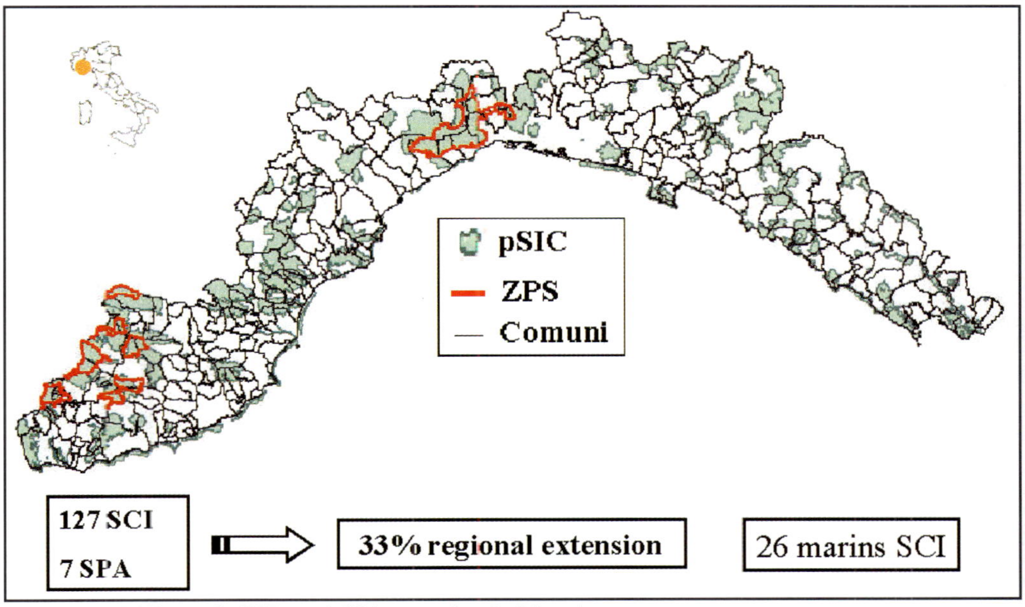

See Page 57. Figure 2. SCIs and SPA extension in Liguria region (Italy).

See Page 60. Figure 12 (below) *Violetta Odorata*, Beigua Park SCIs.

See Page 59. Figure 4. Rocca di Corno, Finalese - Capo Noli SCIs.

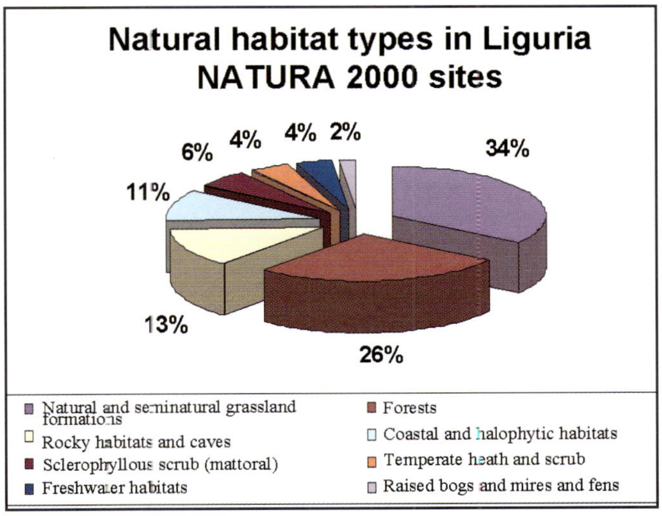

See Page 59. Figure 6. *Cistus Albidus*, Finalese - Capo Noli SCIs.

See Page 60. Figure 10. (above right). Lajone torbiera, Beigua Park SCIs.

See Page 60. Figure 11. (right) Beigua Park SCIs.

See Page 60. Figure 7. The Piana Crixia fungus, an erosion form in Piana Crixia SCIs.

See Page 60. Figure 9. Gargassa valley, Beigua Park SCIs.

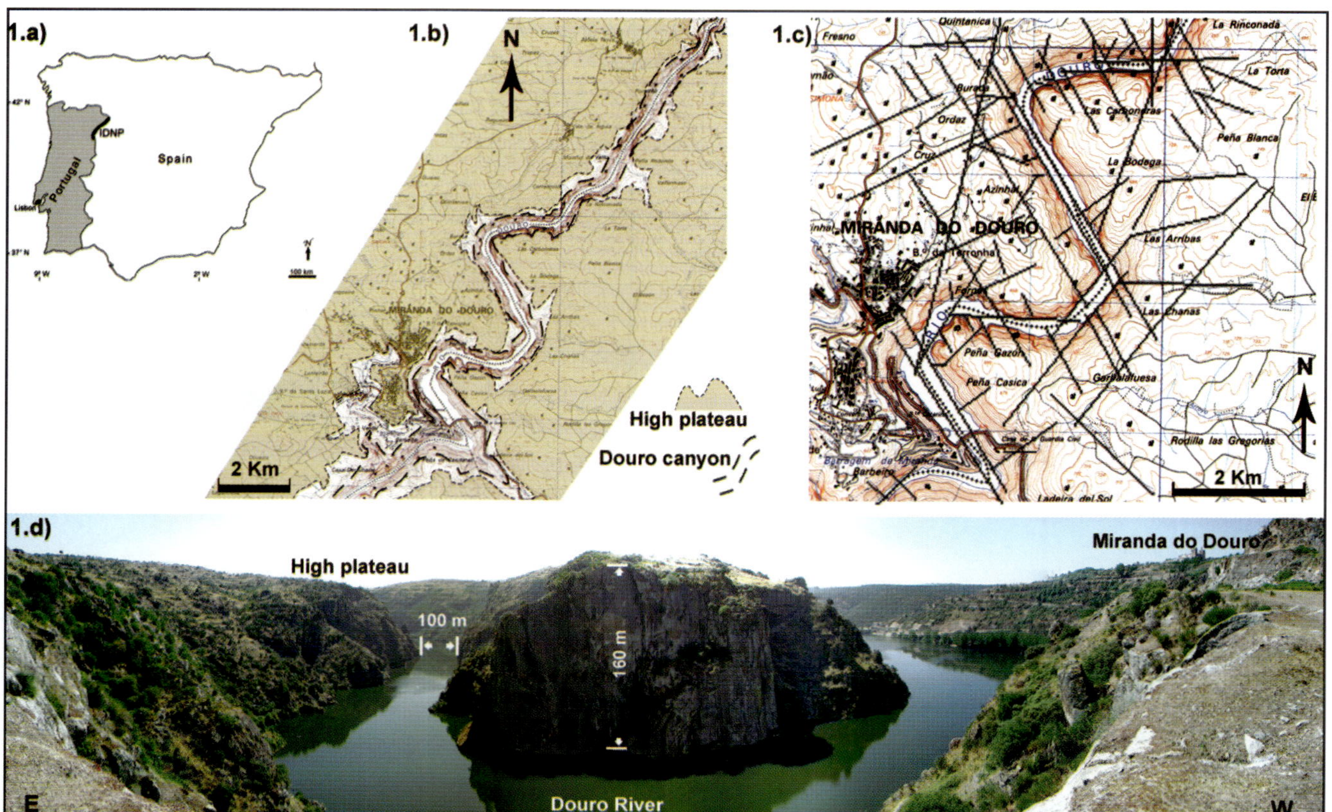

See Page 66. Figure 1. Location of the International Douro Natural Park (IDNP, NE Portugal) (1.a). Geomorphological features (1.b,d) and fracture pattern (1.c) at the Miranda do Douro region (northern IDNP).

See Page 91. Figure 1. Geosites of Kirgyzstan. "Aeolian Castles" in Kongorchek stow. Outcrops of redcolour conglomerates of Paleogene-Neogene as bizarre vertical walls of "castles" and "bastions ".

See Page 91. Figure 2. Cave Chil-Ustun in Visean limestone. It is one of 50 caves of karst region near Osch city.

See Page 92. Figure 3. Geosites of Uzbekistan. Dzharkuduk Geosite "Stone Forest" Upper Cretaceous.

See Page 92. Figure 4. Canyon Charyn river (North Tien-Shan) in Miocene-Pliocene is well known among tourists as "Valley of Castles".

See Page 92. (left) Figure 5. Stone Glacier "Morainic" in Zailiysky Alatau mountain. (above) Figure 6. The Cretaceous and Paleogene sediments are in the background. The famous iridium horizon was revealed at the boundary between them as an evidence of the cosmic disaster which led to the dinosaurs' death.

See Page 92. Figure 7. Monument of Nature "Stone town". Unique landscape complex : granite monadnocks, Alpine lakes and a swamp of the tundra type.

See Page 94. Figure 1. Outcrop Miocene of rocks in Gruschevskian open pit of Nikopolian deposit of manganous ores.

See Page 94. Figure 2. Rapids in place where the river Kamenka cross over Tokovske granite massif.

See Page 95. Figure 3. The sediments of sea Neocene in coastal steep of Kachovskian reservoir (Kachovskian geosite).

See Page 95. Figure 4. Neotectonics in Sarmatian clay-carbonatic sediments.

See Page 95. Figure 5. Geosite " A White rock ".

See Page 96. Figure 6. The nature geological monument " Ribalske open-cast mine ".

See Page 111. (right) Figure 1. Outcrop of loess on the wall of Ravine Malanchin Potik.
See Page 112. (above right) Figure 2.

Outcrop of flat lying thrust in the middle part of Ravine Malanchin Potik. Callovian clay lies below the Glyadov suite glauconitic sands.

See Page 112. (left) Figure 3. Students study geological processes and rock sections in the ravines.

See Page 129. Figure 1. Caher Hill (foreground) and Maghera Mountain (background), Co. Clare.

See Page 132. Figure 7. Kilwaughter iron mines, Co. Antrim, reproduced from a watercolour sketch by G.V. du Noyer.

See Page 132. (left) Figure 8. The sole remaining blast furnace at Creevelea Iron smelting complex, Co. Leitrim.
See Page 134. (right) Figure 12. The Man Engine House, Mountain Mine, Allihies, in early 2002, overlooking Allihies and Ballydonegan strand in the distance.
See Page 135. (below left) Figure 18. Upper Lake, Glendalough, showing spoil scree from 19th C lead mines and below, a natural wetland between mine workings and lake.
(below left) Figure 19. "Avoca Oasis" - natural revegetation in sterile mine spoil, possibly induced by a localised accumulation of organic matter, aided by fertiliser sacks.
See Page 135. Figure 14. Abandoned mine workings and spoil heaps, east Avoca, Co. Wicklow.
See Page 136. Figure 24. The Tankardstown Engine House complex, "Copper Coast" European Geopark, near Bunmahon, Co. Waterford.

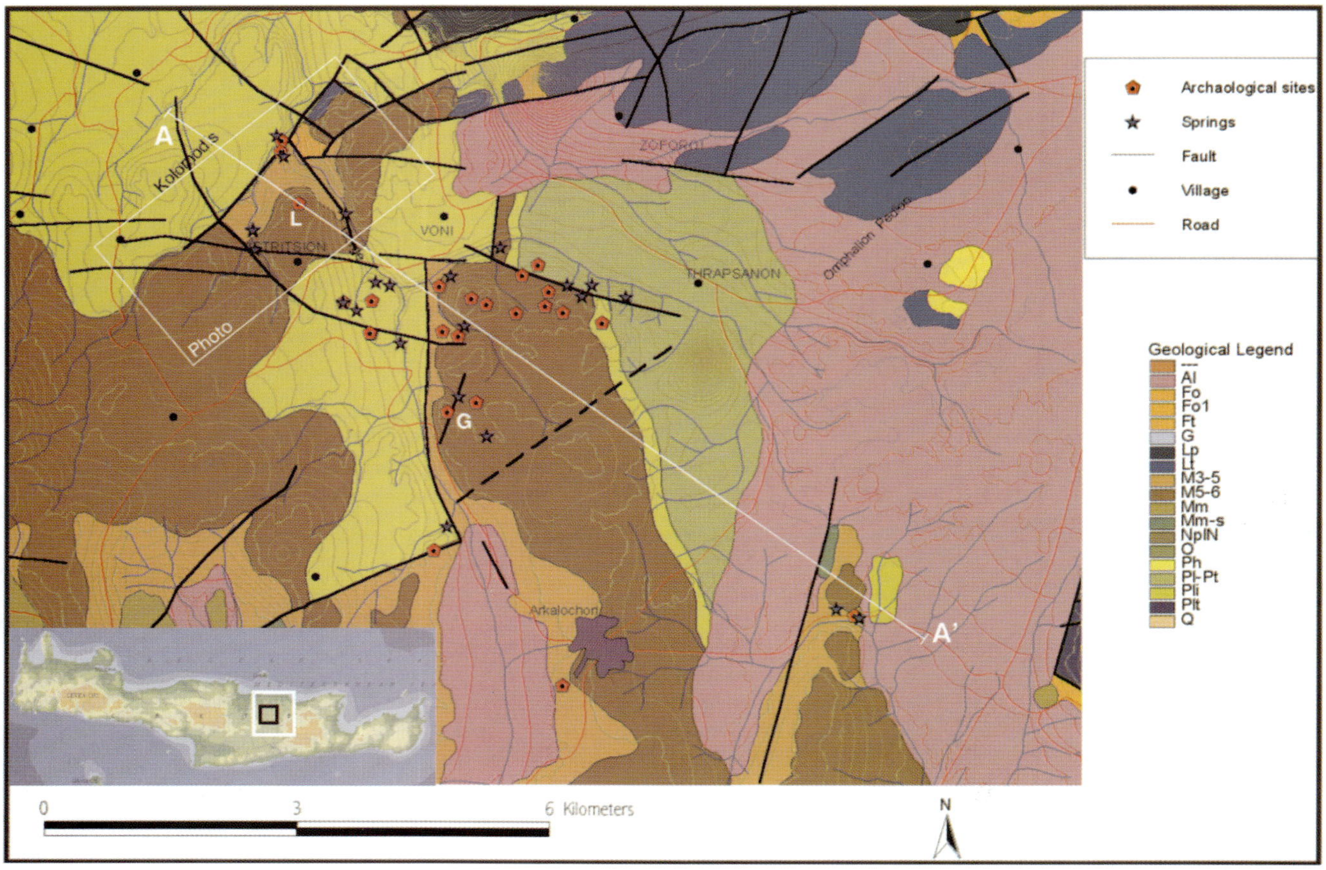

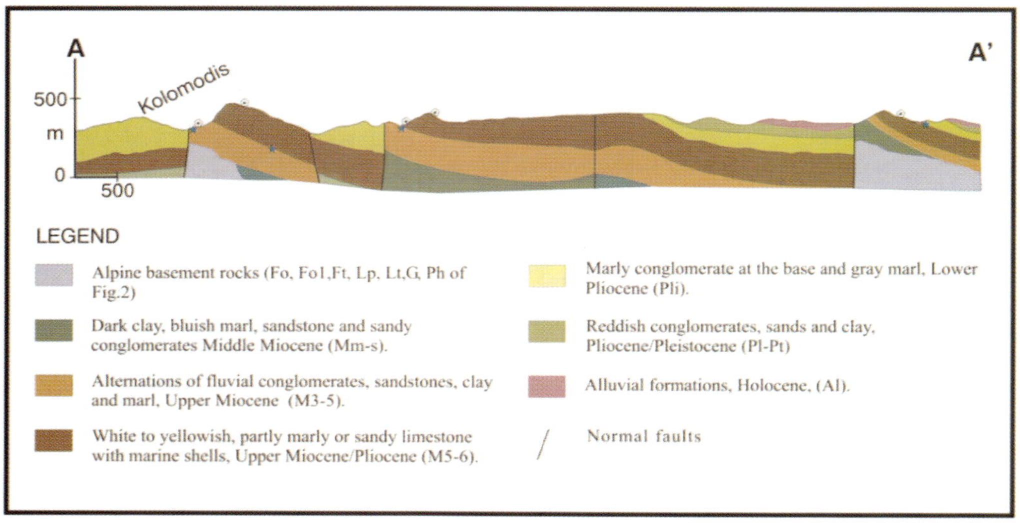

See Page 148. Figure 1. (above) Topographic, geological, and archaeological map of the eastern Pediada area. White line, cross section of Fig. 2, white square the Pediada area, black square study area, L Lykastos, G Galatas. Figure 2. (left) Geological cross section in the study area (Fig. 1). Vertical exaggeration X 2, symbols as in Fig. 1.

See Page 150. Figure 4. Tectonic terraces and ravine development west of Voni. Lykastos on top of central hill. Viewed from the north.

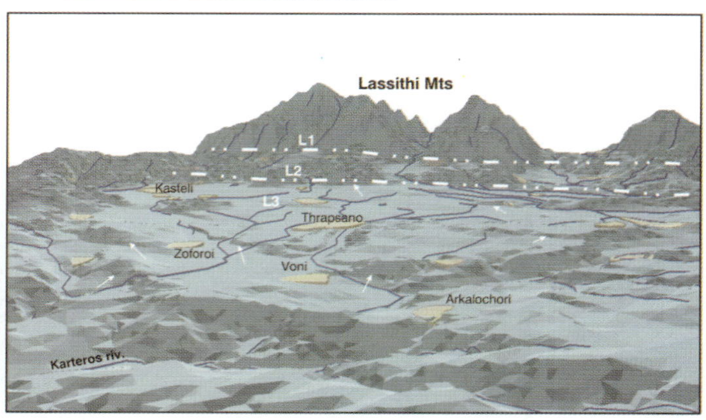

See Page 149. Figure 3. 3-d model of Pediada Region viewed from SW. Vertical exaggeration X 2; white arrowheads fault terraces white dotted lines level borders (L1, L2, L3).

In: Parkes, M.A. (Ed.) 2004 *Natural and Cultural Landscapes - The Geological Foundation*, 113-116, Royal Irish Academy, Dublin

GEOSITES IN PRACTICE - TRANS-NATIONAL COMPARISON AND CORRELATION

Curt Fredén, Lars Karis, Sven Lundqvist, Gunnel Ransed: Geological Survey of Sweden (SGU) and Veli Suominen: Geological Survey of Finland (GSF) and Carl Erik Johansson

Abstract: The GEOSITES project has presented country frameworks for several European countries. However, geology and geological provinces do not comply with national boundaries. It is therefore important to introduce trans-national geological provinces, which are fundamental to the selection of the most typical geological sites. Two of the countries introducing preliminary frameworks and also trying to handle the question of common geoprovincialism, were Finland and Sweden. To illustrate the procedures and methodology of the selection process, this paper presents two examples.

The GEOSITES project was launched by the International Union of Geological Sciences (IUGS) in 1996. Its aims are to compile a global inventory of key geological sites (Geosites) and to create a related database. It will provide a geological input to global conservation efforts, be this for World Heritage purposes or any other conservation programme. The body responsible for this task is the Global Geosites Working Group (GGWG). In Europe, the European Association for the Conservation of the Geological Heritage, ProGEO, has undertaken the operational responsibility for the GEOSITES project. Each national subgroup compiles geological frameworks embracing all aspects of geology. Regional selection and coordination is discussed within ProGEO Regional Working Groups.

A comprehensive review of motives, background and aims was presented by ProGEO (Wimbledon et al. 1998). This paper also presented country frameworks for several European countries. However, geology and geological provinces do not comply with national boundaries. It is therefore important to introduce national and trans-national geological provinces, which are fundamental to the selection of the more, or even most, typical and comprehensive geological site designated to illustrate one or several special geological features. Two of the countries introducing preliminary frameworks and also trying to handle the question of common geoprovincialism, were Finland and Sweden. In order to illustrate the procedures and methodology and how these countries handle the selection process, the paper presents two briefly described examples.

GEOSITES: THE SELECTION PROCESS GENERAL STRATEGY

1 Defining Geological Provinces

The first step is to identify and define coherent geological provinces, regardless of national boundaries, and significant geological formations.

Results: Establishment of national and regional geological frameworks - the reference for the comparison and selection of Geosites.

Examples: Two geological frameworks characteristic of northern Europe are presented (Figures 1 and 2a,b).

2 Selecting National Candidate

The required efforts include creating an inventory of the most representative geological sites within the defined context, creating necessary documentation with reference to established categories of geosites, discussing the sites by relevant authorities, i.e. universities and national and regional governments, and finally selecting one candidate.

Results: Selection of a national candidate to be the optimal geoscientifical representative of a given theme, such as an event, a structure, a province or a time period. This will not exclude other values of minor importance in this context, for example scenic or recreational.

Examples: The distilling process has resulted in the candidates shown in the examples below.

3 Selecting International Candidate

National candidates as well as relevant documentation are compiled and then reviewed. A group of scientists authorised by the countries concerned finalise the selection.

Results: Selection of sites in a comparative and thematic way, regionally and then, lastly, on a global scale.

Examples: When the individual qualities have been weighted, these sites could have been considered to be superior: the Vyborg batholith and the Aurinkovuori ice-marginal formation.

4 Registration in Central Database

A proposed candidate is presented to the IUGS Global Geosites Working Group. If accepted as it is or with any amendments, the selected candidate is registered in the GEOSITES database.

EXAMPLE 1: TERMINAL MORAINES

Framework elements:

Weichselian glaciated areas, Scandinavian ice-sheet, Younger Dryas Ice-Margin Zone.
Type: Landform, terminal moraine.

Geological setting

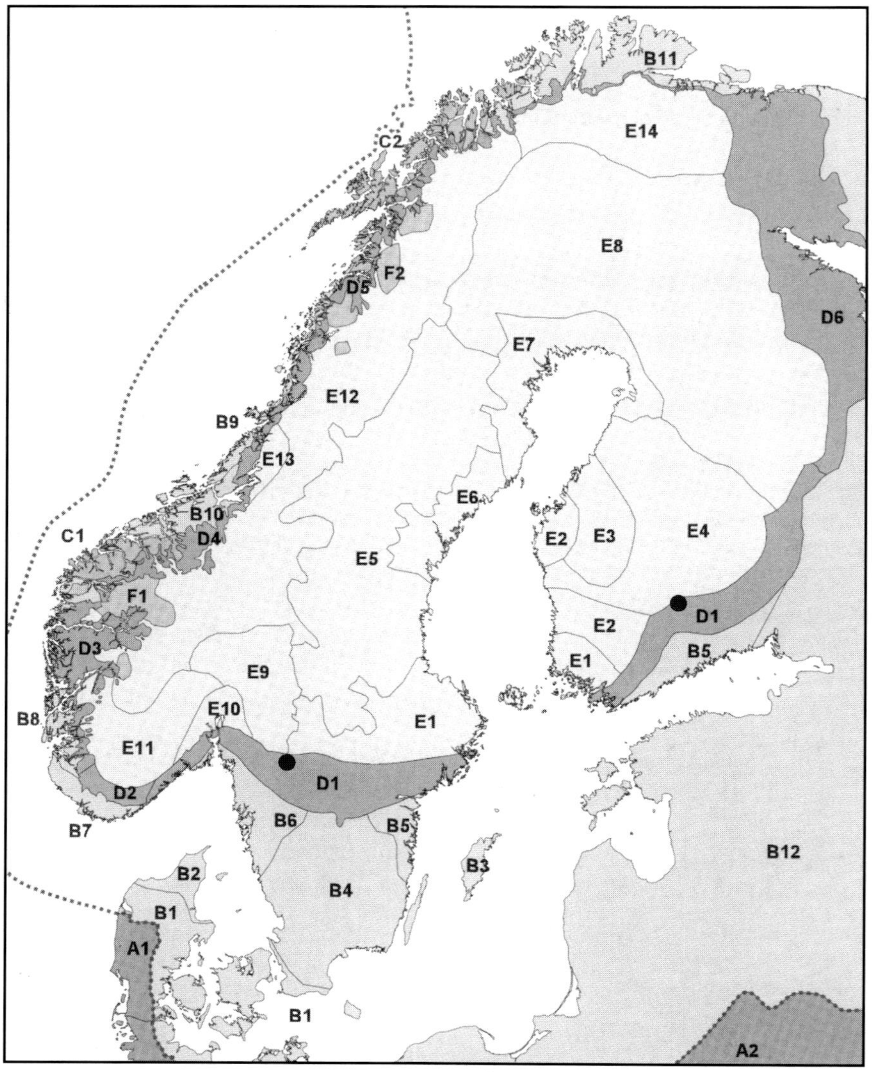

Figure 1. Left. Quaternary geology framework. From "Geodiversitet i nordisk naturvård (2000)".

Figure 2a. Bottom left. Bedrock geology framework. From " Geodiversitet i nordisk naturvård (2000)".

Figure 2b. Bottom right. Distribution of the Rapakivi complexes. Modified from Andersson (1997).

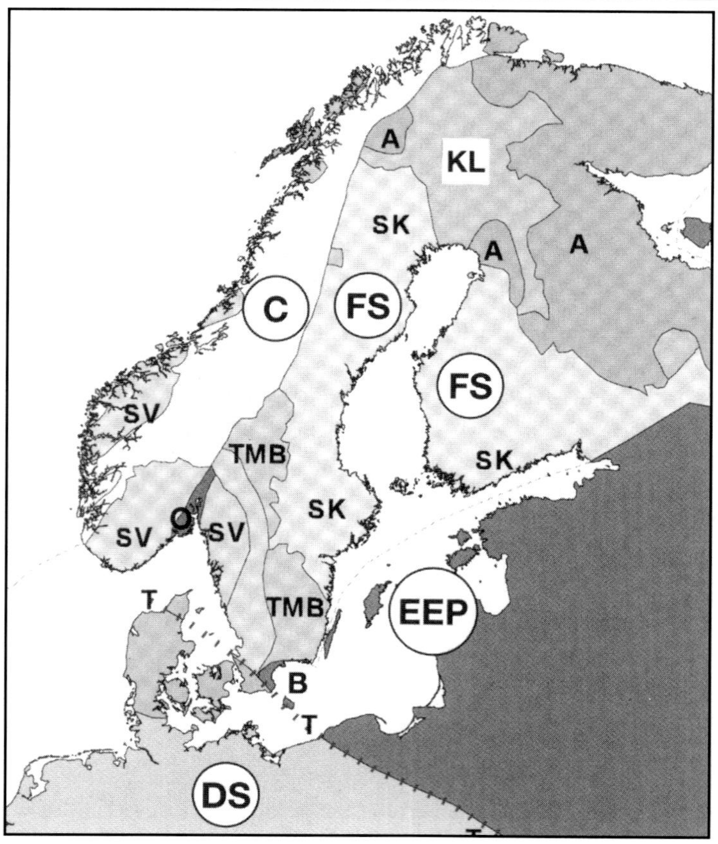

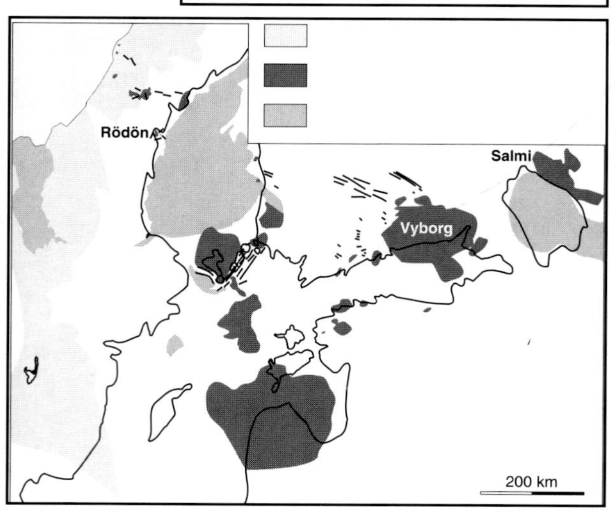

Glacial deposits within this ice-margin zone reflect the stationary or oscillating margin of the retreating Weichselian ice sheet during the Younger Dryas stadial (11.000-10.600 C^{14}-years BP). After this climatic stage the climate rapidly improved and the rate of deglaciation increased dramatically.

General description

The deposits (in Norway - Raerna, in Finland - Salpausselkä) are characterised by large and relatively continuous terminal moraines that also may include glaciofluvial deltas.

Swedish Geosite candidate: Hinden

The elongated point of Hinden reaches 6km out into the lake Vänern, and forms a part of the terminal moraine crossing the lake. In submerged form, it can be followed further in bathymetry in the direction towards the point of Hjorten. The ridge is c. 300m across and the relative height is 15-20m, but younger sediment partly covers its lower parts. A minor parallel ridge is developed only 2km to the north.

Selection motives:

The Hinden terminal moraine is the most magnificent, pronounced and scenic one in Sweden, as a marker of the Younger Dryas event. The site is of great scientific interest for the understanding of the climatic impact and the glaciodynamic connections.

Finnish Geosite candidate: Aurinkovuori

Aurinkovuori is a narrow but morphologically well developed pronounced ice-marginal delta that is part of the Salpausselkä II formation. The delta is connected to the Pulkkila feeding esker.

The delta measures one kilometre along the former Younger Dryas ice-margin and is 300m across. The glaciofluvial deposits are c. 110m thick and rise almost 80m above the surface of Lake Päijänne. On the proximal side there are significant signs of ice contact in the shape of a till bed and hummocky ridges. The top surface marks the youngest and lowest level of the Baltic Ice Lake, 157m a.s.l. A shore terrace at 125-130m a.s.l. formed during the Yoldia sea-stage, is developed on the slope.

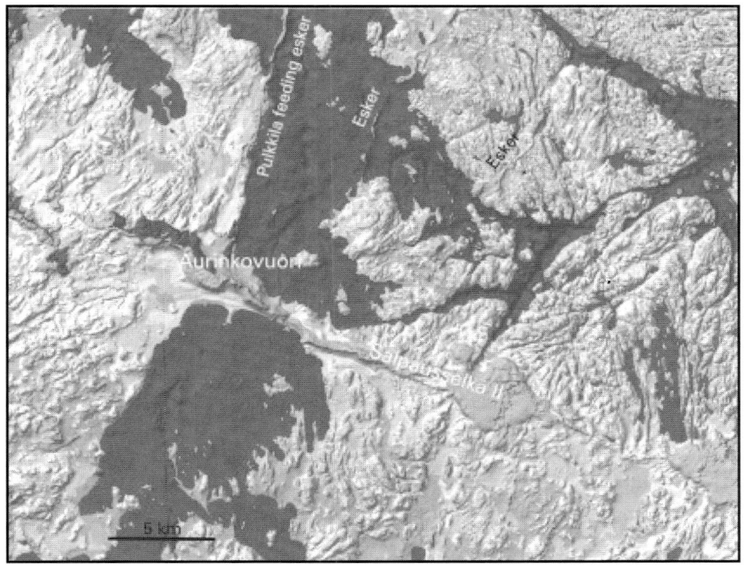

Figure 4. Quaternary geology map draped on terrain model (after Rainio 1990), vertical view.

Figure 3. Geological map draped on terrain model, oblique view. Scale varies; area approximately 50km across (after Karlsson 1870).

Selection motives:

The site is representative in showing the regional and local glaciodynamics and deglaciation pattern, especially as a formation between two ice-lobes and feeding eskers bound to bedrock fissures. It also contributes to the understanding of the development of the Baltic Sea stages. The Salpausselkä-formation, as a unity, represents the most impressive and coherent, complex deposits generated by the inland ice sheet.

EXAMPLE 2: RAPAKIVI GRANITES

Framework elements

Fennoscandian shield, Svecofennian domain, Proterozoic (Palaeo- to Meso-) intrusive rocks, rapakivi complexes. Type: post-tectonic intrusion of rapakivi type.

Geological setting

The classical Fennoscandian rapakivi complexes comprise a number of batholiths and smaller plutons extending from Russian Karelia in the east to central Sweden in the west. Excluding the Salmi batholith in the east, the crystallisation ages of the complexes show a systematic geographical distribution, where the complexes become progressively younger westwards. The ages range from 1.65 Ga to 1.50 Ga.

General description

A typical rapakivi is a coarse-grained granite with alkali feldspar ovoids either surrounded by plagioclase mantles (wiborgite), or not (pyterlite), as well as drop-like phenocrysts of quartz. Since K-feldspar and quartz also dominate the groundmass, they appear in two generations.

Fluorite is a common accessory mineral and anatase is also found. Some other associated granite types (porphyritic with euhedral phenocrysts, equigranular and aplitic) also belong to the rapakivi complexes.

Swedish Geosite candidate: the Rödön Complex

The Rödön granite is a shallow-level intrusion, about 1.50 Ga in age. A predominance of mantled ovoids is typical (wiborgite type). The granite is red, with mantles sometimes not visible, but commonly measuring a few mm. It contains abundant miarolitic cavities, up to about 15cm, partly filled with calcite, quartz, K-feldspar and fluorite. Associated with the small massif are a large number of dykes, including three main categories: quartz-feldspar porphyries, dolerites, and hybrid porphyries. Composite dykes also occur.

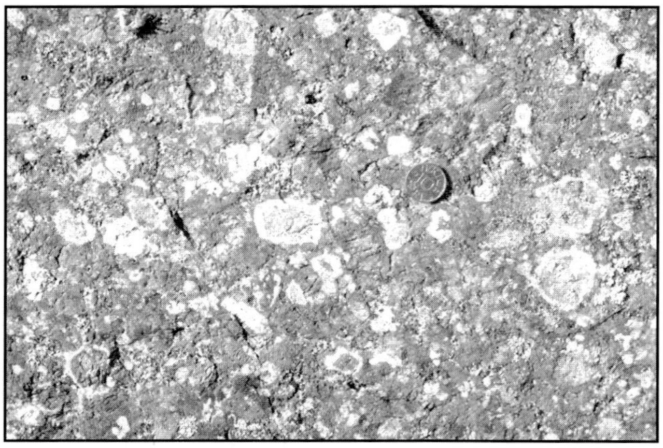

Figure 5. Rödön granite. Photo Ulf B. Andersson.

Selection motives:

The Rödön granite is the only rapakivi granitoid in Sweden that carries the typical ovoidal texture. It is thus the one that most resembles a classical Finnish Rapakivi. Strong fractionation (up to >90%) has resulted in some extremely evolved residual magmas, emplaced as quartz-feldspar porphyry dykes. The basic and the silicic dykes show beautiful examples of magma mixing.

Finnish Geosite candidate: the Vyborg batholith

The Vyborg batholith is a relatively thin, sheet-like body in the upper part of the continental crust and belongs to the oldest group (1.62-1.65 Ga) of the Rapakivi intrusions. It consists of numerous intrusive phases, and therefore consists of several different granite varieties. Most of the batholith, though, is wiborgite. In addition to the granites there are small amounts of minor dyke rocks. Associated with the Vyborg batholith are also mafic rocks that occur as either inclusions or roof pendants.

Selection motives:

The Vyborg batholith is the type area of the classic Finnish rapakivi occurrences - wiborgite is the rapakivi granite proper.

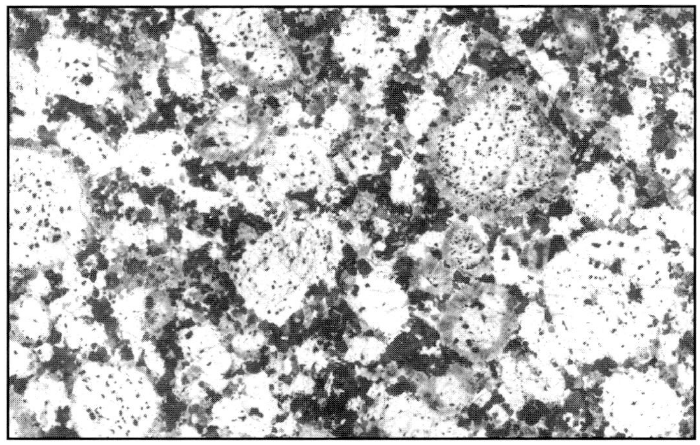

Figure 6. Wiborgite. Photo University of Helsinki, Mineralogical Museum, Helsinki.

ACKNOWLEDGEMENTS

Leif Andersson (SGU) and Hanna Virkki (GSF) prepared the digitally compiled maps, for which we are most grateful. Our deep appreciation is addressed to Jeanette Weihed (SGU) for editing illustrations as well as making the original poster lay-out.

REFERENCES

Andersen, S. & Pedersen, S.S. (eds.), 1998. Israndslinier i Norden. - *Nordisk Ministerråd, TemaNord* 1998:584.

Andersson, U.B., 1997. Petrogenesis of some Proterozoic granitoid suites and associated basic rocks in Sweden (geochemistry and isotope geology). - *Sveriges Geologiska Undersökning, Rapporter och meddelanden* 91, 1-216. [also a reference to figure 2b]

Andersson, U.B., 2002. Petrogenesis of some Mesoproterozoic (Subjotnian) rapakivi complexes of central Sweden: implications from U-Pb zircon ages, Nd, Sr and Pb isotopes. - *Transactions of the Royal Society of Edinburgh: Earth Sciences* 92, 201-228.

Rainio, H., 1990. Geological Map of Finland, Quaternary deposits. *Geological Survey of Finland, sheet* 3112 02 Asikkala.

Johansson, C.E. (ed.), 2001. Geodiversitet i nordisk naturvård. *Nordiska Ministerrådet, Nord* 2000:8. [Figures 1 & 2a]

Koistinen, T. (ed.), 1996. Explanation to the Map of Precambrian basement of the Gulf of Finland and surrounding area 1:1 million. *Geological Survey of Finland, Special Paper* 21, 59-97.

Karlsson, V., 1870. Mapsheet Degerberg. *Sveriges Geologiska Undersökning, Aa 38*. [Figure 3]

ProGEO (Wimbledon, W. et al.), 1998. A first attempt at a GEOSITES framework for Europe - an IUGS initiative to support recognition of world heritage and European geodiversity. - *Geologica Balcanica* 28, 3-4; 5-32

Wimbledon, W.A.P., Andersen, S., Cleal, C.J., Cowie, J.W., Erikstad, L., Gonggrijp, G.P., Johansson, C.E., Karis, L.O. & Suominen, V., 2000. Geological World Heritage: GEOSITES - a global comparative site inventory to enable prioritisation for conservation. *Memorie descrittive carta geologica d'Italia* 54, 45-60.

In: Parkes, M.A. (Ed.) 2004 *Natural and Cultural Landscapes - The Geological Foundation*, 117-120, Royal Irish Academy, Dublin

TOWARDS A GEOSITES FRAMEWORK IN NORTHERN EUROPE - A FIRST ATTEMPT IN THE BALTIC STATES

Jonas Satkunas, Vidas Mikulenas, Jurga Lazauskiene, Lithuania
Rein Raudsep, Krista Taht, Estonia
Aivars Markots, Agris Lacis, Latvia

Abstract: This paper presents the first attempt at a geological framework for the geosites selection of the Baltic States of Lithuania, Latvia and Estonia, created through the ProGEO structure of the Northern Europe Working Group 1. The framework consists of following criteria: pre-Quaternary stratigraphy, geomorphology (pre-Quaternary, and Quaternary), Quaternary stratigraphy, landscape, hydrogeology, and other features. The most representative geosites of each of Baltic (tables 1-3) are placed on this framework now provide an opportunity to determine their comparative value in the regional context.

INTRODUCTION

The ProGEO (The European Association for the Conservation of Geological Heritage) Working Group of Northern Europe (No 3) (http://www.sgu.se/hotell/progeo) currently involves members from Ireland, the United Kingdom, Iceland, The Netherlands, Norway, Denmark, Sweden, Finland, Northwestern Russia, Estonia, Latvia and Lithuania.

The mission of this working group includes: to promote conservation of geological heritage in northern Europe and to aim at a increasing level of awareness concerning geological knowledge and related problems in society; to promote best practice on such matters as databases, criteria, inventory, on-site management, planning, etc.; to strengthen trans-national co-operation and promote initiatives in research, inventory and management of the European elements of the geological heritage; to compile a unified list of geosites representing the geodiversity of northern Europe, contributing to the International Union of Geological Sciences (IUGS) GEOSITES project. The latter goal is one of priorities of current activity of the WG.

This was the first attempt to construct geological framework for the geosites selection of the Baltic States - Lithuania, Latvia and Estonia. The framework consists of following criteria: pre-Quaternary stratigraphy, geomorphology (pre-Quaternary, and Quaternary), Quaternary stratigraphy, landscape, hydrogeology, and other features. The most representative geosites of each of Baltic (tables 1-3) are placed on this framework (Fig) and will be studied aiming to determine their comparative value in the regional context.

Table 1. The most representative geosites of Latvia

No	The name of geotope	Geomorphological feature	Brief description	Geological time
1.	Dauģēni cliffs and caves	cliff, cave, spring	One of the most greatest Middle Devonian outcrops in the Salaca River valley	Q, D
2.	Bērši drumlins	landscape	An impressive and typical landscape of drumlins in one the most extensive European drumlin fields	Q
3.	Vējiņi caves and Elles bedres (Hell Pits)	cave	6 deep and large suffosion funnels with Ezerala - a 46 metre long cave, where the only underground lakes in the Latvia are located	Q, D
4.	Kaltene kalvas (forges)	landscape	A particularly great concentration zone of boulders, "pavement" and bars above the old abrasion terrace of the Litorina Sea	Q
5.	Strante - Ulmale steep coast	landscape	A steep part of the Baltic Sea coast with glaciotectonically deformed deposits	Q
6.	Venta Rapid	waterfall, landscape	The widest waterfall in Latvia, width up to 150 m, height 1.6-2.2 m, formed from Upper Devonian dolomites (Pļaviņas Fm)	Q, D

7.	Kazas valley (with Lībāni - Jaunzemji freshwater lime-stone deposits and Siksparņi caves	landscape, cave	There are outflowing springs at the slope, a freshwater limestone deposit in the NW part, where mining took place in the past, in the N part of the valley (at the upper edge) karst and suffosion features (Siksparņi caves)	Q, D
8.	Dāvida dzirnavas springs	spring	34 springs flow from the Pļaviņas Fm rock	Q, D
9.	Raunis layers	outcrop	Existing information is ambiguous as to whether these are intermoraine deposits. They seem to belong to the end of glaciation and postglacial period	Q
10.	Nīcgale Bigrock (boulder)	boulder	Boulder length 10.5 m, width 10.4 m, height 3.5 m, perimeter 31.1 m, volume above ground ~170 m3 (the biggest in Latvia).	Q
11.	Kraukļi dry bed and underground river	landscape	4 m wide river has formed a ca.10 m deep ravine, where water, at a 200 m long area, is lost underground, after ca. 300 m, at ca. 20 m deep ravine reappearing as a major source.	Q
12.	Lower Zaņa River outcrops	outcrop	Middle Jurassic, Callovian stage; limestone, grey and black clay, sandstone outcrop.	J
13.	Jumprava dolomite outcrop	outcrop	U. Devonian (Stipinai Fm, Bauska Mb) hypostratotype with strong, cavernous, metasomatic dolomite, thin laminar dolomite and domerite, and variegated clay layers, with characteristic fossils.	D
14.	Dolomite prcipice	cliff	25 m high outcrop with dolomite and dolomite marl of the Pļaviņas Fm, overlying siltstone, clay and sandstone of the Amata Fm.	D
15.	Gūtmanis cave	cave	The volume of the cave is 500 m^3 (the biggest cave in Latvia).	D
16.	Sietiņiezis rock	cliff	The greatest outcrop of Upper Devonian white sandstone in Latvia. Gauja Fm, Sietiòi Mb stratotype.	D
17.	Zvārte rock	cliff	U. Devonian (Gauja and Amata Fm) sandstone outcrop. Many fish fossils in the top part of the sandstone section.	D
18.	Lode placoderm deposit	mining	The site with a high concentration of placoderm, crossopterygian, acanthodian, crayfish and plant remains; well studied.	D
19.	Brasla rock	cliff	Gauja Fm sandstone outcrops, caves, niches, seasonal waterfalls and icefalls (during winter).	D
20.	Līči - Laņģi cliff	cliff	Gauja Fm sandstone outcrop (height up to 30 m) with suffosion features	D
21.	Ērģeles (Ērgļi) cliffs	cliff	Gauja Fm stratotype, the most monolithic sandstone cliff wall in Latvia, with the total length 700 m, height up to 22 m, with fractures and niches.	D
22.	Veczemji cliff	cliff	A rocky coastal beach with denudation processes and various petrographic features.	D
23.	Skaņaiskalns rock	cliff	An expressive sandstone outcrop (tectonic slide) at the left bank of the Salaca River.	D

Table 2. The most representative geosites of Lithuania

(recent status of protection is italicised)

GROUP	NAME and *status of protection*	MAIN FEATURES
I. *Landscape landforms (6)*	**1.** Kuršių nerija (Curonian Spit) *Kuršių nerija nacional park*	Active dunes, lacustrine marls.
	2. Biržai karst area *Partly included in Birzai RP*	Sinkholes, notches, cavities, outcrops, springs.
	3. Šeškinė esker *Nature monument*	Esker.
and	**4.** Velnio duobe (Devils Hole) *Nature monument*	Thermokarstic hole.
	5. Puntukas boulder *Nature monument*	The most known erratic boulder.
	6. Puokės akmuo (Puokė Stone) *Nature monument*	The biggest erratic boulder.
hydro-geological features (3)	**7.** Žalsvasis šaltinis (Greenish spring) *Nature monument*	Mineral water (2.4 g/l) spurted out in the karstic sinkhole of 12 m depth, niches.
	8. Kavarsko šaltinis (Kavarskas spring) *Nature monument*	Water spring of biggest yield (up to 18 l/s).
	9. Svilės šaltiniai (Svilė springs) *Nature monument*	Water spring with up to 300 spouts of pure water
II. **Quatern ary type** *sections (9)*	**10.** Rokai outcrop *Included in geomorph. reserve*	Stratotype, lacustrine deposits.
	11. Jonionys outcrop *Included in Dzukija NP*	Sequences of the Eemian Interglacial, etc.
	12. Buivydžiai outcrop	Upper Pleistocene sequence, interglacial gyttja.
	13. Škėvonys outcrop *Nature monument*	Section of Pleistocene deposits(weathering crust
	14. Plikakalnis outcrop *Nature monument*	Section of Pleistocene deposits(postglacial relief
	15. Snaigupėlė outcrop *Nature monument*	Section of Pleistocene deposits
	16. Viriai outcrop	Glacitectonic structure
	17. Butėnai outcrop	Stratotype of Holsteinian Interglacial.
	18. Daumantai outcrop *Included in Anyksciai NP*	Neogene / Quaternary boundary
III. *Pre-Quaternary type sections (8)*	**19.** Varius outcrop *Nature monument*	Lower and Upper Neogene deposits.
	20. Papilė outcrop *Nature monument*	Late Jurassic deposits rich in fossils.
	21. Šaltiškiai quarry	Lower Triassic clay.
	22. Karpėnai quarry	Upper Permian limestone, paleokarstic forms.
	23. Armona outcrops *Included in geological reserve*	Upper Devonian deposits rich in ichtiofauna.
	24. Velniapilis outcrop (niche) *Nature monument*	Upper Devonian gypsum and dolomite.
	25. Muoriškiai outcrop	Upper Devonian dolomite, intensive paleokarst.
	26. Skalių kalnas outcrop *Nature monument*	Upper Devonian dolomite rich in fossils.

Table 3. The most representative geosites of Estonia

No	The name of geosite and *status of protection*	Geomorphological feature	Geological time
1.	North Estonian Klint *Partly protected as landscape reserves*	cliff	Cambrian - Ordovician
2.	Aegviidu glacier marginal formations *Partly protected as landscape reserve*	eskers and kames	Quaternary
3.	Kostivere Karst Field *Mostly protected as Kostivere Landscape Reserve*	karst	Middle Ordovician
4.	Muuga Kabelikivi, an erratic boulder in Muuga *Nature monument*	giant erratic	Quaternary
5.	Tuhala Karst Field *Mostly protected as Tuhala Landscape Reserve*	karst	Middle Ordovician
6.	Helmersen erratic boulders *Nature monument*	giant erratic	Quaternary
7.	Kõpu ancient coastal formations *Mostly protected as landscape reserve*	ancient coastal formation	Quaternary
8.	Kärdla Meteorite Crater *Not protected*	crater	Mid Ordovician
9.	Oil shale outcrop in Kohtla mine *Protected as a part of Oil Shale Museum*	mine	Middle Ordovician
10.	Kurtna Kame Field *Partly protected as landscape reserve*	kames	Quaternary
11.	Vaivara Sinimäed (Blue Mountains) *Vaivara Landscape Reserve*	uplift	Ordovician
12.	Vooremaa Drumlin Field *Partly protected as landscape reserve*	drumlins	Quaternary
13.	Salevere Salumägi cliff *Matsalu Nature Reserve*	cliff	Lower Silurian
14.	Ehalkivi Erratic Boulder in Letipea (Sunset Glow Boulder) *Nature monument*	giant erratic	Quaternary
15.	Ilumetsa Meteorite Craters *Nature monument*	crater	Quaternary
16. 16.1	Piusa Caves and outcrop in Piusa sandstone quarry *Landscape Reserve of Piusa Caves*	Caves, synclinal fold in mine	Middle Devonian
17.	Taevaskoda outcrops *Located on the territory of Ahja Landscape Reserve*	valley	Middle Devonian
18.	Kaali Meteorite Crater *Kaali Landscape Reserve*	crater	Quaternary
19.	The karren (corries) on Vaika islets and Vilsandi Island *Vilsandi National Park, Ramsar area status*	alvars and karst	Lower Silurian
20.	Silurian Klint on Saaremaa Island *Partly protected as landscape reserves*	cliff	Lower Silurian
21.	Kallaste cliff *Kallaste Landscape Reserve*	cliff	Mid Devonian
22.	Karula Upland *Karula National Park*	dome-shaped kames	Quaternary
23.	Härma outcrops *Piusa River Valley Landscape Reserve*	valley	Mid Devonian
24. 24.1	Rõuge ancient valley, springs and Hinni Canyon *Included in Haanja Nature Park*	ancient valley canyon	Quaternary Mid Devonian
25. 25.1	Suur Munamägi (Great Egg Hill) and Vällamägi *Included in Haanja Nature Park*	dome-shaped kames dome-shaped kames	Quaternary Quaternary

SESSION 4

HOW DO HUMANS INFLUENCE THE LANDSCAPE?

Rocks and sediments are the foundation to both urban and rural landscapes, as well as providing the raw materials for construction. Photograph: Matthew Parkes

In: Parkes, M.A. (Ed.) 2004 *Natural and Cultural Landscapes - The Geological Foundation*, 123-127, Royal Irish Academy, Dublin

THE EMERGENCE OF INTEGRATED LAND-CHANGE SCIENCE?

B. L. Turner II, Graduate School of Geography & George Perkins Marsh Institute, Clark University, Worcester, MA, USA. Email: bturner@clarku.edu

KEYNOTE ADDRESS

Abstract: Integrated environmental science is emerging internationally as a major field of research and pedagogy. Among its many versions, that addressing land change has drawn major attention of late because of the recognition of the human impress on the earth's surface, with implications ranging from local ecosystem services and human well being to the functioning of the biosphere itself. Despite the antiquity of this impress, systematic understanding of land change remains underdeveloped, from the character of the basic data to the documentation of the causes of change. Over the past decade and half, however, major advances have been made in monitoring, explaining, and modelling land change worldwide. The advances underway provide the basis for addressing the near- and medium-term trends and trajectory of land changes globally.

GLOBAL CHANGE AND SYNTHESIS SCIENCE

Global environmental change and related activities demonstrate the systemic nature of the biosphere and the unprecedented impress of humankind on its structure and function (Steffen et al. 2002; 2004). This recognition, especially the synergy of human-environment relationships, is not new (e.g., Marsh 1964), attested by, among other examples, Russian attention to studies of the geo-, bio-, and noösphere (Vernadsky 1945). The environmental movement of the middle of the twentieth century helped to foster understanding for specific environmental problems (Worster 1977), but strong and persistent calls for synthesis science awaited global-change problems, launching an era of international and national programmes of integrated environmental science, human-environment science, and sustainability science, (NRC 1999; PNAS 2003; Kates et al. 2000; Schellenhuber and Wenzel 1998), complete with new degree-granting and research programmes and the recrafting of extant environmental and human-environment studies of various kinds, such as forestry and geography.

THE IMPORTANCE OF LAND CHANGE

The condition of and changes in land cover and land use are pivotal to these interests and programmes (Turner 2002; Gutman et al. 2004). They contribute significantly to greenhouse gas emissions (Matson et al. 1998; Watson et al. 2000), changes in the hydrologic cycle and, perhaps, regional precipitation (IGBP 1999), ecosystem and watershed fragmentation, biodiversity loss, and species invasion (Johnson, Revenga and Echeverria 2001; Mooney and Hobbs 2000; Myers 1997; Sala et al. 2000). Land use increasingly usurps large proportions of NPP (DeFries et al 1999; 2000), and changes in land-cover (i.e. afforestation) may account for the missing carbon sink (Fang et al. 2001; Pacala et al. 2001). Land remains the primary source of food and fibre for humankind, and yet this production remains threatened on a large spatial scale by tropospheric pollution (Chamedies et al. 1994) and by assaults on the ability of ecosystem to provide services for this production (Clark et al 1999; Daily et al. 2000). Moreover, the world-wide vulnerability of many people and production systems as well as the larger coupled human-environment systems supporting them are intimately tied to land and the dynamics of its change (Kasperson et al. nd; McCarthy et al. 2001; Turner et al. 2003a,b). For these and other reasons the International Geosphere and Biosphere Programme and the International Human Dimensions Programme have joined two programmes on ecology and land science to form a new "global land Project" to guide international research on land change for the next decade.

SYSTEMATIC COMPARATIVE DATA: CONDITION AND CHANGES

Despite the pivotal role of land, systematic data collection on and understanding of land conditions and their change remain underdeveloped, at least at the global level. Indeed, long-term systematic data on global carbon and understanding of changes in the carbon cycle, no matter the missing carbon sink, are superior to that on land cover and land use. This observation may appear to be counterintuitive, given the centrality of land control in state societies, the history of state cartography and agricultural and resource censuses, and the annual data collected by the United Nations for the past half-century. Nonetheless, a comparison of global data sets, no matter their longitudinal dimension, illustrates the inadequacies noted (Table 1). Significant differences exist among various measures and estimates, especially those dealing with grassland-pasture and the related theme of desertification (arid land degradation).

Several reasons help to explain the quality of data, measures, and estimates of land conditions. The science and application communities historically have not agreed on what to measure and how to do so, and heretofore have not operated within an internationally accepted typology, let alone taxonomy, of land conditions (but see below). Global data sets on several land conditions (e.g., forest and agriculture) are produced by the UN's Food and

Table 1: Global Estimates of Land-Covers and Their Change (10^{-6} km²)

Land Cover	Year(s)	Area	% Change*	Method	Source
Forest-Woodland					
	1990	41.50	-29	expert opinion & Richards 1990	Klein-Goldewijk 2001
	1992	43.90	-20	inventory	Ramankutty & Foley 1999
	1994	41.72		state reported	FAOSTAT 2002
	1997	33.36		adjusted UN	Bryant, Nielsen & Tangley 1997
	2000	41.70			Watson et al. 2000
	2000	45.38		state reports	UNEP-WCMC 2000
Steppe-Savanna-Grassland	1990	17.50	-49%	expert opinion & Richards 1990	Klein-Goldewijk 2001
	1992	26.70	-20%	inventory	Ramankutty & Foley 1999
	2000	35.00			Watson et al. 2000
	2000	53.66		satellite imagery	White, Murray & Rohweder 2000
Cropland					
	1990	14.70	+444	expert opinion & Richards 1990	Klein-Goldewijk 2001
	1992	20.30	+408	inventory	Ramankutty & Foley 1999
	1994	15.10		state reported	FAOSTAT 2002
	2000	16.00			Watson et al. 2000
Pasture					
	1990	3.10		expert opinion & Richards 1990	Klein-Goldewijk 2001
	1994	34.65		state reported	FAOSTAT 2002
	2000	3.50			Watson et al. 2000

* % change for forest and grasslands from pre-human disturbance to date noted; for cropland from 1700 to date noted.

Agricultural Organization (i.e., FAO 2001). These data, however, are generated by individual countries, and despite guidance from the FAO, each country has considerable latitude to interpret what is to be counted and how. Thus, for example, tree plantations to restore degraded lands are apparently counted by China as forest cover, providing the impression that at the country level, China is experiencing forestation and masking the significant loss of extant forests in that country (Fang et al. 2001; Rozelle et al. 2000). International programmes designed to overcome such deficiencies, such the UN's Convention on Desertification, have not necessarily been successful, leading to gross overestimates of human-induced, arid-land degradation (Thomas and Middleton 1994).

Several developments have helped to standardize better some measures of land cover and its change. Foremost are satellite observations to provide consistent information on tree cover (e.g., DeFries et al. 2000). This information is more useful for estimates of biomass and NPP, however, than it is for specific forest types-conditions (e.g., tropical or season tropical forest), and such information says little about land use per se. While improvements are on going, the FAO has taken steps to develop a meta-classification of land-use and land-cover (Gregario and Jansen 2000), and tests are underway to determine its applicability for uniting otherwise disparate typologies.

SYSTEMATIC UNDERSTANDING: CONDITION AND CHANGES

Despite irregularities in the systematic spatio-temporal coverage of land conditions, significant improvement has been made in the understanding the major trends in local to global land conditions, the causes of land change, and the ability to model these changes for different community needs (see Turner 2002).

Trends

Barring some catastrophic changes in the global political economy, the global trends noted in Table 2 are likely to continue for the next several decades. These trends will continue to vary by the economic tier of the controlling polity, aggregated here to the state-level. Third-tier economies will sustain loss and modification of forest and woodlands and degradation of grasslands, owing to the growth in the magnitude of demand for the resources (e.g., timber) and services (cropland) that these lands command and the various reasons that the countries in question will rely on these resources and services. First-tier economies, in contrast, will continue to gain forest cover, peri-urban settlement, and parks-reserves, as land production concentrates and intensifies and resources are secured from distant sources (third tier economies). Second-tier and command economies are more problematic to discern, with high

Table 2: Projected Global Land-Change Trends

Land Cover or Use	Trends-Trajectory
Tropical Forests	Sustained cutting of old growth forest; significant % permanently converted; increased fragmentation, successional growth, burned undergrowth
Tropical-Neo-tropical Woodlands	Sustained conversion; increased fragmentation, successional growth, burned under growth
Mid-latitude Forests	Sustained regrowth in developed economies
Boreal Forests	Significant deforestation in eastern Siberia
Grasslands-Pasture	Local to regional degradation; increases in tropics
Cultivated Lands	Increased intensification on prime lands and in first and second tier economies, and marginal lands taken to other use-covers; sustained semi-permanent cultivation in third tier economies
Dense Settlement	Significantly increased urban, suburban, and peri-urban land; increased ecological footprint

internal variance. Significant deforestation will continue in Siberia and far eastern Russia (21% of extant world forest cover), however, while industrial-urban land uses will consume some prime croplands in southern China (e.g., INAS, CAS, and NAS 2001).

Causes

It is not too far-fetched to assert that the IPAT (Impact = Population x Affluence X Technology) identity dominated much of the thinking during the initial stages of global environmental change research. Its appeal resides in its simple surrogates for resource demand, facilitating linkages to various explanatory themes and theories. While it constitutes one of the few formulations that track with land change over the long-term and at the global scale, it often fails to track well at finer spatio-temporal resolutions (Lambin et al. 2001). This failure follows in part because of the increasing spatial separation of production consumption. For example, forestation in the northeastern United States follows from a service economy that supports suburban/peri-urban settlement replete with tree cover and relies on large-scale importations of material needs otherwise derived from land uses, while deforestation in many parts of Indonesia is driven by international logging activities.

Improved understanding of causes or drivers of land change follows from the new-found attention and support for its study. Such agendas as that of the IGBP-IHDP Land-Use/Cover Change (LUCC) effort focus on the decision-making of the land manager (proximate agent of change), working up the chain of linked factors that influence the manager's decision (opportunities and constraints) as well as undertaking systematic, comparative studies of the causes of certain land covers. Giest and Lambin (2001), for example, demonstrate the multiple array of factors (economic, political-institutional, technological, cultural, and demographic) and their relationships with one another that trigger tropical deforestation by way of linkages among agricultural expansion, infrastructure extension, wood extraction, and other activities. They conclude that 3- and 4-term factors are most common in case studies and that PAT variables are lacking in terms of spatio-temporal fit; that is, even if population increased in the case, it did so owing to policy decisions to promote the occupation an area, not because of local, natural population growth. In another such comparative study of tropical deforestation models, Angelsen and Kaimowitz (1999) conclude with the truism that forest falls when it or the land it covers are profitable to someone or by the ability of some groups to influence policy to enlarge their economic interests.

This complexity notwithstanding, the interaction of major factors influencing various land changes tend to display regional variations, if specific to different land covers and uses (Turner 2002). Again referring to tropical deforestation, the proximate determinants in southeastern Asia-Oceania are largely agricultural expansion and timber extraction, while in Latin America, they are infrastructure and agricultural expansion. The major underlying causes for tropical deforestation, at least in terms of triggering mechanisms, are the economic determinants of timber, market failures, impoverished households, and capital investments which, in turn, are linked to the policy-institutional determinants favouring forestry and development activities coupled with corruption and mismanagement.

Models

A major stimulus for integrated land science is the need for improved land-change models that can explain and project the *kind, magnitude* and *location* of different land covers and uses. Such models are no easy task, given the complexity noted, as well as biophysical feedbacks not addressed above. Nevertheless, large arrays of such land models are under development and major advances have been made (e.g., Veldkamp and Lambin 2001; Parker et al. 2003; Turner 2002). The quality of the models, of course, improves by reducing the covers-uses and spatio-temporal window addressed. The unknowns involved for such critical drivers as policy shifts support the need for agent-based, integrated assessment models.

CONCLUDING COMMENTS

The advances underway in the study of land conditions and change are multidisciplinary, involving the biophysical, social-behavioral, and remote sensing-GIS sciences. Interestingly, an increasing number of on-going projects combine researchers from each of these traditions seeking to understand land change as a coupled human-environment system (e.g., Turner et al. 2001; 2004). The richness of their output promises not only to enhance questions central to each tradition, but also to provide state-of-the art assessments to a wide range of synthesis issues, be it carbon stocks, biodiversity, sustainable production, or household vulnerability. In this sense, integrated land-change science appears well under development.

ACKNOWLEDGEMENT

I thank Alexander Pulsipher for his assistance in preparing this paper.

REFERENCES

Angelsen, A. and D. Kaimowitz. 1999. Rethinking the Causes of Deforestation: Lessons from Economic Models. *The World Bank Research Observer* 14: 73-98.

Bryant, D., Nielsen, D. and Tangley, L. 1997. *Last Frontier Forests: Ecosystems and Economies on the Edge*. Washington, D.C.: World Resources Institute.

Chamedies, W. L., P. S. Kasibhatle, et al. 1994. Growth in Continental-Scale Metro-agroplexes: Regional Ozone Pollution and World Food Production. *Science* 264: 74-77.

Clark, J. S., Carpenter, S. R., Barber, M., Collins, S., Dobson, A., Foley, J. A., Lodge, M., Pacual, M., Pielke, R., Jr., Pizer, W., Pringle, C., Reid, W. V., Rose, K. A., Sala, O., Schlesinger, W. H., Wall, D. H., and Wear, D. 1999. Ecological forecasts: An emerging imperative. *Science* 286: 685-686.

Daily, G. C., Söderqvist, T., Aniyar, S., Arrow, K., Dasgupta, P., Ehrlich, P. R., Folke, C., Hansson, A., Jansson, B.O., Kautsky, N., Levin, S., Lubchenco, J., Mäler, K.-G., Simpson, D., Starrett, D., Tilman, D., and Walker, B. 2000. The value of nature and nature of value. *Science* 289, 395-396.

DeFries, R., Field, C., Fung, I., Collatz, G., and Bounoua, L. 1999. Combining satellite data and biogeochemical models to estimate global effects of human-induced land cover change on carbon emissions and primary productivity. *Global Biogeochemical Cycles* 13, 803-815.

DeFries, R. S., Hansen, M. C., Townshend, J. R. G., Janetos, A. C., and Loveland, T. R. 2000. A new global 1km dataset of percentage tree cover derived from remote sensing. *Global Change Biology* 6, 247-254.

FAO (Food and Agricultural Organization). 2001. *State of the World's Forests, 2001*. Rome, Food and Agricultural Organization of the United Nations.

FAOSTAT. 2002. http://apps.fao.org/ (June 2002).

Fang, J., Chen, A., Peng, C., Zhao, S., and Ci, L. 2001. Changes in forest biomass carbon storage in China between 1949 and 1998. *Science* 292, 2320-2322.

Geist, H. and E. Lambin. 2001. What Drives Tropical Deforestation? A Meta-Analysis of Proximate and Underlying Causes of Deforestation Based on Subnational Case Study Evidence. Luovain-la-Neuve, BE., LUCC International Project Office.

Gregorio, A. and L. J. M. Jansen. 2000. Land Cover Classification System. Rome, Food and Agriculture Organization of the United Nations.

Gutman, G., Janetos, A., Justice, C., Moran, E., Mustard, J., Rindfuss, R., Skole, D., and Turner, B. L., II, eds. 2004. *Land Change Science: Observing, Monitoring, and Understanding Trajectories of Change on the Earth's Surface*. Kluwer Academic Publ., New York, NY (forthcoming).

INAS (Indian National Academy of Science), CAS (Chinese Academy of Science), and NRC (US National Research Council). 2001. *Growing Population, Changing Landscapes: Studies from India, China, and the United States*. Washington, D. C., National Academy Press.

IGBP (International Geosphere-Biosphere Programme. 1999. Special issue on BAHC. *Global Change Newsletter* 39. Stockholm, SW: IGBP.

Johnson, N., Revenga, C., and Echeverria, J. 2001. Managing water for people and nature. *Science* 292, 1071-1072.

Kasperson, J. X., R. E. Kasperson, et al. (nd). Vulnerability to global environmental change. In *The Human Dimensions of Global Environmental Change*. A. Diekmann, T. Dietz, C. C. Jaeger and E. S. Rosa, eds. Cambridge, MA: MIT Press (forthcoming).

Kates, R. W., W. C. Clark, et al. 2001. Sustainability science. *Science* 292, 641-642.

Klien Goldewijk, G. and X. Battjes. 1997. A hundred Year (1890-1990) Database for Integrated Environmental Assessments (HYDE, version 1.1). Bilthoven, NL: National Institute of Public Health and the Environment.

Lambin, E. F., B. L. Turner II, H. Geist, S. Agbola, A. Angelsen, J. W. Bruce, O. Coomes, R. Dirzo, G. Fischer, C. Folke, P. S. George, K. Homewood, J. Imbernon, R. Leemans, X. Li, E. F. Moran, M. Mortimore, P.S. Ramakrishnan , J. F. Richards, H. Skånes, W. Steffen, G. 2001. The causes of land-use and land-cover change - moving beyond the myths. *Global Environmental Change* 11, 5-13.

Marsh, G. P. 1965 (1864). *Man and Nature: or, The Earth as Modified by Human Action*. Cambridge, MA: Belknap Press of Harvard University Press.

Matson, P. A., Naylor, R. L., and Ortiz-Monastario, I. 1998. Integration of environmental, agronomic, and economic aspects of fertilization management. *Science* 280, 112-115.

McCarthy, J. J., Canziani, O. F., Leary, N. A., Dokken, D. J., and White, K. S. eds. 2001. *Climate Change 2001: Impacts, Adaptation, and Vulnerability*. Intergovernmental Panel on Climate Change: Working Group II. Cambridge, UK: Cambridge University Press.

Mooney, H. A. and R. J. Hobbs, eds. 2000. *Invasive Species in a Changing World*. Washington, D. C.: Island Press.

Myers, N. 1997. Mass extinction and evolution. *Science* 278, 597-598.

NRC (US National Research Council). 1999. *Our Common Journey: A Transition Toward Sustainability*.

Washington, D.C.: National Academy Press.
Pacala, S. W., Hurtt, G. C., Baker, D., Peylin, P., Houghton, R. A., Birdsey, R. A., Heath, L., Sundquist, E. T., Stallard, R. F., Ciais, P., Moorcroft, P., Caspersen, J. P., E., S., Moore, B., Kohlmaier, G., Holland, E., Gloor, M., Harmon, M. E., Fan, S.-M., Sarmiento, J. L., Goodale, C. L., Schimel, D., and Field, C. B. 2001. Consistent land and atmosphere-based U. S. carbon sink estimates. *Science* 292, 2316-2320.
Parker, D. C., Manson, S. M., Janssen, M., Hoffmann, M. J., and Deadman, P. J. 2003. Multi-agent systems for the simulation of land use and land cover change: a review. *Annals of the Association of American Geographers* 93, 316-340.
PNAS 2003. Science and technology for sustainable development. *Proceedings of the National Academy of Sciences of the USA* (Special Feature) 100, 8059-8091.
Ramakutty, N. and Foley, J. A. 1999. Estimating historical changes in land cover: North American croplands from 1850 to 1992. *Global Ecology and Biogeography* 8,381-396.
Richards, J. 1990. Land Transformation. In *The Earth Transformed by Human Action*. B. L. Turner II, W. Clark, R. Kates, J. Richards, J. Mathews, and W. Meyer, eds. Cambridge, UK: Cambridge University Press, pp. 163-178.
Rozelle, S., J. Huang, Husain, S. A., and Zazueta, A. 2000. China: *From Afforestation to Poverty Alleviation and Natural Forest Management*. Washington, D. C.: The World Bank.
Sala, O. E., Chapin, F. S., Armesto, J. J., Berlow, E., Bloomfield, J., Dirzo, R., Huber-Sanwald, E., Huenneke, L. F., Jackson, R. B., Kinzig, A., Leemans, R., Lodge, D. M., Mooney, H. A., Oesterheld, M., Poff, N. L., Sykes, M. T., Wlaker, B. H., Walker, M., and Wall, D. H. 2000. Biodiversity: Global biodiversity scenarios for the year 2100. *Science* 287, 1770-1774.
Schellenhuber, H. J. and V. Wenzel, eds. 1998. *Earth System Analysis: Integrating Science for Sustainability*. Berlin, Springer-Verlag.
Steffen, W., Jäger, J., Carson, D., and Bradshaw, C., eds. 2002. *Challenges of a Changing Earth: Proceedings of the Global Change Open Science Conference, Amsterdam, NL, 10-13 July 2000*. Springer-Verlag, Heidelberg, GR.
Steffen, W., Sanderson, A., Tyson, P., Jäger, J., Matson, P., Moore, B., III , Oldfield, F., Richardson, K., Schellnhuber, H.-J., Turner, B. L. I., and Wasson, R. 2004. *Global Change and the Earth System: A Planet Under Pressure*, Springer-Verlag, Berlin, GR.
Thomas, D. S. G. and N. J. Middleton. 1994. *Desertification: Exploding the Myth*. New York, John Wiley & Sons.
Turner, B. L., II. 2002. Toward integrated land-change science: advances in 1.5 decades of sustained international research on land-use and land-cover change. *Challenges of a Changing Earth: Proceedings of the Global Change Open Science Conference, Amsterdam, NL, 10-13 July 2001*. W. Steffan, J. Jäger, D. Carson and C. Bradshaw, eds. Heidelberg, GR:, Springer-Verlag pp. 21-26.
Turner II, B. L., Cortina Villar, S, Foster, D. R., Geoghegan, J., Keys, E., Klepeis, P, Lawrence, D., Macario Mendoza, P., Manson, S., Ogneva-Himmelberger, Y., Plotkin, A. B., Pérez Salicrup, D., Roy Chowdhury, R., Savitsky, B., Schneider, L. Schmook, B., Vance, C. 2001. Deforestation in the southern Yucatán Peninsular region: An integrative approach. *Forest Ecology and Management* 154, 343-370.
Turner II, B. L., Kasperson, R. E., Matson, P. A., McCarthy, J. J., Corell, R. W., Christensen, L., Eckley, N., Kasperson, J. X., Luers, L., Martello, M. L., Polsky, C., Pulsipher, A., and Schiller, A. 2003a. A framework for vulnerability analysis in sustainability science. *Proceedings of the National Academy of Sciences, U. S. A.*, 100, 8074-79.
Turner II, B. L., Matson, P. A., McCarthy, J. J., Corell, R. W., Christensen, L., Eckley, N., Hovelsrud, G., Kasperson, J. X., Kasperson, R. E., Luers, L., Martello, M. L., Mathiesen, S., Naylor, R., Polsky, C., Pulsipher, A., Schiller, A., Selin, H., and Tyler, N. 2003b. Illustrating the coupled human-environment system for vulnerability analysis: three case studies. *Proceedings of the National Academy of Sciences, U. S. A.* 100, 8080-85.
Turner II, B. L., Geoghegan, J., and Foster, D. R., eds. 2004. Integrated Land-Change Science and Tropical Deforestation in the Southern Yucatán: Final Frontiers. Oxford, UK: Oxford University Press.
UNEP-WCMC (United Nations Environmental Programme-World Conservation Monitoring Centre. 2000. http://www.unep-wcmc.org/forest/world.htm. (June 2002).
Veldkamp, A., E. Lambin, et al. 2001. Special Issue: Predicting Land-Use Change. *Agriculture, Ecosystems and Environment* 85.
Vernadsky, W. I. 1945. The biosphere and the noösphere. *American Scientist* 33, 1-12.
Watson, R. T., Noble, I. R., Bolin, B., Ravindranath, N. H., Verardo, D. J., and Dokken, D. J., eds. 2000. *Land Use, Land-Use Change and Forestry*. Special Report of the IPCC (Intergovernmental Panel of Climate Change). Cambridge, UK:, Cambridge University Press.
White, R., Murray, S., and Rohweder, M. 2000. The Pilot Analysis of Global Ecosystems" Grassland Systems. Washington, D. C.: World Resources Institute.
Worster, D. 1977. *Nature's Economy: A History of Ecological Ideas*. Cambridge, U.K.: Cambridge University Press.

In: Parkes, M.A. (Ed.) 2004 *Natural and Cultural Landscapes - The Geological Foundation*, 128, Royal Irish Academy, Dublin

IRISH LANDSCAPE ARCHAEOLOGY: ARCHAEOLOGICAL LANDSCAPES AND HISTORIC LANDSCAPE CHARACTERISATION

Gabriel Cooney, Department of Archaeology, University College Dublin, Belfield, Dublin 4, Ireland. Email: Gabriel.Cooney@ucd.ie

Abstract: In this paper it will be argued that a landscape approach is vital to interpreting Ireland's cultural heritage. There is a need to adopt a landscape approach to the management and sustainability of the archaeological resource in the rural landscape. A key debate regarding approaches to cultural landscape is the compatibility of a whole landscape approach (historic landscape characterisation) with the recognition of specific landscapes (here termed archaeological landscapes). The case for considering these as complementary rather than conflicting approaches will be made.

In: Parkes, M.A. (Ed.) 2004 *Natural and Cultural Landscapes - The Geological Foundation*, 129-138, Royal Irish Academy, Dublin

MINING THROUGH THE MILLENNIA: THE LEGACY OF IRELAND'S MOST ANCIENT INDUSTRY

John H. Morris, Mining Heritage Trust of Ireland [www.mhti.com]

Abstract: This paper provides a brief introduction to the 9,000 year history of mining and quarrying in Ireland, briefly summarises its positive and negative impacts, and outlines how mine heritage is being used as a resource for generally community based sustainable economic development in Ireland, Europe and further afield.

Most people would take for granted that a landscape such as Caher Hill and Maghera Mountain, Co. Clare (Fig. 1) is entirely natural. Nothing could be further from the truth, as apart from its physical shape, it is entirely a man made, cultural landscape - forests, fields, roads, houses, even bogs reflecting slash and burn agrarian activities in pre-historic times.

Figure 1. Caher Hill (foreground) and Maghera Mountain (background), Co. Clare.

Figure 2. Neolithic flint and stone implements.

And, not least, a small quarry, here representing a facet of human industry, which has been intimately associated with the social, cultural and economic evolution of Ireland for thousands of years, since the dawn of human occupation some 9,000 years ago. The contribution that industry has made to Irish society is indisputable, although it may largely go unrecognised and unappreciated, even decried by some. It is manifest, at its most profound, in the names we have chosen, here and in many other parts of Europe, to give to the very earliest human cultural ages, the Stone, (Copper), Bronze and Iron Ages, and it underpins later ages, including the Industrial Revolution era, as well as that of the information knowledge economy of the present age - based, as it is, upon the humble element silicon from which computer chips are currently manufactured (a "silicon" age?). For the older cultural ages, mining and quarrying remains have acquired a patina of respectability for various reasons: their antiquity, or for the cultural value attached to built heritage features, or for the beauty of artefacts skilfully crafted from mined or quarried earth resource materials. But the same is not always the case for more modern examples.

9000 YEARS OF MINING HISTORY

Quarrying, as the first expression of what we now recognise as an industry, may be inferred to have started with the arrival of the very earliest settlers in Ireland, during the Mesolithic period (7,000 - 4,000 BC). Remains of their settlements are scattered around various parts of Ireland, but they are most notably concentrated in the north-east where a ready supply of flint, from which to craft knives, arrow heads and other tools, was readily available for extraction, directly or indirectly, from the flint bearing, Cretaceous age, chalk deposits found primarily in that region only (Stout and Stout 1997, Lyle 2003). The location of some other settlements, for example in the Dingle peninsula, appear to reflect the availability and use of chert and rhyolite as an alternative to flint (Stout and Stout 1997), and, similarly, siliceous concretions in limestone provided a readily available, local substitute for flint, and slate for axeheads, at Boora Bog, Co. Offaly (Mitchell and Ryan 2001).

The Mesolithic extraction and manufacturing activities predate the arrival, or development of agriculture at the beginning of the subsequent Neolithic period (4000 - 2300BC). Some of the most beautifully crafted stone artefacts (Fig. 2), stone axes in particular, as well as a greater range of flint implements, date from this period, as does the development of megalithic tomb construction, of which some 1,350 have been documented throughout Ireland, including Newgrange, an UNESCO World Heritage Site (Fig. 3; Mitchell and Ryan 2001).

About 54% of the total known number of Irish stone axes (c.13,500; 18,000 cited by Mitchell and Ryan 2001)) are composed of porcellanite, a very fine grained, blue-grey coloured rock found, and extracted, from small quarries at only two, relatively obscure locations in NE Ireland: at Tievebulliagh, Co. Antrim and at the Brockley Plug on Rathlin Island (Lyle 2003). Axeheads composed of the

Figure 3. Newgrange megalithic tomb, Co. Meath.

Figure 4. Bronze Age copper mine, Ross Island, Killarney national Park, Co. Kerry. Photo: W. O'Brien.

very distinctive "Lambay Porphyry" form yet another very distinct group, all of which can be related to a single "quarry/axe factory" source on Lambay island, just off the Co. Dublin coast (Mitchell and Ryan 2001).

Megalithic tombs are as equally well known as stone axes, perhaps none more so than the outstanding Newgrange (Fig. 3; c. 3100BC, Harbison 1992), and other Boyne Valley tombs, now designated as an UNESCO World Heritage Site. Most interesting from the perspective of sources of building material is the brilliant white quartz, which has been used as a facing material at Newgrange (Fig. 3). Veins of quartz do occur in the local bedrock in the region, but these are relatively insignificant and unlikely to have been an adequate source for the large volume, and size, of quartz blocks used in the construction. Instead, Mitchell and Ryan (2001) note that they may have been preferentially selected and transported from the fringes of the Wicklow Mountains, where quartz is far more common and abundant, both in situ in bedrock, and as loose boulders.

The advent of the Bronze Age (c. 2300 - 400BC) brought with it an indisputable flowering of mining technology, including a brief period when Ireland was possibly the primary northwestern European producer of at least one of the metals we now admire in the wealth of beautiful copper, bronze, silver and gold artefacts which characterise the age. Most Bronze Age mining sites in Britain and Ireland date between about 1800-1300BC (O'Brien 1996b). A notable exception are the Bronze Age mine workings on Ross Island, Lough Leane, in Killarney National Park, Co. Kerry, which have yielded much earlier dates, between 2400-2000BC - the oldest known such workings so far known anywhere in northwest Europe (Fig. 4; O'Brien 1996a). Indeed, their association with "Beaker Folk" pottery suggests that the sites might have been worked in the very latest part of the Neolithic period (O'Brien 2000). Much about the mining and smelting methodologies, as well as living conditions of the miners, has been revealed by very detailed archaeological excavations initiated in 1992 by Dr. W. O'Brien, and published in summary form in 2000 (O'Brien 2000). The smelting pits are the oldest known examples of their type in Ireland, and the copper metal contains a noticeably high level of arsenic, forming an alloy slightly harder than pure copper, enhancing its usefulness in axeheads and blades. These copper-arsenic alloy artefacts achieved a wide circulation throughout Ireland and further afield, clearly suggesting that for a brief period, between about 2400-2200BC, Ireland was one of, if not the principal source of copper in Ireland and Britain and perhaps northwest Europe - a period which W. O'Brien has informally termed the "Copper Age" (O'Brien 1996a; W. O'Brien, pers. comm. 1997). Available evidence suggests that this specialised mining community was supported by agriculturists in surrounding settlements, though none of the latter has yet been discovered. The settlement appears to have been used only intermittently, perhaps seasonally, or maybe for longer periods, rather than continuously. Evidence of Bronze Age mining is also known from other parts of Counties Cork and Kerry, not least on Mount Gabriel, Co. Cork, where 32 individual mine sites have been documented, all intact and undisturbed by mining activities in subsequent centuries (O'Brien 1994 1996a).

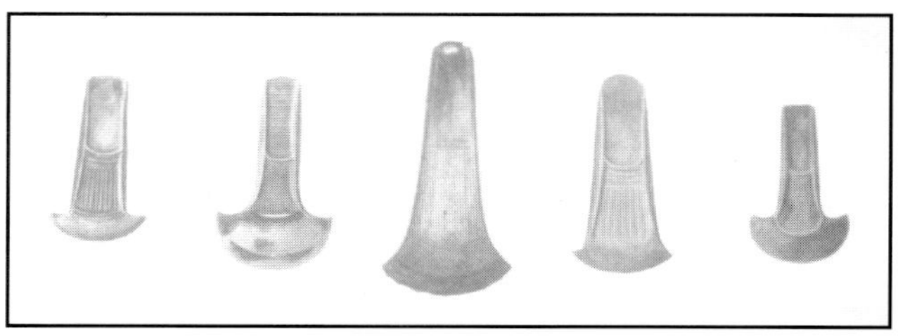

Figure 5. Bronze Age bronze axe heads.

Figure 6. Medieval mining techniques, reproduced from "De Re Metallica" by Georgius Agricola, 1556, reprinted 1950, Dover Publications, New York.

Paramount of all the artefacts that typify the Bronze Age in Ireland, are the elaborate and exquisitely produced range of gold ornaments found in various "hoards" in different parts of the country, for example the Mooghaun hoard found in Co. Clare during the construction of the famous "West Clare Railway" in the 19th Century. The sheer number and richness of these artefacts has frequently been taken to indicate contemporaneous gold mining in Ireland. But, if so, and despite much debate, as well as documentary and metallurgical research, the source of the gold still remains elusive - and an Irish source unproven.

Our knowledge of mining and/or quarrying activities during the Iron Age (400BC - 500AD) presents one of several "information lacunae" between 400BC and the advent of Industrial Revolution related activities from about 1760AD onward. Cowman (2003), in summarising documentary evidence of mining during the first millennium AD, notes reference to a putative copper-iron mine within the hill fort at Freestone Hill, Co. Kilkenny, as well as copper at Bantry. There is a similar dearth of information for the subsequent Early Christian period (c.500 - 1200AD). O'Brien (2000) recounts the 800AD writings of the Welsh monk, Nennius, who, in noting the presence of "circles", of tin, lead, iron and copper around Lough Lein (Leane, Co. Kerry), seems to be reflecting a knowledge of mineral wealth in that part of southwest Ireland. But what might have been all too easily discounted as a rather fanciful documentary record, instead has been corroborated by the discovery of c.700AD metal production furnaces on Ross Island (O'Brien 2000). Evidence of iron and copper smelting sites has been discovered very recently during the construction of the N25 Kilmacthomas bypass road in Co. Waterford (P. Claughton, pers. comm. 2003). The iron site has been dated to about 400-600AD, the copper site to about 600-700AD. An ore sample has been recovered from the latter site but has yet to be matched against an obvious source, the rich copper mineralisation at nearby Bunmahon, which was exploited extensively during the 19th Century. Cowman (2003) notes monastic documentary evidence of knowledge of iron, copper and perhaps silver, as well as evidence of iron smelting at several monastic settlements. The source of the iron ore is uncertain, though "bog iron ore" (ferruginous iron layers found in some of the then vast expanses of blanket bogs), is one possibility. "Old men's workings", of unknown date, are again noted, this time in connection with mines at Milltown, Co. Clare and Ballybrennan, Co. Wexford.

Mining and quarrying activities continued to develop during the Medieval period (c. 1200 - 1760 AD). Norman interest in mining is not documented prior to 1264, when reference to two small mines is made, each employing four miners and a smelting expert, one mine near Waterford, the other at "Oulys" (Oola, Co Limerick?; Cowman and Morris 2003). A subsequent reference in 1289 notes that Florentine and Genoese experts were brought in to operate a silver prospect, almost certainly at Silvermines, Co. Tipperary, which lasted for 14 years before being suspended due to civil unrest (Cowman 1988, Cowman and Morris 2003). Subsequent attempts to re-work the deposit were made, apparently in 1336, and permission to reopen the mines is noted in 1375. Several sources allude to a silver mine and mint at Clonmines, or Barrystown, Co. Wexford (e.g. Cole 1922/1998), although there are now no visible remains which can be assigned definitively to that period of operations. This site may well be that recorded above as "close to Waterford", though the first documentary evidence of it by name dates from 1458, when it was worked by German miners (Cowman 1978, 1986/87, Cowman and Morris 2003).

What might have become the cradle of an Irish "Industrial Revolution" centre, centred upon and around "Slieve an Iarainn", the " Iron Mountains" of Counties Cavan, Leitrim and Roscommon, has had a very long history of fraught, and ultimately unsuccessful developments. Iron works are known to have existed in Drumshanbo as early as the 15th Century, but works expanded rapidly during the 1600s, when by 1621, Sir Charles Coote reputedly employed 3,000 English and Dutch (but no Irish) workers at charcoal fuelled smelting works, which by then had been established at various locations including Ballinamore, Arigna, Garrison (Co. Fermanagh) and Creevelea (McAuley, undated; Cahill 2000). The Drumshanbo works were destroyed in the 1641 rebellion, but restarted in the 1690s along with other works at Dromod, Ballinamore and Creevelea. Most of these works had ceased by the mid-1700s, largely due to the exhaustion of timber supplies for charcoal production. Sir Charles Coote was also directly involved with the development of iron mines and iron works in the Monasterevin - Mountmellick region of Counties Kildare and Laois (Carville 2001). Iron works were established in that area by 1620, using two types of ore, both locally derived: a "white ore" dug out of the ground (bog iron ore presumably) and a "rock-iron" ore mined at Dysert. Operations were, as at Drumshanbo, brought to a halt by the 1640 rebellion. They were not resumed until the 1690s, but after an unspecified closure period during the 1700s, they were active during the 1790s, apparently to produce the distinct Kildare pike used in the 1798 rebellion.

While there are scant references to, or evidence of early medieval mining operations in Ireland, it is reasonable to infer, from the abundance of early Norman period monastic and castle ruins still surviving, that substantial quarrying operations were undertaken in order to supply masonry, as well as slate roofing material. This may be exemplified by the example of Kells Priory, Co. Kilkenny. The Augustinian Priory was most probably founded by Geoffrey fitz Robert in 1193 (Tietzsch-Tyler 1993). Construction proceeded over a 150 year period, starting in 1202, although it was sacked on at least three occasions during the period up to 1350. The precise source of the bulk masonry has yet to be determined, but it appears probable that at least some of the roofing slate was extracted from the Ormonde slate quarry, near Ahenny, Co. Kilkenny, as the Prior of Kells was granted a 49 year lease to extract slate from this site in 1348 (James Power, pers.

comm. 1997). First reference to this substantial and very impressive quarry, and the nearby Victoria slate quarry, dates, however, from 1337, and slate from here was used to roof the Tudor Manor House in Carrick-on-Shannon c.1565, and other houses in that town in the 17th Century. The quarries continued to be actively developed throughout the 18th, 19th and 20th Centuries, winning gold medal prizes at the Dublin exhibition of 1882, and the Crystal Palace Exhibition of 1890. Operations continued up to 1944, although the association with Carrick-on-Shannon was maintained, albeit malodorously, into the 1970s, by the use of the abandoned Victoria quarry as a dump for tannery waste - the sort of fate shared by so many other historic mining and quarrying sites throughout Ireland.

Although the power of steam was known to the ancient Greeks, it was not until the 18th Century that an ironmonger/blacksmith, Thomas Newcomen, first harnessed steam power to serve human needs - and in so doing laid the foundations of the "Industrial Revolution" (1760 - 1900). His invention effectively presaged the end of an era which, at least from the perspective of mining and quarrying technology, had been marked by many centuries of little change, into a new era which witnessed a very dramatic upsurge in demand for earth materials to fuel the very rapid expansion of the new age economy. Indeed the very first application of steam power was to mining, with the development of steam engines to pump water out of mines, which before had been the principal constraint on working mineral deposits to any great depth. This application was applied initially to coal and iron deposits, the foundation materials of the entire Industrial Revolution, and rapidly thereafter to many other types of metallic deposits, such as tin, copper and lead.

The efficiency of this first generation of steam engines left much to be desired however. Considerable improvements were effected by James Watt and Matthew Boulton between 1769 -1800, but the expiry of their patents in 1800 brought with it a rapid flowering of new steam engine technology, not least the inventions of Cornish (mine) engineer Richard Trevithick, inventor of the first steam powered road engine (1801), the first steam railway locomotive (1804, "Penydaren"), but perhaps most importantly, the

Figure 7. Kilwaughter iron mines, Co. Antrim, reproduced from a watercolour sketch by G.V. du Noyer.

Figure 8. The sole remaining blast furnace at Creevelea Iron smelting complex, Co. Leitrim.

development of the use and application of high pressure steam (1812; Hodge 1995). Through these, and other inventions, Trevithick created the foundation for so much that we now take for granted, not least our fossil fuel powered railways and road vehicles. But he also created the foundation of an entire specialist industry - the design, production and installation of Cornish steam engines and Engine Houses (Fig. 9), and related mining technologies and methods, the fruits of which ultimately spanned the globe, from Britain and Ireland, to, amongst other regions, Spain, South America, South Africa and Australia.

The industrial revolution largely by-passed Ireland, apart from the northeast where industrialisation in Belfast and other urban centres was fuelled, at least in part, by the availability of coal and iron resources beneath the Antrim Plateau basalts [Hammond 2000; Fig. 7]. This has bequeathed to us a rich diversity of mine heritage features in that region, for example the evocative caverns of the Glenravel iron mines. Similar attempts were also made to develop the much more marginal coal and iron resources in other parts of Ireland, especially in the Arigna region of Counties Leitrim and Roscommon. There, poor quality coal resources were developed in conjunction with attempts to smelt iron ore derived from nodules in the local limestone sequences, as mentioned above, in particular at Creevelea, Co. Leitrim (McAuley, undated; Fig. 8). Now, only a sole blast furnace remains to remind us of what was a valiant effort to develop the foundation of an "Industrial Revolution" in that area.

After closure in 1768, due to exhaustion of charcoal supplies, the Creevelea works lay dormant until 1851, when it was re-established by a Scottish Company (McAuley, undated). As much as £297,000 is reputed to have been spent on this venture, for which coke was brought from Arigna by cart, and iron ore from an iron mine in Tullynamoyle townland, 13km and 2 km distant from the works respectively. Production only lasted for about 9 months however, with products claimed to have been used in the Tay Bridge, Scotland, the Halfpenny Bridge, Dublin, and cannon for the Crimean war (McAuley, undated). The works were revived in 1861/2, by a consortium of Dublin investors, who instead of using coke as the smelting fuel, instead used a patented form of peat charcoal, the peat for which was extracted from a bog in Gowlaun townland, from banks still known locally as the "Company Banks" (McAuley undated). Despite the success of this production method, and reputed quality of the iron produced, the venture soon collapsed, only to be re-activated once again, using the same process, between 1896-1900. Attempts were made, with varying degrees of success, to mine iron at various other locations in the Republic of Ireland during this period, for example at Gortinee, near Drumsna, Co. Leitrim, and at Redhills, Enaghan, and Cleenrah, all in Co. Cavan (Morris 1984).

Figure 9. (left): Cornish Engine Houses, Wheal Grenville, Cornwall. Figure 10. (right): Williams Engine House, Avoca, Co. Wicklow, as it was in 1956.

Notwithstanding the importance of any of these developments, the most evocative reminder of this age are the gaunt ruins of Cornish Engine Houses and/or associated chimney stacks, which stand guard like lonely sentinels over the sites of 19th Century mining and quarrying sites throughout Ireland: for example at Newtonards, Avoca (Fig. 10), Bunmahon (Fig. 24), Silvermines, Killaloe, Kilbreckin, Glandore, Hope (Castleblayney) and not least at Allihies (Fig. 12). Most of these now roofless shells once housed Cornish design steam engines, used mainly for pumping water out of underground workings, others for winding materials up and down shafts, some for driving ore crushing machinery, and in one exceptional, and unique example, for lifting miners up and down from underground workings. Apart from Killaloe, site of the largest slate quarry in Ireland, material from which was awarded a medallion at the Dublin International Exhibition of 1865, all the others mark the sites of major metal mining complexes: for example copper at Avoca, Allihies and Bunmahon; lead, with or without silver at Newtonards, Silvermines, Glendasan, Glendalough, Hope and Kilbreckan; and manganese at Glandore.

The Allihies, or Bearhaven Copper Mines, located at the very western end of the Beara peninsula, Co. Cork, were the focus of the largest copper mining district in 19th Century Ireland, and are here used to exemplify this entire group of enterprises (Fig. 11). Although Bronze Age mines are documented on the peninsula (see under Bronze Age above), none are evident in the vicinity of the 19th Century mines, the era of which commenced with their initial development in 1812 and continuing uninterrupted up to closure in 1882 (Williams 1993).

After a somewhat slow start from 1812 to about 1820, the mines developed rapidly during the 1820s, principally on the back of the discovery and development of the Mountain Mine, which was to prove the backbone of production up until 1882. The mines entered their most profitable period during the 1830s, with only a minor decline into the 1840s, largely reflecting the increasing cost of production from the ever increasing depth of the mines, coupled with an ongoing decline in the value of copper. Even the Great Famine did not significantly effect production. Earlier potato shortages, or short duration famines are well documented by Williams (ibid) in most instances these being alleviated by importation and sale of food to miners and their families by the Allihies Mining Company, owned by the Puxley family, resident in nearby Dunboy Castle. While this addressed the most immediate problem, taken in conjunction with the contract system of mining used at the mine, the cumulative effect was to increased miners' indebtedness to the Company. This, amongst other factors, contributed to increasing levels of unrest, several strikes and discord between the local community on the one hand, and Cornish miners and mine management on the other.

Figure 11. A copper plate engraving of the Allihies Copper Mines, dated 1838, reproduced from "Rambles in the South of Ireland" by Lady Chatterton, 1839.

Figure 12. (left): the Man Engine House, Mountain Mine, Allihies, as it appeared in early 2002, overlooking the village of Allihies and Ballydonegan strand in the distance. The building has since undergone extensive conservation works by the Mining Heritage Trust of Ireland 2002 - 2004.

The Puxley family sold the mines in the 1860s, and returned to live in South Wales in the 1870s, leaving Dunboy Castle (Fig. 13) in the care of a caretaker, in whose hands it remained until it was destroyed by the IRA in 1921.

The contemporaneous illustration reproduced in Figure 11 shows a number of mine related developments, including several Cornish Engine Houses. Today, only three such buildings remain, all post-dating the time of the engraving, and of these the Man Engine House, Mountain Mine, is the most outstanding example of all (Fig. 12). This building is nationally and internationally unique, now being the sole surviving, most intact example of a purpose built, Cornish design Man Engine House anywhere in the world (Morris and Brown 2001). It was built in 1862, specifically to hoist miners, as well as haul materials up and down from the underground workings. It operated continuously up to mine closure in 1882, after which it lay abandoned until conservation works commenced in 2002 (Morris 2002).

The advent of the Modern era (1900 - 2000 AD) witnessed desultory and generally short lived efforts to maintain mining activities at many of the more significant established metal and coal mining centres, for example at Allihies, Bunmahon, and Glendalough, while more enduring operations continued in other centres, such as Arigna, Avoca, Castlecomer and Kingscourt (gypsum). All of these operations were, however, eclipsed in the latter half of the 20th C by developments at, and subsequent to the discovery, in 1961, of rich lead and zinc ore at a site first recorded in the "Annals of the Four Masters" (Ó Cleirigh et al. 1632) - Tynagh, Co. Galway (Clifford et. al. 1986). This discovery not only resulted in the establishment of one of the single biggest base metal mines in Ireland, but, equally, if not more significantly established Ireland as a mining centre of world wide importance. Exploration venture capital

Figure 13. Dunboy Castle, Castletownbere, as it was in 1997 (author R.A. Williams in the foreground).

poured into Ireland during the subsequent decades up to the 1990s, resulting in the discovery of a number of base metal deposits, some of which were subsequently developed into very large mines (year of discovery in brackets): Gortdrum (1963), Silvermines/Ballynoe (1962 - extension of known resources), Galmoy (1986), Lisheen (1990), and not least, Navan (1970), now the largest zinc-lead mine in Europe, and the sixth largest such mine in the world. These successes spawned exploration for many other commodities, including, for example, gold, with significant resources defined at Lecanvey, Cregganbaun, Clontibret, Curraghinalt and Lack, the latter two both in Co. Tyrone, and the last named currently the only producer of gold anywhere in Ireland; silver at Mallow and Aherlow; lithium at Aclare House, Co. Wicklow; platinum group metals at Carlingford; and diamond indicator minerals in Co. Donegal.

THE LEGACY

Successive generations of Irish schoolchildren have been brought up on the mantra that "Ireland is a land without significant mineral resources". Nothing could be further from the truth, as the brief history above demonstrates. But this mineral endowment, and more particularly its extraction, has left both positive and negative impacts, the positive impacts the rich and diverse built heritage and artefacts noted above, the negative impacts that of environmental degradation, both visual as well as chemical.

To many people, modern, large scale mining operations have created a veritable "blot on the landscape", for example the large, sterile, spoil heaps at Avoca and Silvermines (Fig. 14). In other areas, heavy metal rich, wind blown dust has created health hazards, as in Silvermines, while in other areas, such as Avoca, and in Cornwall, acidic, metal rich waters have caused serious damage to river water quality and adversely affected fish stocks (Figs. 15, 16).

The mining heritage community does not deny the gravity of this particular dimension of the mining legacy, and to that end it has worked positively and constructively with mitigation programmes planned for implementation at Silvermines and Avoca. This pro-active approach has

Figure 14. Abandoned mine workings and spoil heaps, east Avoca, Co. Wicklow.
Figure 15. (centre): metal rich, acid mine drainage, County Adit, Cornwall.

Figure 16. Banner headline article, Sunday Times, identifying the Avoca River as the most polluted river in Ireland.

ensured that while mitigation will be undertaken, it will be done in such a manner that will not result in the destruction of the mining heritage landscapes or built features in either area. Indeed, the communities in both areas have fought very actively and vigorously for the protection of their mine heritage, as they consider it part of their community history - their sense of place. This very welcome outcome also contributes towards a heightened understanding of another, little appreciated aspect of such mining sites: that many are developing ecosystems and habitats worthy of conservation in their own right.

This dimension is perhaps better appreciated in other countries, for example the UK, where sites such as the oil shale "bings" in the Midland Valley of Scotland, the Mona Mine buildings and environs, Parys Mountain, Anglesey, and the spoil heaps at Wheal Poldice near St. Day, Cornwall are listed as SSSIs for their scientific and/or ecological interests: the Mona Mine for lichens, the Wheal Poldice spoil for its natural heathland interest, contrasting very markedly with adjoining crofters field enclosures, both generated directly from the mining industry, the spoil from the mines, the filed by enclosure of pre-existing commonage (Fig. 17). Downstream, at Devoran Bridge, swards of sea pinks have recolonised metal rich mud flats (Fig. 29), and further out into the estuary, a new sub-species of nematode has been discovered in the same metal rich sediments.

Efforts have been made in Ireland to raise awareness of the ecological value of historic mining sites, most notably in the form of a thematic workshop on the subject in 1997 and subsequent publication by the Royal Irish Academy (Parkes 1999). Studies presented ranged from naturally established wet lands filtering mine waters at Glendalough (Fig. 18), to definition of mycorrhizal associations and lichen ecosystems, natural heathland re-generation (Fig. 19) and bat colonies in old mine workings, to artificial wet land experiments. More recently, Moles and Moles have documented the natural lead and heavy metal extraction capacity of bracken at Newtonards, as part of an analysis of the destruction of natural re-vegetation and consequent environmental damage and pollution being caused by off-road bikers on the site. Similar, though unquantified damage is evident at Glenmalure, Co. Wicklow.

PRESENTING THE LEGACY

Mine heritage as an educational, tourism and leisure resource is still a relatively novel idea in Ireland, but one which has, sadly, drawn forth sarcastic comment in the press. But in other countries, the value of this resource is better appreciated and valued - both as a heritage resource in its own right, as well as a sustainable resource to contribute to community based economic regeneration. The scale and style of such presentations is very variable, from the "theme park" style, but commercially very successful approach of centres such as Barkerville in British Columbia, Canada, Gold Reef City, Johannesburg, South Africa and Sovereign Hill, Ballarat in Western Australia, to more technologically focussed presentations, such as the Michell Whim Engine House and the St. Edward Mine in Cornwall, UK. In between there is a vast array of other,

Figure 17. Wheal Poldice spoil heaps, foreground, with natural heathland regeneration, now protected as a SSSI. Note the contrast with the mine crofter's enclosures in the background.

Figure 18. Upper Lake, Glendalough, showing spoil scree from 19th C lead mines on left, and below, a natural wetland between the mine workings and the lake.

Figure 19. "Avoca Oasis" - natural revegetation in otherwise sterile mine spoil, possibly induced by a localised accumulation of organic matter, aided by fertiliser sacks.

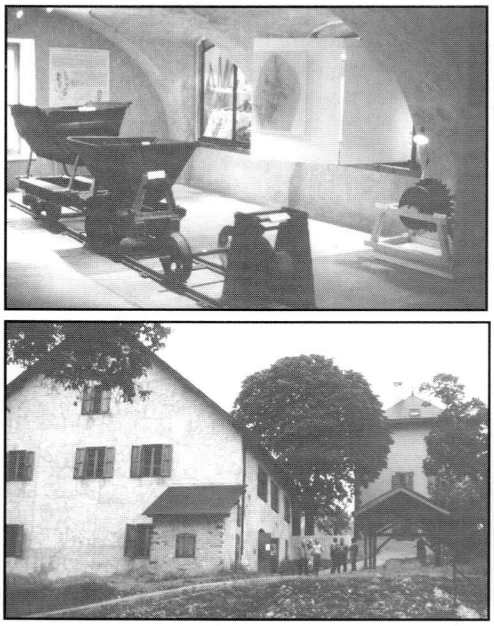

Figure 20. (left): The Lady Isabella water wheel, Laxey Mine, Isle of Man. Figure 21. (centre); internal and external views of the Mine Museum and Mining Archaeological Centre, L'Argentière-la-Bessée, France. Figure 22. (right): visitor entrance to underground mine tour, Miniera Paola, Italy.

more general style presentations, such as the Laxey Mine, with the Lady Isabella water wheel now virtually a national icon for the Isle of Man (Fig. 20); the award winning Killhope Mine, Co. Durham, and Poldark Mine, Cornwall, UK; the Llechwedd and Gloddfa Ganol slate quarries and mines, and "Big Pit" coal mining centre, Wales; and, further afield, the UNESCO World Heritage Site Rammelsberg Mine, Germany, the Clausthal-Zellerfeld museums and mines complex, Germany; Zloty Stok gold mine, Poland; Les Mines d'Argent des Gorges du Fournel, L'Argentière-la-Bessée, France (Fig. 21); Miniera Paola, Perosa Argentina, Italy (Fig. 22); and the Coronation Mine, Canada. Many other sites centres are being developed actively, such as Linares, Andalucia, Spain and salt and sulphur mines in Sicily.

In Ireland, there are now two such tourist mines, the Glengowla silver mine Oughterard, Co. Galway, the first in the country, which opened in 1998 (Fig. 23), and more recently, the Arigna Coal Mine, museum and miners trail, Co. Roscommon which opened in 2003. Other centres are being actively developed by various, largely community based groups: Castlecomer, based upon the Deerpark Colliery, Co. Kilkenny; Allihies Mine Museum, Co. Cork, as well as conservation of the Man Engine House (Fig. 12); the Copper Coast European Geopark, Bunmahon, Co. Waterford (Fig. 24); a proposed national Mine Heritage Centre at Silvermines, Co. Tipperary; and a Mining Park and conservation of all built mine heritage features at Avoca, Co. Wicklow.

At an international level, presentation initiatives range from a bid to have *Cornish Mining* listed as an UNESCO World Heritage Site (Figs 25 - 29), to revitalisation of a nascent European Mine Heritage Network (Europamines) to promote conservation and presentation of mine heritage throughout Europe, to EU ERDF funded projects which incorporate mine heritage conservation and presentation actions. *Cornish Mining* seeks to integrate Cornish mines and mining technology within its wider social, cultural,

Figure 23. Glengowla silver-lead mine office and workshop buildings, with reconstructed shaft top in foreground, Oughterard, Co. Galway.

Figure 24. The Tankardstown Engine House complex atop the cliffs, at the centre of the "Copper Coast" European Geopark, near Bunmahon, Co. Waterford.

Figures 25 - 29: Cornish Mining - a total mine, cultural and natural heritage concept, and candidate site for designation as an UNESCO World Heritage Site. 25 (top left), conserved Cornish Engine House ruins at Botallack; 26 (centre), close up view of part of the conserved Cornish Steam Engine at Parkandillick; 27 (top right), a terrace of conserved miners cottages, St. Day; 28 (bottom left), the Methodist preaching pit, Gwennap; 29 (bottom right), swards of thrift ("Sea Pinks") at Devoran Bridge.

religious and economic framework, to create a total mine, cultural and natural heritage concept (Figs 25 - 29). This all-embracing concept is very far ranging, including not only technical dimensions, such as Engine Houses, steam engines (Figs 25, 26) and related processing and smelting dimensions, but also the life and times of the people, for example their houses (Fig. 27), religion (Fig. 28), and schools, but also the natural environment framework, including new, developing ecosystems and habitats (Fig. 29).

In Ireland, the Copper Coast "European Geopark" (Fig. 24) is one of three participants in a c. Euro 4M INTERRREG 3B project which received conditional approval in late 2003. The "Copper Coast" strand includes provision of c. Euro 0.6M specifically to undertake a complete archaeological excavation of the Tankardstown Engine house site (Fig. 24), conservation of all built heritage features and provision of visitor facilities. Mine heritage, its conservation, presentation and use as an educational, leisure and visitor resource has finally come of age: and it is now rightfully starting to take its place as the most ancient component of industrial cultural heritage in Ireland.

REFERENCES

Cahill, M. 2000. Sliabh an Iarainn's iron works. *Technology Ireland*, November, 7.

Carville, G. 2001. The iron-works of Co. Laois. *Journal of the Mining Heritage Trust of Ireland*, 1, 55-56.

Clifford, J.A, Ryan, P. and Kucha, H. 1986. A review of the geological setting of the Tynagh orebody, Co. Galway. In: Andrew, C.J., Crowe, R.W.A., Finlay, S., Pennell, W.M. and Pyne, J.F. (eds), *Geology and Genesis of Mineral Deposits in Ireland*, Irish Association for Economic Geology, 419 - 439.

Cole, G.A. 1998. *Memoir of Minerals of Economic Importance and Metalliferous Mines In Ireland. Memoir, Geological Survey of Ireland*, 1922, reprinted 1998 by the Mining Heritage Society of Ireland, 155pp.

Cowman, D. 1978. The Royal Silver Mines of Bannow. *Decies* IX, 28-42

Cowman, D. 1982. Bronze Age Copper Mining At Danes Island. *Decies* XX, 22-27

Cowman, D. 1986/87. The German Miners at Bannow. *Wexford Historical Society Journal*, 11, 67-82.

Cowman, D. 1988. Sporadic Workings at Silvermines. *Tipperary Historical Journal*, 36-115.

Cowman, D. and Morris, J.H. 2003. An history of quarrying and mining in Ireland up to 1700. *Journal of the Mining Heritage Trust of Ireland*, 3, 25-32.

Harbison, P. 1992. *Guide to National and Historic Monuments of Ireland.* Gill and Macmillan Ltd, Dublin, 377pp.

Hammond, F. 1991. *Antrim Coasts and Glens, Industrial Heritage.* Department of the Environment, HMSO, Belfast, 103pp.

Hodge, J. 1995. *Richard Trevithick, 1771 - 1833.* Shire Publications, Princes Risborough, UK, 48pp.

Lyle, P. 2003. *Classic geology in Europe - The north of Ireland.* Terra Publishing, UK, 214pp.

McAuley, J. Undated. Creevalea Iron Works, 1600 - 1898. Pamphlet, North Leitrim Glens Development Company.

Mitchell, F. and Ryan, M. 2001. *Reading the Irish Landscape*. Townhouse, Dublin, 392pp.

Morris, J.H. 2002. The Man Engine House, Allihies, County Cork: a pictorial record of conservation works completed in November 2002. *Journal of the Mining Heritage Trust of Ireland*, 2, 35 - 39.

Morris, J.H and Brown K. 2001. The Man Engine House, Mountain Mine, Allihies, Co. Cork. *Journal of the Mining Heritage Trust of Ireland*, 1, 39 - 48.

Ó Cleirigh, M., Ó Cleirigh, P., O'Duigenan, P. and O'Mulcrony, F. 1632. *Annals of the Four Masters*.

O'Brien, W. 1994. Mount Gabriel: Bronze Age Mining in Ireland. *Bronze Age Studies*, Number 3, Galway University Press.

O'Brien, W. 1996a. *From the Fiery Furnace: copper sources and early Irish metallurgy*. Royal Dublin Society, Occasional Paper 12, 16pp.

O'Brien, W. 1996b. *Bronze Age copper mining in Britain and Ireland.* Shire Publications.

O'Brien, W. 2000. *Ross Island and the Mining Heritage of Killarney*. Department of Archaeology, NUI, Galway, 30pp.

Parkes, M. (ed.) 1999. The Ecology of Old Mine Sites. Proceedings of the Royal Irish Academy, *Biology and Environment*, 99B, No. 1, 77pp.

Stout, G. and Stout, M. 1997. Early landscapes: from pre-history to plantation. In: Aalen, F.H.A., Whelan, K. and Stout, M (eds), *Atlas of the Irish Rural Landscape*, Cork University Press, 31 - 63.

Tietzsch-Tyler, D. 1993. *The Augustinian Priory of Kells, Co. Kilkenny - an exploration*. Kells Region Economic and Tourism Enterprise Ltd, 40pp.

Williams, R.A. 1993. *The Berehaven Copper Mines*. British Mining No. 42, 1991, reprinted 1993 by A.B. O'Connor, Kenmare Book Shop, 228pp.

In: Parkes, M.A. (Ed.) 2004 *Natural and Cultural Landscapes - The Geological Foundation*, 139-142, Royal Irish Academy, Dublin

LANDSCAPE IN AN AGRICULTURAL CONTEXT

John A. Finn and Noel Culleton, Teagasc, Johnstown Castle Research Centre, Wexford, Ireland.
Email: jfinn@johnstown.teagasc.ie; nculleton@johnstown.teagasc.ie

Abstract: Given the extent of agricultural activity (both farming and forestry) within Ireland, the Irish rural environment consists of a variety of cultural, semi-natural landscapes. Economically motivated agricultural practices have been, and continue to be, strong driving forces of environmental impacts (negative and positive) within Irish landscapes. The adoption of modern intensive practices has seen a reduction in both the quality of water and the biological diversity of habitats and species. Traditional farming practices are needed to support habitats of highest conservation value, and maintain the character of traditional landscapes. Farm-level decision-making about agricultural practice is driven strongly by both policy and economics. Agri-environmental schemes that are based on sound policy and economic principles can be the driving force to entice and pay farmers to deliver a combination of environmental goods, farm income and agricultural produce that is a necessary part of most Irish landscapes. However, it is important that agri-environmental schemes are evaluated for cost-efficiency, and monitored to prove that there is added value to the environment.

INTRODUCTION

'Landscape' may be defined as the pattern of individual surface patches belonging to each of the land attributes such as rocks, topography, soil and vegetation. Landscapes typically comprise a blend of geology, landform, land use systems, social culture, archaeological sites, agricultural practices, biodiversity (habitats and species) and environmental attributes such as air, water and soil. In a more holistic sense, landscapes also include the visual and aesthetic aspects, as well as being ecosystems with dynamic biological interactions, and energy and nutrient flows. Here, we focus on the interplay between agricultural practices, environmental impacts and biodiversity within landscapes. In particular, we outline the dominant land use system that agriculture represents in Irish landscapes, its negative and positive contribution, and the importance of agricultural and environmental policy in promoting 'living landscapes'.

HOW HAS AGRICULTURE INFLUENCED IRISH LANDSCAPES?

Of the 6.8 million ha in Ireland, about 5 million ha is currently devoted to farming and forestry. Of the 4.4 million ha of farmland, about 50% is pasture, about 30% is grass conserved as silage and hay, about 11% is rough grazing and about 9% is under arable crops (1997 figures; from Emerson and Gillmor 1999). However, the current land use composition of the Irish landscape is the result of a quite dynamic history in terms of changes in the extent and nature of agricultural land use.

Historically, major land-use changes in Ireland include the extensive removal of the woodland that covered almost the entire mineral soil area of the country, so that by 1600, approximately 12% of Ireland was remaining wooded. By 1905, only 1% of the country was under trees, but a programme of State forestry during the twentieth century has increased forestry to about 9% of Ireland's land cover at the beginning of the twenty-first century. An intensive programme of afforestation proposes to increase forest area to about 17% of the country by 2030 (Anonymous 1996). Such a considerable increase in afforestation will result in a significant change in land use in many parts of the Irish landscape. While there may be social, economic and environmental benefits to this increase in forestry, care will need to be taken that existing sites of high conservation importance (from a biodiversity and/or landscape perspective) are not damaged, as has happened in the past. To this end, more stringent site evaluation and management policies are being implemented by the Forestry Service.

Most of the present lowland landscape of Ireland was laid out in more recent history (18th and 19th Century), when plantation landlords forced their tenants to enclose common land by establishing earth banks, stone walls and hedgerows. Enclosure of millions of hectares of land created the 'patchwork-quilt' appearance of the present rural landscape. Thus, the present rural landscape is a cultural, semi-natural landscape, and is largely the result of a dynamic interplay between forces of nature, and the imposed agricultural and economic forces, (Culleton, Tunney, and Coulter 1994).

The Green Revolution during the twentieth century has resulted in significant changes in agricultural practices and land-use. Following the major developments in agricultural science, government grant-aid resulted in extensive drainage of land during the 1950's as a precursor to intensified land use. Water was removed from land at a much greater rate, which allowed more grazing days in the year, enhanced productivity, permitted a wider variety of crops to be grown, and was a prerequisite for the effectiveness of fertilisers. This period of extensive drainage was accompanied by the widespread destruction of wetland habitats, the agricultural improvement of other marginal lands (wildlife habitats), increased soil fertility and the use of a wide range of fertilisers and pesticides. After World War II, soil fertility (especially lime and phosphorous status) was typically very low, and successive Governments encouraged the use of lime and phosphorus to improve productivity. This process has been highly effective. The average soil P level of samples analysed at Johnstown Castle was about 1 mg/l in 1950, and about 9 mg/l in 1993

(Coulter and Tunney 1996). Other changes that have occurred include increased mechanization, silage rather than hay conservation, and a widespread housing of livestock in the winter months.

ENVIRONMENTAL IMPACTS OF AGRICULTURE

The above changes dramatically increased agricultural output over the past several decades. However, the EU, government and the Irish public gradually realised during the 1980's and 1990's that the increases in agricultural productivity and profitability were coming at a heavy price to the environment. Here, we consider briefly the deterioration in water and biodiversity as two examples of the environmental impact of agriculture.

The Environmental Protection Agency calculated that up to 50% of the pollution in Irish water can be attributed to agricultural sources (Lucey et al. 1999). Water quality is now regarded as a national asset. There are three major sources of nutrient movement from the agricultural landscape to water. The first is the loss of phosphorus from agricultural land. It is argued that the phosphatic fertiliser applied to land can be carried by overland flow of water into watercourses, although the exact pathways of phosphorus (P) loss have not been fully quantified. Nonetheless, there is strong evidence that the higher the soil P levels, the greater the risk of P run off into water, if overland flow occurs. Recent reports suggest that it is possible to have soil P levels that allow optimum, or near optimum, agricultural output, and yet pose minimal threat to the environment; fertiliser P usage has dropped by 30% from 1995 to 2000 (Coulter et al. 2002). The key is to apply fertilisers to exact specifications and to spread according to the codes of good environmental practices. A second nutrient source is seepage of slurry into water bodies. It has been common practice to spread slurry during the winter months, often on fields with slopes leading to water bodies. It is inevitable that some of this slurry will be washed by heavy rain into the water, which poses a serious source of pollution. Shifting the applications of slurry from winter to summer would help resolve this issue. While there are some practical and physical problems with doing this, it is now happening on a widespread scale and it is hoped that incidences of pollution caused by slurry spreading can be minimized into the future. A third nutrient source is leakage of soiled water from farmyards. Water that seeps from farmyards nearly always contain nutrients that have enormous potential to pollute (Dunne et al. 2002), and every effort must be made to prevent these nutrients reaching watercourses.

As well as affecting the quality of potable water, agricultural practice may also affect the biodiversity component of landscapes. In the aquatic environment, the drainage of fields, canalisation of streams and rivers, and decline of water quality have all affected aquatic biodiversity. The quantity and quality of aquatic and semi-aquatic habitats have declined significantly e.g. wet grasslands, reed beds, streamside vegetation with consequent impacts on plant, invertebrate, fish and bird species. In terrestrial habitats, significant landscape features such as stone walls, hedgerows and field margins have been removed in order to facilitate bigger fields. In many parts of the country, the high species diversity of our traditional semi-natural grasslands has given way to more productive monocultures of perennial ryegrasses, a process facilitated by ploughing, reseeding, high soil fertility and herbicide use. A Department of Environment/Countryside Commission (1986) survey in England and Wales showed a 22% decrease in hedgerow length between 1947 and 1980, and a 25% decline in semi-natural habitats. Whilst we are unaware of corresponding data for Ireland, there is little doubt that considerable quantities of wildlife habitats have disappeared in Ireland since the 1950's. One environmental impact of agriculture has been the over-stocking of sheep on the peaty mountainous areas in the West of Ireland. The sheep have been consistently overgrazing, leading to the degradation of the indigenous vegetation. The combination of trampling and low soil cover has led to the erosion of the peaty soil, with consequent impacts on the visual landscape and nearby salmonid rivers.

While fully acknowledging the prominent role of agriculture in some of the recent environmental problems in Irish landscapes (above), it is important to point out the positive and, in many cases, necessary role of agriculture in maintaining the "pattern of individual surface patches" of Ireland. In particular, traditional agricultural practices over the past several hundred years have developed the diversity of plant species in various grasslands (and other habitats) around the country. In species-rich meadows and pastures, these traditional practices are characterised by regular cutting or grazing regimes, lower inputs, low soil fertility and lower productivity. Traditional agriculture is responsible for much of the biodiversity of the Irish landscape e.g. The Burren, Shannon Callows, semi-natural vegetation of hay meadows, pastures and machairs. By maintaining the processes that created such species-rich habitats, Irish farmers from the past to the present have largely determined, and will continue to determine, the biodiversity of the Irish landscape.

AGRI-ENVIRONMENTAL POLICY, PRACTICE AND PAYMENTS

In discussing the role of agriculture in influencing the landscape, it is somewhat simplistic to hold farmers solely responsible for some of the environmental problems (although there is no excuse for irresponsible or illegal practices). Ultimately, it is agricultural and environmental economic policy that dictates decision-making by farmers about agricultural practices that may have positive or negative environmental impacts- "The managed agricultural landscape is a product of hard, marketing economic forces, and a place where people make a living." (Hindmarch and Pienkowski 1997). Thus, top-down delivery and implementation of policy results in farm-level, bottom-up delivery of environmental benefits or problems. For example, the environmental impacts related to overstocking in upland areas has largely been a consequence of headage payments as part of EU subsidy and price incentives.

Looking to the future, we believe that the maintenance and delivery of desirable cultural landscapes (that are visually pleasing, biologically diverse, of high environmental quality and culturally valuable) will not be feasible without the participation of farmers (e.g. Wakeford 1999). The maintenance of such desirable landscapes requires certain forms of land management practiced by farmers- thus, Irish landscapes are very much 'living landscapes' that are reliant on their cultural (and agricultural) components. Policy objectives at national and EU level suggest that there will be considerable changes in the nature of agricultural land use in the future. Projected future farm numbers suggest a more polarised practice of agriculture that is composed of a relatively small number of intensive farms, and relatively large numbers of more extensive farms. Any declines in farm incomes and economic viability may threaten the essential cultural component of the landscape. Thus, an important question is: Can agricultural and environmental policy deliver environmental benefits, farm incomes and agricultural production?

These questions are at the heart of developing a sustainable agri-environment that delivers a desirable landscape. As production-related subsidies disappear, there will be an increasing emphasis on the role of agri-environmental payments. We believe that such payments will form a central role in answering the above question. If farming provides environmental goods that benefit society (e.g. quality water, quality air, pleasing landscape, high biodiversity, maintained soil condition), then farmers should receive payment. Whereas traditional subsidies paid for an agricultural market product, the new agri-environmental payments are paying farmers to manage land for non-market, environmental goods. Note that it is important that these payments are not perceived as 'new money for old rope' i.e. that farming can conduct business as usual and collect agri-environmental payments. Agri-environmental payments are intended for a service provided (the delivery of environmental goods), and are not simply intended as compensation for lost profit. In practice, the difference between the monetary value of the payment and the lost profit will dictate the incentive level for farmers to participate in agri-environmental schemes.

THE RURAL ENVIRONMENT PROTECTION SCHEME AND LANDSCAPE POLICY

The Rural Environment Protection Scheme (REPS) was initiated in 1994, and its objectives are:

- to establish farming practices and production methods which reflect the increasing concern for conservation, landscape protection and wider environmental problems.
- to protect wildlife habitats and endangered species of flora and fauna.
- to produce quality food in an extensive and environmentally friendly manner.

Farmers who comply with the conditions of the scheme, based on an individually tailored agri-environmental plan, receive a basic annual payment of □151 per hectare, up to a maximum of □6040 per annum. Farmers who undertake supplementary measures are entitled to additional payments. Supplementary Measure A 'Conservation of Natural Heritage' applies to Natural Heritage Areas (NHA's), Special Areas of Conservation (SAC's) and commonages.

By October 1998, there were about 37,500 participants across 1.27 million ha of land, with a total payment of □161 million. There is some opinion that the REPS is not acting sufficiently as a disincentive to harmful practices (e.g. Hickie 1997; Bohnsack and Carrucan 1999). On the other hand, the management recommendations within the REPS would be expected to have improved nutrient management and farms, to have afforded protection to many hedgerows and other wildlife habitats, and to have increased environmental awareness among the farming community. An objective evaluation of the environmental performance of the scheme is not possible at the moment, in part reflecting the lack of baseline data and the absence of an ongoing monitoring program.

No single model of agri-environmental scheme can deliver the wide variety of environmental goods that are desired. From a landscape perspective, the voluntary and spatially non-targeted structure of the REPS means that there is no protection specifically directed at the landscape scale; the REPS only operates at the level of individual farms. This is not to say that the REPS does not contribute to landscape protection- in fact, many of the measures would be expected to do so. More generally, Hickie (1997) indicated that "there is no national overview of landscape protection" in Ireland. For example, an Area of Outstanding Natural Beauty (AONB) in the UK is designated for its landscape and scenic beauty. AONB's are not necessarily areas of high nature conservation value, but often include areas that are. An AONB usually has special funding to help promote good management and sustainable development within it, and may have planning restrictions.

The following landscape policy guidelines were suggested by Jeffrey (1994) and seem as relevant now as when they were first presented:

1. Large landscape units, and an integrated approach to them, should prevail.
2. Landscape policy should be discussed on an interdepartmental and interagency basis
3. Rural people should be able to comment on central policy and initiate local action.
4. Tourism and amenity resources should be carefully protected in defined areas.
5. High quality and high added value should characterise all products of the rural landscape.

6. Monitor progress and social, environmental and economic change in rural areas.
7. Maintaining a viable rural population is a key challenge.
8. Share solutions with other regions in Europe.

ACKNOWLEDGEMENTS

This paper is dedicated to the memory of our late colleague Finnain MacNaeidhe We thank John Lee, Austin O'Sullivan and Rogier Schulte for comments.

REFERENCES

Anonymous 1996. *Growing for the Future. A strategic plan for the development of the forestry sector in Ireland.* Department of Agriculture, Food and Forestry, Ireland.

Bohnsack, U. and Carrucan, P. 1999. *An assessment of farming prescriptions under the Rural Environment Protection Scheme in the uplands of the Burren Karstic region, Co. Clare.* The Heritage Council.

Coulter, B. and Tunney, H. 1996. The background to the P debate. In: *Phosphorus recommendations for grassland: good agronomic practice.* Carton, O.T., Ryan, M. and Magette, W.L. (eds). Teagasc.

Coulter, B, Murphy, W.E., Culleton, N. Finnerty, E. and Connolly, L. 2002. *A survey of fertiliser use in 2000 for grassland and arable crops.* Teagasc.

Countryside Commission. 1986. *Monitoring landscape change.* Cheltenham.

Culleton N., Tunney H. and Coulter B. 1994. Sustainability in Irish Agriculture. *Irish Geography* 27: 36-47.

Dunne, E., Culleton, N., Harrington, R. and O'Donovan, G. 2002. Integrated Constructed Wetlands. In: *Principles of Successful Organic Farming.* Culleton, N., Fox, R. Schulte, R. and Finn, J.A. (eds.)

Emerson, H.J. and Gillmor, D.A. 1999. The Rural Environment Protection Scheme of the Republic of Ireland. *Land Use Policy* 16: 235-245.

Hickie, D. 1997. *Evaluations of environmental designations in Ireland.* The Heritage Council.

Hindmarch, C., and Pienkowski, M. 1997. *Land management: the hidden costs.* Blackwell Science.

Jeffrey, D.W. 1994. The future of the Irish countryside. In: *Science, Technology and Realism.* Moriarty, C., McCloskey, B. and Power, C. (eds.). pp. 49-55.

Lee, J., Culleton, N., Tunney, H. and Coulter, B. 1994. Sustainable agriculture with particular reference to soil fertility and nutrient usage. In: *Science, Technology and Realism.* Moriarty, C., McCloskey, B. and Power, C. (eds.). pp. 31-41.

Lucey, J. Bowman, J.J., Clabby, K.J., Cunningham, P., Lehane, M., MacCárthaigh, M.L. McCarrigle, M.L. and Toner, P.F. 1999. *Water quality in Ireland. 1995-1997.* Environmental Protection Agency, Wexford.

Wakeford, R. 1999. The English countryside: making character count. In: *Policies and priorities for Ireland's landscape. Conference proceedings.* The Heritage Council.

In: Parkes, M.A. (Ed.) 2004 *Natural and Cultural Landscapes - The Geological Foundation*, 143-146, Royal Irish Academy, Dublin

THE MAKING OF THE BOYNE LANDSCAPE

Geraldine Stout, Archaeological Survey of Ireland, Department of Environment, Heritage and Local Government, Ely Place, Dublin 2

Abstract: This paper looks at the role of rock in the formation of a lowland cultural landscape. The study area is the Bend of the Boyne in County Meath, known in ancient times as Brugh na Bóinne and since 1993 designated as a World Heritage Site. Continuous use of local stone as a building material in this valley has rooted its man-made features in the landscape and provided a harmonious link between prehistoric tomb, medieval castle and modern house. The monumental value of stone, its sheer durability has ensured exceptional survival of a vast legacy of prehistoric and historic monuments. Current geological research into the make-up of these monuments is providing, significant insights into these early communities. Features of local geomorphology have also been responsible for influencing historical events. Today, the challenge lies in protecting the Boyne valley when the commercial value of rock and its heritage value come into conflict.

The Bend of the Boyne lies in the lower Boyne valley in the Central Lowlands region, a landscape type, which is characterised by its undulating drift-covered relief. The bend was formed at the end of the Ice Age c.12000 years ago. The Boyne incised itself into glacial till, left by an ice sheet retreating to the north-west and wound a course, dictated by a gorge cut through a ridge by glacial meltwaters and a fault line in the underlying bedrock. [1] The lower Boyne valley opens onto an area of the coastline where the near-continuous upland perimeter of the island is absent; here the lowlands reach to the Irish sea and the Boyne valley provides one of the few easy eastern avenues of access to Ireland's interior. Moreover, this area also lies within the 'eastern triangle', a part of Ireland which receives less rain, and contains less bog and mountain than any other compact area of similar size in the country. [2] The 'eastern triangle' in general, and the Boyne Valley in particular, traditionally represents the geopolitical focus of Ireland. Accessibility and environmental resources have combined to make the area desirable to a succession of groups and conducive to striking indigenous cultural developments.

The Boyne landscape has developed on ancient rocks influenced by the folding and faulting of the Caledonian mountain-building phase and largely blanketed by drift deposits derived from the last Ice Age (fig.1). The area consists mainly of rocks of Carboniferous age and lies between two inliers of Lower Palaeozoic rock, the Louth massif to the north and the Balbriggan massif to the south. The oldest rock types in the area are the Lower Palaeozoic rocks (slates) which form an east-west ridge on the northern edge of the valley, dominating its northern prospect; they lie close to the north bank of the Mattock river and extend northwards as far as Nobber in north-west Meath. They make contact with Carboniferous limestones to the north of the Boyne on a fault line running through the townlands of Monknewtown and Littlegrange. They are strongly folded and consist of steeply dipping beds of green quartzites or greywackes, hard greenish-grey and buff mudstones, slates and black-brown slates. [3]

The lower Boyne valley is essentially a Carboniferous syncline with a dominant northeast/ southwest axis reflected in the local topography. [4] The core of this syncline is formed of Namurian shales and sandstones. The Carboniferous shales in the area lie stratigraphically above the limestone. The Platin limestone is the best exposed of all the Carboniferous limestone in the area. It lies in a belt stretching from the southwest near Duleek, through Platin cement works and Sheephouse to Oldbridge in the north. It is exposed as limestone outcrop with low crags and hollows visible north of Duleek and at Cruicerath. In general it is thick bedded, medium to coarse grained, mainly pale limestone. The top of the Platin limestone is found at Mullaghcrone House quarry where the limestone is exposed. There is a still younger limestone in the area which is dark, well bedded with some bands of chert. It outcrops in two places; to the north of Newgrange and Dowth, and also in a north-east to south-west belt from Oldbridge to as far south as Cruicerath. [5] The presence of limestone is mirrored in the building stone used locally and in the names of houses such as Whiterock House near Duleek. North of the Boyne River lies an elongated ridge of Carboniferous shale, overlying the earlier Carboniferous limestone. Exposure is mainly confined to stream beds and banks (such as that beside Rock House opposite the Bru na Boinne Visitor Centre). On the south side of the river, this shale forms a still higher rock ridge between Donore and

1	G. Mitchell, 'The landscape' in G. Eogan, *Excavations at Knowth I* (Dublin, 1984), pp 9-11; R.T.Meehan & W.P.Warren, *The Boyne Valley in the Ice Age* (Geological Survey of Ireland, Dublin 1999)
2	J. Andrews, 'A geographer's view of Irish history' in T. Moody and F. Martin (ed.), *The course of Irish history* (Cork, 1967), pp 17-29, see p. 21.
3	A. Bowden, 'The Carboniferous geology of P.L. 1039 and P.L. 1040 and adjoining application areas near Drogheda, county Louth' (unpublished GSI report AB1/71, 1971), pp 1-30.
4	A. McCabe, 'Directions of late-Pleistocene ice-flows in eastern counties Meath and Louth, Ireland', in *Irish Geography*, xx (1972), pp 443-61, see p. 443.
5	Bowden, 'Carboniferous geology', pp 11-2.

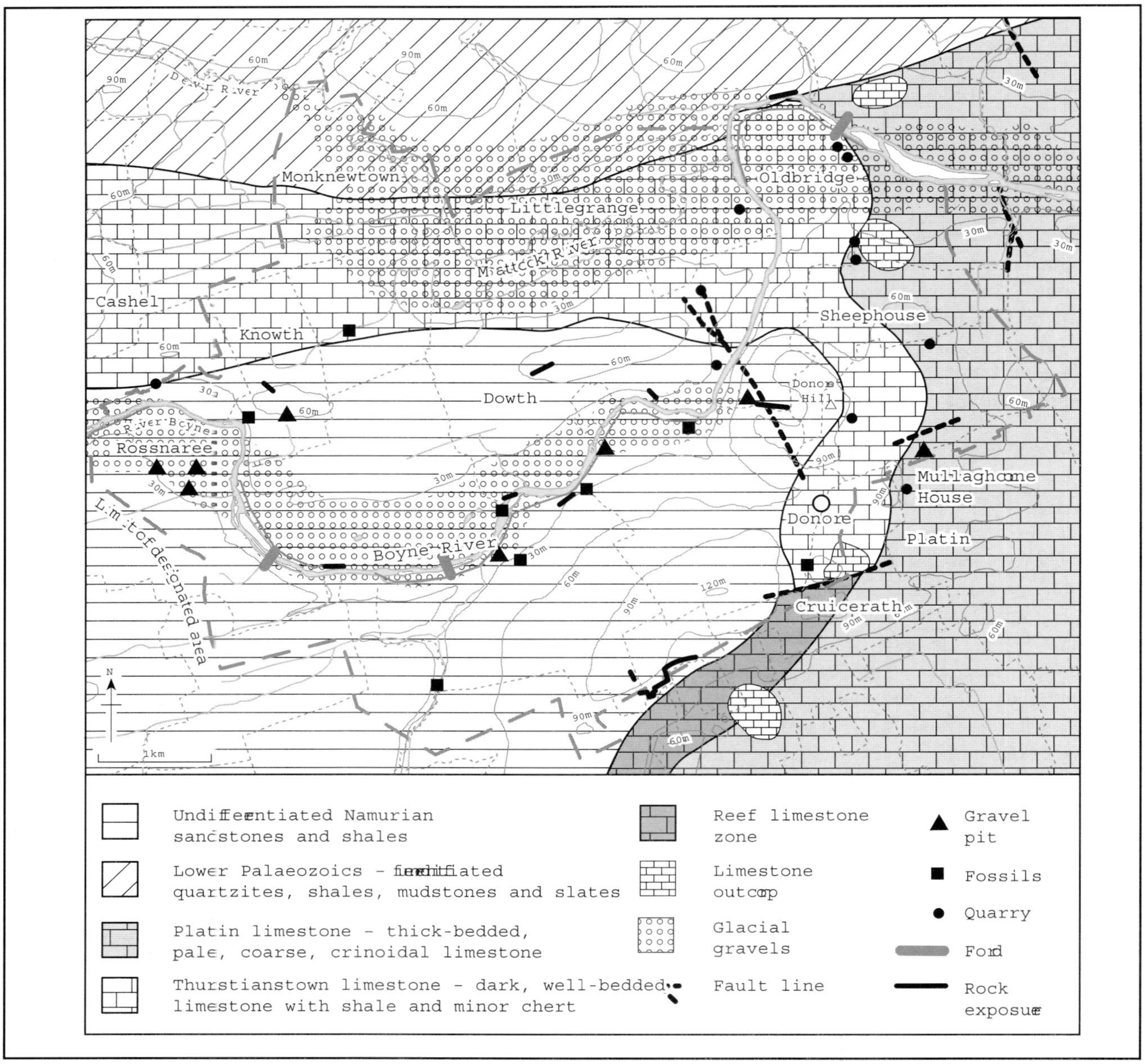

Figure 1. Geological map of the Bend of the Boyne (after Bowden 1971).

Redmountain. The Redmountain shale group consists of black shales and massive sandstones. The eastern extremity of these shales is being quarried at Donore by Irish Cement.

Passage tombs are the oldest surviving monuments in the Boyne Valley today, built some 5,000 years ago they testify to the monumental value of stone. Their round, stone-kerbed mounds dominate the ridge top and south-facing slopes, which run between the rivers Boyne and Mattock. Thirty-one definite and nine possible passage tomb sites can be identified in the Bend of the Boyne. They appear in clusters or cemeteries built on the most prominent knolls. Their densest concentration is around the main mound at Knowth at the western edge of the ridge. [6] They incorporate over 600 decorated stones, representing two-thirds of the megalithic art of Europe and more than four-fifths of the decorated passage tomb stones in Ireland. [7] These megalithic tombs have had a more profound impact on the landscape than any other ideological force in the history of this valley, creating a 'sacred landscape' which still attracts an international audience five thousand years after they were built.

The Boyne tombs embody the natural resources of their immediate environs. At Knowth local sandstones and limestones were used for the structural stones whilst some glacial erratics from drift cover were used in the kerbstones of the smaller tombs. Better quality stone was used for the orthostats of its chamber and geologists are indicating a stone source north along the coast at Clogherhead. At Newgrange, the majority of the stones, including most of

6 G. Cooney, 'The place of megalithic tomb cemeteries in Ireland' in *Antiquity*, lxiv (1990), pp 741-53.
7 G.Eogan, *Knowth and the passage tombs of Ireland* (London 1986) p.186

the kerbstones, were Lower Palaeozoic greywacke, a green coarse slate that outcrops to the north of Newgrange (see above). This type of rock weathers well and provides a good surface for decorating; it is the most used stone type of the tomb-builders. . The tombs were covered by mounds composed of the local boulder clay interspersed with sods removed from nearby fields and water rolled stones taken from the riverbed and local drift. Some of the stones used, especially in the Dowth cairn, were quarried and could be a by-product from quarrying for kerbstones. Stone settings within the make-up of the tomb are present in a large number of the excavated tombs.

Figure 2. The use of local stone forms a link between prehistoric tomb and stone cottage at Newgrange.

Given the quantity of stone used there must have been organised extraction of this material. The precise location of these prehistoric quarries has yet to be discovered. The tomb builders probably used natural outcrops where the stratification of the bedrock would have facilitated simple quarrying. The maximum depth of these deposits does not exceed 3m. so exploitation of many sites would have been necessary. Primitive quarries, if back-filled, would not be noticeable today. In France much research has taken place on the prehistoric quarries used by megalithic tomb builders. At Bougon in Deux-Sèvres quarries were discovered, using resistivity survey techniques, close to tumuli. The rock was extracted from conical pits, c.3m deep. They were almost completely filled in after use. Quarrying tools, including several picks of deer antler, were used to scrape away the split layers of rocky bands to expose blocks of solid rock. These were then loosened with axe hammers. Stones extracted in this way (without the use of fire) are fairly regular and parallel-sided. They can be used to form kerbs both inside and outside the mound. [8]

What is particularly enlightening is the information geologists are providing on the specific origin of the non-local material in these tombs and what it is telling us about the extent of communication and transport amongst communities in Neolithic Ireland. Five types of non-local 'collectable cobbles' were used to embellish the facades of the tombs at Newgrange and Knowth. [9] Dark, oval, naturally rounded cobbles of granodiorite were found around the entrance to Newgrange, Newgrange Site Z and Knowth. There are also well, rounded cobbles of granite at Newgrange which came from the eastern Mournes and microgranite cobbles from Slieve Gullion or Carlingford. The gabbro cobbles used are from Carlingford Mountain, the most likely source being a stretch of shore of Dundalk Bay, between Giles Quay and Rathcor. The white quartz, which occurs as angular fragments in large quantities near the main entrance to Newgrange and in smaller quantities at Newgrange Site Z, came mainly from the Wicklow mountains, and some may also have come from the Newry area. All this material would have been brought to the Boyne manually, probably carried by boat. Presence of these imports, highlight the extent of communications with regions north and south of the valley where physical evidence survives for tomb-building communities.

Features of local geology and geomorphology have influenced historical events in the Boyne valley. Prominent hilltops have always been a focus of assembly and ceremony in Ireland. Tradition records that the Hill of Slane was the starting point for St Patrick's Christian mission in Ireland and the Hill of Tara was for many centuries the political centre and Seat of the High kings of Ireland. From a military standpoint terrain can offer strategic advantage or disadvantage and it had a bearing on events in the famous Battle of the Boyne in 1690, which decided the destiny of the crowns of England, Ireland and Scotland. With an outflanking movement led by the Duke of Schomberg upstream at Rossnaree and a frontal attack at Oldbridge on 1 July 1690, King William of Orange defeated James II, and strengthened his hold on the English throne. The Jacobite army were helped by a natural combination of kame, moraine and a long shale ridge at Roughgrange and Donore Hill which screened observations from the north for the full extent of its length. King William's Glen provided concealment for the victorious Williamite armies. However, it was the physical barriers posed by the area's glacial history that impeded a major engagement between Williamite and Jacobite forces on the south side of the river. Meltwater flows had cut spectacular channels in the

8 J.-P. Mohen, *The world of megaliths* (New York, 1990), pp 158-68.

9 F. Mitchell, 'Notes on some non-local cobbles at the entrance to the passage-graves at Newgrange and Knowth, county Meath' in *Journal of the Royal Society of Antiquaries of Ireland*, cxxii (1992), pp 128-45. I. Meighan, D. Simpson,. and B. Hartwell, 'Newgrange: Sourcing of its granitic cobbles' in *Archaeology Ireland* , xvi (2002) pp 32-5.

10 G.Story, *An impartial history of the wars of Ireland* (London 1693), p. 75.

area at the end of the Ice Age and at Roughgrange an extensive stream system, which drains much of this escarpment. This area is shown as hilly and inhospitable on Story's contemporary map, which depicts the difficult terrain, which both armies experienced. [10] Generals' Sarsfield and Maxwell reported that a cavalry charge was impossible, as there were two double ditches with high banks and a little brook between them in the valley that separated the two forces

We drew up to enclose the enemy's whole army but a damned deep bog lay between us, we could not soon pass. [11]

The Williamite army had succeeded in diverting Jacobite forces from Oldbridge and at least at Roughgrange no body got hurt!

Local stone continued to be used as a building material through prehistoric and historic times. The post-Williamite period in the valley witnessed the creation of a system of estates leading to a rural landscape of demesnes, farms and fields and with this came, improvements in trade and communications. The Boyne, one of the earliest navigation schemes in Ireland, was canalised between 1748 and 1790 in order to encourage trade with the Dublin market and to transport corn to the port at Drogheda from inland markets. [12] The Boyne Navigation in particular left behind an impressive legacy in stone of locks and lock houses, gates, bridges, towpaths, ramparts, weirs and walls. There were twenty locks built on the navigation and eight of these locks were built between Drogheda and Slane in the Lower Boyne Navigation. The locks built along the canal were to be made of the largest 'rude stones' available and the 'walls of lime and stone [made] in the best manner'. [13] Major excavations were involved including cutting through stone cliffs such as that at the Fennor arch near Slane. The stone cutters and masons employed were mostly locals such as William Norris of Oldbridge, and Patrick Begg and John Elliott of Dowth and some of these have given there names to places such as Norris Hill near Slane. The carved label stones, which appear, on the locks are the work of Simon Austin and John Colwell.

Use of clay rather than stone is largely responsible for the disappearance of over half of the Ninteenth housing stock in the area. However the enactment of the Labourers (Ireland) Acts from 1893 to 1919 resulted in the replacement of some of the poor mud built houses in the Boyne Valley by stone cottages. This was part of a major public re-housing scheme in Ireland, which sought to provide suitable dwellings and half-acre allotments for landless labourers who were inadequately housed. [14] In the Boyne Valley and elsewhere, these cottages came to be commonly known as 'Parnell Cottages' because the parliamentarian Charles Stewart Parnell who lobbied for the improvement of the housing of the disadvantaged in rural Irish society. [15] These labourers' cottages form a distinctive house style in the Boyne Valley. In this area a local architect, P.J. Dodd of Drogheda, designed them. [16] For the most part they are detached dwellings, one and a half stories high with unplastered stone walls, built of limestone from the local quarry in Sheephouse. Features include dressed quoins and slated roofs with brick detailing on doors windows and chimney. These solid houses are remarkably well preserved a tribute to their architect and the materials used.

The current boom in the Irish economy and housing market has caused a huge demand for raw materials nationwide and many new quarries have been opened and old quarries re-activated in the Boyne valley. Increased commercial exploitation of shale and limestone resources in the Boyne valley poses a real threat to its natural beauty. Quarrying causes visual scarring; whole farms and fields are being cleared and replaced by large industrial scale buildings. Levels of exploitation continue to rise at an alarming rate in a countryside, designated as an *Area of High Natural Beauty, High Amenity and Special Archaeological Interest.* A series of planning decisions have highlighted the weakness of these Local Government landscape protection orders in our pro-development economy. In 1989, Meath County Council gave permission to Irish Cement for a shale quarry and settling ponds in Donore within the Boyne Valley Protected Area. In a further planning appeal, Bord Pleanála considered the proposed site marginal in terms of its visual quality and in terms of its amenity value, which was narrowly, identified as the passage tombs of Knowth, Dowth and Newgrange only. [17] The significance of this ruling for landscape conservation is that the Meath Development Plan did not set out to limit development in these areas identified as protected and that a precedent has been set to allow quarrying in the protected area.

11 Quoted in Beresford-Ellis, *Boyne Water*, p. 82
12 C. Ellison, *The waters of the Boyne and Blackwater* (Navan 1983),p.12 13 Ibid. p. 13.
14 F. Aalen, 'The re-housing of rural labourers in Ireland under the Labourers (Ireland) Act, 1883-1919' in *Journal of Historical Geography*, xii (1986), pp 287-306,
15 F. Aalen, 'Public housing in Ireland' in *Planning Perspectives*, ii (1987), pp 175-93, see p. 179.
16 J. Geraghty, 'P.J. Dodd of Drogheda: architect and civil engineer' in *Journal of the Old Drogheda Society*, ix (1994), pp 7-37, see pp 24-5.
17 Meath County Council, PL 17/5/74394.

In: Parkes, M.A. (Ed.) 2004 *Natural and Cultural Landscapes - The Geological Foundation*, 147-150, Royal Irish Academy, Dublin

COULD THE RISE AND DEVELOPMENT OF PAST CIVILIZATIONS DEPEND ON THE GEOLOGICAL ENVIRONMENT?

Fassoulas Charalampos, Natural History Museum, University of Crete, Knossou Av., Heraklion 71409, Greece. Email: fassoulas@nhmc.uoc.gr
Panagiotakis Nikolaos, D. Solomou 25, Heraklion, Crete, Greece

Abstract: The Pediada region in central Crete lies between the Lassithi Mountains and the Yiouktas ridge. The largest part of it, the Upper Pediada, is a plateau (altitude 300 - 400 metres) that extends south of the two Minoan palatial sites Knossos and Malia. The rocks of the region are mainly Neogene sediments resting over carbonate basement rocks. The sediments are intercalations of massive marly limestone, siltstone, conglomerates and marls. Intense neotectonic fragmentation and selective erosion have caused the formation of gorges that start from the fault planes leading to the hinterland, also cliffs and steep slopes related to fault planes and rock sliding, as well as high river terraces. Many springs appear at the contacts of the carbonate rocks and streams run along the impermeable formations. Protected by the mountain ranges from east and the "natural walls" of the tectonic and river terraces from west, by its altitude and controlled routes through river beds and gorges from the north sea, rich in fertile soils and water, the Pediada kindled several cultures from the Neolithic Period to the Present Day. During the Minoan Period four palatial sites developed within it or at its edge: Knossos, Malia, Archanes and Galatas; during the Classical and Hellenistic Periods seven important cities; during the Venetian Period eleven fortresses and at Present it is the hinterland of Heraklion, the largest town in Crete. We believe that the characteristic geomorphology and regional geology played a very crucial role in the development of the economy and the rise of the aforementioned civilizations.

INTRODUCTION

The Pediada region in central Crete extends south from the north sea and west of the Lassithi mountains up to the Yiouktas mountain and is the inland part of the Heraklio Prefecture. It is the region that gave life to and sustained the two major Bronze Age palatial sites of Knossos and Mallia lying at its northwest and northeast corners respectively, as well as the newly identified palatial site of Galatas at its southwestern part. It is not a flat plain as its name suggests, but, being an extension of the Lassithi mountains, it forms a plateau that rises immediately south of the strip of the north coast line. It consists of small plains and undulating countryside, watered by the Karteros River and its tributaries as well as some tributaries of Anapodaris River to the south. It is a very fertile region forming a vital source for the present economy of the island.

The Pediada has been occupied continuously since the Neolithic Period (c.3500 B.C.), but it was during the Bronze Age (especially between 1900-1450 B.C.) that it reached its highest peak of development as part of the Minoan Civilization. In this study we examine how the geology and geomorphology of the region may have affected the civilizations that developed within it.

HISTORY OF THE REGION

The Pediada was first settled in the Neolithic Period (c. 3500 B.C.) but it was at the end of the 20th and the beginning of the 19th centuries B.C. that an explosion in the number of settlements occurred (Panagiotakis 2004a,b). Most of the hillsides were occupied, but spots where water and building stone outcrops existed were very much preferred. In the next period (1700-1425 B.C.), which is the period when the Minoan Palaces of Knossos, Mallia and Galatas were enjoying great power and prosperity, the Pediada was densely populated; settlements were found everywhere, but more so in areas that were adjacent to fertile plains and valleys, water and building stone as well as along major routes that lead to the Palaces. It is thus obvious that the Pediada was fully exploited by the Palatial economies which needed its natural resources for their sustenance and prestige. Towards the end of the Bronze Age (and especially after the destruction of the Palaces) a gradual decrease in site density occurred but during the next periods (Geometric, Orientalizing) large settlements existed especially along riversides (mainly along Karteros). In the Classical and Hellenistic periods seven City-states existed (Knossos, Lyktos, Lykastos, Chersonissos, Arkades, Eltynaia, Choumeri Kefala) and were sustained by the Pediada's natural resources. During the Roman Period large aqueducts were constructed to supply large towns with water and many bath installations have been identified together with a large number of towns, all suggesting that the Pediada was enjoying peace and prosperity. The Byzantine Period is evident in a large number of churches and settlements. The full exploitation of the Pediada's resources during the Venetian period is evident in the large numbers of fortresses and towers, some of which are still partly standing, and the existence of hundreds of large wine pressing installations. During the Ottoman period the presence of hundreds of water mills (for grinding cereals) and the archives relating to produce show that the Pediada continued to be a major source of wealth. In the periods that followed (20th century- beginning of 21st) the Pediada still forms a major asset for the economy of the island (the main products being: olive oil, raisins, table grapes, wine).

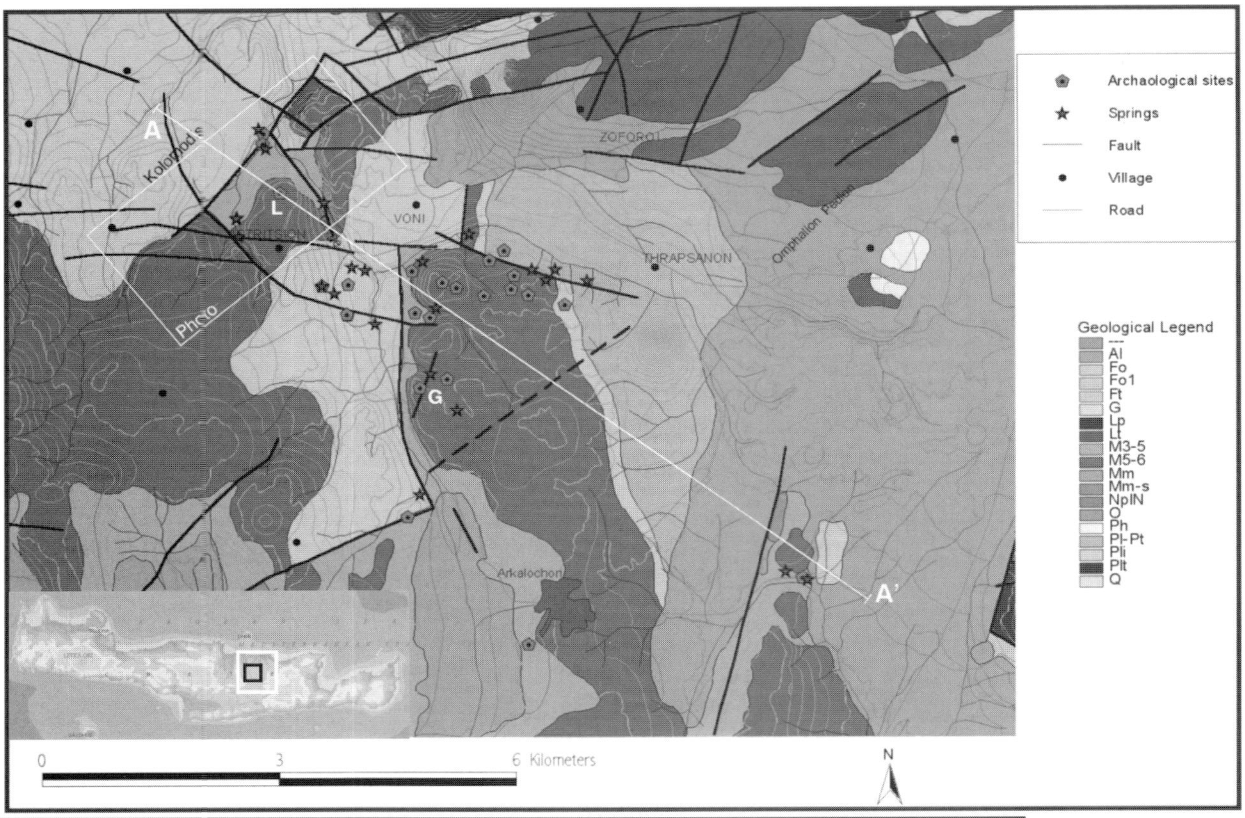

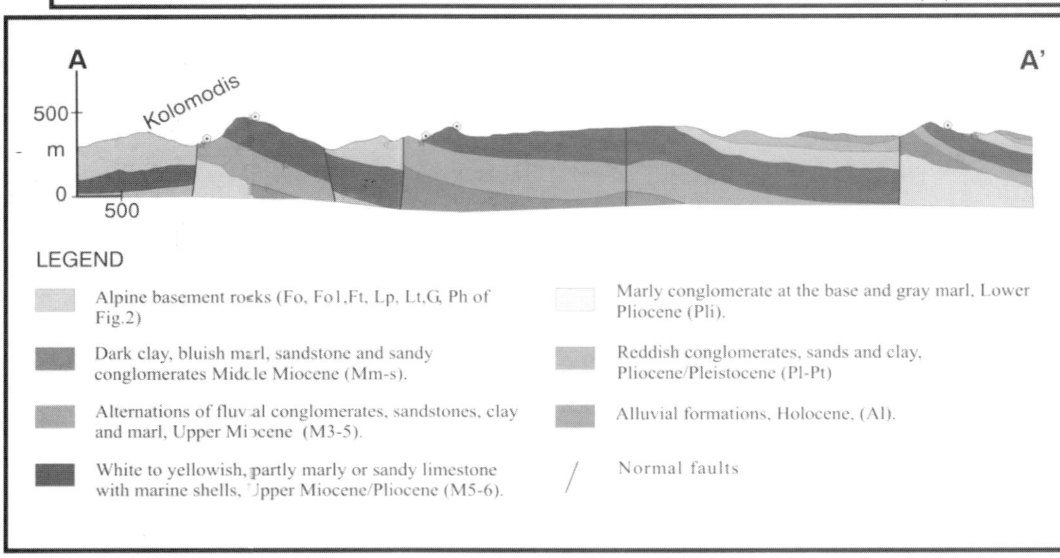

Figure 1. Above Topographic, geological, and archaeological map of the eastern Pediada area. White line, cross section of Fig. 2, white square the Pediada area, black square study area, L Lykastos, G Galatas. Figure 2. Left Geological cross section in the study area (Fig. 1). Vertical exaggeration X 2, symbols as in Fig. 1.

GEOLOGY - GEOMORPHOLOGY

It is well known that Crete has a complex geological structure as a result of the Alpine and present-day orogenic processes. Mesozoic marine sequences occurring in nappe structures constitute the main body of the island over which Tertiary and Quaternary sediments have been deposited (Creutzburg et al. 1977). Syn- and post-depositional tectonism fragmented the island forming grabens and horsts (Fassoulas 2001). In the Pediada region three nappes outcrop at the northern and eastern ridges (IGME 1994): the lower, Phyllite-quartzite nappe, and the overlying, Tripolitsa and the Pindos nappes (Fytrolakis 1980).

During Neogene times intense fragmentation and subsidence affected Crete as a result of the back-arc extension. Many faults developed fragmenting the island into smaller islands and marine basins (Dermitzakis and Papanikolaou 1981). Several sedimentary formations were deposited during Miocene - Pleistocene times over the basement rocks most of which are present today in the Pediada region (Figures 1, 2). Sediments are composed mainly of the lower, Middle Miocene sandstone, clay and marls, the Tortonian limestone and the upper, Pliocene marls and Quaternary conglomerates and sands. Alluvial sediments and soils cover the present day valleys. The Neogene and Quaternary tectonism formed mainly normal faults in several directions (Fassoulas 2001). The older faults trend E-W and are related with the formation and development of the Miocene basins. These were cut during late Miocene-Pliocene times by N-S trending faults that caused significant vertical offsets. Younger, NE-SW trending faults fragmented the area forming new basins and ridges.

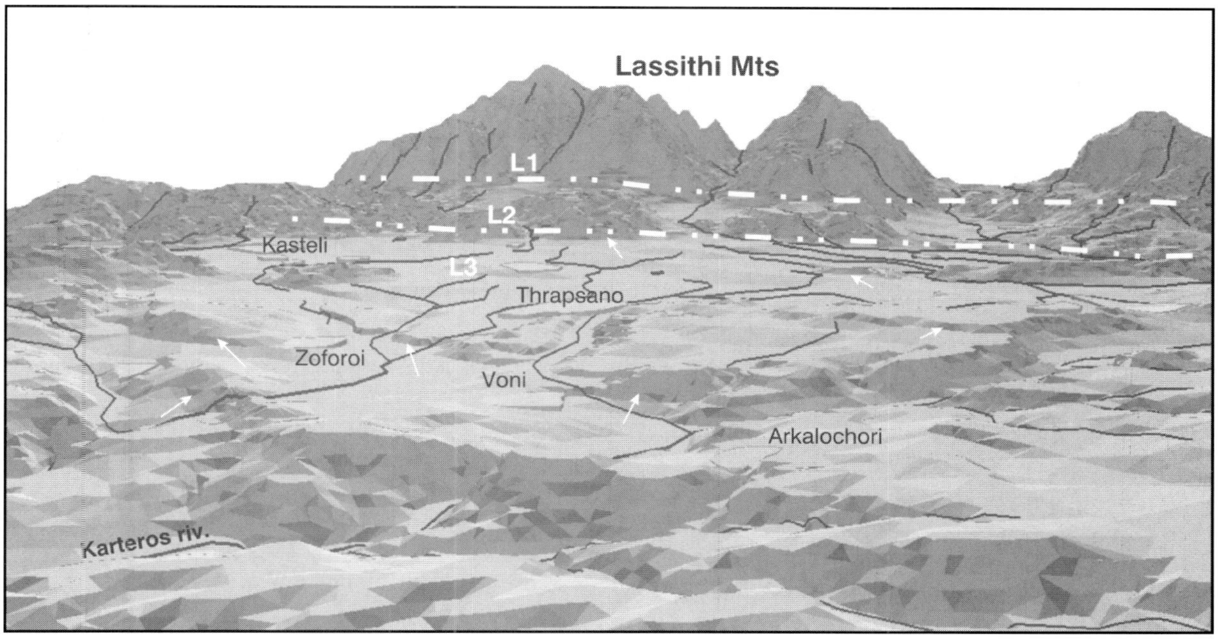

Figure 3. 3-d model of Pediada Region viewed from SW. Vertical exaggeration X 2; white arrowheads fault terraces white dotted lines level borders (L1, L2, L3).

Although precipitation is moderate many streams develop over the impermeable sediments draining mainly to the north. The most important is Karteros (having its own sources west of Arkalochori) and it's tributaries that run from south and south - east towards the northern coast, flowing into sea near the archaeological site of Amnissos. The river forms a long and deep valley cutting al the lithologies of the area. Another important stream, Anapodaris, is formed at the southern edge of Pediada running southwards to the Libyan sea. It's tributaries form deep valleys and steep river terraces. Furthermore, many springs also occur as a result of the alternation of permeable and impermeable sediments. Many of them (as Thrapsanon, Choumeri etc.) have huge capacities and today supply Heraklion town with drinking water. The role of faults to the underground water movement is crucial. Many older faults react as pathways to the underground water, whereas, other (as the Kasteli fault at the foothill of Lassithi mountains) form a natural diaphragm to the underground water of Lassithi mountains forcing it to move northwards to Mallia area.

The landscape of the broad Pediada region is characterized by a step-like mode that declines gradually from the high Lassithi mountains (2000m) to the Karteros river valley to the west (Figure 3). Normal faults trending in several directions are responsible for the creation of three main levels (steps): the high Lassithi mountains (L1), the intermediate zone (L2) and the western lowlands (L3) which consist of the Omphalion Pedion and the Karteros valleys; they are also responsible for the many abrupt, minor relief changes. Most of these lay in general NE-SW, parallel to the main fault directions A big active fault (Kastelli fault.) occurring at the foothill of Lassithi mountains caused a relief change of about 800 m and the formation of the L1 and L2 levels (Figure 3). Another N-S trending fault with a vertical offset of about 200 m borders the L2 and L3 levels. Minor other faults occurring at the lowlands of the Pediada area fragment the plateau forming the flat area of Omphalion Pedion (Figure 3).

Deep erosion has been accelerated due to vertical movements. In carbonate rocks gorges were developed normal to the fault alignment as a result of Karstic weathering and uplift of the footwall (Fassoulas 2000). Each of these gorges is a route to enter the inland of the footwall area. Selective erosion, and weathering contrasts between carbonate and pelitic sediments resulted in rockslides, gliding and landslides. This was more intense along fault zones where tectonic terraces were finally formed exposing steep cliffs and slopes (Figure 4). At the same time the Pediada region is naturally protected from the east by the Lassithi mountains, at the west and south by lower ridges and from the north by its naturally raised position above the sea. Due to these geomorphologic structures, access to the 'hinterland' of the Pediada region (the part raised above the north sea) was possible from the north only through the river valleys, especially Karteros, and through naturally protected paths, that often run through gorges and ravines.

A CASE STUDY

To show the relationship and interdependence of landscape and development of civilizations we present as an example a small area in the south-west part of the Pediada Region where as is evident in Figures 1 and 2 many normal faults occur (between the modern villages Astritsion and near Thrapsano, as well as between Voni and near Arkalochori, in North East-South West directions). The area is undulating and the hills consist of Tortonian limestone with general dipping to the east covered by the Pliocene and Pleistocene fertile sediments. Several small gorges were developed normal to fault zones as in the areas of Astritsi, Sambas, Galatas, and Zoforoi. The hydrology of the area is thus governed by the structure and appearance of the limestone in between the impermeable pelitic sediments over- and underlying it. Large numbers of springs exist either as

Figure 4. Tectonic terraces and ravine development west of Voni. Lykastos on top of central hill. Viewed from the north.

contact or drainage springs at the contact of the Tortonian limestone with the overlying or underlying sediments, or as tectonic over-flooding springs along faults that fragment the limestone. Many of the springs appear in the above-mentioned gorges where erosion has exhumed either the contact of the Tortonian limestone or the fault crosscut.

It is significant that a great number of archaeological sites of all periods (some multi-period sites) exist, among which the Minoan Palatial site of Galatas, and the Minoan as well as Classical (multi-period site) and later important city of Lykastos; they are found on the above mentioned tectonic terraces, thus taking advantage of the natural protection provided by the terraces (some of which rise very abruptly, as in the case of both Galatas and Lykastos), of the building stone available and of the water of the springs that are found along these tectonic terraces. The archaeological sites are always found within walking distance from the aforementioned springs, the Palatial site of Galatas, for instance, has a spring (now dry) at its northwest corner and another (still rich in water) is 600 m. south east and shared with Mandrakia, another contemporary site that existed immediately east. Lykastos too had two main springs very near it and many more within walking distance.

It is thus evident that in the small area of several km^2 there are 20 archaeological sites of great importance (and a few hundreds of lesser importance) built by or near 18 of the richest springs. The archaeological sites in this area always overlook the fertile Omalion Pedion and the valleys along the tributaries of Karteros. The presence of a great number of important archaeological sites and the Palace of Galatas point to the importance of the region as a major agricultural and pastoral centre necessary for the sustenance not only of the palatial society of Galatas but mainly of that of Knossos. Both the Galatas palace and Lykastos are situated along the main route leading from the Pediada to Knossos (Lykastos also guards a gorge through which the route runs), perhaps protecting this route and easing the mobilization of the Pediada produce to Knossos and the harbours along the north sea, especially that of Amnissos (with which there was direct connection).

DISCUSSION - CONCLUSIONS

The Pediada Region taken as a case study has shown that the geomorphology and geology of a region affect enormously the development of a civilization. The geomorphology of the area (due to the vertical motions and selective erosion) makes it a plateau accessible only through easily controlled passes, providing thus a natural protection to its inhabitants. The geological features of Pediada, the marly and sandy sediments, as well as the wealth in water (all due to the special structure of the Tortonian limestone that is fragmented in many parts and tilted to the east) make it a fertile region, suitable for every kind of crop.

All these factors (landscape, fertile soils, plenty of spring water) together with the protection provided by its landscape against invaders and at the same time its connection with the north sea - a major route to other cultures within and around the Aegean - must have been responsible for its diachronic importance. They all contributed to the development of civilizations in the Pediada Region for the past 5000y. It cannot be thus coincidental that the Pediada cradled and kindled a series of civilizations from the 3rd millennium B.C. onwards.

ACKNOWLEDGMENTS

The authors express their sincere thanks to M. Kritsotakis and M. Panagiotaki for their critical comments and suggestions and to M. Nikolakakis for provision of the maps, as well as to the INSTAP for funding this study.

REFERENCES

Creutzburg et al. 1977. *Geological Map of Crete* (scale 1:200.000). IGME. Athens.

Dermitzakis M & Papanikolou D. 1981. Paleogeography and geodynamics of the Aegean region during the Neogene. *Ann. Geol. Pays. Hellen, Hors*, 3, 246-290.

Fassoulas C. 2001. The tectonic development of a Neogene basin at the leading adge of the European margin: The Heraklion basin, Crete, Greece. *J. Geodynamics*, 31, 49-70,.

Fassoulas C. 2000. *Field guide to the geology of Crete*. Natural History Museum, Heraklion Crete

Fytrolakis N. 1980. *Contribution to the geological structure of Crete*. Habil. Thesis, Nat. Techn. Uni., Athens.

IGME 1994. *Geological map of Greece* (scale 1:50.000). Epano Archanai, Mochos shts. Athens.

Panagiotakis N. 2004a. *Le développement archéologique de la Pediada Région de la Crète Centrale*: Un Résume. B.C.H. (in press).

Panagiotakis N. 2004b. Contacts between Knossos and the Pediada Region in Central Crete. In: G. Gadogan (ed), *Palace, City, State*. B.S.A. supl. Vol. (in press).

In: Parkes, M.A. (Ed.) 2004 *Natural and Cultural Landscapes - The Geological Foundation*, 151-152, Royal Irish Academy, Dublin

GRAVES AND GRAVE STONES: GEOLOGICAL INFLUENCE ON CULTURE

Matthew Parkes, Geological Survey of Ireland, Beggars Bush, Haddington Road, Dublin 4, Ireland. Email: matthew.parkes@gsi.ie

Abstract: The choice of stone or other material to mark the burial place of an individual often varies according to cultural tradition, but these cultural choices are themselves heavily influenced by geological factors, sometimes only apparent after consideration of local geology. The geodiversity of easily available rock sources has lead to vernacular variations in gravestones.

INTRODUCTION

Disposal of dead bodies varies considerably according to religious and cultural practices in different countries. Approaches such as cremation (in a variety of ways) are common, but others such as mummification or cannibalism are recorded. However, burial has been widely practised in a majority of cultures, and is probably the most common procedure for dealing with human bodies. Since prehistoric times, people have chosen to mark the last resting place of physical remains with stone monuments. Indeed, archaeology as a discipline has had to often focus on these burial monuments as the most enduring and tangible link to the lives of our predecessors. Until modern times, with the ability to move heavy and bulky rock materials around the world, people normally used whatever the most durable local sources of rock that was available. This has led to some interesting and rarely noted variation from one area to another in terms of the gravestones, which is essentially a very small aspect of vernacular architecture. Three examples are used to illustrate this point.

COPENHAGEN

The city of Copenhagen in Denmark lies in an area of 'soft rocks'. The geology of the region is lacking in hard durable stone types that make good gravestones. However, the extensive glacial till deposits over the landscape include large rounded boulders of exotic metamorphic and igneous rock types brought from further north in Sweden and Norway. This limited availability of durable stone has led to the traditional use of these stones as headstones on graves in the city cemeteries (Figures 1, 2).

Figure 1. Traditional boulder style gravestones with minimal dressing.

Figure 2. Traditional boulder style gravestones in Copenhagen cemetery.

Figure 3. Display area of available stones in the Stone Mason's yard. Almost all of these had been cut and polished to give one smooth face.

Figure 4. Another display area, with mostly rounded 'boulder' style stones, even though many were imported and rounded off.

At one particular graveyard in a city centre location, although the stonemason now imports a great variety of stone types from the world over, they are worked to create the traditional local style for the majority of customers. They are imported in rectangular blocks, but they are dressed by burning to produce the rounded boulder shapes that are the preferred headstone type. A walk around the graveyard is sufficient to show that it is not only older gravestones that are of the exotic boulder type, but many recent graves are also this type. In fact, the majority of the stones on offer in the stonemason's are already rounded and polished to replicate the traditional sources.

ICELAND

In Iceland, the entire landscape is dominated by volcanic lavas, but there is a traditional pattern of using sections of columnar basalt as gravestones. In some sections of a main cemetery in Reyjavik, entire rows of older graves are marked by basaltic columns, with small carved inscriptions upon them. They vary considerably in height, but this author speculates that the height of column used may be related to a persons wealth! Although basalt would be the principal rock type of Iceland, columnal outcrops are perhaps sufficiently uncommon to have caused some value to be placed on substantial pieces, in the past. However, there appeared to be a break from the past in that few new or recent graves were noted with columnar basalt gravestones, but a variety of more generic crosses, metal markers and different stone types (presumably imported).

Figure 5. This shows usage of larger columnar basalts in a spectacular mountainside memorial.

Figure 6. Columnar basalt gravestones mixed with metal crosses and other slab styles in an old Icelandic graveyard.

HONG KONG

It is clear in some areas that lack of suitable locations leads to alternative strategies. In many mountain villages in Sardinia, for example, the rocky substrate does not allow easy burial, and bodies may be buried in concrete cells above ground. In Hong Kong, the steep and often unstable slopes make space at a premium, and a person may be buried in underground tombs, but then after a year or so their bones are collected and then kept in the family home, so that the tomb can be reused again and again.

Figure 7. A Hong Kong tomb.

CONCLUSION

All of these burial rituals and use of gravestones are influenced by the underlying geology and an analysis of many cultures, now and in the past, would possibly show comparable geo-influence on cultural behaviour.

ACKNOWLEDGEMENTS

Helgi Thorfassen is thanked for providing the picture of the Icelandic monument. Ann Laursen is thanked for taking the Copenhagen cemetery pictures after my own failed to work properly.

SESSION 5

PROTECTING AND MANAGING OUR LANDSCAPES: INTERNATIONAL AND NATIONAL SCALE

Giant's Causeway, Co. Antrim, Northern Ireland - a World Heritage Site for the geology. Photograph: Sophie Preteseille.

In: Parkes, M.A. (Ed.) 2004 *Natural and Cultural Landscapes - The Geological Foundation*, 155-157, Royal Irish Academy, Dublin

PROTECTING AND MANAGING OUR LANDSCAPES: THE POLICY CONTEXT.

Michael Starrett, Chief Executive, The Heritage Council, Kilkenny

KEYNOTE ADDRESS

Abstract: The Irish landscape has been the focus of attention for generations and holds a significant place in the culture of Ireland. Its development and future now face unprecedented change as Ireland's rapid economic growth places stresses and strains on the systems which are there to protect its landscape. Protecting the landscape in this time of change demands partnership and co-ordination between all those agencies and activities which have a potential impact on it. Furthermore people must be involved in the decisions which affect their landscape. This is the approach which the Irish Heritage Council has promoted since its establishment in 1995. This short paper outlines the approach adopted and the policy which has emerged as a result.

The Irish landscape has been the focus of attention for generations. Whether through publications, such as Irish Geographical Studies (Stephens and Glasscock, 1970) or the Atlas of the Irish Rural Landscape (Aalen et al, 1997), or through the myth and legend of storytellers, our landscape holds a significant place in the culture of Ireland. The sense of place in Ireland is part of that culture, epitomised in the retention of townland and place names, many of which, in the native Irish language, reflect the relationships between people and their environment. The influence of humankind on Ireland's wild and savage natural beauty is pervasive. Ireland enjoys no wilderness, and even its highest peaks and its deepest bogs can yield marvellous examples of the influence which people have had upon the shaping of that landscape. Its development and future now face unprecedented change as Ireland's rapid economic growth places stresses and strains on the systems which are there to protect its landscape. Protecting the landscape in this time of change demands partnership and co-ordination between all those agencies and activities which have a potential impact on it. Furthermore people must be involved in the decisions which affect their landscape. This is the approach which the Irish Heritage Council has promoted since its establishment in 1995. This short paper outlines the approach adopted and the policy which has emerged as a result.

THE EMERGENCE OF HERITAGE AND LANDSCAPE POLICY IN IRELAND

Ireland's National Development Plan; 2000-2006 (Government Publications 2002) was framed after an extensive consultation process to reflect the broad consensus on future national development needs. The plan lays the foundation for Ireland's continued economic and social development, and sets out a coherent development strategy supported by a quantified commitment to invest circa □48 billion over the plan period. Funding will be mainly directed towards infrastructure development, education and training, the productive sector and the promotion of social inclusion.

While the Council was not involved in the preparation of the National Development Plan, it commented on it. Council recognised the need for substantial infrastructure improvements throughout Ireland, but argued that much of the information which decision makers required in order to plan for these was not available (The Heritage Council, 2000a). This could lead to lengthy delays in the planning process, and thus the targets set in the plan might not be realised. The Council believed that an integrated and effective landscape policy could help achieve the objectives of the National Development Plan by providing a context within which to plan infrastructure schemes. Moreover, the incorporation of heritage concerns, especially landscape, into development planning would encourage awareness of the value of the national heritage and so help improve the quality of life for all. The Council therefore advocated its emerging landscape approach and argued that planning policies should be made heritage proof through heritage appraisal of development plans.

Heritage appraisal is a methodology developed by the Heritage Council to raise awareness of the significance of the national heritage within the development planning process. It is targeted at assessing the policies contained within local authority development plans, and at the evaluation of specific developments. The methodology, which was developed in partnership with local authorities, was published in July 2000. Its application is seen as complementary to the Council policy on landscape (The Heritage Council 2002).

The Heritage Council takes a holistic and integrated approach to landscape. It believes that landscape policy should embrace all landscape elements, and recognise their significance. This calls for a multidisciplinary approach to the way in which Ireland manages and develops landscapes. Recent work in the forestry sector undertaken through the Forestry Inventory Planning System, initiated by the Department of Communications, Marine and Natural Resources, is one positive illustration of such an approach by a specific sectoral interest. Similarly the recent consultative draft guidelines on landscape characterisation, published by the Department of the Environment & Local Government, aimed specifically at local authorities, will assist in the ongoing debate and the raising of awareness on landscape issues.

THE VISION FOR LANDSCAPE

In order to inform the development of its landscape policy, and consolidate international experience, the Heritage Council organised an international conference on landscape at Tullamore in 1999. This meeting heard a challenge from Prof. Michael Ryan: 'I wonder is it possible to write a simple short statement, a vision of what we want for the Irish landscape? If we can write that and accommodate within it all the social, economic, developmental, protective and other needs everything else will flow from that vision' (The Heritage Council, 1999).

Agreement on such a vision could be starting point. It should then be possible to develop a framework for actions that would help achieve the vision, including the negotiations which are required with different interests. Thus the vision becomes the goal to which everyone's efforts are directed. Lessons drawn from best practice from abroad, the planning of new initiatives and the adjustment of ongoing practical work in a variety of sectors can all be harnessed to the achievement of that vision.

The challenge to formulate a vision for Ireland's landscape was accepted by the Heritage Council, which referred this task to a working group on policy development for Ireland's landscape. On being asked for their vision for the Irish landscape, the members of the group identified the relationship between society and environment as a central theme. The link between cultural and natural elements of the landscape heritage was self-evident. The underlying implication was that people recognised that they had an influence over the 'shape' of the landscape and wished to maintain their power to exercise that influence in future. Landscape was seen as dynamic and changing over time. The group believed that people wished to retain a dynamic landscape.

Any work on landscape in Ireland has to recognise that there is at present a low level of public awareness. For example, the Council policy on heritage awareness (the Heritage Council 2000b) shows that less than 3 per cent of the population considered landscape as part of their national heritage. It was nonetheless felt necessary to validate the group's views about landscape through further market research (Landsdowne Market Research 2000). As a result, the group felt that its initial views had been confirmed.

It put forward this vision for the Irish landscape: *The Irish landscape will be a dynamic landscape, one that accommodates the physical and spiritual needs of society with the needs of nature in a harmonious manner, and as a result brings benefits to both in kind.*

This vision forms the core from which Council's policy is developed

THE PILOT PROJECT - COUNTY CLARE

The pilot project (ERM 2000) was initially proposed at the Tullamore conference. It was designed to test and resolve a number of the issues surrounding the development of policy for Ireland's landscapes, including the availability, accessibility and quality of baseline information. The pilot was seen as a way to inform the Heritage Council on the development of an approach towards integration of landscape policy across sectoral divides.

In developing the pilot exercise on landscape character, the Council drew on the work on landscape characterisation undertaken since the early 1980s elsewhere in Europe. This work has been fraught with problems in its initial conceptual stages and many lessons have been learnt by the agencies involved in promoting what is now clearly emerging as a most useful tool. The Council looked particularly to experience in Scotland and the emphasis on involving stakeholders in the process.

The emergence of the European Landscape Convention (Council of Europe 2000) and its aim to promote landscape protection, management and planning and to organise European co-operation on landscape issues, supported and encouraged the approach adopted. As a specific measure the convention sees every citizen as having a share in landscape and requires campaigns to raise awareness as a component part of its implementation. The Irish Government ratified the convention in March 2002 and is to date one of three countries to do so. Another eighteen countries have signed the Convention and their ratification is awaited.

Allied to the specific aims of the Convention, characterisation, for the Heritage Council, provided the opportunity to fully involve all stakeholders in the process of policy development. The approach recognises that all agencies and individuals can identify with the physical attributes of landscape. What is more difficult is to impart the value and significance which landscape has to our overall quality of life. Because characterisation work requires access to a wide range of information and the co-operation of the various bodies that control this, the process has the potential to bring together many different agencies and organisations and to place their work in a landscape context. This allows a direct connection to be made between their work and the maintenance and enhancement of over all quality of life for all.

THE NEXT STEPS

There are several elements of follow-up to the Clare pilot. For example, Council has currently extended work within the pilot area. This work aims to produce a full County-based Landscape Character Assessment (CLCA) in association with partner agencies.

The Heritage Council has also started to work closely with

local authorities elsewhere in Ireland on the application of characterisation to help assess major development proposals which would affect designed landscapes.

In the light of Council's policy and the ratification of the European Landscape Convention by the Irish Government, all follow-up work includes a strong element of stakeholder participation. This not only reflects best international practice but also allows Ireland to demonstrate how it is meeting its requirements arising from the Convention.

THE HERITAGE COUNCIL POLICY

As the major output of its work on landscape in, the Heritage Council (July 2002) published proposals for integrated policies for Ireland's landscape. This policy articulates the rationale behind the Council's thinking on landscape and contains recommendations which support the production of a standardised and consistent national landscape policy, which in turn will support the development of County landscape characterisation. In preparing its recommendations, the Council wished to allay fears of a top down approach, and to avoid encumbering local authorities and others with burdensome work; at the same time it was determined to raise the profile of landscape issues in shaping development policy in Ireland.

The landscape policy now proposed by the Heritage Council, and the recommendations which flow from that policy, are designed to allow those with responsibility for the management and development of landscape (and whose use of land impacts on it) to assess the policies they are implementing against specific indicators. They also propose a fully integrated approach to the landscape to be adopted by the Government. If acted upon, these recommendations will Council asserts help to ensure that the quality of life becomes an integral part of the implementation of the National Development Plan. The Heritage Council now awaits a full response from government on the detail of the proposals contained in its policy document.

CONCLUSION

In the absence of the consistent and standardised approach to landscape as proposed by the Council, there is a danger that agencies will act in an ad hoc and ill-informed way. In the current climate of rapid economic development, this would represent a real threat to the well being of the Irish landscape. The work of the Heritage Council therefore seeks to place the national heritage (including landscape) much higher on all agendas. The Council has sought to raise awareness of the significance of that heritage to the quality of life of all those who live in, work in and visit the Irish landscape. It believes that, through the integrated approach to landscape characterisation developed for Clare and now advocated for the whole of Ireland, all those with an interest in the future of Ireland's landscape heritage can have a say in how it will evolve.

Indeed the greatest danger to the future well being of the Irish landscape is that the focus and vision are too narrow. The landscape is where people live and represents many of the aspects of Ireland that people take pride in sharing with others. While it allows the Irish people to enjoy a quality of life which is the envy of many, its future cannot be taken for granted. The well-being of the landscape is entrusted to the Irish people and it is they who must value it if its qualities are to survive. To rely on one group or another to look after it will inevitably fail. The responsibility is that of society as a whole.

REFERENCES

Aalen et al (1997) *Atlas of the Irish Rural Landscape.* Cork University Press

Council of Europe (2000) - *The European Landscape Convention*

David Tyldesley & Associates (in association with Donegal County Council) (2000): *Heritage Appraisal of Development Plans: A Methodology for Planning Authorities*

Environmental Resource Management (ERM) and ERA-Maptec Ltd (2000) *Pilot Study on Landscape Characterisation in County Clare.* Unpublished report to the Heritage Council, Kilkenny

The Heritage Council (1999) *Policies and Priorities for Ireland's Landscape: Conference Papers, Tullamore, Co. Offaly. Ireland, April 1999.* The Heritage Council, Kilkenny

The Heritage Council (2000a) *Submission on the National Development Plan 2000-2006.* The Heritage Council, Kilkenny

The Heritage Council (2000b) - *Policy Paper on Heritage Awareness in Ireland.* The Heritage Council, Kilkenny

The Heritage Council (2002) - *Policy Paper on Ireland's Landscapes and the National Heritage.* The Heritage Council, Kilkenny

Landsdowne Market Research (2000) *The Irish Landscape.* Unpublished report to the Heritage Council

National Development Plan 2000 - 2006 Government Publications Dublin (2000)

Stephens and Glasscock - in Honour of E. Estyn Evans, Irish Geographical Studies, Department of Geography Queen's University of Belfast 1970.

Footnote: The European Landscape Convention which is now fully ratified by 11 Countries, will be fully in force in March 2004.

In: Parkes, M.A. (Ed.) 2004 *Natural and Cultural Landscapes - The Geological Foundation*, 159-160, Royal Irish Academy, Dublin

THE ROLE OF THE WORLD HERITAGE CONVENTION IN LANDSCAPE PROTECTION

Tony Weighell, Joint Nature Conservation Committee, UK

Abstract: The World Heritage Convention is unique in recognising cultural and natural heritage, including biological and geological diversity. In the narrowest sense the Convention offers the opportunity to recognise some of the best geological and geomorphological sites in the world. Within the scope of the Convention itself there are broader opportunities to recognise those Earth heritage values that underpin biodiversity or influence human settlement and land use. The Convention, through the Cultural Landscapes concept also offers a potentially powerful model for landscape scale conservation initiatives that go beyond the protected area approach, and integrate cultural, biological and geological heritage values across entire landscapes.

THE WORLD HERITAGE CONVENTION

The 'Convention Concerning the Protection of the World Cultural and Natural Heritage' was signed in 1972. There are currently 174 State Party signatories and 730 properties have been inscribed on the World Heritage List. The Convention is unique in recognising cultural and natural heritage, including biological and geological diversity. Of the sites recognised, 563 are cultural, 144 natural and 23 are mixed properties.

According to the World Heritage Convention, cultural heritage is a 'monument, group of buildings or site of historical, aesthetic, archaeological, scientific, ethnological or anthropological value'. Natural heritage designates 'outstanding physical, biological, and geological features; habitats of threatened plants or animal species and areas of value on scientific or aesthetic grounds or from the point of view of conservation'. Although two distinct sets of criteria currently exist for identifying sites that qualify for World Heritage status - one set for cultural the other for natural sites - there is a growing trend to recognise the close relationship between the two, particularly at the landscape scale. Revision to the operational guidelines for the convention are likely to result in a single set of criteria which will reduce the distinction between the two sets of values.

THE WORLD HERITAGE PROCESS

The operation of the World Heritage Convention is overseen by the World Heritage Committee, elected from signatories to the Convention. The Committee is advised on natural heritage (biology and geology) issues by the IUCN who evaluate all site nominations. Acceptance by the Committee of new nominations depends not only on the natural heritage values of a site, based on the concept of 'outstanding universal value' (sites must be of global and not merely national or regional significance), but also on site condition and the existence of an effective management plan. Scientific value is not enough - a site must be properly protected.

Nominations are only accepted if sites are already on a national Tentative List which acts as a preliminary screening process to ensure a reasonable chance of success for a nomination. To assist national and regional experts in the formulation of these lists the IUCN has undertaken a range of thematic reviews, including reviews of Forest Protected Areas , Wetland and Marine Protected Areas and Human Use of WH Natural Sites. In 1994 a Global Strategy for a balanced and representative World Heritage List was adopted by the World Heritage Committee. Its aim is to ensure that the List reflects the world's cultural and natural diversity of outstanding universal value. The thematic reviews are an essential part of this strategy and a review of the role of geology in the World Heritage Convention is currently underway.

GEOLOGY IN WORLD HERITAGE

The Convention refers specifically to 'geological features' and the operational guidelines make clear the type of feature that the convention seeks to recognise. Natural Criteria (i) states such features should be 'outstanding examples representing major stages of earth's history, including the record of life, significant on-going geological processes in the development of landforms, or significant geomorphic or physiographic features'.

The Convention is therefore capable of recognising a wide range of geological and geomorphological features and at a range of scales. Small sites important for their fossils lie at one end of the scale, landforms and the processes that produce them at the other. Although strictly site based, the Convention can recognise significant landscapes by inscription on the World Heritage List for geological as

Sites included on the World heritage List for their natural values should:

i. be outstanding examples representing major stages of earth's history, including the record of life, significant on-going geological processes in the development of landforms, or significant geomorphic or physiographic features; or
ii. be outstanding examples representing significant on-going ecological and biological processes in the evolution and development of terrestrial, fresh water, coastal and marine ecosystems and communities of plants and animals; or
iii. contain superlative natural phenomena or areas of exceptional natural beauty and aesthetic importance; or
iv. contain the most important and significant natural habitats for in-situ conservation of biological diversity, including those containing threatened species of outstanding universal value from the point of view of science or conservation.

well as biological and cultural reasons.

The Convention's strict criteria in terms of outstanding universal value, site integrity and management requirements mean that the Convention can only recognise a small number of sites worldwide. No upper limit has been set to the number of sites that should be inscribed on the World Heritage List but it could double to around 1500 in the next thirty years. The relevance of the Convention to wider landscape conservation, beyond the boundaries of a small number of protected sites, might therefore be questioned. In some cases (for example the Grand Canyon, the volcanoes of Kamchatka in Russia, the Aeolian Islands of Italy) large elements of the landscape of geological significance are recognised through the application of Criteria N(i).

Although inscribed on the basis of other natural or cultural criteria many sites have important geological and geomorphological values that underpin biodiversity patterns or influence human settlement patterns and landuse. Mount Kinabalu in Malyasia is a 'biological site' but the extraordinary plant biodiversity on the mountain is in part due to the diversity of rock types in the area. Iguazu in Argentina/Brazil is also inscribed on the World Heritage list for its biodiversity values, but at the heart of the site, and of overwhelming landscape significance, are a set of waterfalls developed over a series of basalt flows. Again a strong geological control is evident.

The Convention certainly offers the opportunity for special geological sites to be given world status through Criteria N(i), but it offers an opportunity to recognise wider Earth heritage values across the full range of World Heritage site types - cultural and natural. Sites may be inscribed on specific cultural or biological criteria but there is considerable scope for the Convention to recognise the underlying role of geology and geomorphology in creating or sustaining these values. World Heritage status can therefore emphasise not only the headline values, but the full range of cultural and natural heritage values that make sites special. Management plans and educational programmes designed around World Heritage sites should be encouraged to recognise this full spectrum of values, and draw out where appropriate the Earth heritage role. This approach goes beyond the simple application of Criteria N(i) to determine if a site has Earth heritage values.

Sites included on the World Heritage List for their cultural values should:

i. represent a masterpiece of human creative genius; or
ii. exhibit an important interchange of human values, over a span of time or within a cultural area of the world, on developments in architecture; or technology, monumental arts, town-planning or landscape design; or
iii. bear a unique or at least exceptional testimony to a cultural tradition or to a civilization which is living or which has disappeared; or
iv. be an outstanding example of a type of building or architectural or technological ensemble or landscape which illustrates (a) significant stage(s) in human history; or
v. be an outstanding example of a traditional human settlement or land-use which is representative of a culture (or cultures), especially when it has become vulnerable under the impact of irreversible change; or
vi. be directly or tangibly associated with events or living traditions, with ideas, or with beliefs, with artistic and literary works of outstanding universal significance.

CULTURAL LANDSCAPES

Recognition of landscapes with a range of outstanding values is becoming increasingly important under the Convention and this is often now achieved through use of Cultural Criteria (v) that recognises landscapes that are 'outstanding example(s) of a traditional human settlement or land-use'. These represent landscapes that are the combined works of nature and of man, semi-natural landscapes that reflect long associations between human society and the landscapes in which they live. Approximately twenty sites have been listed on this basis with significant opportunities in the future for recognition of world class landscapes that involve a wide range of natural and cultural values. The role of geology and geomorphology in shaping such landscapes, influencing biological patterns and determining human land use will often be very important.

The Cultural Landscape concept is powerful, its development brings together the full range of heritage values and recognises the links and interdependencies. In many parts of the world - certainly in the United Kingdom - we have effectively separated cultural and natural conservation efforts, and even within nature conservation there can be some separation of effort between geological and biological work. In other parts of the world such distinctions are less obvious. The trend now is towards greater integration of cultural and natural heritage efforts, a move anticipated by the World Heritage cultural landscapes approach.

The idea of specifically recognising landscapes that result from the 'combined works of nature and of man' fits with the wide recognition of the semi-natural character of many landscapes, particularly in Europe, and the need to mange these for the benefit of wildlife and people. The World Heritage Convention itself provides an opportunity to recognise only a limited number of properties worldwide but it offers a useful model for wider application. National and international landscape programmes for landscape protection (including the European Landscape Convention) can learn from the UNESCO approach.

By definition the World Heritage List should only contain world class phenomena, including landscapes, but most landscapes do not fall into this category, most are more mundane, less spectacular. The World Heritage model, in particular the Cultural Landscape concept, is relevant to all landscapes. Not just as a direct mechanism for designation of protected areas, but as an example of how cultural and natural heritage values can be integrated at landscape scale. The Convention therefore offers a lesson for conservation of all landscapes, the good, the bad and the merely ordinary.

In: Parkes, M.A. (Ed.) 2004 *Natural and Cultural Landscapes - The Geological Foundation*, 161-164, Royal Irish Academy, Dublin

INCLUSIVE MANAGEMENT - GEODIVERSITY AND BIODIVERSITY WORKING TOGETHER

Dr Jonathan Larwood and Dr Michael Murphy, English Nature, Northminster House, Peterborough PE1 1UA, UK.
Email: jonathan.larwood@english-nature.org.uk; michael.murphy@english-nature.org.uk

Abstract Environmental conservation has been typically channelled in two directions: the conservation of biodiversity, which has received much global support, and the conservation of geodiversity, which has been the 'Cinderella' of the conservation community.

This paper examines the differing scales at which the interaction between geodiversity and biodiversity can be considered. On the landscape scale of Natural Areas and on a scale of detailed interaction between geology, geomorphology, habitats and species. The importance of holistic management and integration is stressed as essential to achieving successful long-term management of our natural environment.

GEODIVERSITY AND BIODIVERSITY - LANDSCAPE

The diversity of British geology underpins the diversity of landscape, associated habitats and species we see today and is an association that has long been recognised (Fortey 1993).

For example, the Upper Cretaceous Chalk of southern and eastern England typically produces a rolling landscape of chalk downland. This is characteristically intersected by a network of dry valleys, has an associated calcareous chalk flora and supports a rich agricultural economy. In contrast, the Carboniferous Limestone of northern England is harder and forms extensive upland areas. The calcareous vegetation has similarities to that seen on the chalk but differing altitude, aspect and land-use history have produced a distinctive landscape, fauna and flora seen at its most dramatic in the karstic limestone pavements.

Despite this strong link, an integrated approach to the management of our natural resources is rare. Recognition of this link in England drove the establishment of the Natural Areas concept by English Nature (Duff 1994). This is a holistic approach to conservation, based on geology and geomorphology, that seeks to establish wider objectives and targets based upon both geological and wildlife conservation. Natural Areas are tracts of land unified by underlying geology, landforms and soils. They display characteristic vegetation types and wildlife species, and support broadly similar land uses and settlement patterns. They are influenced by climate, altitude and aspect and, importantly, reflect land use and characteristics that are recognisable and familiar at a local scale.

Geology and geomorphology are central to the Natural Areas concept and the Natural Area map (97 terrestrial and 23 maritime Natural Areas) clearly mirrors the underlying geology and geomorphology. For example, the distribution of Cretaceous Chalk and Carboniferous Limestone (described above) produces characteristic landscapes, habitats, species and land-use patterns, which are recognised as discrete Natural Areas. The Natural Areas concept continues to be important to the work of English Nature (King 1999) and similar concepts have been developed by the Countryside Agency - the Countryside Character Initiative - and English Heritage - the Historic Landscape Characterisation programme (Corfield 2002).

GEODIVERSITY AND BIODIVERSITY - HABITATS AND SPECIES

There are, however, limits to the broad concepts established through Natural Areas. At a landscape and countrywide scale the associations between geology, geomorphology and wildlife are reasonably well demonstrated. In detail, however, little work has been done on the finer associations (and the limits of these associations).

Since the general demise of the Victorian naturalist there seems to have been little dialogue between ecologist, geologist and geomorphologist and the disciplines have become increasingly specialised and separated. As a result the extent of the research on the detailed interactions between geology, geomorphology and wildlife is limited.

There are, however, a number of unusual associations, which bear witness to the detailed influence that the underlying geology and geomorphology has on modern biodiversity (Hopkins 1994, 2003). There are a surprising number of rare and local species, particularly plant species, which are associated with unusual lithologies.

The toxicity of heavy metal mineralisation has plant (eg leadwort and alpine penny-cress), moss and lichen associations virtually restricted to these situations (Humphries and Donnelly 2002). For example, lichens, which are particularly successful colonisers of extreme habitats, have been found to have specific associations with uranium minerals, concentrating uranium within their tissues (McLean

Figure 1. Dungeness SSSI, Kent, England. Britain's largest shingle foreland reflecting over 5000 years of coastal development today supports a diverse fauna and flora in the shingle beaches, ridges and associated brackish and freshwater pools demonstrating a direct link between geodiversity and biodiversity. Photograph: Peter Wakely

Purvis and Williamson 1996). Also extremely localised are the unusual serpentinite-plant associations of the Lizard in Cornwall. Unusually, both acid and calcareous species are found together. Strikingly, the most abundant species found on the Cornish Lizard is the Cornish Heath (*Erica vagans*) which, apart from a small site in County Fermanagh, Northern Ireland, is found nowhere else in the UK.

Faunal associations tend to be more general (Humphries and Donnelly 2002). Examples include invertebrates associated with eroding cliffs (eg the rare tiger beetle *Cicinedela germaninca* associated with coastal cliffs and the mason bee associated with inland cliffs), bats with caves and mines, and birds with rocky nesting habitats in coastal areas, as well as sand and gravel pits.

Geology and geomorphology clearly play an important role in delineating the distribution of habitats and species, but this association is far from simple. Climate, altitude, aspect and geographical location have a fundamental ecological influence. Weathering, and the development of soil profile, is also central to habitat and species distribution. Underlying geology may influence the chemistry of local soils but may also be unrelated, for example, where local geology is covered by thick layers of glacial till and sand such as the sandy soils of the Breckland Heaths, which overly and conceal the chalk of parts of Norfolk.

The relationship between geodiversity and biodiversity is further complicated by the influence of man. Much of the environment seen in England today is semi-natural. The diversity of habitat has been influenced by a long history of management including grazing, burning, cultivation, and woodland management. Our own interaction with the natural environment and the way we manage this environment is an important consideration.

GEODIVERSITY AND BIODIVERSITY - MANAGEMENT

Recent research (Humphries and Donnelly 2002) has looked at the techniques used for geological conservation and considered the benefits these may bring to associated habitats and species.

Typical geological conservation techniques include scaling, reprofiling and stabilisation of rock faces, clearance of rock debris and scree and the management of vegetation. All these actions can have a tangible benefit for both geological and biological conservation. Maintaining exposed rock faces can benefit lower plants and ferns, by establishing open areas and surfaces for colonisation, or birds, by establishing suitable rocky nesting sites. Similarly removal and management of vegetation can benefit a wide range of fauna and flora through the maintenance of open areas. In particular, the establishment and maintenance of bare ground, often considered unsightly, is essential for many plants and animals (Key and Gent 1993, Key 2000). Some of our rarest reptiles and amphibians need bare sand, as do many grassland and heathland plants and invertebrates.

Conversely, management of habitats can directly benefit geology and geomorphology. For example, the removal of conifer plantation from karst landforms in Cumbria (Webb 2001) has not only re-established the valuable limestone pavement habitat but has also revealed limestone pavement, an important element of the natural landscape. Similarly, management of encroaching scrub in Cheddar Gorge (Corns 1988) has benefited habitat, species, geology and geomorphology alike.

The mutual benefits from adopting a holistic approach to conservation management are clear but are rarely considered. There is no doubt that a better understanding of the interaction between geology, geomorphology, habitat and species will bring more successful and sustainable management of our natural environment.

GEODIVERSITY AND BIODIVERSITY - INTEGRATION

Establishing integration between geodiversity and biodiversity is a significant challenge. The major framework for the delivery of biological conservation in the UK is the UK Biodiversity Action Plan (UK BAP). This has established a long-term strategy for the protection and enhancement of biological diversity, which is co-ordinated and delivered through Local Biodiversity Action Plans (LBAPs). There are examples of LBAPs that do encompass geology, however, these are the exceptions.

A similar strategic approach does not exist for geological and geomorphological conservation. In response to this, the feasibility of Local Geodiversity Action Plans (LGAPs) is currently being investigated (Burek and Potter 2002, Larwood 2002) and are now being developed for a number of counties.

The success (and failings) of the BAP process has been looked at in detail in order to guide the possible development of LGAPs. If they work, LGAPs will provide the opportunity to establish a framework for geological conservation that considers not only site based objectives at local, regional and national levels but also the context of those sites in the surrounding landscape. Importantly, LGAPs should provide an opportunity for integration with existing LBAPs and the establishment of management policies and practice that encompass both geological and biological conservation.

THE FUTURE

At all scales, there are tangible links to be drawn between geology, geomorphology, landscape, habitats and species. However, this is a complex relationship that needs to be better understood. The Natural Areas concept draws these links generally at a landscape scale. In detail there is less understanding of this complex relationship. Initial research has demonstrated the value of a holistic approach to conservation management but better understanding of the detailed interaction between geodiversity and biodiversity is needed. Research into this relationship is currently being undertaken on behalf of English Nature.

Delivery of an intergrated approach to conservation is also a major challenge. LGAPs may provide an integrated approach for geological and geomorphological conservation but could also provide an opportunity for integration with other aspects of environmental conservation. Pilot LGAPs, now being established, will hopefully develop this integration.

Geological and geomorphological conservation may be considered a 'Cinderella' of conservation. However, because of the cross-cutting nature of the geological and geomorphological resource this Cinderella may offer the best opportunity for integrated conservation management between geodiversity and biodiversity.

REFERENCES

Burek, C. and Potter, J., 2002. *Local Geodiversity Action Plans - setting the context for geological conservation.* Unpublished report for English Nature, 64pp.

Corfield, M., 2002. Conserving character. *Earth heritage*, 17, 16-17.

Corns, R.G., 1988. Cheddar Gorge and the scrub control programme. *Earth Science Conservation*, 25, 36-40.

Duff, K., 1994. Natural Areas: an holistic approach to conservation based on geology. In: O'Halloran, D, Green, C., Harley, M, Stanley, M and Knill, J (eds) *Geological and Landscape Conservation.* Geological Society, pp 121-126.

Fortey, R., 1993. *The Hidden Landscape*. Jonathan Cape, 310pp.

Hopkins, J., 1994. Geology, geomorphology and biodiversity. *Earth heritage*, 2, 3-6.

Hopkins, J., 2003. Some apects of geology and the British flora. *British Wildlife*, 14 (3), 186-194.

Humphries, R.N and Donnelly, L., 2002. *Geological conservation: benefits for biodiversity*. Unpublished report for English Nature, 23pp.

Key, R. and Gent, T., 1993. Bare but not barren. *Enact*, 1 (2), 15-16.

Key, R., 2000. Bare ground and the conservation of invertebrates. *British Wildlife*, 11, 183-191.

King, A., 1999. Natural Areas: an opportunity taken for a holistic approach? *Earth heritage*, 11, 23-24.

Larwood, J.G., 2002. Local Geodiversity Action Plans (LGAPs) - setting a context for geological conserva-

tion? *Nature's Place*, 25, 12-13.
McLean, J., Purvis., O.W. and Williamson, B.J., 1996. Role of lichen melanins in uranium remediation. *Nature*, 391, p649.
Webb, S., 2001. Life after conifers. *Enact*, 9 (2), 18-22.

In: Parkes, M.A. (Ed.) 2004 *Natural and Cultural Landscapes - The Geological Foundation*, 165-169, Royal Irish Academy, Dublin

(RE)THINKING IRELAND'S ATTITUDE TO LANDSCAPE

Dr John McDonagh, Department of Geography, National University of Ireland, Galway

Abstract: The physical landscape has moved further up the National and International agenda, yet it still does not receive the attention it deserves. Ireland's attitude towards landscape management remains very top-down and interventionist rather than bottom-up or community driven. Indeed current practice and attitude towards landscape management and conservation in Ireland is still overly concerned with individual features, historic sites and designated areas. This paper argues for a change in this "thinking". Using brief examples of different developments on the landscape the paper argues for a greater understanding and appreciation of the 'window on the past' that Irish landscapes provide and, a more inclusive and holistic management practice which recognises that landscapes are a continually evolving product of human interaction with nature. Overall, the paper calls for a change in Irish thinking, planning and attitude toward altering the physical landscape. A change that can only emerge from a greater awareness among the wider public and policy makers, that landscape is dynamic, and that landscape once it is altered/removed, can never be recreated.

INTRODUCTION

Geomorphological conservation has been described as the Cinderella of nature conservation (Gray 1997). While there is continued debate on protecting our environment, the current dominant focus is one of protecting archaeological sites, heritage buildings, habitats etc., with limited recognition of landform conservation. Indeed debates on conservation of the physical landscape demonstrate a lack of perspective both from community and official policy. The alteration, and in some cases obliteration, of geomorphological features for roadways, housing developments or leisure pursuits like golf course development, is depriving future generations of a repository of knowledge on their past. So why is this the case - is it due to the lack of a national landscape policy or is it because of our attitude toward the landscape? This paper argues that it is a combination of both and, as such, argues for a change in Irish thinking, planning and attitude toward altering the physical landscape. Overall the paper advocates a greater understanding and appreciation of the 'window on the past' that the Irish landscape provides and calls for greater awareness among the wider public that landscape, once altered, can never be recreated. Further, the paper calls for landscape to be viewed holistically, recognising physical and human interaction and natural change. Lastly, the need for people to be placed at the centre of the 'environment versus development' debate is also argued, as is the need for a more determined policy from government to conserve our landscape narrative for future generations.

THE IRISH LANDSCAPE

Ireland's countryside is a rich mosaic of physical landscapes, numerous habitats, varieties of flora and fauna, changing landuse patterns, agricultural practices and different cultural traditions. This mosaic is coming under increased pressure from the large-scale development and the unprecedented economic prosperity that Ireland has gone through in the last 5-8 years. The country has emerged from a period of large-scale unemployment, poor growth and high emigration to a prosperous economy, almost full employment and unprecedented growth in industry, tourism and services.

While there has been a pre-occupation with development in terms of economics, creating employment etc. it is now realised that there is urgent need to consider the impacts of development on landscapes and to approach landscape appraisal in a systematic way - a way which will enable a proactive view of development to be taken while maintaining that respect for the environment and heritage generally, which is required by sustainability (DoELG 2000). To this end landscape considerations need to become central to national policy particularly in terms of the National Development Plan and the National Spatial Strategy.

SO WHY BOTHER ABOUT OUR PHYSICAL LANDSCAPE?

Perhaps one of the first problems we have in dealing with the physical landscape is the problem of definition. What do people regard as landscape? Can it be categorised for the purposes of protection? Is one person's landscape another's visual obstruction? For a term that trips easily off the tongue, landscape creates a great deal of discussion when one attempts definition. The German geographer Carl Troll used the term landscape ecology meaning an approach that interprets landscape as supporting both natural and cultural ecosystems (cited in Goudie 1997). Landscape is also dynamic and, with increasing economic development taking place, it is inevitable that landscape will change. My argument is not against change. More-so it is that landscape must be viewed as central to the sustainability debate currently dominant. Landform conservation must be given more importance and, as part of ensuring sustainability, there should be a 'fundamental importance of incorporating community issues in (to) landscape planning and management' (Hughes 1999, p.74).

Landscapes are a continually evolving product of human interaction with nature and while it may be possible to describe the distinctness of one landscape type from another, it may be foolhardy to attempt to determine whether one landscape is better or deserves more protection than another. In the Irish system, classification has been more akin to an approach of mapping and identifying landuse changes.

What we must also embrace is that there is a narrative within our landscape that is more easily told by some features than others and this should not be summarily dismissed by the JCB. How much 'protection' these features are afforded is invariably determined by the aesthetic, social and economic factors, the weightings these are given and, their recognition in the wider policy environment. This weighting must be informed by the historical chronology that different features can provide, an element often over-looked due to a lack of knowledge on the part of the developer or road designer.

Landscape 'embraces all that is visible when one looks across an area of land' (DoELG 2000) and more - it is the linkage of human and physical environments, each dependent on the other for its survival. What we must also grasp is that overconfidence in our ability to manage our physical landscape is proving an ultimate threat. The severe depletion of natural habitats and the consequences this has for the food chain, allied to other concerns such as those raised by dam constructions, regraded slopes, straightened rivers and other 'feats of engineering', are now major threats to future sustainability.

POLICY APPROACHES!

Some recent approaches to landscape, appraisals that have been done extensively in Britain and Northern Ireland, are based on different understandings and uses of landscapes than that currently underway in Ireland (DoELG 2000). By contrast in Ireland, there have been few documents produced on landscape bar the quasi-official document Outstanding Landscapes (An Foras Forbartha 1976), the more recent DoELG (2000) document Landscape and Landscape Assessment - Consultation Draft of Guidelines for Planning Authorities, and the Heritage Council's (2002) Policy Paper on Ireland's Landscape and the National Heritage. The Heritage Council (2002) describe its vision of the Irish landscape as being 'a dynamic landscape, one that accommodates the physical and spiritual needs of society with the needs of nature in a harmonious manner ... (which) ... as a result bring benefits to both' (p.13). The DoELG (2000) document, in its vision, sought to develop a more systematic approach to landscape moving away from 'concepts such as sublime, beautiful, outstanding etc. as criteria or as a means of categorisation' as these were seen as 'categories which ... (gave) ... rise to a view of landscape which was unnecessarily restrictive, protectionist and conservationist' (p.3). In fact the DoELG suggested that it was seeking to develop new methods which favoured characterisation (from landcover to historical and cultural significance) with the end result allowing landscapes to be viewed in relation to their 'ability to accommodate developments' and to 'give indicators as to which developments might be most suited, under what conditions and using what design criteria' (DoELG 2000, p.3). The absence of any linkages to the 'narrative' or historical chronology in this desire is worrying however.

Most other aspects of the policy environment dealing with landscape in Ireland are very much linked to EU Directives such as the EU Birds Directive (79/409/EEC) which requires Member States to designate Special Protection Areas (SPAs- approx. 111 areas) for wild birds etc.; The EU Habitats Directive (92/43/EEC) which requires the designation of Special Areas of Conservation (SAC - approx. 362 areas) and includes Blanket and Raised bogs, sand dunes, machairs, turloughs, limestone pavement, fens, grasslands, woodlands, salt marshes/meadows, estuaries/mudflats/cliff, shallow bays and lakes/rivers. There is also the Rural Environmental Protection Scheme (REPS) that encourages farmers to farm in an environmentally sustainable way with enhanced levels of payment applying in SPAs and SACs (in total approx. 990,000ha are areas under protection out of a land bank of 6.9m ha). The Local Government (Planning and Development) Acts 1963-1998 have also been amended and now require the inclusion of objectives in relation to the protection of wildlife habitats in county and city developments under new regulations of the Wildlife Acts (www.enfo.ie).

Thus, while the physical landscape has moved further up the national (and international) agenda, it still does not receive the attention it deserves. In fact Ireland's attitude towards landscape management remains very top-down and interventionist rather than bottom-up or community driven. Aalen et al. (1997) highlighted this when they suggested that: 'at present, we (Ireland) have a designation policy rather than a landscape policy, a concern with sites rather than landscapes, with elements rather than wholes. The chief strategy has been to establish enclaves encompassing valued resources and to safeguard them by limiting human use. Landscape is treated as a series of discrete elements to be planned in isolation; the outcome is a fragmented and confused perspective, inimical to the landscape, which is holistic and requires broad strategic approaches. Nowadays, whole landscapes are the issue, not sites. The ... protective policies applied to special areas need urgent complementing by others which address the realities of constant unpredictable change on a nationwide scale. The radically changed objective must be accompanied by a fundamental shift in approach' (p.255). This lengthy quote sums up how current practice and attitude towards landscape management and conservation in Ireland is still overly concerned with individual features, historic sites and designated areas, rather than the more inclusive and holistic management that landscapes require.

This viewpoint is reiterated by the Heritage Council's Policy Paper (2002), which suggested that it was important that landscape be considered as a whole rather than in the compartmentalised manner in which it is often treated. The Council argue that environmental agencies, archaeologists, farmers, the forestry and tourism sectors and people in general, must follow an integrated approach which recognises the significance, interdependence and interrelationships between these actors.

So how are these issues manifest 'on the ground'? What kind of landscape changes are we talking about and is it possible to have a holistic view of landscape? To illustrate, albeit briefly, the lack of a holistic view, the lack of integration and the often conflicting debate that surrounds landscape and development, one can look at the examples of house construction and golf course development. These 'developments' illustrate the attitudes towards landscape and further reinforce the need for a national landscape policy and the urgency with which it is required.

LANDSCAPE CUSTODIANS?

The idea that one person's landscape is another's visual obstruction perhaps best sums up the difficulty of dealing with the dynamic concept that is our landscape. For a term that trips easily off the tongue, landscape creates a great deal of discussion and conflicting views when one attempts to determine what activities or developments can take place on any given landscape. The 'attitude' we have toward landscape plays a major role in the 'value' that becomes attached, be it in symbolic landscapes, landscape aesthetics, landscape and ideology or landscape and tradition. It is to this latter point, landscape and tradition, which I want to turn to first. The attachment to the land and the tradition of building houses on family farms is very much entrenched within Irish culture and tradition. This tradition is however coming under increased strain as those in the Heritage Services (Dúchas for example), argue for a reduction in this practice arguing for greater protection of the natural environment. But what does this mean? Are landscapes not intricate combinations of human and physical environments? This is where the first example of conflicting views of the landscapes emerges.

'I want to build my home on the site that my father gave me. I am sincerely asking for permission. I am not a stranger. This is our valley. Our family has farmed and lived here for centuries. These fields carry the memory of our family from Famine times, landlord times and Black and Tan times until the present time. The very imagery and contours of these fields is at the very heart of my imagination. I am asking for is as a member of this family to live here where we have lived for centuries ... to continue our family on these fields that we have minded for centuries ... I wish in no way to injure the environment. I will do everything possible to avoid this ... the memory, aesthetic and mystical presence of this landscape really moves me. I do not have to imagine myself into the way of life here, I was born within it. I belong within it and know it from the inside' (quotes from John O'Donoghue as reported in the Clare Champion, March 15th 2002).

These words from John O'Donoghue conjure up images of tradition, culture and the closeness to 'mother earth' that is often linked to Irish rural space and people. The talk of identity and links with landlord and Black and Tan times evoke the nationalist spirit and the attachment to the land that is often uppermost in Ireland's history since independence. So why are these images so important and why are they used in conjunction with house building? This is where the understanding of landscape and the attitude towards landscape reaches another level. This is where landscape becomes more than a physical space and where the dynamism that is created through human interaction with the physical landscape is perhaps under estimated by those wishing to preserve (what they see as) the character and integrity of a landscape. It is somewhat ironic in this particular example (and while this is just one example, I feel that is opens the debate on landscape quite significantly), that the applicant (Mr O'Donoghue) had been at the fore front of ensuring the preservation and protection of the Burren and was central to the successful opposition of plans by Dúchas and the government Department for Arts and Culture to build an interpretative centre at Mullaghmore in the Burren. In viewing both sides of the respective arguments, it is easy to see in the above quote Mr O'Donoghue's passion and desire for the landscape. What becomes interesting within this debate is how this passion and desire is challenged by others who feel equally as passionate about preventing this happening lest it create a precedent for further development within the area that 'would be contrary to the County Development Plan ... and... would impinge significantly on the character and integrity of the landscape' (Clare Champion March 15th 2002). Other contradictions (conflicting views) that surface in this debate are reflected in the EIS (Environmental Impact Statement) conducted for the site by Mr O'Donoghue which suggested that the development would have no impact on priority habitat such as limestone pavements and that the site was located on disturbed grassland of minimal conservational value. That the same development was described by Dúchas as potentially destroying habitats such as limestone pavements and orchid-rich semi-natural grasslands, protected under Annex I of the European Habitats Directive, seemed to further reinforce the gap between those wanting to be part of the landscape and those making specific distinctions between human and physical landscapes. The latter claims ring hollow in view of the comments that the report 'which claims that the site for the house is orchid-rich was drafted on foot of an inspection last November where it would have been impossible to say if orchids grew there or not' (Laffery, cited in Clare Champion March 15th 2002).

So how can two groups that champion the preservation of the landscape be on opposing sides? And who should have a say in the type of development that takes place in the landscape? This is where the attitude of people towards the landscape and, what people want from the landscape in which they live, comes to the fore. This is also where the dynamic that is the landscape has to be viewed as the intertwining of both physical and human strands. It also indicates the need for viewing whole landscapes with policies that address the realities of these landscapes allied to a fundamental shift in people's attitudes to, and understanding of, the landscape. The notion of 'ownership' is also central to this debate and the need to challenge the idea that where a person owns land, they have a right to build on that land what ever the implications for the environment (due to the limited scope available, the opportunity to pursue this in greater depth is not available). Whatever the 'thinking' and whatever the challenges facing the landscape are, it is abundantly clear that people must become central to the equation of balancing between environment and development.

THE EXAMPLE OF GOLF COURSES

The next example worth exploring with regard to conflicts between environment and development is that of golf courses. With an 18-hole golf course taking up in the region of 45-55ha, and with almost 100 courses within a 50km radius of Dublin city for example, golf courses are becoming an extremely significant part of our landscape. What we are seeing in this expanding golfscape is a large scale change from agricultural landuse to recreational landuse and the physical alternation of the landscape that this brings with it - raised tees, bunkers, tiered fairways etc. There are at least 3 broad groups into which the golfscape can be divided (1) courses which convert and landscape farmland; (2) courses which adapt already existing parkland centred on a former or still-existing big house and (3) coastal dunes that can be adapted for a links-type course (Horner, 1993, p.155). There was a time, Gray (1997) suggested, that golf courses were designed with the existing landforms being utilised (ex's of Turnberry, St. Andrews's) etc. However, this use of the natural landscape has given way in recent years to feats of design and engineering and over-confidence in our ability to manage the landscape. A recent headline in the Irish Times newspaper is testament to this where it signalled the triumph of man [sic] over nature in the creating of the Castleisland Golf course in Co. Kerry. Some quotes from this report highlight this 'success':

'the experts claimed (this course) would prove impossible to build because of the inhospitable terrain'

'When the initial plan to build a golf course on hilly terrain just north of (Castleisland) was mooted there were those who wondered about the sanity of those making the proposal'

'Built on 180 acres, .. the course is blessed by the presence of two valleys which have been integrated to form two spectacular - and potentially punishing - holes, the 7th and the 18th'

'the land was difficult to work with, which made the work of the contractors all the more admirable ... the back nine uses the landscape exceptionally well and nowhere better than on the 12th where there is a preponderance of wildlife all the way down the left ...'

'what was little more than bog has been transformed - in some cases by the addition of copious amounts of sand - into genuine golfing turf' (all Reid 2002).

In other cases like the Doonbeg Links in Co. Clare, the course designer Greg Norman spoke of the challenge of integrating the dunes and natural landscape of the area into the course and lauded the opportunity for him as a course designer to have to work with the terrain as it was, rather than mould an artificial landscape (this was due to the restrictions imposed and the initial objections to the project. What was also somewhat ironic was that while the development was opposed on the grounds of 'protecting a fragile environment' (among others) it was discovered that part of the area had been degraded due to sand extraction that should have been prevented many years previous). In fact Norman suggested that such was the uniqueness of this particular way of approaching a golf course design that it was among one of the last remaining opportunities in Europe to take such an approach. The argument here is not that we shouldn't have golf courses, but that the 'exception' that Norman refers to in Doonbeg where the natural terrain has to be left mostly undisturbed, should be the norm.

These examples underline what I suggest is the cavalier way we view our landscape. The Irish landscape is probably one of the best 'fieldsites' in the world as it contains numerous examples of features and landforms sculpted by our recent glacial history. The moulding of the landscape during the Pleistocene has left a rich and irreplaceable landscape with a variety and complexity of deposits from drumlins and eskers to rocky hillocks and rolling morainic topographies. If we accept that in order to understand what is happening in current day processes - be it climate change, mass movements or increased flooding - we must understand our past landscapes and the narrative they provide, then there is surely a need for a greater understanding and appreciation of the 'window on the past' that the Irish landscape provides. On the one hand we have those who wish to triumph over the landscape and on the other are those wishing to leave the landscape untouched. Neither is realistic nor sufficient. Landscapes provide a historical narrative that must be decoded. Landscapes also create new narratives that are moulded by human and physical dynamics. The challenge is to recognise and accommodate both.

SO WHAT CAN BE DONE?

It is not my argument that changes to the landscape must be prevented. This is neither realistic nor feasible. What I do argue however is that we have a duty to future generations to preserve our landscape for their use. Further if we do not document those features of our landscape which have an important 'story to tell' (prior to their excavation should their demise not be preventable), then we are destroying an irreplaceable part of our past history and depriving future generations of a rich well of knowledge. One example of a step in the right direction is that of the GSI project and Landscape Tourism. Recognising the importance of the scenic landscape to Ireland's tourism sector the GSI have sought to use their expert knowledge of the physical features of the countryside to enhance people's (visitors and locals) understanding of the scenic landscapes around them. One of the successful projects undertaken by the GSI (in association with the Geological Survey of Northern Ireland) is the 'Landscapes in Stone' project begun in 1997.

The GSI has also undertaken, as part of the Geological Heritage Programme, the objective of conserving and promoting the best of Irish geology with karst being one of the first areas chosen.

What I would also argue for, is:

(1) The need to preserve not only key geomorphological features but to document those features where their demise is not preventable;

(2) Site-led design and the need for an integrative framework between developers and the landscape on which developments take place. The question here is whether people involved with planning, policy and design, are sufficiently trained or aware of the narrative landforms can provide;

(3) Increased public awareness - a greater awareness by the wider public that landscape once altered can never be recreated and the need to inform the general public that vital past histories and chronologies can be determined from the complex make-up of glacial deposits;

(4) Community involvement - The Draft Convention on Landscape proposed by the Council of Europe suggests that landscape is 'an important part of people's lives - giving individuals and communities a sense of identity and belonging and bestowing a sense of place on our surroundings'. The more people understand about a landscape/landform and the more people want to conserve a landscape/landform then the easier it will be to achieve this. Conserving/protecting the landscape can only be viable if public opinion is there and hence the importance of local community involvement in managing our landscapes. There is a notable lack within the DoELG Guidelines on the part to be played by local communities in this process. The concentration placed on planning and management is outlined through key concepts of landscape character, values and sensitivity, the implications for policy responses with only the smallest reference to 'achieving local consensus'. Landscape cannot be viewed purely in terms of topography and terrain, but must incorporate the local communities that inhabit and manage the landscape;

(5) Policy response - There is a need for a more determined policy from government to conserve key landforms and landscapes within Ireland and keep in tact a narrative of our evolution, both as a people and as a country, for future generations. The quality of the environment largely determines the quality of life that can be experienced - it also determines the quality of our economy; the quality of our culture and the quality of our heritage.

Finally, no '*development can take place which does not change the landscape. As has happened throughout history, the landscape is constantly changing both through the actions of nature and human intervention. The challenge to our generation is to bring about change in a way that is respectful and creative. Our landscapes are therefore a resource which is to be used prudently and sustainably*' (The Draft Convention on Landscape). However while it is prudent to develop strategies and polices to protect the landscape, one must also remember that while the discussion is on-going on what should be done, where houses can and can not be built for example, more destruction of the physical landscape is continuing apace, more landscape narratives disappearing, their stories having never been told.

REFERENCES

Aalen, F.H.A., Whelan, K. and Stout, M. (1997) *Atlas of the Irish Rural Landscape*, Cork University Press.

DoELG (2000) *Landscape and Landscape Assessment - Consultation Draft of Guidelines for Planning Authorities*, Department of the Environment and Local Government.

Foras Forbartha (An) (1976) *Outstanding Landscapes*, An Foras Forbartha.

Goudie, A. (1997) *The Human Impact on the Natural Environment* (4th Edtn), Blackwell Publishers.

Gray, J.M. (1997) 'Planning and Landform: geomorphological authenticity or incongruity in the countryside?' *Area*, Vol. 29 (4) pp-312-324.

Heritage Council (1999) *Policies and Priorities for Ireland's Landscape*, Conference Papers, The Heritage Council of Ireland Series.

Heritage Council (2002) *Policy Paper on Ireland's Landscape and the National Heritage*, The Heritage Council of Ireland Series.

Horner, A. (1993) Changing Ireland - Dublin's Expanding Golfscape. *Irish Geography*, Vol 26 (2) pp.151-157.

Hughes, E. (1999) 'Community Participation in landscape Planning and Management', in, Heritage Council, Policies and Priorities for Ireland's Landscape, Conference Papers, The Heritage Council of Ireland Series, pp.73-87.

Reid, P. (2002) 'Top class work in the valleys', *Irish Times* Aug 27th p.23

www.enfo.ie

In: Parkes, M.A. (Ed.) 2004 *Natural and Cultural Landscapes - The Geological Foundation*, 170, Royal Irish Academy, Dublin

ASSESSING RAPID GEOLOGICAL CHANGES IN LANDSCAPES AND ECOSYSTEMS

Antony Berger and Jonas Satkunas,
IUGS Geoindicator Initiative, Victoria BC, and Vilnius, Lithuania
Email: aberger@uvic.ca; Jonas.Satkunas@lgt.lt

Abstract: The geological foundation of thelandscape changes due both to human (technogenic) actions and non-human causes. Rapid geological changes, driven by exogenic and endogenic processes, may be hazardous (e.g. landslides, sinkhole collapse, earthquakes) or slower and more pervasive (variations in groundwater level or chemistry). These and other such phenomena, including glacier flucutations, dune movement, soil erosion, and stream sediment transport, also affect living organisms in a variety of ways. To assist in understanding ecosystem condition and resilience, the International Union of Geological Sciences has developed the geoindicator concept. This brings together in a convenient format various measures for assessing significant geological change that can occur on human time-scales (see www.gcrio.org/geo/title and www.lgt.lt/geoin). Technogenic) activities commonly retard or accelerate natural processes, but in order to understand and predict the landscape impact of any such activity, it is necessary to establish the background rate of natural change. The crucial issue is to identify the key indications of geological change and to carry out systematic monitoring. The geoindicator approach is now being used in State-of-the-Environment reporting in several countries and in a new program of environmental assessment in national parks in Canada and the US. Since some of the most valued natural landsapes are protected in parks and other protected areas, it is especially important to identify and track changes to both ecosystems and their abiotic foundation. The presentation will include examples where understanding rapid geological change is essential to the resolution of ecological issues.

In: Parkes, M.A. (Ed.) 2004 *Natural and Cultural Landscapes - The Geological Foundation*, 171-174, Royal Irish Academy, Dublin

'LAND FORM' RATHER THAN 'LANDFORMS': GEOMORPHOLOGICAL CONSERVATION OUTSIDE PROTECTED AREAS

Murray Gray, Department of Geography, Queen Mary, University of London, Mile End Road, London E1 4NS, UK. Email: j.m.gray@qmul.ac.uk

Abstract: This paper argues the case for conservation of 'land form' in the general sense rather than specific, well-developed 'landforms'. In many areas the natural form of the land is still basically intact and has value in various ways, including its contribution to landscape and biodiversity. Some landforms are highly sensitive to human impacts and too many morphological changes in the countryside show little respect for local topographic distinctiveness or geomorphological concepts. Examples include, golf course remodelling and waste landraising schemes.

The 'Natural Areas' and 'Natural Heritage Zone' approaches in the UK, supported by 'Landscape Character Assessment' and planning policies, have the potential to describe, value and conserve geomorphological character, though the full potential has yet to be realised. These approaches should apply to the 'wider countryside' beyond protected areas such as National Parks and Sites of Special Scientific Interest.

There is also scope for landscape enhancement by authentic restoration, including quarry, coastal and river restoration. What is needed, but very difficult to achieve, is a greater understanding of our geomorphological heritage amongst landscape architects, planners and the wider public so that we value and conserve the natural land form and achieve geomorphological authenticity in land development schemes and landscape restoration.

VALUING LAND FORM

The three-dimensional form of the land surface is arguably the most important element in producing landscape variation, in influencing biodiversity and in many other ways. This applies at all scales. At a global scale, topographic diversity is provided by mountain ranges and ocean trenches, island arcs and mid-ocean ridges, drainage basins and coastlines, ice-sheets and deserts. At the local level rivers and ponds, beaches and coasts, hills and valleys all give variety to local landscapes and in themselves show micro-morphological variation. For example, a river has its flood-plain and banks, bedforms and materials, planform and gradient, water and sediment discharge, all of which may vary in space or time.

This diversity of land form and geomorphological processes has value in several ways apart from its intrinsic value free of human valuations (see Table 1).

THREATS TO LAND FORM

There is, perhaps, a general tendency to think of the biological world as fragile and vulnerable and therefore in need of conservation, whereas the abiotic landforms and rocks are stable, static and much too prolific ever to be endangered. This is a gross oversimplification though much depends on the sensitivity or robustness of the area/site concerned. Tasmanian geomorphologists have compiled a 10-point geosensitivity scale
(e.g. Kiernan, 1997) ranging from:
1 = Sensitive to inadvertent damage by free-ranging pedestrian passage (e.g. brittle lava forms that can be crushed underfoot)
2 = Sensitive to effects of more focused human pedestrian passage (e.g. coastal dune disturbance by footpaths)
3 = Sensitive to damage by collecting, sampling or deliberate vandalism or theft (e.g. some fossil and mineral collecting, rock coring and speleothem sampling)
to:

8 = Sensitive to major removal or addition of geomaterial (e.g. quarrying, golf course remodelling, waste disposal by landraising)
9 = Sensitive only to very large-scale contour change (e.g. large open-cast quarries; major river channelisation)
10 = Sensitive only to catastrophic events (e.g. major landslides & tsunamis, human-induced sea-level change).

The point here is that many human activities alter the Earth's surface form or interfere with geomorphological processes, that different systems have different sensitivities, and that anthropogenic change is an insidious process whose cumulative impacts can be serious.

CONSERVING LANDFORMS BY PROTECTED AREA DESIGNATION

Given these values and threats, and on the basis of the simple relationship:
Value + Threat = Conservation Need
it is not surprising that attempts have long been made at geomorphological conservation. For example, Yellowstone National Park was established in 1872 largely for its geothermal features and scenic beauty. And in 1880, Edinburgh City Council took measures to enclose striations identified earlier that century by Louis Agassiz as part of the evidence that Britain had been glaciated.

Type of value	Sub-class	Examples
Intrinsic value	1. Intrinsic value	Abiotic nature free of human valuations
Cultural value	2. Folklore	Giant's Causeway, NI; Devil's Tower, USA
	3. Archaeological/Historical	Cave shelters; hill forts
	4. Spiritual	Uluru, Australia; N.American Indian sites
	5. Sense of place	White Cliffs of Dover, UK; Rock of Gibraltar; local places
Aesthetic value	6. Local landscapes	Sea-views; countryside walks
	7. Geotourism	Grand Canyon, USA; Norwegian fjords; Canadian Rockies
	8. Leisure activities	Rock climbing; caving; whitewater rafting
	9. Remote appreciation	Landforms & processes in magazines and TV
	10. Artistic inspiration	Literature (Hardy); music (Sibelius); painting (Turner)
Functional value	11. Platforms	Building construction; soil formation; plant growth
	12. System functions	River flow; longshore drift; soil formation
	13. Habitat provision	Biodiversity
Research & Education value	14 Scientific discovery	Geoprocesses
	15 Earth history	Origin of landforms
	16 History of research	Early identification of glacial activity in Britain
	17 Environmental monitoring	Sea-level change; glacier shrinkage
	18. Education & training	Field studies; professional training

Table 1 Some values of topographic diversity

Since then many other important landforms, landform assemblages and landforming processes have been conserved through their designation as protected areas of one sort or another. This traditional approach is evident in many countries and also at the international level. For example, World Heritage Sites include the Grand Canyon, USA and Uluru (Ayers Rock), Australia. Geoparks include the Cuilcagh Country Park (karst topography) in Northern Ireland and the Zhangzhou Littoral Volcanoes Geopark in China. National Parks include the Yellowstone geothermal features in the USA, and the Caldera of Taburiente on the Spanish Canary Island of La Palma. National Monuments include the Devil's Tower in Wyoming, USA and 226 catalogued erratic boulders in Estonia. And many other forms of protected area have been established by countries or even parts of countries. For example, Sites of SpecialScientific Interest (UK), National Natural Landmarks (USA), and Areas of Natural and Scientific Interest (Canadian Province of Ontario).

Although these designations play a very important role in the conservation of landforms and processes, several criticisms have been made of specific national or sub-national systems or of the protected area approach in general. For example:

- The systems are very diverse and do not always conform to the IUCN classification of protected areas. The value of protected areas also varies or is unclear. Some are protected for their research/educational value whereas others appear to be mainly for recreational use, thus increasing the threats to their geomorphological integrity;
- Site selection criteria vary. In some cases the emphasis is on protecting the spectacular or unusual landforms whereas in other cases attempts are made to ensure that a representative selection of a nation's geodiversity is protected by establishing systematic scientific inventories of the landforms present;
- It is relatively easy to designate an area or site but less easy to protect it from degradation or manage its long-term health. The "paper parks" phenomenon is well known in conservation;
- The protected area approach may give the impression that effective conservation can be achieved solely in such areas, and that it can be ignored elsewhere. This may make it harder to achieve integrated policies.
- Protected area boundaries are often fairly arbitrarily drawn and may not reflect the wider landform assemblage or operation of geomorphological processes, for example on coastlines, in drainage basins or through karstic systems;
- Much of the land surface does not consist of specific named landforms that are easily protected by drawing lines around them. Yet much topography is essentially natural, having been the subject of only limited disturbance by humans. There is therefore a case for changing the focus of geomorphological conservation to "land form" rather than "landforms";
- The net result of protected areas is a fragmented approach to nature conservation rather than one that integrates geological and biological conservation or wider land use management.

As a result of these criticisms a number of new approaches are being adopted to ensure a wider approach to nature conservation in general, and this gives an opportunity to review our objectives in geomorphological conservation.

CONSERVING LAND FORM IN "THE WIDER COUNTRYSIDE"

In order to conserve the wider geomorphological resource we need, first of all some system of classifying landscape, secondly a description of the land form and processes within each unit, and thirdly methods of protecting and enhancing the natural land form. The general approach used by some countries can be termed "regionalisation" whereby the country is divided into regions on the basis of common physical attributes (geology, land form, drainage, soils, vegetation). Although this approach has also been criticised as being effectively just another boundary-based network, it does give a way of assessing the land form on a continuous basis.

In Australia, the Land Systems Mapping approach developed by the CSIRO in the 1950s and 1960s and subsequently applied in Africa, Latin America and Asia, as well as Australia, was an early attempt at this approach though the purpose was essentially to assess agricultural potential rather than conservation. However, the principle of spatial classification of physical landscapes has been adopted by conservation agencies in various countries.

For example, Parks Canada has divided the country into 39 Natural Regions on the basis of landscape types. Some Canadian provinces have taken this further and divided the Natural Regions within their provinces into Natural Subregions and are using the resulting spatial systems as a framework for managing the wider georesource. For example, in Alberta the six Natural Regions represented in the province have been further mapped into 20 Subregions. Below this are three further levels of subdivision, referred to as Natural History Themes, and these are being used for nature conservation management, state of the environment reporting and integrated resource planning, though the application to geoconservation has yet to be fully developed.

Similarly, English Nature has divided the country into 120 Natural Areas (97 terrestrial and 23 maritime) on the basis of similarities in landscape, landform and other natural characteristics. The advantages of using these Natural Areas as the basis for developing national and local conservation strategies are seen as:

- offering a more effective framework for planning and achieving conservation objectives than administrative county boundaries;
- providing a framework for appraisal and evaluation of what is important and why, and for setting objectives to protect local landscape diversity;
- enabling conservation to take place in an integrated way rather than focusing on protected sites, and therefore achieving a less fragmented approach to nature conservation. Sites can be better related to each other by seeing them in a wider contect;
- increasing the integration of geological and biological conservation;
- acting as a framework for involving local communities in valuing the distinctiveness of their local areas and developing a sense of place and a place for nature.

Natural Area Profiles have been produced which describe the character, threats and management objectives for each area (English Nature 1998). Although the system can be supported in principle, the implementation as far as geomorphology is concerned means that an opportunity has been missed to crystallise what is valued in local landscape character, how sensitive landscapes are to geomorphological change and what needs to be done to protect landscapes from inappropriate development (Gray 2001).

Scottish Natural Heritage (2002) has followed a similar approach, but with a more systematic description and assessment of the geomorphology of the country and the threats to it. Scotland has been divided into 21 Natural Heritage Zones (now called Natural Heritage Futures). For each of these 21 areas Scottish Natural Heritage has prepared a "Local Prospectus" which describes the natural heritage for the area and the processes that have produced it, discusses the changes taking place, including human impacts, and presents a series of key goals and specific actions which aim to close the gap between the vision and current trends over a 25 year period. The publication includes a CD-ROM describing the earth science interests of each area and the pressures, threats and vulnerability to change.

Landscape Character Assessment has been used in the UK at a finer scale than Natural Areas/Natural Heritage Futures. It subdivides landscapes into Landscape Character Types which share common physical characteristics (e.g. chalk river valleys) and Landscape Character Areas which are unique and discrete geographical areas of a landscape type (e.g. Itchen Valley as an example of a chalk river valley). In Scotland, Scottish Natural Heritage has co-ordinated a comprehensive national programme of Landscape Character Assessment in partnership with local authorities. Twenty nine separate assessments have now been published in which 366 local landscape character types and nearly 4000 individual local character areas have been identified (Swanwick and Land Use Consultants 2002). A similar approach aimed at integrating natural and cultural landscape characterisation on a systematic basis has recently been proposed for Ireland (Heritage Council 2002).

These regionalisations then allow us to do several things. First they allow us to know and understand the character of the landscape and land-forming processes. Secondly, they help us to appreciate the value of local landscape character and what makes it distinctive. And thirdly, on the basis of this knowledge and understanding, they allow us to plan and control future changes more effectively. This can be done on 3 fronts:

- we can try to retain the natural land form or ensure that changes are sensitive to the scale, gradients, morphology, etc. of the existing landscape. There needs to be more authentic geomorphological design in landscape change (Jarman 1994; Gray 1997).
- we should try to allow natural processes to operate

except where human life and property is seriously endangered. For example, we need to learn to work with nature rather than against it by always channelising rivers or building hard coastal defences;

- we need to extend efforts to restore static and dynamic landscapes, in ways that are already being initiated, for example in river restoration schemes, managed coastal realignment or quarry restoration. But geomorphologists need to be involved so that geomorphological principles and authentic design are adhered to.

These objectives can be achieved through a range of measures such as physically excluding the public from highly sensitive areas by fencing, using land-use planning systems to regulate landform change, developing policies, charters or action plans for land form conservation, and educating landscape architects, planners and the wider public of the value of our geomorphological heritage. These ideas on valuing and conserving land form are developed further by Gray (2004).

REFERENCES

English Nature. 1998. *Natural Areas: nature conservation in context* (CD-Rom). English Nature, Peterborough.

Gray, J.M. 1997. Planning and landform: geomorphological authenticity or incongruity in the countryside. *Area*, 29, 312-324.

Gray, J.M. 2001. Geomorphological conservation and public policy in England: a geomorphological critique of English Nature's 'Natural Areas' approach. *Earth Surface Processes & Landforms*, 26, 1009-1023.

Gray, J.M. 2004. *Geodiversity: valuing and conserving abiotic nature*. Wiley, Chichester.

Heritage Council. 2002. *Integrating policies for Ireland's landscape*. The Heritage Council, Kilkenny.

Jarman, D. 1994. Geomorphological authenticity: the planning contribution. In Stevens, C., Gordon, J.E., Green, C.P. and Macklin, M.G. (eds) *Conserving our landscape*. English Nature, Peterborough, 41-45.

Kiernan, K. 1997. Landform classification for geoconservation. In Eberhard, R. (ed) *Pattern & process: towards a regional approach to national estate assessment of geodiversity*. Australian Heritage Commission, Canberra, 21-34.

Scottish Natural Heritage. 2002. *Natural Heritage Futures* (6 theme volumes, 21 area descriptions & CD-ROM). Scottish Natural Heritage, Edinburgh.

Swanwick, C. and Land Use Consultants 2002. *Landscape character assessment: guidance for Scotland and England.* Countryside Agency and Scottish Natural Heritage.

In: Parkes, M.A. (Ed.) 2004 *Natural and Cultural Landscapes - The Geological Foundation*, 175-178, Royal Irish Academy, Dublin

GEOLOGY AND RELIEF OF POLAND IN LANDSCAPE PROTECTION

Zofia Alexandrowicz, Polish Academy of Sciences, Institute of Nature Conservation, al. A Mickiewicza 33, 31-120 Kraków, Poland, Email: alexandrowicz@iop.krakow.pl
Kaj Romeyko-Hurko, Polish Academy of Sciences, Wladyslaw Szafer Institute of Botany, ul. Lubicz 46, 31-512 Kraków, Poland, Email: romeyko@ib-pan.krakow.pl
Jan Urban, Polish Academy of Sciences, Institute of Nature Conservation, al. A Mickiewicza 33, 31-120 Kraków, Poland, Email: urban@iop.krakow.pl

Abstract: Regarding the genesis and its reflection in the relief, three main types of landscape units should be distinguished in Poland: a) young-glacial area of relief connected with Vistulian (Weichselian) ice-sheet, b) old-glacial area characterized by morphology more or less transformed by denudation in Pleistocene and Holocene, c) structural area reflecting lithological properties of the pre-Quaternary formations. The areas are spread as belts parallel to latitude and present different landscapes: lowlands, uplands and mountains. A National project of the landscape protection was concluded in 1971 and enacted in 1991. Typical and the most picturesque landscapes of Poland are covered by 23 national parks and very numerous nature reserves. Specific categories of landscape protection are: landscape parks and areas of protected landscape, as well as natural-landscape complexes. Totally about 33% of Polish territory representing different types of relief is legally protected.

INTRODUCTION

Landscape reflects processes and phenomena existing on the Earth's surface and - in some cases - within the Earth's crust. On the other hand landscape represents an important element of the environment of human life, which influences human consciousness forming our impressionability and imagination. Nowadays landscape also becomes a matter and result of right or improper human activity. Thus the present-day state of landscape is differentiated, because landscape is controlled by natural and cultural elements, which form a mosaic of types.

Particular regions are featured by specific, unique landscapes. Consequently, protection of landscape ought to be individual and regional. The main objective of it is conservation of the peculiarity of features documenting natural, cultural and historical development of the region. Definition of individuality and diversity of landscapes must regard - besides other factors - the elements conditioned by geological structure. These elements determine in great part, the possibilities and ways of human activity and economic use of the areas.

LEGAL BASE OF LANDSCAPE PROTECTION IN POLAND

Landscape protection in Poland started with the first edict on protection of monuments formed in 1918, although the informal traditions had been much older. The edict declared - among others - protection of elements of morphology, whereas subsequent laws enabled conservation of large areas and all elements of landscape. Polish laws on nature conservation enacted in 1934 and 1949 introduced the term of landscape, that was the cause of creation of the type "landscape reserve" in nature reserves classification. In the national parks being established, genesis and transformation of the landscape have been more and more appreciated as an ecological agent controlling biodiversity. A National project of landscape protection was concluded and openly presented in 1971 (Kozlowski 1973, Czemerda 1990). Since this moment landscape parks as well as areas of protected landscape have been systematically established by provincial authorities based first on the law on environment protection and subsequently on the law on nature conservation enacted in 1991 and amended in 2001.

Currently the landscape parks and the areas of protected landscape, together with national parks and nature reserves form the national system of nature conservation which covers about 33% of territory of the country. Moreover local features of the landscape can be formally protected as natural-landscape complexes - a new category of conservation introduced by a law enacted in 1991 with the aim of stimulation of activity of local communities (Alexandrowicz et al. 1992, Bujno 2000).

The importance of the particular landscape parks and areas of protected landscape is confirmed by nature reserves, nature monuments (among them numerous ones of geological and geomorphological value) as well as sites of ecological interest and documentary sites of inanimate nature located within these large-scale areas of nature protection.

National parks and landscape parks play a significant role in the system of landscape protection, because their vast areas represent key features (values) of the regions, most characteristic for Polish territory (Fig. 1). Legally established by Polish government 23 national parks cover 3145.08 km^2 (0.98% of the country area). Five of them have been formally established as international areas of nature protection - biosphere reserves (MAB UNESCO). National parks and their buffer zones are the most representative and picturesque areas of original landscape of Poland. They are concentrated in the southern and eastern part of the country.

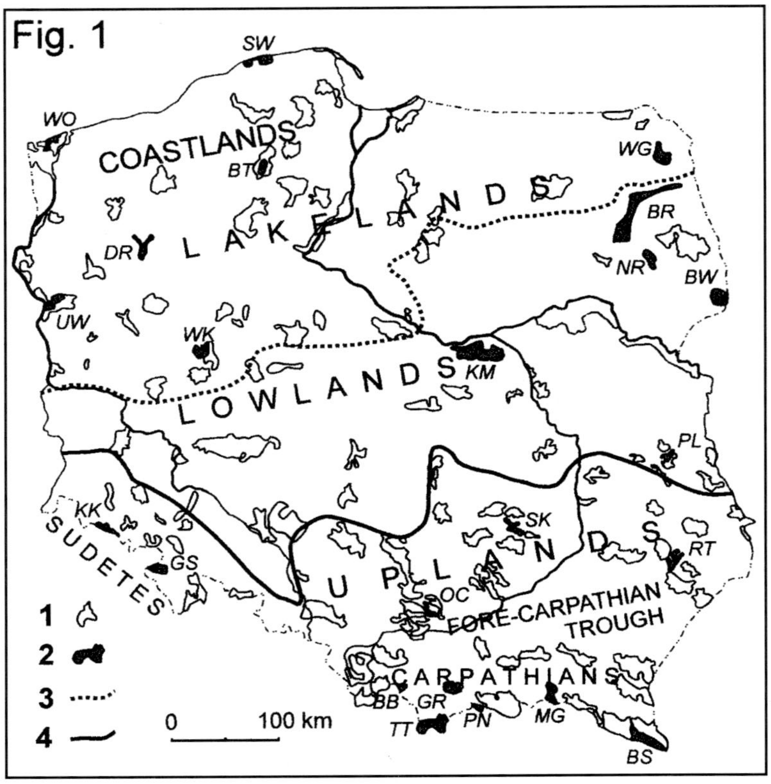

Figure 1. National parks and landscape parks in Poland:
1 - landscape park (after Kurzynski, Mielnicka 2002); 2 - national park: BB - Babiogorski N. P., BR - Biebrzanski N. P., BS - Bieszczadzki N. P., BT - Bory Tucholskie N. P., BW - Bialowieski N. P., DR - Drawienski N. P., GC - Gorczanski N. P., GS - Gory Stolowe N. P., KK - Karkonoski N. P., KM - Kampinoski N. P., MG - Magurski N. P., NR - Narwianski N. P., OC - Ojcowski N. P., PL - Poleski N. P., PN - Pieninski N. P., RT - Roztoczanski N. P., SK - Swietokrzyski N. P., SW - Slowinski N. P., TT - Tatrzanski N. P., UW - Ujscie Warty N. P., WG - Wigierski N. P., WK - Wielkopolski N. P., WO - Wolinski N. P.; 3 - extend of Vistulian (Weichselian) glaciation - young glacial area; 4 - extent of continous cover of Quaternary sediments - old glacial area.

Landscape parks represent areas of very varied types of natural and cultural landscape, modified in different scales by human activity. 119 landscape parks covering 25320,36 km^2 (7.8% of the country area) have been established by provincial authorities so far. Developed network of these parks and their buffer zones represents an essential element, which prevents the fragmentation of natural landscapes caused by current changes of land use.

Strict, rigorous orders are observed in nature reserves situated in the landscape parks and out of them. The nature reserves represent typical landforms, well exposed in the landscape. Approximately 160 nature reserves (some 350 km^2) from total number 1346 (1635,47 km^2) represent great landscape values. The national parks and landscape parks (or their complexes) are managed by special, local administrative centres with scientific and didactic units. Seasonal plans for the parks are made so as to guarantee proper management of these areas.

PROTECTED LANDSCAPES IN RELATION TO GEOLOGICAL AND MORPHOLOGIC REGIONALISM OF POLAND

The landscape of Poland is considerably differentiated. Many processes acting on the various backgrounds have formed the relief of Polish area. In the lowlands of the northern and central part of Polish territory Pleistocene glaciations and subsequently Holocene changes of environment represent the main agents determining the relief. In the mountains and uplands of the southern and central part of the country old geological formations and structures are reflected in the morphology, which has been framed during Tertiary denudation as well as Quaternary denudation and local accumulation. From north to south three main genetic types of the landscape are distinguished: young-glacial, old-glacial and structural (Kondracki 2000). The areas belonging to these types form belts parallel to latitude and characterized by disparate assemblies of landforms. Differentiation of the protected landscape sufficiently reflects geological and geomorphological key features of the particular regions (Fig. 1).

Young-glacial area

This spreads from the Baltic shoreline to the southern margin of the Vistulian (Weichselian) ice-sheet extent. Differentiated relief of coastal lowlands, lakelands with morainic hummocks and ridges cut by numerous river valleys and glacial troughs with lakes characterizes these areas. Along the sea-shore three protected areas: Wolinski National Park with the highest cliff, Slowinski N.P (Biosphere Reserve), with large shifting dunes and Nadmorski Lanscape Park on the Hel Peninsula, also with aeolian forms, represent different types of coastal landscape. The North-Polish Lakelands extend parallel to frontal moraines and sandrs of the last glaciation. Four national parks and several large landscape parks are situated within this area. A few of them are noteworthy: Ujscie Warty N.P., Mazurski L.P. with the Lake Luknajno (Biosphere Reserve) and the largest Polish lake Sniardwy (113 km^2), Suwalski L.P. with the deepest Polish lake - Hancza (108 m) and large fields of erratic boulders.

Old-glacial area

This spreads between the uplands and the southern margin of the Vistulian ice-sheet extent. Glacial relief has been more or less transformed by denudation mostly related to periglacial conditions. In denudation flats fragments of old frontal moraines are conserved as local elevations. Large-space protected areas are located along rivers and depressions covered with swamps and peat-bogs. Kampinoski N.P. near Warszawa enclosing a fragment of ice-marginal valley with wide terraces covered by dunes is interesting owing to these geomorphological features and Quaternary

Figure 2. The castle medieval ruin Ogrodzieniec harmonising with Jurassic limestones as one of the protected landscape component. Photo by Z. Alexandrowicz.

(Holocene) stratigraphy. Several areas of the east part of Poland are of high values, too. There are: Biebrzanski N.P. - a large territory of Holocene swamps and peat-bogs, Narwianski N.P. - a beautiful anastamosing river valley, Poleski N.P. and surroundings landscape parks with covered karst forms developed in Cretaceous limestones and reproduced in Quaternary deposits.

Structural area

This represents uplands and mountains of the southern part of Poland. Its morphology reflects a differentiated geological basement. Polish Uplands enclose three regions spreading from west eastward: the Silesia-Cracow Upland, Malopolska Upland and Lublin Upland. In the first of them folded Upper Palaeozoic formations are covered with Mesozoic ones, which are inclined to north-east as a large monocline. Patterns of morphology are distinctly determined by this structure, particularly along the belt of Upper Jurassic limestones, exposed as hardrocks rising above the Tertiary planation surface. The surface is segmented by rocky valleys and crowned by fantastic limestone tors with ruins of medieval castles (Fig. 2). Phenomena of karst relief are protected in Complex of Jurajskie L.P. and Ojcowski N.P. In the southernmost part of upland the Vistula river valley passes along tectonic troughs between rocky hills (Fig. 3).

Figure 4. Babia Gora National Park - MAB Biosphere Reserve (Fig. 1 - BB) - the range elevated above the timberline; on the first plan the northern slope transformed by several landslides. Photo by S.W. Alexandrowicz.

Figure 3. The area of landscape park near Krakow - the gap of the Vistula river bordered by Jurassic limestones with the medieval Benedictine Abbey on the right side. Photo by Z. Alexandrowicz.

In the Malopolska Upland the Holy Cross Mountains with bordering zone represent the whole geological history from Cambrian till Quaternary (Wróblewski 2000). Large blocks fields of Cambrian quartzite formed during the Pleistocene periglacial climate are the main feature of the Swietokrzyski N.P. Five landscape parks rich in instructive outcrops of different geological formations as well as interesting, mainly denudational, often karst landforms are dislocated around this national park. One of them, Checinsko-Kielecki L.P. was established because of the geological and geomorphological values of the area. The Nida Basin upland situated to the south is famous for Tertiary gypsum with primary depositional structures and secondary karst forms. They are protected in Complex of Ponidzie L.P.

The Lublin Upland is a large plateau formed of Upper Cretaceous carbonate rocks, covered with loess, and in the southern part also with Miocene limestones. At the south border the edge is crossed by some epigenetic river gorges with numerous small waterfalls. Together with a range of limestone hills, the gorges form specific features of the subregion called Roztocze, passing up to Ukraine (Buraczynski 1997). It is protected in Roztoczanski N.P. and five landscape parks.

Figure 5. Bieszczady National Park - International MAB Biosphere Reserve (Fig. 1 - BS) - a typical structural relief formed on resistant sandstones; the highest parts of the range covered with mountain meadows. Photo by S. Michalik.

The southern edge of this range is generated by Tertiary tectonic movements and framed the Fore-Carpathian Trough situated to the south-west. This last region separates the above described uplands from the Carpathians. It is filled with thick complex of Miocene clays and covered with loess, locally also with relicts of fluvioglacial sediments of Sanian ice-sheet.

Mountains elevated during the Alpine orogenic movements - the Carpathians and Sudetes - spread in the southernmost part of the country. The largest mountainous area of Poland and Central Europe - the Carpathians are divided according to the diversity of geological structures into several parts (in Poland): Flysch Carpathians (Outer Carpathians), Pieniny Klippen Belt, Podhale Basin and Tatra Mts (Inner Carpathians).

The Outer Carpathians are formed of thick, folded flysch deposits of uppermost Jurassic to Lower Miocene age. A few facial-tectonic units are distinguished as nappes overthrust northward. Typical litho- and biostratigraphical sequences, tectonic structures and interesting landforms are protected in four national parks and 13 landscape parks (Alexandrowicz and Poprawa 2000). Babiogorski N.P. (Biosphere Reserve) is noteworthy owing to the huge landslide modelling the whole northern slope of the range (Fig. 4). It is the highest range in the Polish flysch Carpathians (1725m a.s.l.). Bieszczady N. P. (International Biosphere Reserve) represents the typical landscape of Eastern Carpathians with mountain meadows above the forests and well visible structural elements of the relief developed on most resistant sandstones of the flysch formations (Fig. 5).

The Pieniny Klippen Belt is a zone of steep and thrust folds as well as strongly dislocated overthrust nappes. Jurassic and Cretaceous limestones and marls are main components of several lithostratigraphic units. Rocks of different resistance form a very picturesque landscape with numerous deep valleys and gorges. The famous antecedent Dunajec River gap is particularly interesting. All these elements are situated within the Pieninski N.P.

The Tatra Mts, the highest chain of Carpathians with the culminating point of Poland (Mt Rysy 2499m a.s.l.) are separated from the above described belt with the Podhale Basin filled by Palaeogene flysch. The core of the Tatra consists of Precambrian metamorphic rocks and Variscan granite. They are overlain by sedimentary formations of Triassic - Cretaceous age, composed mainly of limestones and dolomites, overthrust as nappes to the north. Almost the whole chain represents high-mountain landscape and is protected in Tatrzanski N.P. (Biosphere Reserve).

The Sudetes (Sudety Mountains) ranging along the south-western border of Poland belong to the Bohemian Massif. It is a mosaic of chains formed of very differentiated geological formations: Precambrian metamorphic rocks, sequences of Lower and Upper Palaeozoic deposits, Variscan granites, Permian volcanic rocks, Upper Cretaceous sandstones and Tertiary basalts. In the north-eastern part they are limited by the dislocation zone clearly reflected in the landscape as the edge of mountains. The downthrown part is covered by Tertiary and Quaternary sediments. Two national parks and 9 landscape parks are located in this area (Gawlikowska 2000). Karkonoski N.P. (Biosphere Reserve) encloses the highest part of mountains (Mt Sniezka 1602m a.s.l.). An old planation surface crowned by granite hardrocks and landforms related to mountain glaciers are the most characteristic features of this park. Stolowe Gory N.P. represents the platy mountains, unique in Poland (900m a.s.l.), formed of sandstones, bounded with high rocky scarps. A lot of isolated tors and rocky walls diversify this relief. Slezanski L.P. and Snieznicki L.P. are particularly valuable because of occurrence of old metamorphic rocks and differentiated associations of minerals.

The review of geological structures and geomorphological features of Poland indicates great variety of landscape values corresponding to geodiversity. The main and most interesting components of relief and geological structures are included in the system of landscape protection in Poland. This system composed of national parks, landscape parks and other categories represents an important element of a European network of nature conservation.

REFERENCES

Alexandrowicz Z., Kucmierz A., Urban J. and Oteska-Budzyn J. 1992. *Evaluation of inanimate nature of protected areas and objects in Poland* (English summary). Pol. Geol. Inst., Warszawa, pp. 140.

Alexandrowicz Z. and Poprawa D. (eds). 2000. *Geodiversity conservation of the Polish Carpathians* (English summary). Pol. Geol. Inst., Warszawa, pp. 142.

Bujno S. 2000. *Conservation of nature in Poland. The map 1:750 000* (English explanation). PPWK, Warszawa.

Buraczynski J. 1997. *Geology, relief and landscape of Roztocze Upland* (English summary). Univ. M. Curie-Sklodowska, Lublin, pp. 189.

Czemerda A. 1990. Evolution of the conception of landscape protection in Poland. *Ochrona Przyrody* 47: 359-379.

Gawlikowska E. 2000. *Geodiversity conservation of the Lower Silesia* (English summary)). Pol. Geol. Inst., Warszawa, pp. 72.

Kondracki J. 2000. *Geografia regionalna Polski* (in Polish). Wyd. Nauk. PWN, Warszawa.

Kozlowski S. 1973. The plan of saveguarding Poland's landscape and its initial accomplishment (English summary). *Ochrona Przyrody* 38: 61-83.

Kurzynski J., Mielnicka B. 2002. Tourist function of the landscape parks (English summary). In: Partyka J. (ed.): *Tourism in national parks*: 485-498. Inst. Nat. Conserv. Pol. Acad. Sci and Ojcow Nat. Park, Ojcow.

Wroblewski T. 2000. *Geodiversity conservation of the Gory Swietokrzyskie region* (English summary). Pol. Geol. Inst., Warszawa, pp. 88.

In: Parkes, M.A. (Ed.) 2004 *Natural and Cultural Landscapes - The Geological Foundation*, 179-182, Royal Irish Academy, Dublin

GEOLOGICAL AND LANDSCAPE CONSERVATION - FUNDING AT LAST !

Colin Prosser, English Nature, Northminster House, Peterborough, PE11UA, United Kingdom.
Email: colin.prosser@english-nature.org.uk

Abstract: In Britain, there is a long history of geological and landscape conservation, with legislation allowing for the conservation of important geological, geomorphological and landscape features having been in place since 1949. The subsequent 53 years have seen the methodology used to systematically identify the most important sites become increasingly refined, and the strength of the planning and nature conservation legislation protecting these sites increase. We are now at a point in England, where there is a range of conservation designations working to protect our geology and landscape. These include Sites of Special Scientific Interest, National Parks, Areas of Outstanding Natural Beauty, a World Heritage Site, National Nature Reserves and Local Nature Reserves. All of these attract, or have potential to attract, some government resourcing. A further major development in geological conservation over the last 10-15 years has been the rise of the geological conservation voluntary sector led by the RIGS (Regionally Important Geological/geomorphological Sites) movement. Here a considerable amount of volunteer time and effort has gone into identifying locally important RIGS and planning work to record, manage and interpret them for public use. Unfortunately, funding to drive this work forward has not been easy to secure. However, this is about to change, as large sums of money are now available for aspects of geological, geomorphological and landscape conservation as a result of a new levy on the extraction and sale of primary aggregate. Funds raised through the levy are to be distributed through a Sustainability Fund, targeted at achieving a number of environmental improvements in areas where aggregate has been extracted. One target area of the fund is to conserve and manage geological features. Sustainability Fund grants should enable millions of pounds to be made available for geological and landscape conservation in England, and provides an opportunity not to be missed.

CONSERVATION HISTORY AND LEGISLATION

In Great Britain, there is a long history of geological, geomorphological and landscape conservation. This is largely due to the fact that the diversity of geology (used throughout this paper to include geomorphology) and landscape which occurs in Great Britain, and the significance of British geology in terms of the historic development of the science of geology, have long been recognised by those interested in conservation. Furthermore, increasing development pressure, in what is a relatively small and densely populated country, has provided enough threat and damage to important geological sites, to convince those responsible for making conservation legislation, that conservation of geological features and landscape is a necessary activity.

In Great Britain, the origin of geological conservation, founded primarily on scientific importance, and landscape conservation, based on more aesthetic and cultural importance, goes back more than fifty years. It was 1947, when the Wild Life Conservation Special Committee, created by the Government of the day to examine how they could better engage in a national nature conservation effort, published their report "Command 7122". This set the direction for nature conservation in Great Britain, and led to the passing of the The National Parks and Access to the Countryside Act of 1949. This Act required local authorities to be notified of the location of important "geological or physiographical" (now called geomorphological) features, as well as making provision for the creation of National Parks and Areas of Outstanding Natural Beauty. These three 'designations' still form the backbone of geological and landscape conservation in Great Britain.
The designation of land as an SSSI (Site of Special Scientific Interest), provides the basis of geological conservation in Great Britain. This designation flows from The Wildlife and Countryside Act (1981), and was strengthened recently by The Countryside and Rights of Way (CROW) Act (2000). All candidate SSSIs were identified through the Geological Conservation Review, a systematic and scientific assessment of the geological and geomorphological resource of Great Britain (Ellis (ed) 1996). There are approximately 1240 SSSIs with a geological interest in England alone, making a significant contribution to geological and implicitly to wider landscape conservation. Biological SSSIs too, help protect important areas of landscape, especially in upland areas where large areas of land may be notified.

Broader landscape designations in England form part of the remit of the Countryside Agency, and include National Parks run by National Park Authorities (there are seven accounting for 7% of the land), Areas of Outstanding Natural Beauty (there are 35 of these 'jewels of the English landscape' covering 15% of England) and Heritage Coasts, scenic stretches of coastline managed to conserve their 'natural beauty'.

A number of other designations also exist which can help with the conservation and management of geology and landscape. These include:

- NNRs (National Nature Reserves), where an area of land of SSSI quality is managed primarily for nature conservation (most SSSIs are managed primarily for other purpos-

Figure 1. Aggregate extraction, Derbyshire, England. The Aggregates Levy, a tax on the extraction and sale of primary aggregate, is generating a Sustainability Fund, providing funding for geological and landscape conservation projects in areas of aggregate extraction. (Photo. Colin Prosser, English Nature).

es) it can be declared and managed as an NNR. The Stiperstones NNR in Shropshire, illustrates a geomorphological feature managed in this way.

- LNRs (Local Nature Reserves), declared by a local authority on any land they wish to manage for nature conservation. LNRs tend to be small but still make a significant contribution to geological conservation.

- RIGS (Regionally Important Geological/ geomorphological Sites), provide a conservation option that has appeared in Great Britain over the last 12 years. RIGS sites can be designated for a range of features, including, geology, geomorphology, or landscape. Although not afforded statutory protection, many RIGS are brought to the attention of local planners, and thus inform decision making with respect to development proposals that may impact on a RIGS.

- A recent, and much publicised development has been the World Heritage Site status granted to the Dorset and East Devon Coast. This adds to the SSSI and international wildlife designations that already apply to this stretch of coast, and provides a new 'tool' with which to deliver geological and landscape conservation.

In summary, the last fifty years has seen the methodology for the systematic identification of the most important geological and landscape sites become increasingly refined. There are now a wide range of designations protecting England's geology and landscape, most of which have some associated government funding, or in the case of the World Heritage Site, potential to attract funds.

ENERGY AND IDEAS BUT NOT ENOUGH CASH

Over the last 10-15 years, a major development in geological conservation in Great Britain has been the rise of the voluntary sector. Led by the RIGS movement, a considerable amount of volunteer time and effort has gone into identifying locally important sites and planning the work needed to record, manage and interpret them for use by geologists, schools and the general public. The concept behind RIGS, and the aims of the RIGS movement can be gleaned from Harley (1994), Prosser and King (1998) and UKRIGS (2001). In short, RIGS groups are made up of geologists, planners, teachers, conservationists and others, working at a county level at identifying, managing and promoting locally important geological sites.

RIGS groups provide a strong impetus for local conservation activity, and many projects aimed at managing and interpreting geological sites have been delivered or proposed by RIGS groups (eg. Campbell and McCall 2002). These projects, however, often require resources if they are to happen, and funding to take forward this kind of work has not traditionally been easy to secure. This situation has changed dramatically, since April 2002, as a result of a new tax on extraction of primary aggregate, and large grants are now available for geological and landscape conservation work where a link to aggregate extraction, past or present, can be demonstrated.

THE AGGREGATES LEVY - A TAX ON QUARRYING

On April 1 2002, the Aggregates Levy, introduced by the Finance Act 2001, came into force in Britain. This is a tax on the commercial exploitation of aggregate, and has been set at an initial figure of £1.60 per tonne. The argument for the Levy is that there are significant environmental costs associated with quarrying that are not already covered by regulation. These include noise, dust, visual intrusion, loss of amenity and damage to biodiversity. It is intended that the Levy will bring about environmental benefits by making the price of aggregates better reflect these costs, and by encouraging the use of recycled materials and certain waste products which are now subject to the Levy. This is one of a number of 'green taxes' being considered or implemented by the Government. It has been broadly welcomed by environmentalists and is strongly opposed by the aggregates industry. The Levy applies to primary extraction of gravel, sand and rock to be crushed as aggregate. At present, clay, coal, slate waste and china clay waste etc is exempt from the Levy.

Figure 2. Dryhill Quarry, SSSI and LNR, Kent, England. A disused quarry that is well managed and provides open access for scientific, educational and recreational use. The Sustainability Fund should lead to more sites providing a geological resource of this type. (Photo. Colin Prosser, English Nature).

THE SUSTAINABILITY FUND

The major significance of this Levy for those with an interest in geological and landscape conservation is that part of the money raised through the Levy, approximately 10%, will be used to create a Sustainability Fund. Although only guaranteed for an initial period of two years, the Fund is aimed at delivering a wide range of environmental improvements, and at helping to promote and develop a more sustainable aggregates industry. In England, a total of £ 29.3 million per year will be distributed through English Nature, English Heritage, the Countryside Agency, three local authorities and other bodies concerned with the running of a sustainable minerals industry. English Nature's grant scheme relates to increasing biodiversity and conserving/ enhancing geological features in any area affected by aggregates extraction. Given the strong link between geology and mineral extraction, this provides a great opportunity to make the most of the geological features and sites exposed through aggregate extraction. The Countryside Agency, with its remit for landscape conservation, will be able to use Sustainability Fund grants to support projects relevant to broader landscape conservation in areas affected by aggregate extraction.

WHAT CAN BE FUNDED?

In terms of geological and landscape conservation, grants will be available from the Sustainability Fund, distributed by English Nature, The Countryside Agency and English Heritage, to fund geological and landscape conservation work including site management, interpretation, access and promotion, and specimen rescue - providing that the project proposal has a direct link with aggregate extraction, past or present. Although only guaranteed for two years, this Fund should enable millions of pounds to be made available for geological and landscape conservation and promotion in England. These funding levels are vastly in excess of anything available previously, and provide an opportunity not to be missed. The potential to support the work of the voluntary sector relating to geological and landscape conservation is now greater than ever, and should result in a whole range of new projects being delivered. A similar scheme has been developed in Wales but as yet nothing is in place in Scotland.

WHAT WILL WE GET?

The hope is that the guaranteed funding for two years will enable many geological groups, working with a large number of partners in industry, local government, professional and amateur groups and societies, deliver projects that will both conserve and promote geology and landscape. Imaginative projects that celebrate the link between aggregates extraction and geology are particularly welcome, whether it is on site management, interpretation, or access work, or projects that deliver these ends through off-site interpretation (eg. leaflets, displays of material or web based information). Applicants will need to emphasise the links between their projects and aggregate extraction, and demonstrate how the local community will benefit as a result of the project.

Although already in place, and not funded through the Sustainability Fund, Dryhill SSSI and LNR, a disused quarry near Sevenoaks, Kent (Figure 2), illustrates the sort of end result expected from many projects. Here the disused quarry is well managed, has interpretation panels, fencing and paths, and provides safe and open access to the public, allowing recreational, educational and scientific use.

The Sustainability Fund offers a golden opportunity for geological and landscape conservation in Great Britain, and it is an opportunity that must be exploited to the full. It also provides a model to be considered in other countries around the World.

REFERENCES

Campbell, M. and McCall, J. 2002. RIGS' first decade.

Geoscientist, 12 (8), pp. 12 - 13. The Geological Society, London.

Ellis, N.V. (Ed), Bowen, D.Q., Campbell, S., Knill, J.L., McKirdy, A.P., Prosser, C.D., Vincent, M.A. and Wilson, R.C.L. 1996. *An Introduction to the Geological Conservation Review*. GCR Series No. 1, Joint Nature Conservation Committee, Peterborough.

Harley. 1994. The RIGS (Regionally Important Geological / geomorphological Sites) challenge - involving local volunteers in conserving England's geological heritage. In: (O 'Halloran, D., Green, C., Harley, M., Stanley, M., and Knill, J. ed. *Geological and Landscape Conservation*. Proceedings of the Malvern International Conference 1993. pp. 313-317, The Geological Society, London.

Prosser, C.D., and King, A.H. 1998. Regionally Important Geological and Geomorphological Sites: the origin and a forward view. In: Oliver, P.,G. (ed.) 1998. *Proceedings of the first UK RIGS conference*, pp. 1-8. Hereford and Worcestershire RIGS Group, Worcester.

UKRIGS. 2001. *UKRIGS development strategy 2001-2005*. pp. 1-27. Association of UK RIGS Groups.

Postscript: In December 2003, it was announced by the UK Government, that the Aggregates Levy Sustainability Fund grant scheme would be extended for a further three years until March 2006.

In: Parkes, M.A. (Ed.) 2004 *Natural and Cultural Landscapes - The Geological Foundation*, 183, Royal Irish Academy, Dublin

ITALIAN LANDSCAPE PHYSIOGRAPHIC UNITS

Marisa Amadei, Roberto Bagnaia, Lucilla Laureti, Francesca R. Lugeri, Nicola Lugeri - Dipartimento per i Servizi Tecnici Nazionali - Servizio Geologico Nazionale, via Curtatone 3, 00185 Roma, Italy. Email: nicola.lugeri@dstn.it

Abstract: The "Italian landscape physiographic units map" allows to identify - to a regional scale (1:250.000) - homogeneous portions of the land surface characterized by an arrangement of structural elements of the landscape and by a typical pattern from a morphological, lithologic, and land-use point of view [1].

The subject is part of a very ambitious project (Map of the Nature) defined by a state law, that aims to providing an assessment and planning tool in the field of environmental quality and protection. At the chosen scale (1:250,000), it arises that the main information on the territory is provided by the "landscape units", as defined from a synthetic approach mainly based on the physiographic aspects of the territory.

The project is completed by the map of Biotopes (scale 1:50,000) that is in progress (about 20% of the national territory is planned to be mapped by the end of the year), to which the assessment of natural value and threat will be referred.

The realisation of the map has principally followed remote sensing methods, mainly working on aerial photographs (Volo Italia, 1989; ~1:75,000). Moreover, field surveys have been performed and ancillary data used, too.

The observed shapes have been organised in a number of landscape types. They are theoretical typical associations of land features recognisable and are not referred to a geographically specific portion of the land.

The landscape units, that do refer to specific portions of the land, have been then recognised starting from the types, but defining their local and unique features. Finally, the landscape types and units, with related descriptions, have been mapped and organised in a GIS.

References

[1] - Amadio V., Amadei M., Bagnaia R., Di Bucci D., Laureti L., Lisi A., Lugeri F.R., Lugeri N.: "The role of geomorphology in Landscape Ecology" in Applied Geomorphology: Theory and Practice R.J: Allison, ed. John Wiley & Sons (2002) and references therein

GEOLOGICAL ASPECTS IN ECONOMIC EVALUATION OF A REGION (HUNGARY)

Dr Géza Chikán, geologist, cartographer, economist
Geological Institute of Hungary, H-1143 Budapest, Stefánia út 14. Hungary.
Email: chikan@mafi.hu

Abstract: The registration of raw materials of Hungary is in progress in different institutions from many years. Specifications of our raw material's stocks are made annually. This work helps the mining and the planning of obtaining.

My opinion is that the economics in geology means not only the registration of raw materials but it means all things coming from economic perspectives caused by geological characteristic of formations too. The joining to European Community demands a special regional view in appreciation of economic potential of geological formations. Therefore I would like to show the possibilities and importance of a Hungarian region: the Western Mecsek area.

Main tasks of my presentation are:

1. **The place and role of Western Mecsek area in the South Transdanubian Region**
2. **Geological formations of Western Mecsek**
3. **Economic potential of geological formations**
    - **Raw materials**
    - **Utilization of geological characteristic**
    - **Tourist value of geological formations**
    - **Other possibilities of utilization**
4. **Mineral raw materials**
    - **Types**
    - **Quantity, role, quality**
5. **Role of geological situations**
    - **Types of deposits**
    - **Tectonics**
6. **The perspectives of utilization of geological formations in the Region**

SESSION 6

HOW SHOULD WE MANAGE LANDSCAPES AT A LOCAL SCALE?

Cliff retreat on coasts, in unconsolidated sediments is a problem in many countries. "Managing" such dynamic natural processes is often possible with engineering solutions but may have consequences elsewhere along the coast which are not easily modelled. Photograph of cliffs near Greystones, Co. Wicklow: Matthew Parkes

In: Parkes, M.A. (Ed.) 2004 *Natural and Cultural Landscapes - The Geological Foundation*, 187-192, Royal Irish Academy, Dublin

GEOSITE MANAGEMENT - A WIDELY APPLICABLE, PRACTICAL APPROACH

W.A.P. Wimbledon, A.F. Barnard and A.G. Peterken, Countryside Council for Wales

KEYNOTE ADDRESS

Abstract: Geoconservation is the protection of key geosites or terrains for use, in research, training, education and understanding. This paper introduces a simple method to manage both the geological resource of geosites and the factors which conceal or damage the resource.

Many countries manage reserves and parks. Some conserve smaller or distinct geological monuments or areas by law, recognising the wider value of geological features. Most of what is called "management", however, is about managing people, providing paths and signs, educational facilities, controlling or informing visitors, a touristic experience, managing the infrastructure, or land-use other than geological usage.

Some countries combine legal protection with a functioning town and country planning system: the essential for protection. In Britain, conservation, including geoconservation, has by statute been an established governmental function only since 1949 (NP&AC Act 1949: W&C Act 1981; CROW 2000: Black, 1995). In this, management has been largely ad hoc, it has consisted of management of people and sometimes cleaning of geology. However, only in recent times has coherent site management planning been started, to examine the needs of tangible scientific features, which are the fundamental basis and justification for conservation of sites (in Britain as Sites of Special Scientific Interest (SSSIs)). The Countryside Council for Wales has pioneered a new method of "minimum-format management plans" (MFMP) for geological SSSIs: to put sustainable management of this irreplaceable resource first, before other considerations.

Here we detail this site management method. It depends on a full understanding of the scientific features and sub-features and a methodical assessment of their needs, so that use (scientific and educational) may continue. No meaningful monitoring of change, of damage or improvement can happen without the existence of such a management plan.

INTRODUCTION

The diverse geological heritage of Britain is one of the features that make it unique. Internationally, sites have been given increased recognition and priority by creation of the GEOSITES programme by the International Union of Geological Sciences (IUGS). Its aim is to create a "representative selection of internationally significant sites". The first European framework (Ishchenko et al. 1998) includes an important inventory of British sites. This British Geosite listing is based on the Geological Conservation Review (GCR) selection of SSSIs (Wimbledon et al.1995), completed in 1991, and still being described in the GCR published volumes. In this project, Britain was divided into selection categories, GCR 'blocks', and sites selected in these contextual frameworks.

Within wider landscapes sit small and large geosites, windows into the geology, which underlies and controls that landscape. Many geosites are of the highest scientific importance. In Wales, individual GCR sites total more than four hundred, and, although we have no active volcanoes or deserts, there is otherwise great geological variation, seen in coasts, caves, karst, igneous, stratigraphical and fossil sites.

The geology of Wales, and Britain, has played a great part in global scientific history, and its diverse range of sites and their treasures present great possibilities for the future, in science, education and geotourism. Such would benefit both local communities and site protection. But before sites can be managed and protected in a coherent manner, or exploited in any way for education or tourism, they need a management plan to provide a methodical framework for action.

This paper introduces a simple method to manage the geological resource provided by geosites. Only in recent times has coherent site-specific management planning been implemented to examine the needs of features, which are the justification for conservation of sites as SSSI's. The Countryside Council for Wales has pioneered a new method of "Minimum Format Management Plans" (MFMP) for geological SSSI's: to, for the first time, address the sustainable management of this irreplaceable resource. The Countryside Council for Wales (CCW) is a front-runner in management planning, their guidelines being endorsed at the 1993, 5th conference of the RAMSAR contracting parties, Kushiro, Japan. A modified form of this approach to suit geological needs is described here.

AIMS AND APPROACH

Britain has a formal planning control system (by local authority councils, central government) of major schemes and developments, and this acts as a 'safety net' to stop damage to SSSI's. But a management plan is an essential internal tool for the day-to-day management of a site, to

Figure 1. Panoramic (180 degree) view of U.K. site.

monitor if it is in "favourable conservation status" (FCS = assessment based on current state of the feature plus assessment of its long term prospects). Geoconservation is one land use amongst many, but when considering geosite management there is an imperative that local uses do not take priority before geology of national and international significance.

A management plan is a site specific, comprehensive and influential document aiming to further the interest of the site and determine the factors affecting it. In Wales, the management plan writer starts the process with the site's GCR interest already defined, and delimited on a map. Management planning then deals with the localisation and documentation of the site's primary features of interest on site. The start of the management plan contains a very brief outline of the salient interest e.g.
"Wern Ddu claypits is a disused opencast clay and coal working situated to the Southeast of Caerphilly. Its rocks are very important for understanding the Westphalian history of the SE South Wales Coalfield".

Further documentation of the site must be based upon field observations, real geology, not a generalised description culled from literature. Descriptive documentation may be enhanced by the use of sedimentary logs, sketches and photographs (or even commissioned research). This information forms the kernel of the management plan.

HOW DOES MANAGEMENT PLANNING WORK?

The process is:

- Identify features/sub-features of primary interest
- Set objectives for the interest(s)
  (what is "FCS" for the site - a vision)
- Identify performance indicators, which will allow assessment of the features against the objective - these include:

Identify factors affecting the interest(s) of the site
Splitting down the site into compartments, based on features and their physical distribution (some compartments will have no interest)
Define quantifiable "attributes" for sub-features. Simplest method: to define x sq metres of rock, or % of total area, so as to measure change, for good or bad

- Assess overall "conservation status" of the site
- Write management rationale and management projects

IDENTIFICATION OF FEATURES

First we identify the features of wider interest and importance. It is important not to base the management plan on literature, but to know the site and general area thoroughly. All British sites were selected by the GCR and in a geological selection category ('block'), which is taken as the "feature" for the purposes of management. The feature is the primary interest, for instance the GCR 'block' Westphalian stratigraphy, and this gives a descriptive setting for the sub-features, the individual marker beds, fossil-bearing levels, that collectively make up the feature, conceptually and in tangible geographic sense on the ground. The "sub-features" are the geology of the site that is recorded in detail, and which it is important to have accessible.

The scale of the sub-features will obviously depend upon the diversity and extent of the geology. It may not be possible to find millimetre sized remains within an obscure exposure on site. Not all things may be determined in the field and therefore literature and research results must be relied upon. Although in the majority of cases it is possible, with a little survey work, to define tangible and more obvious sub-features. Once the valuable sub-features have been identified, locations for these primary interests can be definitively surveyed.

Photographic coverage of feature and its sub-feature locations is a standard requirement of the documentation process. With the advent of the digital photographic darkroom, it has a wider, seemingly limitless, application (Figure 1).

CONSERVATION AIMS - SETTING OBJECTIVES

The objective is always to keep the feature in a condition that is beneficial for future understanding of the site, in "favourable conservation status". Intrinsic in keeping the feature in FCS is the maintenance of all the constituent sub-features. A site is in FCS when its state enables a full understanding of the primary interest (GCR interest) and will do for the foreseeable future. It may be expanded upon, to contain extent of the interest or access route to point of interest. A typical example of an objective for the interest would be:

> "The exposures of rock within the site boundary will contain sufficient quantity and quality of outcrop so as to enable a full understanding of the feature. This amount will be sustainable in the long term."
>
> "Management will aim to improve exposure and access of the rock outcrop on this site. At least 50% of the 6ha site will have clean and accessible exposures of Westphalian strata. This includes a complete exposed section through the stratigraphy. Coals A and B, and marine bands 1 and 2, will be kept exposed."

This is a generic statement, made as a reference point.

Figure 2. Subfeatures from left to right: 1) exposed bedding plane with ammonites, 2) formation boundary - marine to non-marine, 3) fossil soil, 5) erosion surface on charophyte-bearing lake micrites, 6) marine shell bed, marking transgression. Project: 4) concealed interval to be exposed.

FACTORS

Factors are the processes and pressures that affect a site. These are the issues that management of the site aims to control. The Factors section of the management plan may include past and current land management and it forms the 'backdrop' to the site's current condition and usage.

Factors impact upon the condition of a site and may be good or bad. Anything from the owner's prerogatives to the growth of

rare plants may be included. In general, factors are defined succinctly and not expanded upon. For example, "growth of scrub or vegetation on rock face/ledges and quarry floor" is clear explanation of a factor that adversely affects the geology. In certain specific locations, for example, quarries or public open spaces, it may be relevant to highlight hazards due to deterioration as separate safety factors.

"Factors" may be internal to the site or external, good or bad. We tend to concentrate on the bad ones. Factors may be natural or Man-made effects. For instance:

- Natural decay, erosion, sedimentation, soil wash, talus accumulation
- Grazing by livestock, trampling damage
- Vegetation/root growth
- Tipping of waste
- Change of drainage, flooding

Factors fall into two main categories, those that directly affect a site and those that cause a site to be more vulnerable.

SUB-FEATURES

Sub-features are the core areas, which are central to understanding and valuation of the locality (the feature being the GCR interest for which it was selected). They are also physical entities. A mineral vein, a suite of stratigraphic units in a larger site, or a single fossil-bearing bed or a volcanic pipe can be measured.

In all cases, it is important to establish a baseline for monitoring such important areas. This is where digital photographic coverage is invaluable. For most sites it is possible to acquire images of the sub-features and combine, manipulate and annotate them accurately, giving a means to document and assess all change. Figure 2 shows a set of subfeatures.

Quite often sub-features occur in close proximity and it is not unusual for a collection of defined sub-features to all have the same factors impacting upon them. Management issues being the same, there is no reason to delimit and assess each one singly. Areas, termed "compartments", may be defined which consist of collections of sub-features.

In Figure 2 the key stratigraphic horizons have been highlighted to define important areas to be given "attributes". An attribute provides a scientific and areal basis for a condition assessment of any given sub-feature.

ATTRIBUTES

- They are measurable characteristics or qualities of the site's interest (sub-features for instance)
- They can be recognised and defined on the ground, on maps and in digital photography to give the exact extent of defined elements of geology/geomorphology
- Attributes are the characteristics, qualities or properties of selected sub-features.
- It may be decided that all the sub-features should be attributes if the interest(s) are limited and vulnerable, only some sub-features may be need this treatment.

"Attributes" are central to this planning technique. Individual features may be given one or many attributes. They are employed as a convenient 'yardstick' to measure change. For instance, a specified area of a coal seam or a marine mudstone in a coal-measures cycle may be used as an attribute: its extent being employed as a measure or indication of condition of the whole stratigraphic section. In Figure 2 the sub-features are used as attributes.

When choosing attributes, the following should be noted: an attribute is a convenient means to measure the presence or absence of sub-features and it measures also the impact of factors. Attributes should be good indicators of the feature, and should be measurable. They can be economical surrogates, for example, the presence of a number of marker bands across a site is sufficient to identify the key localities.

Table 2 shows a hypothetical example.

Sub-Feature	Targets	Attribute % of outcrop
Bed 27	Current Condition 03/02	70
	Lower Limit	50

Table 2

In this example, 70% of the potential rock exposure, of the sub-feature - bed 27, a coal seam - is exposed. It is decided that there should be a 20 % "buffer zone" before reaching the set lower limit, and it is the manager's judgement that 50% of the outcrop is the minimum acceptible level to meet FCS, When the extent of exposure deteriorates to 50%, 'alarm bells' start ringing and a response will be triggered. This response is likely to take the form of a management project such as cleaning and clearing of identified sub-standard areas.

Where the current condition of a sub-feature is above a selected lower limit, no action is needed: the site is in acceptable condition (FCS). But if the condition falls below the specified lower limit, a project is required to bring the site back up to "favourable" status.

The measured attributes indicate that this sub-feature is in within the limits of acceptable condition, and work is unlikely to be needed in the near future. (It has to be remembered that attributes are chosen as indicators of condition, and are not ncessarily the whole site: all subfeatures present on a site may be chosen to be attributes, but this is not always the case.)

CONSERVATION STATUS

Having identified an area of importance (sub-feature(s)) and defined attributes for that entity (for instance, x sq metres of outcrop, delimited on photographic coverage), it is possible to assess if the sub-feature is in the required condition or not.

A statement can be made defining the overall condition of the site based on assessment of the attributes. The conclusion of this process is to give the site a "status", at a point in time (the last site inspection). Status ranges through:

Favourable maintained
Favourable recovered
Favourable recovering
Unfavourable no change
Unfavourable declining
Destroyed

If a site has an unfavourable conservation status, projects are defined to bring the offending attributes (relating directly to individual areas and specific sub-features) into a favourable status. Projects undertaken would control factors, undo damage and decay, and enhance key elements.

Unless a site is kept clean by natural erosional processes, it is likely that it will fall into the unfavourable declining category. Even recently cleaned sites will soon start to decline. It is only on the day that work is being carried out that the site is, in condition terms, actually improving. If decline is not yet threatening our "full understanding" then the site has a "favourable maintained" status: if decline takes it below the limits set, then it is "unfavourable declining". It is possible that a small number of sites may show no change, being static or imperceptibly changing, although these are rare examples.

RATIONALE

The final statement of the management plan, is the rationale. The rationale summarises the key management issues. The management issues include those factors it is necessary to manage to ensure that favourable condition is sustained. The rationale is also used to add detail, and to focus in on other important site-specific problems, such as the impact of visitors and users. This section of the management plan follows the vision the writer has for the site, the status of the feature is now known and the rationale has highlighted the key management issues and the factors to be controlled. Next, these management issues will be addressed in the management projects.

PROJECTS

Projects may be specific to part of or the entire site. Management projects attempt to combat the effects of negative factors and bring the site back into favourable condition. For projects to be effective they need to target both the factors that negatively affect the site and the individual attributes/sub-features stating precisely intentions: what needs to be done, where and how, costs, methodology etc.

Management projects are easily documented with diagrams. A simple example (fig. 3) is a small, disused quarry situated in a residential inner-city area. A path running along the top of the cliff presents a danger to the public. Fencing the path would result in the deterioration of access to the site: whereas excavation of a recessed path (in compartments 1a and 1b) would provide safe access and new exposures. Another project (in compartment 5) would be to expose a concealed bed containing primitive fossil plants. Completion of projects will undoubtedly hinge upon good relationships with owner/occupiers. In many

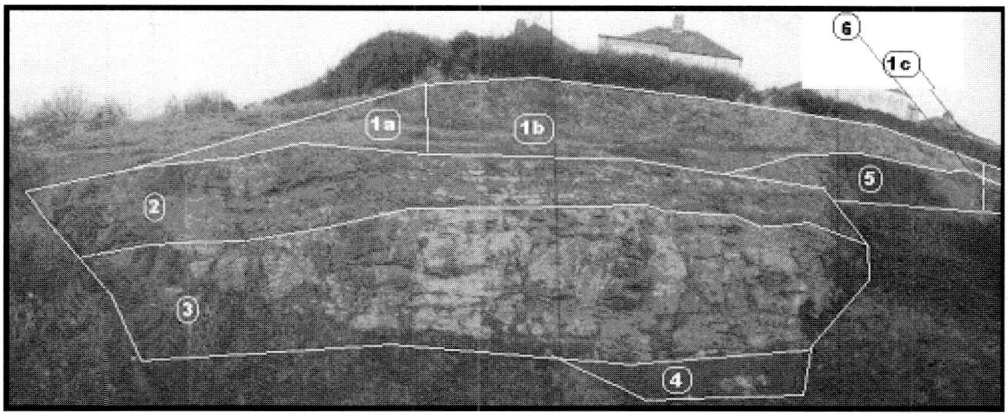

Figure 3

casework issues there will be a compromise between what is of benefit to the site and the prerogatives of the owners and/or occupiers.

The MFMP may require a formal Agreement with an owner, and this would allow financial payments to be made to the owner as long as specified beneficial site use and management occur.

MONITORING

Implementation of a management plan requires subsequent monitoring, the frequency of which is determined by the degradation potential of the site, which in turn is dictated by the factors affecting the site. The initial photographic coverage is used as a baseline for subsequent and on-going photographic monitoring and definition of sub-features. For monitoring to be meaningful it must relate directly to the sub-features and their condition as assessed in the MFMP.

FOLLOW UP AND CYCLE

This management planning methodology assesses factors touching a site, their control, and the best management for the benefit of the primary geological resource. It can be used as a practical and widely applicable tool to control the condition of a site, decay, use and misuse. The aim is to achieve sustainable management, and therefore must be shared with site owners and managers, as essential partners in protection. This is because there can be no action on a site that is not affected by the management plan, as there can be no monitoring without a management plan having been produced as that defines absolutely the baseline for substantive assessment of improvement or decay.

There is a clock built into this process and a cyclical method, it is:

1. Write the management plan
2. Carry out the projects
3. Monitor the site
4. Review the plan, then re-write if necessary

Importantly, there is no such thing as a finished management plan, it is constantly evolving, and will need revision to remain useful.

In the examples given there is a bias to stratigraphic sites. The geological diversity of Britain is expressed in a vast array of protected sites (in 100 GCR 'blocks'), with varying management issues. But it is still possible, through minor modification, to fit this MFMP method to all categories of site.

CONCLUSION AND EUROPE

To keep sites in favourable condition (FCS), we need to manage them. The Minimum Format Management Plan (MFMP) is the mechanism that delivers "management planning", pulls together on-site observations and key information on the primary scientific interest of the site, and all matters concerning its condition and use. Our method depends on a full understanding of the geological features and sub-features and a methodical assessment of their needs, so that use (scientific and educational) may continue. No meaningful monitoring of change, damage or improvement can happen without the existence of such a management plan. The corollary is that monitoring is an essential in managemnt planning.

A management plan format that is applicable across Europe appears to be a paradox, due to the geodiversity, size and location of sites. However, it is not unreasonable to consider implementing a standard European method and approach to management planning for the geological resource. This would be an extremely useful tool in the protection of important sites.

For more information on the details of this management planning technique, contact: b.wimbledon@ccw.gov.uk, a.barnard@ccw.gov.uk or a.peterken@ccw.gov.uk

ACKNOWLEDGEMENTS

We take full responsibility for our adaptation of the method, but are very happy to acknowledge the great and fundamental contribution of our CCW colleague Mike Alexander to the development of management planning.

REFERENCES

Black, G.P. 1995. pp 179-272 In: Barrett-Hall, R. Environmental Law. Sweet & Maxwell, London 1168 pp

Countryside & Rights of Way Act, 2000. Her Majesty's Stationery Office.

Ishchenko, A.A., Gerasimenko, N.P., Alexandrowicz, Z, Vinokurov, V., Liscak, P., Vozar, J., Vozarova, A., Bezak, V., Kohut, M., Polak, M., Mello, J., Potfaj, M., Gross, P., Elecko, M., Nagy, A., Barath, I., Lapo, A.,Vdovets, M., Klincharov, S., Marjanac, L., Mijovic, D. Dimitrijevic, M., Gavrilovic, D., Theodossiou-Drandaki, I., Serjani, A., Todorov, T., Nakov, R., Zagorchev, I., Perez-Gonzalez, A., Benvenuti, M., Boni, M., Brancucci, G., Bortolami, G., Burlando, M., Costantini, E., D'Andrea, M., Gisotti, G., Guado, G., Marchetti, M., Massoli-Novelli, R., Panizza, M., Pavia, G., Poli, G., Zarlenga, F., Satkunas, J., Mikulenas, V., Suominen, V., Kananoja, T., Lehtinen, M., Gonggrijp, G.P., Look, E.R., Grube. A., Johansson, C. E., Karis, L.O., Parkes, M., Raudsep, R., Andersen, S., Cleal, C.J., Bevins, R., Wimbledon, W.A.P. 1999. A first attempt at a GEOSITES framework for Europe - an IUGS initiative to support recognition of world heritage and European geodiversity. *Geologica Balcanica* 28, 5-32.

National Parks & Access to the Countryside Act, 1949. Her Majesty's Stationery Office.

Wildlife & Countryside Act, 1981. Her Majesty's Stationery Office.

Wimbledon, W.A.P., Benton, M.J., Bevins, R.E., Black, G. P., Bridgland, D. R., Cleal, C. J., Cooper, R G. and May, V.J. 1995. The development of a methodology for the selection of British sites for conservation. Part 1. *Modern Geology* 20, 159-202.

In: Parkes, M.A. (Ed.) 2004 *Natural and Cultural Landscapes - The Geological Foundation*, 193-195, Royal Irish Academy, Dublin

LOCAL LANDSCAPE - LET PLANNING DECISIONS BE BASED ON GEOSCIENCE ARGUMENT

Krzysztof Miskiewicz, Institute of Nature Conservation, Polish Academy of Sciences, Al. Mickiewicza 33, 31-120 Krakow, Poland. Email: dex16@wp.pl

Abstract: The natural values are both of great significance and very difficult to preserve in areas of intensive agricultural use, where only small pieces of land have remained untouched, thus forming enclaves of natural landscape. In such cultural landscapes where arable fields, housing areas, roads and power transmission lines prevail, any landscape, scientific and didactic values of nature decline due to man-made transformations. Thus, it is of utmost importance to preserve such small landscape components to prevent environmental devastation due to our ignorance and to gain new significant sites of an inanimate nature. An appropriate planning and conservation of such small landscape features forms a base for large-scale protection activities and often involves the successful protection of unique geological sites of great scientific importance. Local planning decisions can have a formal basis to lay the foundations for conservation of key geological features. This can raise awareness among people and enhance tourist attractiveness. Landscape features are formed from a geological basement that documents the events from history of the Earth. One of such events in the history of life was the appearance and evolution of graptolites, early ancestors of chordates, i.e. humans. Graptolite bearing deposits are very rare in Poland and can be found in the Swietokrzyskie Mts. and Sudety Mts. Such fossils crop out in some exposures in ravines, trenches, cross-cuts and closed down quarries etc. thus adding an unusual scientific value to such spots. Due to protective activities based on applicable legislation (nature reserve, nature monument and documentary site) such important geological sites will become a natural refuge in an agricultural landscape. Such enclaves formed in compliance with environmental interest based on geoscience argument provide the only possible way to preserve natural landscape components in highly anthropogenic areas.

The local landscape is the nearest surroundings for humans, a place of their everyday life - work and rest. For a long time, it has been noticed and reflected by musicians, poets, painters or photographers in works of art. To a little extent, the features related to beauty and views are subject to very rapid changes and are discerned first - that is why local governments can and should react to them by competent decision-making. Man transforms the landscape in an uncontrolled way and often leaves out the fact that local changes to the environment add up and contribute to global changes. Therefore, while planning the landscape surrounding us, we should start with the nearest places.

Natural values are very difficult to be saved in any land subject to intensive agriculture use; there, in the open space of arable crops, only some parts of the land, not useful for business purposes, constitute enclaves of natural landscape. Those are: ravines, ditches, cross-cuts, decommissioned quarries, river valleys etc, often tree-covered and serving as a refuge for a large number of floral and animal species. Bedrock can often outcrop at those places; in many cases the bedrock itself is of unique value. In such a culture-related landscape, dominated by fields, boroughs, roads and power transmission lines, the landscape value is lost due to anthropogenic transformations.

The investigations of changes in the vegetal cover as well as the use of the land indicate agriculture as the main causative agent of global changes in the environment. The alterations caused due to the land use, taking place in Europe after the year 1700, have been the reason for today's division of the landscape: 15% natural landscape, 20% landscape changed in part, 65% appreciably altered (Mannion 1997). Intensification of agriculture disturbs the biological balance, and in consequence, causes the forest surface to dwindle, brings about habitat fragmentation, intensification of floods, water and eolian erosion, stepping and desertification. Thus, the agricultural landscape requires special attention while working out a plan of sustainable development.

The landscape protection may be implemented in many ways, for example, by saving the valuable, clear-cut objects of that landscape. Their proper development is the foundation for taking protective steps nationwide. At the local level, the protection plans are provided with all formal attributes so as to construct the basis for saving the objects which are essential for both the geological heritage and landscape protection.

In Poland, single geological objects can be protected as nature monuments and documenting sites, whereas larger objects - as nature reserves (Alexandrowicz 1991, Alexandrowicz et al. 1992). The purpose of each form of protection is quite different. The inanimate nature monuments are objects standing out in the landscape, of scientific, didactic, aesthetic and historical value. The documentation sites of inanimate nature are places of occurrence of specific rocks, fossils, mineral associations, tectonic structures etc; although they have no beauty values, their merits are substantial. Inanimate nature reserves are places of unique value; they occupy larger areas than the preceding protection forms, and in the hierarchy they are placed just after national parks.

The basement complex is the landscape-shaping element. It

documents the events from the Earth's history. One of such events in the development of the biosphere was the speciation and development of graptolites. They are possibly evolutionary ancestors of the vertebrates - animals who dominate nowadays.

Fig. 1

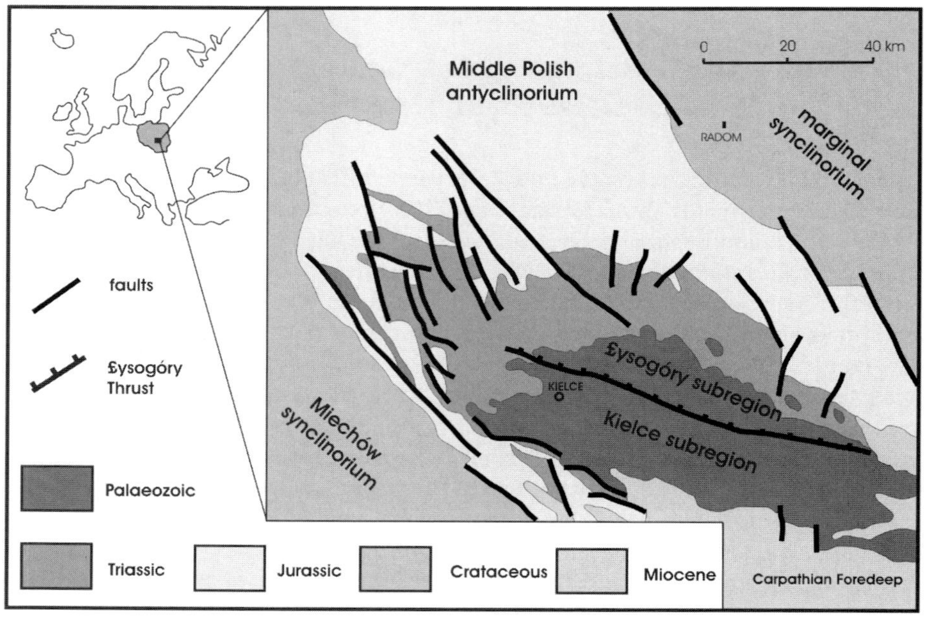

Fig. 2

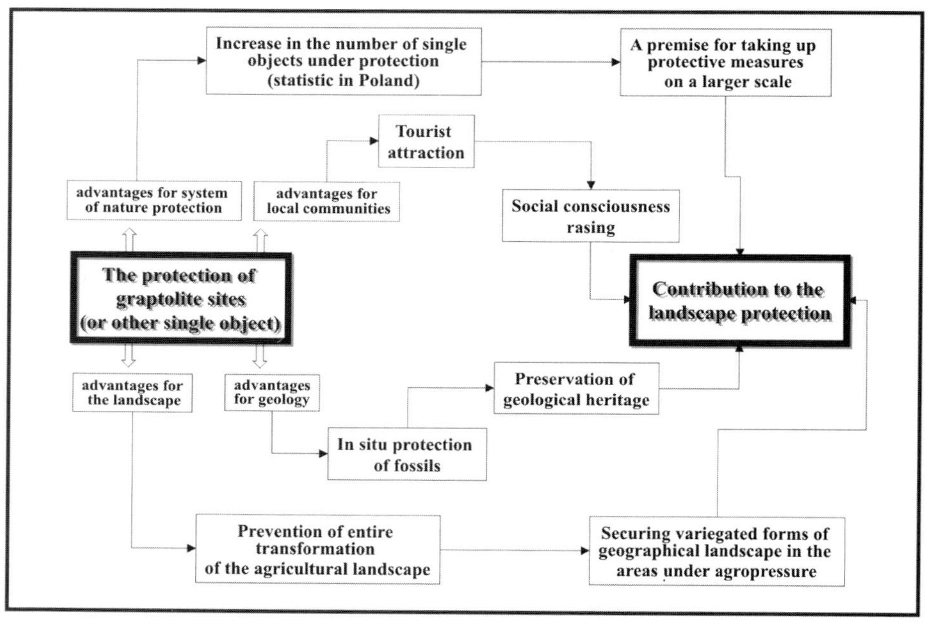

Graptolites came to exist in the middle Cambrian and became extinct in the Lower Carboniferous period. Their acme of development took place in the Ordovician and Silurian - and in most cases they are found exactly in those deposits. They were evolutionary progressive forms, used in the orthobiostratigraphy of the Ordovician and Silurian. Graptolites were colonial animals; respective individuals lived in 1-2 mm long, funnel-shaped integuments called thecas. The integuments were clustered like twigs, constituting colonies. Little houses were built in a characteristic way - they consisted of single strips, combined one with another, making a fusellar structure. The nature of such a seam was the chief argument indicated by Roman Kozlowski, a Polish palaeontologist, as a proof for a relationship between the Pterobranchia and the graptolites (Kozlowski 1948, 1966). And the Pterobranchia are unquestionable ancestors of the vertebrates. Summing up the similarity between the Pterobranchia and the graptolites: macroscopic similarities in the internal structure of thecas, approximately identical composition of the house structural material, organisation of the individuals within a colony (Bates and Kirk 1986). However, the difference is the collagen composition - theca structural material - that is why the relationship between the Pterobranchia and the graptolites is still an open question (Urbanek, Dilly 2000). Graptolites are usually preserved in clay deposits; they have given another name to those stones - graptolite shales. They occur in the layer surfaces, and seem to be a delicate drawing in a rock - hence the genesis of their name [graptos - drawing, lithos - stone], and are an asset for collectors. Graptolites are also preserved in calcareous and silica rocks; they can be extracted by dissolving the mineral parts in acids.

As far as palaeography of the Palaeozoic is concerned, Poland lies in the borderline between two big sedimentation environments - the West European platform and the East European platform - separated with a great fissure zone, called Teisseyre-Tornquist line. In the Polish territory, the 'graptolite -bearing' rocks occur in the surface in situ in Saint Cross Mountains [Góry Swietokrzyskie] and Sudety Mountains and ex situ as erratics.

The Saint Cross Mountains belong to the faulted mountains of the Hercynian orogenesis - they are limited with faults along which they were lifted (Figure 1). They consist of Palaeozoic rocks and a Mesozoic buffer zone. The biggest dislocation - the Saint Cross overthrust - divides the Palaeozoic part into two subregions: Kielce and Lysogory. Ordovician and Silurian deposits occur in both these parts. The protection plan described here, regards the southern [viz. Kielce] part of Professor R Kozlowski's exploratory area.

There are 10 documented graptolite sites in that part; however, only 6 of them are accessible for observations, and they are still in danger:

Telegraf Mount - the outcrop is subject to geologists' intensive penetration, and this fact corroborated by many pits - a product of a large number of exploration holes. Exposures are no longer available for observations.
Bukówka - the exposure was little and sited in a scarp by the road. Nowadays, it is entirely covered with waste-mantle.
Otrocz Mount - nowadays, the access to the Silurian rocks is very difficult. The old outcrop is no longer in existence, and the entire area is occupied by the tree stand.
Widelki - like in the case of the sites aforementioned, the exposure of graptolite shales in Widelki is presently covered with waste-mantle. It visualises the scale of the phenomenon of exposure disappearance in consequence of natural processes of erosion and plant succession.
Czyzów-Koziel - the outcrop of graptolite shales has been levelled off in consequence of the man-made land use. All of the hill is covered with arable crops.
Pragowiec - an example of fossil gophering in a picturesque ravine. Graptolites occur there in mass, and are easily accessible. The protection of this place is indispensable.
Zalesie - objects already under protection; nevertheless, there is still a danger that the outcrop is endangered by overgrowth of plants and soil layer creep.
Zbelutka - a cover of waste-mantle and forest litter enables observations of graptolite shales.
Wysoczki - an outcrop of Ordovician chalcedonites, a decommissioned working is being gradually overgrown and gradually covered with leaf litter.

What purpose should the protection serve? The plan of protection for graptolite sites in the Saint Cross Mountains is complex so as to allow their rational use (Gonera and Miskiewicz 2003). Two sites are already under protection; for the most important ones, protective measures have been postulated in order to reach documentary, scientific and didactic goals. Some, instead, have been left for collecting purposes. Due to the aforesaid, the sites which are precious from the scientific viewpoints will be preserved, whereas it is still possible to keep on scientific investigations and to collect fossils. The protection of graptolite sites (or other single objects) can offer advantages for landscape protection (Figure 2).

REFERENCES

Alexandrowicz, Z. 1991. Stanowisko dokumentacyjne jako nowa kategoria ochrony przyrody nieozywionej (The site of scientific documentation - a new category of the inanimate nature conservation). *Chronmy Przyrode Ojczysta* XLVII, 1/2: 5-9.

Alexandrowicz, Z., Kucmierz, A., Urban, J. and Oteska-Budzyn J. 1992. *Waloryzacja przyrody nieozywionej obszarów i obiektów chronionych w Polsce* (Evaluation of inanimate nature of protected areas and objects in Poland). Pan. Inst. Geol., Warszawa.

Bates, D.E.B. and Kirk, N.H. 1986. Mode of secretion of graptolite periderm, in normal and retiolite graptolites. In: C.P. Hughes, R.B. Rickards (Eds.) *Palaeoecology and Biostratigraphy of Graptolites*, Geological Society Special Publication 20: 221-236

Gonera, M. and Miskiewicz, K. 2003. Graptolite-bearing sites in Poland - concept of protection. In memory of Professor Roman Kozlowski. Nature Conservation 59.

Kozlowski, R. 1948. Les graptolithes et quelques nouveax groupes d'animaux du Tremadoc de la Pologne. *Palaeontologica Polonica* 3: 229 s.

Kozlowski, R. 1966. On the structure and relationship of graptolitides. *Journal of Paleontology*. 40,3: 489-501.

Mannion, M. A. 1997. *Global environmental change. A natural and cultural environmental history*. Addison Wesley Longman Limited.

Urbanek, A. and Dilly, P.N. 2000. The stolon system in Rhabdopleura compacta (Hemichordata) and its phylogenetic implications. *Acta Palaeontologica Polonica* 45,3: 201-226.

In: Parkes, M.A. (Ed.) 2004 *Natural and Cultural Landscapes - The Geological Foundation*, 196, Royal Irish Academy, Dublin

INTEGRATING GEODIVERSITY AND BIODIVERSITY IN LOCAL LANDSCAPE AREAS OF NORTHERN IRELAND.

R.W. Tomlinson, E. Muise, G. Mallon and P. O'Hare, School of Geography, Queen's University Belfast, Belfast BT7 1NN.

Abstract: Northern Ireland has been divided into 130 Landscape Character Areas (LCAs). Each is defined by its distinctive character, based on local geology, land forms, land use, cultural and ecological features. Threats to each LCA are also considered. LCAs may be seen as a mechanism by which landscapes may be conserved and new, appropriate development opportunities grasped. National and Local Governments are also required by international agreements, to conserve geodiversity and biodiversity and to encourage local communities to become involved. The poster shows by examples how geodiversity and biodiversity in Northern Ireland have been linked to LCAs; it suggests that this linkage will provide a framework for planners and will enable local communities better to appreciate the landscapes and habitats of their area. Difficulties in the linkages are also discussed

In: Parkes, M.A. (Ed.) 2004 *Natural and Cultural Landscapes - The Geological Foundation*, 197-200, Royal Irish Academy, Dublin

INTEGRATION ASSESSMENT AND MANAGEMENT OF IRELAND'S LANDSCAPES

Martin Critchley, ERA-Maptec Ltd., 36 Dame Street, Dublin 2
Julie Martin, Julie Martin Associates, The Round House, Swale Cottage, Station Road, Richmond, North Yorkshire, DL10 4LU
Sean O'Riordain, Environmental Resources Management Ireland Ltd (ERM), 22 Earlsfort Terrace, Dublin 2

Abstract: Landscape character assessment is vital tool in the management of Ireland's countryside resources. It provides a well defined mechanism for mapping, classifying and describing the landscape units of an area and the process of making judgements about landscape character to inform different decisions. Using County Clare as model, a staged approach has been developed for the assessment of landscape character which includes analysis of the physical (stage 1); cultural/historical (stage 2) and sensitivity (stage 3) parameters. Crucially this approach is underpinned by consultation at a local level covering bodies from a wide socio-economic base. Landscape character assessment is currently being undertaken for a number of counties in Ireland and this assessment needs to be co-ordinated at a national level to ensure compatibility and consistency of results. This is especially important if the landscape character areas are to be used to assist policy advice at a regional and national level.

INTRODUCTION

Ireland's landscape is regarded as one of our most precious national assets, contributing to the overall uniqueness of our country and to people's sense of identity and place.

The Landscape is a dynamic component of our environment, typically comprising of a blend of geology, landform, biology (habitats and biodiversity), archaeology, land use, agricultural practices, settlement patterns; biodiversity and environmental attributes such as air, soil, inland and marine waters. In a wider context the landscape embraces visual and aesthetic aspects. The landscape is also regarded as one that accommodates the physical and spiritual needs of society and the needs of nature in a harmonious manner.

In addition and perhaps most critically, the landscape has strong economic, social and community values and can be viewed as an asset attracting for example tourism development and other business activities to local communities.

LANDSCAPE CHARACTER ASSESSSMENT

Landscape characterisation is a complex process but one that is increasingly used to classify our countryside resources, and to help guide their future conservation, development, management and enhancement. The overall objective of Landscape Character Assessment is to analyse the character, value and sensitivity of the landscape. Landscape Character Assessment is a way of understanding the character of the whole landscape ie:

- pattern of elements that occurs consistently in a particular landscape (mapping and GIS)
- specific combinations of landform, landcover, landscape features
- recognisable sense of place, identity

Above all it is a systematic approach which can be documented, is repeatable and applicable to the country as a whole.

In Ireland, Draft Landscape and Landscape Assessment Guidelines have been issued by the Department of the Environment, Heritage and Local Government (DoEHLG). These reflect, as would be expected, the strategic concerns of the Department to ensure an integrated approach to planning and landscape management. In addition however Ireland has signed the European Landscape Convention. The principles enshrined in the Convention provide a wider interpretation to the management of the landscape. These seek to;

- reflect an integrated approach to landscape development and management that characterises all types of landscape in natural and cultural terms, and not solely designated or special landscapes;

- to help government, departments and organisations create development programmes which preserve and enhance the integrity of existing landscapes and townscapes; and

- help identify and preserve priority views along the coast, assisting in coastal policy formulation and decision-making and project inception in both coastal and marine environments.

The Convention therefore pushes the issue of dynamic landscape management into all Government Departments and Agencies and not just those that fall within the immediate remit of the Department of the Environment, Heritage and Local Government and local authorities.

SEASCAPE CHARACTER ASSESSMENT

Landscape assessment also applies to marine areas. Seascapes are an important element in any maritime nation's sense of identity and culture. The objective of a

Seascape Assessment is to analyse the character, value and sensitivity of Seascapes. The long-term objective of seascape assessment is to maintain the character and quality of the coastline, enhancing key seascape elements, where possible. This requires consideration of;

OBJECTIVES OF LANDSCAPE/ SEASCAPE CHARACTER ASSESSMENT

The main objective therefore of the Landscape/Seascape Character Assessment process is to demonstrate the significance and value of landscape to all activities which impact on that landscape or seascape. This Assessment should then inform both the policy-making and services delivery of local, regional and national authorities and agencies, and provide a policy framework for individuals and developers that is open and transparent.

APPLICABILITY AT DIFFERENT LEVELS

It is widely accepted that Landscape Character Assessment can be applied at a number of different levels from national or indeed European to regional or local level. Of particular importance is the uniformity in the methodology by which such landscape characterisation is undertaken.

Ideally assessments at different scales should be consistent so that they can fit together as integrated systems or as a hierarchy of landscape character types and/or areas so that assessment at each level adds more detail to the one above. For example all landscape assessments currently undertaken at county level in Ireland should be able to feed into the overall or national assessment of Ireland and vice versa. International experience indicates that a failure to seek a horizontal integration of landscape related policies results in the creation of inherent policy conflicts across a range of areas including physical planning, agricultural development, transport and communications.

It is widely accepted that co-ordination and integration of landscape characterisation will be best achieved by identifying one particular agency to develop, co-ordinate and continuously update the system at the national level. Through this agency, State and local bodies can, along with private development interests and community groups/environmental bodies and others, have regard for the management of the landscape and the creation of a dynamic approach to population settlement, social and economic development, environmental protection and maintenance of a strong cultural perspective and local identity. Clearly the adoption of the National Spatial Strategy can underpin such an approach and it may be appropriate to have the DoEHLG through its NSS steering committee take on this role. It does however indicate that there is a need for a national model that goes beyond the current draft guidelines and yet retains a clear focus on local as well as national delivery of dynamic landscape management that adds to better public administration.

In recognition of this the Heritage Council commissioned ERM Environmental Resources Management Ireland Ltd. in association with ERA -Maptec to undertake a pilot landscape assessment in County Clare. The aim of the Council was that the assessment would;

- fully reflect the principles enshrined in the European Landscape Convention and the DoELG Draft Guidelines on Landscape Characterisation;
- fully reflect the recommendations contained in the Council's Policy;
- fully reflect quality of life aspects of the landscape throughout the County and the views and concerns of communities living and working in the County;

- incorporate greater analysis and characterisation of the ecological and archaeological components of landscapes, which is considered an integral part of any landscape characterisation process; and most critically;
- establish a framework for the development of a model for landscape characterisation that can be applied at a national level.

One of the purposes of this paper is therefore to highlight the methodology and experiences that have emerged from the Landscape Character Assessment of County Clare and to demonstrate how this approach can be applied at a national level.

OVERALL APPROACH TO THE LANDSCAPE ASSESSMENT PROCESS

The concept of landscape characterisation has emerged as an appropriate way to deal with landscape and has become an increasingly important and widely used tool. Three basic stages of assessment are involved in the landscape characterisation process. These are:

Stage 1: Definition of physical
Stage 2: Cultural and historical landscapes and definition of landscape values.
Stage 3: Assessment of sensitivity of landscapes to change

The first stage of assessment characterises the landscape in terms of land cover -trees, vegetation, settlements, water etc and landform which results from geological and geomorphological history.

Added to this first stage of assessment is a second stage referred to as values in the DoEHLG Guidelines. This stage takes historical, cultural, religious and other understandings of the landscape into account. The final stage takes into account the sensitivity of the landscape to accommodate change or intervention without suffering unacceptable effects to its character or value.

Critically this approach was underpinned by a consultation process at local level covering bodies from a wide socio-economic base in the County. The base was provided by the Community and Enterprise Directorate of Clare County Council. Local representatives of the State sector and the County Planning Authority were also actively consulted. Presentations to the planning policy committee and the full

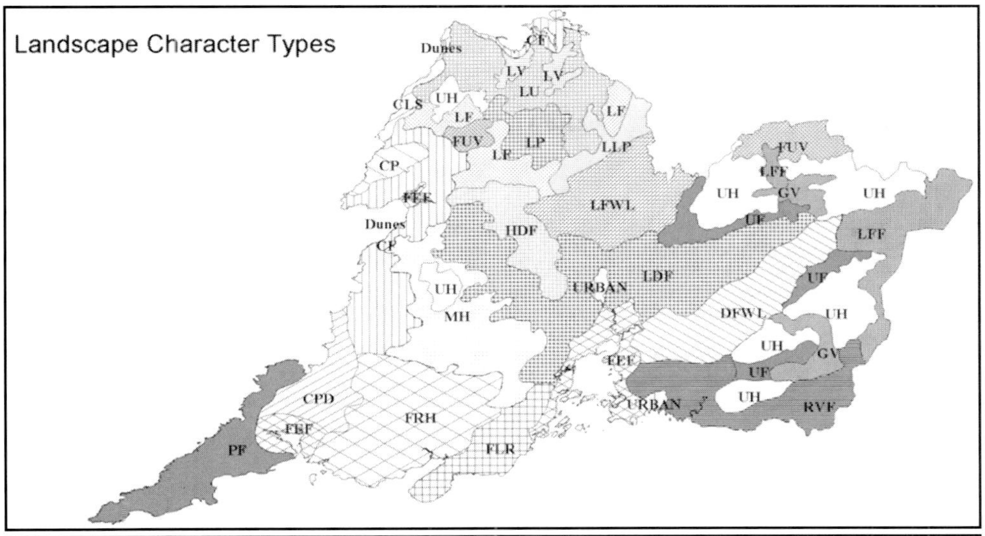

Figure 1. Landscape character types for County Clare. Character types are generic units having particular combinations of data sets which can be repeated in different parts of the county or country. For example, Upland Hills (UH) occur in the north-west of the county and the east.

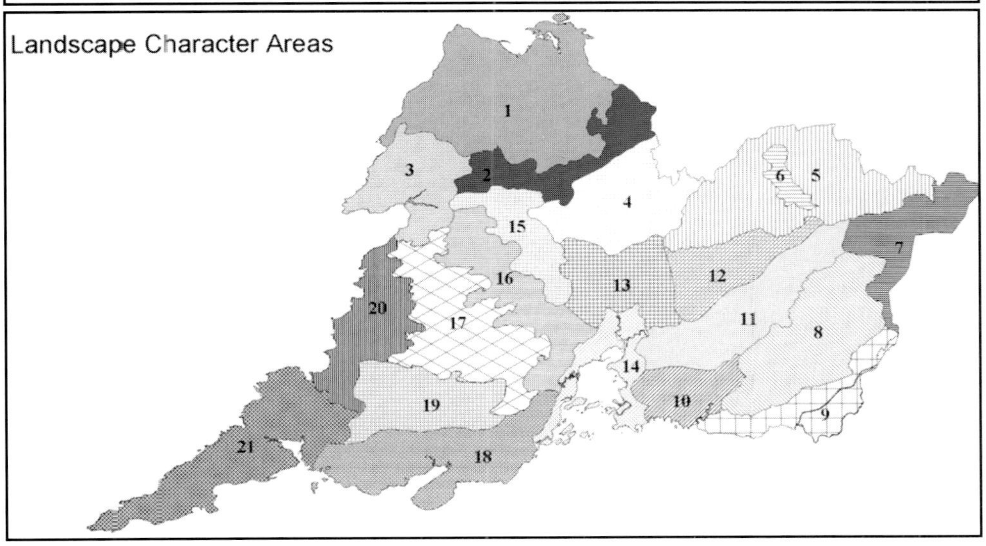

Figure 2. Landscape character areas for County Clare. Character areas are unique to a specific area and can contain one or more character types and may cross-cut character type boundaries. For example, the Burren (1) is comprised of at least four landscape character types as shown in figure 1 (UH - upland hills, LP - Limestone Plateau, LU - Limestone Upland, LV - Limestone Valley).

membership of Clare County Council also took place.

The actual Landscape Characterisation process embraced mapping, classifying and describing the landscape character, and the process of making judgements about landscape character to inform different decisions. The following are the main outputs as a result of the assessment process: Landscape Character Types (LCT's) which are distinct types of landscape that are relatively homogenous in character. They are generic in nature in that they may occur in different localities throughout the Country. Nevertheless, where they do occur, they commonly share similar combinations of geology, topography, landcover and historical land use, for example, limestone river valleys.

Landscape Character Areas (LCA's) which are units of the landscape that are geographically specific and have their own character and sense of place. Each LCA has its own distinctive character based upon patterns of geology, landform, landuse, cultural, historical and ecological features.

Policy advice covering a range of key forces for change including forestry, renewable energy, agricultural practice, residential development, transport planning and others.

Experience also revealed that both Seascape Character Areas (as described above) and Historic Landscape Types (HLT) would assist in the overall landscape assessment process and should therefore also be identified and mapped. Cooney (2000) describes a HLT as a discrete landscape based on the 'scale and integrity of the archaeological features [that] reflect significantly on the human history and land use of that area (Cooney et al,2000, 22).

Thematic maps showing landscape/seascape character areas and types were produced to illustrate the outputs. Maps illustrating Historic Landscape Types were also produced. These were accompanied by written descriptions of Landscape and Seascape Character Areas including an outline of values and sensitivities and principles for landscape management.

APPLICABILITY FOR NATIONAL MODEL

The European Landscape Convention notes that developments in agriculture, forestry, industrial and mineral production techniques and in regional planning, town planning, transport, infrastructure, tourism and recreation and, at a more general level, changes in the world economy are in many cases accelerating the transformation of landscapes (The Heritage Council 2002). Therefore there is a greater necessity for Landscape Character Assessment to be undertaken and for a national model to be finalised in Ireland and indeed elsewhere in Europe.

Quite a number of departments, agencies and individuals could benefit from the practical applications of Landscape Character Assessment. In particular landscape assessment could help in processes which:

- identify what environmental and cultural features are present in a locality and thus provide the basis for local economic development opportunities ;
- monitor change in the environment;
- understand a locality's sensitivity to change and development; and
- inform the conditions for any development and change.

More specifically Landscape Character Assessment can be used in a wide range of situations. Some of the main applications include:

- informing development plan policies at strategic and local level to help prepare Landscape Management Strategies and Forestry Strategies;
- contributing to landscape capacity studies relating to the supply of land for housing, minerals, renewable energy or other land uses;
- providing an input to Environmental Impact Assessment
- informing work on special areas, including identification of areas for designation, mapping of boundaries, justification for special application of policies, justification for special treatment by designation, and input to management plans and other management initiatives;
- helping to guide land use change in positive and sustainable ways i.e programmes of programmes of woodland expansion;
- informing the targeting of Agri-environmental schemes;
- contributing to wider environmental initiatives such as biodiversity plans, heritage plan and policies; and
- identifying areas where a strategy of enhancement or regeneration is appropriate for a particular landscape area or type.
- identification of priority geographically based economic actions through the regional assemblies/ regional development agencies.

This could happen through the completion of landscape assessments by local authorities at one level in line with the methodology set out above. For those that have completed assessments a simple bench marking exercise to identify gaps/differences is all that may be necessary to bring them into line with the model. More critically at national level the drawing together of the assessments could be facilitated through, in the first instance, the redrafting of the current draft guidelines to ensure that all local authorities apply the model in a coherent manner. Equally, the drafting of similar guidelines to be applied by national departments and their operational agencies should be considered given the fact that these also have direct impact upon Ireland's landform.

The actual tasks associated with the model i.e. data collection and interpretation, landscape analysis and consultation can replicate the suggested model at the national level through it is suggested an inter departmental committee along the lines of the Task Force for the Integration of local government and local development or the NSS steering group could be appropriate to oversee this particular work. This would place landscape management at the core of corporate planning in each department in much the way of social inclusion and rural proofing has been in recent years.

CONCLUSION

In essence, landscape character assessment aims to improve our understanding of landscapes and provide a framework that allows landscape considerations to be taken into account in all aspects of decision-making and promotes a unified approach to landscape management. Co-ordination and integration is imperative to the success of landscape character assessment. This requires a national approach which is based on local delivery.

The Government recognises the central and positive role that integrated landscape policies play in achieving national and international objectives by acknowledging the value of having a consistent model for landscape characterisation (The Heritage Council 2002). This places a greater emphasis on finalising the development and implementation of a national model. The national model should provide a baseline against which change can be gauged and monitored and reflect the principles enshrined in the European landscape convention, the DoEHLG Landscape guidelines and heritage policies.

As mentioned above there are a number of agencies who will benefit from the implementation of a national model and therefore the challenge is to facilitate the development of an understanding of landscape issues within these bodies and to create a momentum within the different departments and agencies to assist in the overall development and implementation of the model.

ACKNOWLEDGEMENTS

We are grateful to the Heritage Council for permission to publish this paper and for its active support of the County Clare pilot landscape characterisation project. We would also like to expression our gratitude to all the stakeholders who contributed their time, knowledge and expertise to the project.

REFERENCES

Cooney, G., Condit, T. and Byrnes, E. 2000. 'The Archaeological Landscape' In N. Buttimer, C Rynne and H. Guerin (eds) *The Heritage of Ireland: natural, man-made and cultural heritage, conservation and interpretation, business and administration*, 18-28, The Collins Press.

Draft Guidelines for Planning Authorities on Landscape and Landscape Assessment. 2000. Department of the Environment, Heritage and Local Government

Heritage Council 2002. Policy Paper on Ireland's Landscape and the National Heritage.

In: Parkes, M.A. (Ed.) 2004 *Natural and Cultural Landscapes - The Geological Foundation*, 201-204, Royal Irish Academy, Dublin

FROM "ROCK DESERT" TO FOREST - AN EXAMPLE FROM KRAS

Andrej Kranjc, Karst Research Institute, Titov trg 2, SI-6230 Postojna, Slovenia.
Email: kranjc@zrc-sazu.si

Abstract: Natural vegetation on the limestone plateau Kras (Karst) used to be Euro-Siberian type of forest. Man started to exploit the land during the Neolithic period. Clearing of forest continued and due to overgrazing, accelerated erosion and wind erosion the Kras plateau started to change into a bare limestone country. This is the reason why travellers and scholars started to equate the word karst with bare rocky landscape. Since the 11th century there had been attempts to preserve the forest. During the 18th and in the beginning of the 19th century Kras was practically a rocky desert and then afforestation started. First they tried with oak, then Austrian black pine. The first successful attempt occurred in 1859 in the background of Trieste. Planting saplings was a good job and for many people it was the main income. The process continued to the middle of the 20th century. As changes in the economy took place, less and less people worked on the land and they ceased to breed sheep and goats. Now overgrown pastures are changing into forest. If the process will continue Kras will be completely covered by shrubs and forest by the year 2013.

Kras (Karst, Carso) is a limestone plateau above the Gulf of Trieste in the Adriatic Sea. The 45^0 45' N parallel and the 14^0 00' meridian east crosses in the middle of it. The Mediterranean influences are modified by its position and altitude (200 - 500 m). The climate is Mediterranean: hot and dry summers, cold winters with precipitation, strong NE wind burja. There are no olive trees as winters are too cold. The Kras plateau is the northwesternmost part of Dinaric karst. It is 40km long and up to 13 km wide, covering little more than 600 km^2, sloping down towards the NW (Kranjc 1997).

From the name of this country the international term karst is derived, due to: bareness of the surface, lack of soil, special hydrological features, and climatic phenomena (Kranjc 1997). Valvasor (1689) wrote "*In some places it is possible to see for some miles, but everything is only grey, nothing green, because all the country is covered by stones.*"

Figure 1. Bare landscape of Kras as seen by the 17th century traveller (Valvasor 1689).

(Figure 1). Illustrations from the same time show the cultivated land at the bottom of dolines only (Valvasor 1679). Natural vegetation in Slovenia is forest and before the appearance of man the forest covered at least 90 % of its surface (Anko 1989). Kras belongs to the Euro-Siberian and North-American (and not to the Mediterranean) region, where approximately 5000 years BP beech tree (*Fagus*) and fir tree (*Abies*) were common (Šercelj 1996).

Deforestation can be traced from the early periods. Intensive phases of deforestation occurred during prehistory (Neolithic and Iron age), due to grazing (Slapšak 1995; Culiberg in: Kranjc 1997), during medieval colonisation and during the phase of population (agrarian) maximum in

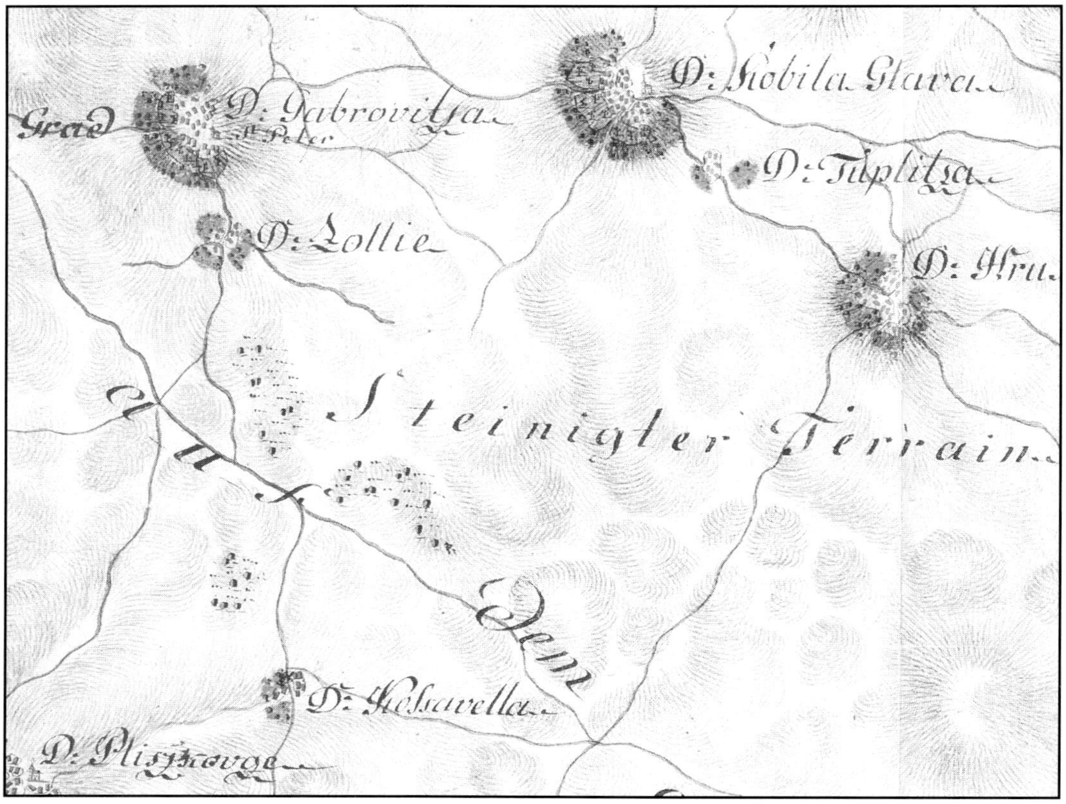

Figure 2. Sector of a military map (Rajšp & Trpin 1997) showing rocky surface of the central part of Kras.

the 19th century. To confirm the prehistoric deforestation Gams's (1991) researches on Rillenkarren are very illustrative. He found out that they started to be formed on Kras 3000 - 3500 years BP, when the forest was destroyed and bare rocks began to appear. In the Roman period cultivated land was in the form of islands around the settlements and quite a lot of forest still remained inbetween (Slapšak 1995). The "military maps" (Josephinische Landesaufnahme) and accompanying descriptions of the 18th century show very few forests on Kras (Fig. 2). For example the text of the village of "Gabrovitza" (Gabrovica pri Komnu) says: "*Forests: There are no forests in the neighbourhood, only small bush one hour away ...*" (Rajšp & Trpin 1997).

In general in Slovenia there was a minimum of surface covered by forest around the year 1875. Forest then covered 37 % of the surface, while in the Postojna district, which included also a part of the Kras, the forest covered only 26 % in 1880 (Azarov 1994). In 1989 the share of forest had increased to 51 % and nowadays it is still increasing, its share being estimated to be over 60 % of the whole surface of Slovenia. On the Kras itself the share of forest was even smaller, about 20 % in 1900. The changed microclimatic conditions even enabled the spreading of some steppe species. But in 1985 forest increased to 42 % (Gams 1991) and in 1998 to over 50 %.

The reasons for deforestation were clearing for new farming land reclamation, cutting of wood for different purposes, for fire wood, for ship building and for construction. Some popular sources mention that "Venice stands on the forest of Kras", meaning that piles for the foundations of the buildings on Venice marshy land are from Kras. The last but not least important reason was the grazing of sheep and especially goats.

Man began to be aware of the deforestation quite early and the owners tried to safeguard the remaining forest by different ways. Very old "forest acts" exist, for example protective regulations by the town of Trieste of 1150. For Istria, Friuli and Karst the "Waldordnung" of 1541 is known, as well as Venetian forest acts from 1452 on (Gašperšic and Winkler 1986). In 1583 even an armed guard had to be organised by the town of Trieste (Guttenberg 1881) to protect the forest. The forest was preserved as the primary material for shipbuilding (state or domain forests) as can be seen from Ressel's "Geschichte der k.k. Marine-Wälder" (1855), to sell the timber to Venice and to use it in mines (as timber) and in iron foundries (for charcoal) (Gašperšic and Winkler 1986).

J. Ressel tried the first to reafforest Kras. He made a reafforestation plan in 1842 (Die Wieder bewalderung der Gemainde Gründe Istriens) in which he proposed to use oak acorns, stating the oak being an autochthonous sort with a good wood and therefore the best tree for reafforestation. But as the oak grows slowly this was not the best choice. The first successful reafforestation occurred in 1859 in the vicinity of Bazovica by J. Koller, who used the young seedlings of black pine (*Pinus nigra* var. *austriaca*) (Azarov 1994). As the first attempts proved to be successful (Fig. 3), the "Reafforestation Act for Kranjska" (Carniola) was issued in 1885 (Azarov 1994). The reafforestation was really successful and from all over Europe specialists came to Kras to see it (Rubbia 1912). Finally, to prevent the deforestation and to help the forest to grow Yugoslavia banned goat breeding in 1953 (Gams 1991).

Figure 3. The beginnings of the reafforestation on Kras (Goll 1898).

The present situation on the Kras shows over 50 % of the surface covered by forest. It is mostly the monoculture pine tree forest. The course of reafforestation is now going on by itself: pastures become overgrown by shrub and this slowly transforms into forest. That means that the surface is being overgrown first with tall herbs, then with shrub species, and finally with trees and forest ground flora. The front of pine forest progresses without human help at a maximum of 17 m per year; computer modelling forecasts that the whole region will be overgrown by shrub by 2013 (Favretto and Poldini 1986).

There are two main types of problems due to the monoculture pine tree forest. The first is hypersensitivity to different insect pests and to forest fires (Gams 1997). The second type of problem is caused by too great share of the surface covered by forest. This is shown in processes such as disappearing of cultural landscape, decreasing of biodiversity (especially floral) (Piskernik 1990), and changing of microclimate (affecting vineyards, ham drying, etc.).

After 150 years of work and efforts, including large sums of money for the reafforestation of Kras, the specialists and laymen started to ask: how to reasonably stop the overgrowing of Kras? The foresters suggest slowly replacing pine by oak (Mlinšek 1993), while farmers, land-owners and inhabitants in general suggested increasing grazing of goats and sheep. The foresters also suggested that Kras should be a field experimental laboratory of international importance, where it would be possible to study the revitalisation of completely degraded country. Specially important has to be the study of revitalisation of thermophile associations, which are the most affected and at the same time the most suppressed and neglected by the world's public (Mlinšek 1993).

REFERENCES

Anko, B., 1989. Gozdnatost.- Enciklopedija Slovenije, 3, 341-342, Ljubljana

Azarov, E., 1994. Crni bor na Krasu.- Kras, 4, 18-21, Ljubljana

Favretto, D. and L. Poldini, 1986. Extinction time of a sample of Karst pastures due to bush encroachment.- *Ecological Modelling* 33, 85-88, Amsterdam

Gams, I., 1991. *The origin of the term Karst in the time of transition of Karst (Kras) from deforestation to forestation.* Quaderni del Dipartimento di Geografia, 13, Proceedings of the International Conference on Environmental Changes in Karst Areas, 1-8, Padova

Gams, I., 1997. Gozdni po□ari na Krasu in vloga borovih sestojev (Forest Fires in the Karst and the Role of Austrian Pine Plantations). *Ujma* 11, 113-117, Ljubljana

Gašperšic, F. and I. Winkler, 1986. Ponovna ozelenitev in gozdnogospodarsko aktiviranje slovenskega krasa. *Gozdarski vestnik* 5, 169-183, Ljubljana

Goll, W., 1898. *Die Karstaufforstung in Krain.* III-VII, 1-120, Laibach

Guttenberg, H. v., 1881. Der Karst und seine forstlichen Verhältnisse. *Zeitschrift des Deutschen und oesterreichischen Alpenvereins* B. XII, 24-62, München

Kranjc, A., 1994. About the Name and the History of the Region Kras.- *Acta carsologica* 23, 81-90, Ljubljana

Kranjc, A. (ed.), 1997. *Slovene Classical Karst "Kras"*. 254 pp., Ljubljana

Mlinšek, D., 1993. □ivjenjski prostor "nizki kras", primer clovekove destruktivnosti, energije □ivljenja, upanja v cloveka in trajen raziskovalni laboratorij. *Gozdarski vestnik* Vol. 51, No. 5-6, 280-293, Ljubljana

Piskernik, M., 1990. Kaj se dogaja v gozdovih na Krasu po zadnji vojni? *Proteus* 52, 6, 214-217, Ljubljana

Rajšp, V. and D. Trpin, 1997. Slovenija na vojaškem zemljevidu 1763-1787 (1804) - Josephinische Landesaufnahme 1763-1787 (1804) für das Gebiet der Republik Slowenien. 3, V-XL, 1-436, Ljubljana

Rubbia, K., 1912. *Petindvajset let pogozdovanja Krasa na Kranjskem*. 1-85, Ljubljana

Slapšak, B., 1995. Mo□nosti študija poselitve v arheologiji. *Arheo* 17, 2-90, Ljubljana

Šercelj, A., 1996. *Zacetki in razvoj gozdov v Sloveniji* - The Origins and Development of Forests in Slovenia.- Dela - Opera 35, SAZU, Razr. IV, 1-142, Ljubljana

Valvasor, J. W., 1679. *Topographia Ducatus Carnioliae Modernae*. Wagensperg in Crain (fascimile edition, Ljubljana and München 1970)

Valvasor, J. W., 1689. Die Ehre deß Herzogthums Crain. I. Th., pp. 696, Laibach - Nürnberg

In: Parkes, M.A. (Ed.) 2004 *Natural and Cultural Landscapes - The Geological Foundation*, 205-208, Royal Irish Academy, Dublin

MYSTICAL MYTHICAL IRELAND; GROUNDWATER AND THE ADVANTAGES OF SUSPENDED DISBELIEF

David M. Ball, Hydrogeologist

Abstract: The purpose of the conference is to make linkages between people, their landscape and the foundations of the landscape. Geologists instinctively understand the role and fundamental importance of the subsurface to the surface landscape, because most geologists have been trained to read the landscape looking for clues to understand the subsurface. The situation that the conference has to face head on, is that most people are not interested in the subsurface unless they have a need to know about it. Geologists can discuss the subsurface amongst themselves with great enthusiasm, but for most people, "rocks are boring", unless they can relate them to their individual circumstances, or the community within which they live. There have been many imaginative attempts by geologists to hook the interest of the general public in the subsurface, but most efforts have started from the basis of making rock, fossils, minerals and geomorphology interesting. Whilst the layman can see that these facets of the subsurface are a legitimate interest, they nevertheless usually see them as something old, static and of little immediate importance or relevance to their own personal lives. The purpose of my paper is to introduce a very important part of the subsurface that does have a general, immediate and often very personal interest to the layman; water. I will then go on to look at the cultural landscape that, I have found from my fieldwork, blocks or inhibits understanding of the relationship between water in the subsurface, the people and the surface landscape.

The geological foundation of the landscape, in Ireland consists of two components; a static framework of solid rock and soils, through which moves a dynamic fluid; groundwater. "Water is essential for life", and for most rural people in Ireland and elsewhere, the groundwater resources in the subsurface, are the source of their water supply. The ubiquitous presence, the seasonal replenishment, and the variable movement of this hidden water resource, have underpinned the natural environment on the island, and consequently, the historical and cultural development of the island of Ireland. Yet its importance in creating and sustaining the landscape is seldom dwelt upon, and its presence seldom considered or properly understood.

The subsurface, and the fluid moving through it, are often held by both the public and professionals to be a realm of mystery. It is hidden and therefore difficult to measure or observe. Discourse concerning the subsurface and groundwater is often filled with myth, hearsay, partial truths, misapprehensions, misnomers and mysticism. The divine, God and holy are often woven into people's perspectives of groundwater. It is exceedingly rare to have a conversation in modern Ireland concerning groundwater without reference to water diviners and holy wells. Most people believe that water flows up hill, as if by magic, to emerge as springs. The word 'spring' for many people actually implies a jump from the earth. This belief in water flowing up hill is, in effect, a denial of Newton's Law of Gravity which has been widely accepted for about 300 years. One of the roles of a hydrogeologist in Ireland is to dispel pre-Enlightenment thinking, and show that mankind can understand, and can influence, conditions in the subsurface. What happens underground is not just God's will.

The widespread lack of understanding relating to the underground and groundwater is a major impediment to understanding the environment, and to taking responsibility for actions that will inevitably, eventually, degrade the surface (or superficial) environment. It is not widely recognised, for example, that the water seen in rivers during summer is almost wholly groundwater; that our streams and rivers are essentially 'groundwater drains', and that if we impair groundwater quality, we mess up the rivers, and hence, the aquatic life contained therein. In short; fishes depend on groundwater.

There appear to be three reasons why the general population wish to adhere to pre-Enlightenment ideas. These reasons are not rooted in science or technology, but are rooted in personal beliefs reinforced by culture and even agencies of the State. Geologists usually do not feel inclined, or trained, to deal with people problems. But, if we are to reach out, hook, and engage people, so that they can see that the subsurface belongs to them, and is the foundation of their environment, then we have first to identify the personal blocks. The three reasons I encounter most frequently in my field work are: -

1. Myths regarding springs and prolific wells were absorbed by Christianity and recognised by the church. They became venerated and sometimes places of worship. They became holy wells. Therefore to challenge these myths means challenging the church, and an ancient sense of spirit of place. Most people voluntarily do not feel inclined to mount this challenge.

There is a sense of mystery about the subsurface prevalent in society. A second component of this mystery is that many decision makers, professionals and ordinary people wholly, or partly believe in "water diviners". Even though individuals may profess that they don't believe in that "old rubbish", they still retain a sentimental hope that there is something in it. In a sense they are always wanting it to work. There is probably a child in everyone that wants to believe in magic. The name 'Diviner', through the word divine, links back to God's omniscience. It is as if the people who put themselves forward as 'diviners' have been blessed with the skill to interpret God's understanding of

Figure 1. A Water Diviner (left) using a stopwatch on his daughter's velvet ribbon informing a polite driller (right) that the water supply borehole he has just successfully drilled is wrong.

the subsurface. The presence of water diviners in every county, and even parish in Ireland, acts as an intervention between individual people and the ubiquitous groundwater flowing beneath their feet. Every 'apparent' success by a local diviner keeps alive the feeling that magic can work. The message received by ordinary people who need a new water supply is that "there's no need for you to try to understand the subsurface - only diviners can".

There are also many people who partially or wholly believe that God created the world in six days. A common problem for geologists in the 19th century is still a problem in the 21st. A hydrogeologist in modern Ireland explaining the simple facts about the history of the subsurface and that all water flows downhill, is secretly regarded by many as a heretic.

A final factor to reinforce the link in peoples' minds between groundwater and the sacred is a choice by a state authority to inscribe in red colour the words "Holy Well" on the most up to date topographical maps of Ireland. The position of these old, often shallow, wells and springs, that by their nature, are frequently contaminated, could have been described, accurately, with just the words "Well", "Old Well" or "Spring". Just as, for example on the same maps, a rock where Mass was held in the past is accurately described as a "Mass Rock", and not as a "Holy Rock".

The adjective 'Holy' to the uncritical mind implies that the well, and its water, is sacred, blessed, safe and pure. The words "pure spring water" are commonly used by people in association with water from these wells. Yet the water from these wells is often neither safe nor pure. With this in mind, to direct people, for example, to the site of a shallow well, described as "holy", in an actively used graveyard is irresponsible. A graveyard is essentially a low intensity landfill site for organic waste. Modern coffins made from spot glued, chipboard or plywood will collapse rapidly in wet or humid graves under the weight of soil. The soft parts of a human corpse can decompose and flow out of a broken coffin within two weeks. Therefore a dilute version of the decomposed deceased can be in a down-slope shallow well or spring, within days.

Ireland is not unique. The same cultural and religious misconceptions about the subsurface, and the fluid moving through it, are found in most countries in Europe and the Americas.

2. It is comforting to the psyche to leave the subsurface as a mystery. Most people are still buried in Ireland. Therefore the only time anyone can be sure that they will have an intimate contact with the subsurface is when they are dead. Therefore thinking about the subsurface and understanding the bacteria and water and processes below the ground is unsettling for many people. Thinking about

Trinity Well - Co. Limerick June 2001

Figure 2. Trinity Holy Well, Co. Limerick, with a slatted animal house and slurry spreading in the adjacent field. June 2001.

Figure 3. A Holy Well in an active graveyard at St. Coleman's, Mayo. From a hydrogeology and water quality perspective, a graveyard is a place for burying organic waste. In fact it could be described as a 'low intensity landfill site' The probability that the groundwater at this site contains the dilute juices from the decomposition of the recently deceased, begs the question of whether the Church and the State should be encouraging its citizens to revere the well and perhaps drink its water. Photograph from* Fish Stone Water, *2001. Copyright © Anna Rackard and Liam O'Callaghan by permission of Attic Press Ltd, Crawford Business Park, Crosses Green, Cork, Ireland.

the subsurface means facing their own mortality, and the reality of their inevitable decomposition. Subconscious denial of the unpleasant is a natural human reflex. It is also a powerful block to the use of common sense when considering the linkages and processes that take place underground.

3. Avoidance of rational thought about the presence of water, the movement of water, and the processes that take place in water underground, means that an individual, or a group of individuals, can avoid responsibility for prejudicial actions they may take at the surface. By maintaining ignorance, it is possible, without conscience, to pollute the surface and watch the pollutants seep into the ground. People without knowledge of the consequences just say "It is nature taking its course". This 'surface down approach' and 'Out of sight, out of mind', has been very convenient in the context of agri-business waste disposal and slurry spreading, or poor design and site selection of housing and septic tanks. It costs money to dispose of waste properly, or to plan infrastructural developments properly. It appears that it saves money, and a nagging conscience, to remain ignorant of the subsurface.

These three broad reasons indicate why society and individuals find it comforting, convenient and financially pragmatic to ignore the subsurface. However ignorance is not bliss. People are not stupid and they know that ignorance is often dangerous, damaging, inconvenient and expensive. Therefore to get people to think of the underground and understand that it is not a mystery (gravity still applies), geologists need to explain the danger or cost of ignorance in a personal way. Geologists for example need to explain that cheap water supply borehole construction often leads to shallow (easily contaminated) groundwater bypassing the upper casing and cascading down the hole to mix with the better-protected deeper groundwater. Or put in another, even more personal, way; explain that the sound of the waterfall inside their borehole, when they turn on the pump, is probably a dilute version of either their own or, even worse, their neighbours septic tank effluent. When common sense and the vertical dimension underground are explained in these terms, people sit up and listen. People become very interested. They readily put aside their myths and misconceptions. It is the personal nature of a water supply that can be an effective 'hook' to get people to want to understand the subsurface. On a larger scale the same can apply with government and local government or community projects. It can be shown that flawed water supply borehole design and construction, costs money. Either an expensive 'end of pipe' treatment system has to be bolted on to the delivery system to purify the water, or a new source needs to be found or constructed properly.

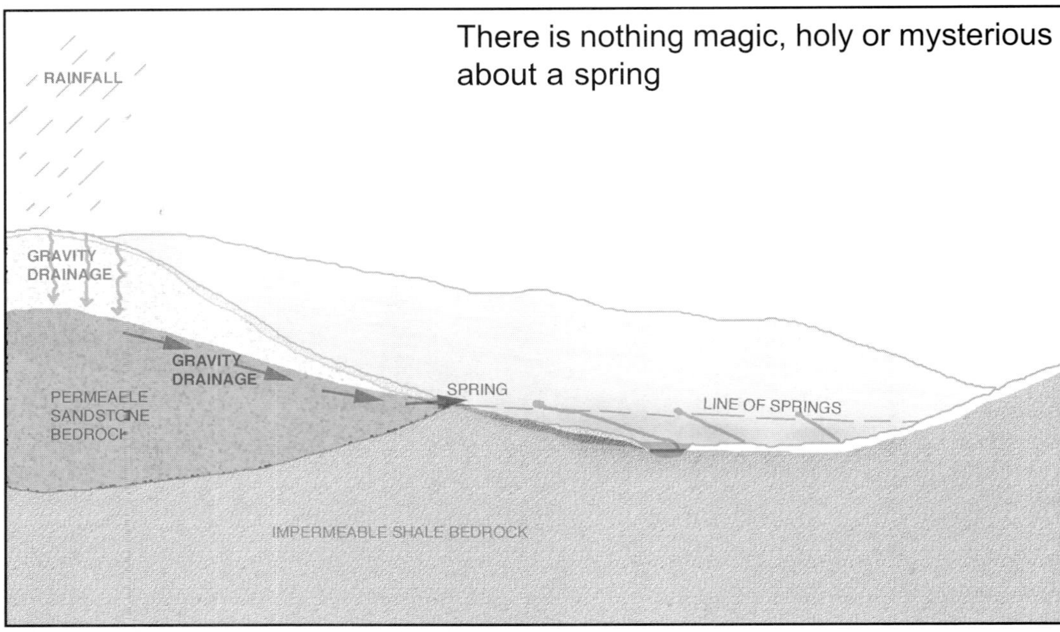

Figure 4. Above. There is nothing magic, holy or mysterious about a spring. This diagram illustrates one scenario of rocks and groundwater resulting in springs at the contact of permeable and impermeable rocks.

Figure 5. Left. Tubber Maura, Co. Louth, close by an active graveyard. This spring is shown on the latest O.S. 1:50,000 scale 'Discovery Series' maps as a 'Holy Well'. The same red lettering used shows the location of the ancient religious centre of Monasterboice, with its Churches, Roundtower and High Cross.

In summary; geologists and geomorphologists will probably continue to find it difficult to communicate the fundamental importance of the geological foundation of the landscape to non geologists, if they start by explaining the wonders of the age, the evolution and the composition of the static, solid rock framework. A rock that has been there for 300 million years is not personal, immediate or expensive to the general public or decision-makers. "It will probably be still there tomorrow", and the consequences of ignoring it are slight. By contrast, the dynamic fluid moving through the framework of these static, "boring" rocks, often has an immediate, financial or personal relevance. Once the ordinary person has become interested in the water, and the common sense way that the water moves through the rock, then they will begin to realise that they need also to understand the nature of the rock, in order to obtain and maintain both the quantity and quality of their water. In other words; use a person's real interest in water, to get them to think about the underground. Once they realise that the water moving underground is controlled by gravity and the structure of the rock, they will start to understand the relevance and importance of the rock in their day-to-day lives.

The mythical mystical Ireland image was sold abroad to attract visitors, but if we continue to collude, and believe this image ourselves, then it is a real block to our national understanding of the foundation of the real landscape that we, as a nation, are trying to sell to attract visitors. Encouraging others and ourselves to suspend disbelief is 300 years out of date. Portraying our subsurface in terms of myth and mystic beliefs helped create a short-term gain of tourist dollars in the 1950's, that is now turning to a long-term national embarrassment on water quality in the European Court.

In: Parkes, M.A. (Ed.) 2004 *Natural and Cultural Landscapes - The Geological Foundation*, 209-210, Royal Irish Academy, Dublin

' UNBUNGLING ' IRISH LANDSCAPE: A GAIA LOGICAL FOUNDATION

Paul Leech: BE B Arch (Hons) RIBA MIEI, Principal GAIA ecotecture, 11 Upper Mount Street, Dublin 2. Email: paulleech@eircom.net

Thirty years ago the late Oisín Kelly, sculptor, impressed me with the remark that the literal meaning of any iconography is unimportant: what matters is what we do within that iconography.

Poetry subsists in the spaces between the words, rather than the words themselves. Culture subsists within a landscape: it is not the landscape itself.

The use of 'scape' is challenging: indicating "a scene or view of something". Geology is hard science: I have studied and enjoyed it at University: 'views' of things touch on aesthetics: hard science and aesthetics intersect in culture. Culture requires a constructive critique of what is happening with indications for future progress. 'Gaia-logical' issues have an inner necessity and are of ultimate concern at this time in Ireland: we Irish have been taking cultural possession of our national landscapes while economic processes have reached a state of unprecedented change. Social and ecological processes have sadly been relegated while the vulgarity of the powerful prevails, both in public and private domains.

Some of us who consciously decided to stay in this country thirty years ago, to make a career and raise a family have bitter-sweet feelings about a burgeoning economy, (in some senses, thankfully, raising the absolute level of economic well being), while at the same time witnessing a 'sell out' of many reasons for staying, both in social and environmental terms.

At one time, lack of economic resources was used as an excuse for poor stewardship of our environmental resources or social responsibilities: we no longer can accept these excuses. The results since the foundation of the State in landscape terms can be described as 'poor', or at best 'fair': in short I believe that our landscape has been badly 'bungled' in many senses: copious platitudinous words have not been matched by actions, a pattern I recognise as nationally characteristic. Our landscape condition expresses who we have been. It is full of potential for expressing who we might become.

For example the new-found affluence of Ireland and failure of governance has recently led to uncontrolled inflation: in house prices this has lead to appalling suburban sprawls, remote from jobs and services: these have created an unacceptable quality of life and 'landscapes' which store up problems for the future generating absurd levels of "private" transport; unsustainable burdens on services infrastructure.

Taking the right directions now is crucial. Serious 'landscape' damage already done is remediable ("unbungling" Ireland is possible) and future damage avoidable, while caring for landscape in culturally creative ways, within economic, ecological and social dimensions.

Issues of social policy, spatial strategy, transport, energy, waste management, natural and man made conservation, universal access, safety etc all require difficult decisions where no one parameter can be optimised but a balance must be struck between conflicting demands within a mature culture.

To the horror of the more environmentally and aesthetically aware our rural landscapes continue to be defaced by ugly building of individual, badly designed and sited, 'bungalows' (18,000 of the 50,000 housing unit output in any one year), especially damaging in treeless landscapes of our western seaboard counties.

This island has an extraordinary landscape in cultural and natural terms in urgent need of remedial therapies: almost no building however ugly is beyond salvation. I would like to suggest a new form of national forest, completely diffused through the country, specifically targeting ugly buildings with appropriate species especially, in otherwise vulnerable locations: these could be associated with other remedial measure concerning sewage treatment, shared services etc: 'ecologising' the bungalow with transitional spaces etc is one key task in the macro landscape: there are many others in the urban landscape.

Greedy Architects have been complicit in the bungling process: Institutions which apply 'greenwash' to 'business as usual' do no service to the nation: Architects did not generally design the bungalows but from the sixties forward they have been the lackeys of development capital resulting, often, in poor results.

Synergies exist between for example the requirements of the Kyoto protocol: Carbon fixing and afforestation interrelate: programmes by Coillte, and the like could envision this dispersed 'forest' around all the ugly buildings, rail and motorways, that already or soon will 'afflict' our landscape: "Free trees" would receive wide uptake: microclimate creation, primary energy use reduction, rainfall management, and merciful screening can together heal the damage in many dimensions.

1. A living grass covered form in a wild landscape: (Iorras, Mayo) as a live / work environment, replacing an ugly mass concrete house and galvanised steel sheds.

2. A main street location where a credit union has been grafted into the visual ecology of the urban landscape (Longford). Many of the elements of the replaced building have been recycled literally and culturally.

Complaining, alone, is not creative: creativity is life.

In conclusion and not to complain unduly but to challenge myself as to what I have done about it all, I can only offer examples which address these issues, being four illustrative interventions from our own work.

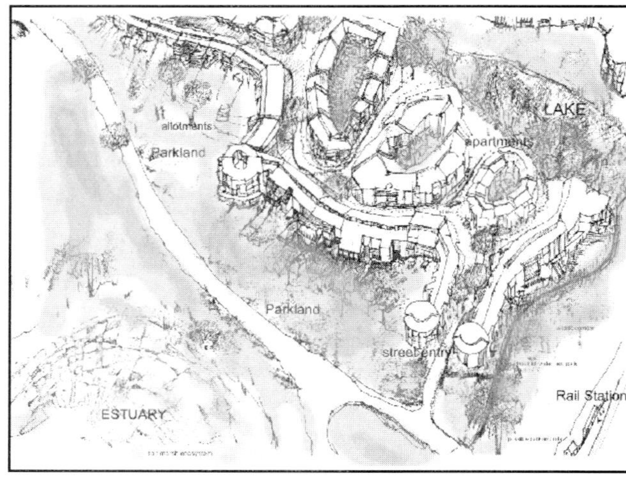

3. Eco urbanism at the Estuary of the Nanny river, Laytown on the east coast . A recently adopted land zoning of "eco residential business" is the first in Ireland. What the market will make of such a zoning in time (5 years only) will tell.

4. The tallest timber office building (30m) in the World? :an eco office tower project in a sprawling midland Irish town, relating to two fine Church towers as transcendental signs in landscape. An Bord Pleanála has recoiled from such height despite precedents such as church campanile, round towers and many other urban examples including that in Youghal, wonderfully straddling a street. Timidity is not wise where banality flourishes. A revised version is proceeding to site.

In: Parkes, M.A. (Ed.) 2004 *Natural and Cultural Landscapes - The Geological Foundation*, 211-214, Royal Irish Academy, Dublin

"PRACTICAL CONSERVATION IN SUPPORT OF SUPERLATIVE SCIENCE" - THE RE-INTERPRETATION OF THE TYPE LLANDOVERY AREA

John H. Davies, Countryside Council For Wales, Y Gelli, Stryd fawr, Llandysul, Ceredigion, Cymru, SA44 4DP, Email: sion_cwm_hir@hotmail.com

Abstract: At the beginning of the 19th century, Murchison named the Llandovery Series as a major sub-division of his Silurian. O.T. Jones mapped the type Llandovery area in the late 1920s, and published the classical papers: Southern Llandovery area 1925, Northern Llandovery area 1949. In the 1970s the International Stratigraphical Subcommittee was establishing stratotype sections for the Silurian, but rejected the small, isolated exposures in the Southern Llandovery area, and began to favour other parts of the world. At the beginning of the 1980s new forestry access roads near Llandovery provided the required continuous sections. Consequently the Silurian Subcomission chose these cuttings, which show all sub-divisions of the Llandovery Series. These were subsequently chosen as SSSI sites. The new roads were constructed for access, not for the geology, and loose debris and vegetation soon obscured the rock sections. By 1991 (in less than 10 years), they were so overgrown and degraded, that they were no longer usable as geological sites. The Countryside Council for Wales decided to clear them of loose debris and vegetation and extended them. From 1994 to 1999, the Scrâch, Trefawr, Cwm-coed-aeron, and Coed Glyn Môch track sections; in the northern area and Cilgwyn in the south, were considerably enhanced. Cefn Cerrig quarry, part of the Fron road section SSSI, in the southern area was also cleared. As a result, these sections have never before been so clearly exposed, enabling detailed study. New fossiliferous horizons were located, and many parts of the stratigraphical sequence, which were not known in detail, have been exposed for the first time. Some sequences which were thought to have been transitional from one formation to another, are now seen to change much more abruptly. At both Scrâch and Cilgwyn track sections, the Ordovician-Silurian boundary sequence was exposed and the character of the sequence was different to what was expected: it was also found to be much more fossiliferous. The Coed Glyn Môch section is now shown to be deficient as a stratigraphical standard, due to its deformation in a thrust imbrication zone. As a result of this clearance work, Wales, which has for many years been considered the home of lower Palaeozoic stratigraphy, now has cleanly exposed standard sections, which will enable renewed international study.

INTRODUCTION

In the northern area, the four GCR sites, Scrâch, Trefawr, Cwm-coed-aeron and Coed Glyn Môch, are forestry road cuttings, in Crychan Forest, NE of the town of Llandovery. In 1980 a network of forest roads was constructed, to provide access to the plantations, and exposed the sub-divisions of the Llandovery. In the southern area, the Fron GCR site, shows Silurian in cuttings along the road from Llandovery towards Myddfai. The site includes two quarries at Cefn-cerig, which show the Aeronian/Telychian stage boundary. The Cilgwyn - Ydw GCR site is in the valley of the river Ydw. The site includes four exposures near to the road from Myddfai to Llangadog. The Cilgwyn forestry road section is the western-most of these.

HISTORY OF RESEARCH

At the beginning of the 19th century, Murchison named the Llandovery Series as a major sub-division of his Silurian. O.T. Jones mapped the type Llandovery area in the late 1920s (1925, 1949). In the 1970s the International Stratigraphical Subcomission was establishing stratotype sections for the Silurian, but rejected the small, isolated exposures in the Southern Llandovery area, and began to favour other parts of the world. At the beginning of the 1980s new forestry access roads near Llandovery provided the required continuous sections. Consequently the Silurian Subcomission chose these cuttings, (four in the northern area, and one in the southern area) which show all sub-divisions of the Llandovery Series. All the cuttings subsequently became (GCR) SSSI sites. The new roads were constructed for access, not for the geology, and loose debris and vegetation soon obscured the rock sections. By 1991 (in less than 10 years), they were so overgrown and degraded, that they were no longer usable as geological sites. The Countryside Council for Wales began to clear them of loose debris and vegetation and extended them. From 1994 to 1999, the Scrâch, Trefawr, Cwm-coed-aeron and Coed Glyn Môch track sections; in the northern area and Cilgwyn in the south, were considerably enhanced. Cefn Cerrig quarry, part of the Fron road section, in the southern area was also cleared.

As a result, the sections have never been so clearly exposed, enabling them to be studied in much greater detail. New fossiliferous horizons were located, and parts of the stratigraphical sequence have been exposed for the first time. Some sequences thought to have been transitional from one formation to another, are now seen to change much more abruptly. At Cilgwyn and Scrâch the uppermost Ordovician sediments were exposed, and at Scrâch the Ordovician-Silurian boundary. The sedimentology and structure was different to previous descriptions and found to be much more fossiliferous. The Coed Glyn Môch section is now shown to be deficient as a stratigraphical standard, due to its deformation in a thrust imbrication zone. As a result of this clearance work, Wales, for many years considered the home of lower Palaeozoic stratigraphy, now has cleanly exposed standard sections, to enable renewed international study.

DESCRIPTION OF WORK

The northern Llandovery area

Work commenced (1994), on both Scrâch and Trefawr tracks, in the northern Llandovery area. More of Scrâch, and the remainder of Trefawr tracks were cleared in 1996. Work on Coed Glyn Môch commenced in 1997, Scrâch and Coed Glyn Môch were completed in 1999, with Cwm-coed-aeron the last remaining cutting in the northern area. Two of the sections showing the Llandovery stratotype in Crychan Forest were partially cleared in a joint programme of work financed by the Llandovery unit of Forest Enterprise and CCW.

The southern Llandovery area

Cefn Cerig quarry, which shows the Aeronian/Telychian stage boundary, was cleared by hand in 1994 by OUGS students, supervised by CCW. Further work in 1996 exposed a complete section linking all the quarry faces. In 1999, the Cilgwyn forestry road section also in the southern area was cleared revealing a new sequence.

The northern Llandovery area - Crychan Forest road sections

The Scrâch track section: This forestry road cutting stretches for 700m ascending southwards, across the hillslope which forms the valley side of a tributary Cwm Crychan. The hill is formed by a NW dipping Ashgill and Llandovery sequence. The chief interest of the quarry, which lies outside the GCR site, and the adjacent road cutting is the sequence from the base of the Scrâch Formation. There is then a continuously exposed section up hill through the lower part of the Llandovery, including the Bronydd and Crychan Formations.

Before clearance: The quarry and the section up the road to the next junction, was overgrown, and obscured by soil and rotting branches from felled trees. The various units along the section were traced by making small hand excavations, and interpretation of the relationships of beds, or detailed tectonic structure was impossible without clearing the whole section. The boundary between the Scrâch and overlying Bronydd formations was excavated by hand. In parts, the cutting was generally less than 1.5m high and obscured with grass and small trees. Where these trees had blown down their roots showed muddy sandstones of the Crychan Formation.

After clearance: The clearance of vegetation and scree revealed the whole sequence, from the Scrâch Formation, through the Bronydd Formation into the Crychan Formation, with each of the formation boundaries. A measured section was compiled for the whole sequence. A number of significant features were revealed, including many bands of bentonite clay. Several bedding-planes had trace fossils and current structures, many of which had been destroyed by bedding decollement, which caused slickensides. Bioturbation is common within the mudstone layers; sandy layers commonly show ripple-drifting.

The Scrâch Formation, shows a fining upward sequence commencing with massive conglomerate, in the quarry, which passes up into alternating turbidite sandstones and slates, and finally slates. This whole sequence is very similar to the section recently cleared at Cwm Clyd in Breconshire, but the section at Scrâch is highly faulted and folded. The boundary with the overlying Bronydd Formation is unfortunately faulted. The Bronydd Formation is composed of (30m) massive pale-grey mudstones, dipping at 25°NW. The angle of the cutting steepens, with the gradual onset of alternating (up to 0.15m) storm-surge sandstones and mudstones (22m thick) sequence. The sandstone bed thickness and frequency decreases upward and the top of the formation is marked by the uppermost (0.03m) of these sandstones. Conformably above the Bronydd Formation are continuous uniform muddy-sandstones and sandy-mudstones of the Crychan Formation, which continue for the remainder of the section. These beds, which occasionally show a coarser base, are frequently bioturbated. The dip steepens up the section (to 35°W) before a small stream flows across the face indicating the location of a number of faults. In the cutting W of the fault, the dip of beds of the Crychan Formation has increased to 40°W.

Structure: The Scrâch Formation is highly folded and faulted, and highly sheared close to the Bronydd Formation boundary. The lower part of the latter formation also shows many faults, and bedding decollement is present throughout the section.

Trefawr track section

The GCR site consists of 5 continuous cuttings, stretching for almost 1km along the NW side of the forestry road near the top of Cefn Trefawr. The cuttings are generally no more than 6m high and almost all occur on the SE side of the road, but at the SW end, the cuttings occur on both sides. The chief interest of this section is the almost continuous sequence through the lower half of the Aeronian Stage and particularly, the outcrop of the boundary between the Aeronian and Rhuddanian stages.

Before clearance: The road sections were only poorly exposed, with large areas obscured by scree and moss, and vegetation including small conifers. The exposure of the base of the Aeronian stage was visible.

After clearance: The whole Trefawr track section was cleared and is now almost continuously exposed. The important Aeronian stage boundary was also cleared re-enforcing its importance as an international stratotype. The section at the NE end commences a few tens of metres above the base of the Trefawr Formation and continues through that formation into the lower (20m) of the overlying Rhydings Formation. The Trefawr Formation can now be subdivided into several units, which dip gently towards the SW, being the plunge of the core of the Cefn-y-garreg Syncline. Some of the beds are thus faulted and many of the bedding-planes show decollement. In the upper half of the Trefawr road cuttings are not continuous, but it is possible to trace the beds from the upper Trefawr Formation into the overlying Rhydings Formation.

Cwm-coed-aeron track section: This forestry road cutting stretches for 300m along the SE side of a ridge, trending approximately NE-SW, formed by the NW dipping Llandovery escarpment. The chief interest is the transition from the Trefawr Formation at the E end to the Rhydings Formation at the W end.

Before clearance: There was a gap between the lowest beds of the Trefawr Formation and the main site. Westwards along the section, though the Trefawr Formation silty mudstones were continuously exposed, they were not visible or accessible for large sections because of dense gorse bushes. The lowest metre was also buried by scree. The sandy-mudstones of the Rhydings Formation formed a much better exposed part of the cutting, where a crag on the hill has been transected by the road cutting, but the lower parts of the section were masked by a large amount of scree and young pine trees.

After clearance: A continuous section was cleared, showing the transitional coarsening-upward nature of the Trefawr-Rydings Formation boundary. The sequence dips generally at 15°W. The oldest shaley part of the Trefawr Formation lies outside the GCR site boundary to the E. After gorse clearance the section was found to be largely un-faulted, but the Rhyding's Formation has been penetrated by a strong cleavage, dipping at 80°SE, associated with a large fault, which crosses the hill 50m NW of the site.

Coed Glyn Môch road section

This 600m long, L shaped, forestry road cutting, occurred on the SE, and SW sides of a NE-SW trending ridge, forming the escarpment of the steeply dipping upper Llandovery. The chief interest of this section is the sequence through the Rhydings and Wormwood Formations (Aeronian Stage), to the Cerig Formation (Telychian Stage). These rocks should contain Monograptus turriculatus and Monograptus segwickii biozones, and consequently, the Aeronian and Telychian Stage boundary.

Before clearance: From the NW end of the site, eastwards, for approximately 90m, the cutting did not show any exposed rock, due to vegetation, and for the next 20m, only intermittent outcrop of the Rhydings Formation occurred. The first continuous section then extended E for 70m consisting of muddy sandstones of the Rhydings Formation. There was some sign of faulting which may have disrupted the stratigraphy. SE from there, the Rhydings Formation was only poorly exposed for 100m, almost to a bend in the track. It was suspected that vegetation obscured major tectonic structures, implied by the complexity of the rocks in the SE half the cutting. In addition, the beds of the Rhydings Formation at the beginning of the SE exposure are overturned dipping NW, indicating that there was a major change of dip within this poorly exposed area. Around the road bend, and continuing NE to the end of the site, the rocks were more continuous, but with much vegetation and scree obscuring critical areas of outcrop. These showed deformed rock sequences, with faults and a thrust.

After clearance: Vegetation and scree was cleared and completely exposed the section, for almost its whole length (Figure 1). The structure is now revealed as an overthrusted anticline, the core of which occurs at the W end of the section, and the dip steepens towards the SE, until the beds are overturned. For the last 50m towards the road bend, the section shows a complex mixture of faulted Rhydings and Wormwood Formation, and along the SE half of the cutting, all the formation boundaries have been disrupted by faulting. At the bend itself, a strongly foliated shale, with lens-shaped, sheared, sandy-mudstone boudins, occurs in the lower part of the cutting, below a NW dipping thrust-plane. Immediately above this in the face, are the sandstones of the overturned Rhydings Formation, which dip at 65°NW. The sequence of foliated shale and sandy-mudstone boudins is equated with the Wormwood Formation and extends for 150m NE in the lower part of the cutting. The thrust has an undulating outcrop in this oblique strike section, frequently branching into numerous separate structures. Above the thrust are larger bodies of inverted, steeply-dipping, sandy-mudstones resembling the Rhydings Formation, which are isolated from each other by zones of sheared shales and boudins. These are parallel, listric-faults, and repeat slices of Redding's Formation above a sole thrust. The last 50m at the NE end of the section is composed of Cerig Formation shales below the thrust, in thrusted contact with the the Rhydings Formation sandy-mudstones, locally cutting out the whole of the Wormwood Formation. Thrust faults splaying from the main dislocation can be traced onward through the Cerrig Formation shales. Although fossils from the Rhydings, Wormwood and Cerrig Formations have been collected from the section, in consecutive, inverted order, the section is very complex and is of no stratigraphical use since all the boundaries are dislocated, and many of the beds are repeated several times by faults. The outcrops show a much more complex relationship between the stratigraphical units than might be expected from the published descriptions of this important site, possibly with the exception of O.T.Jones (1947).

The southern Llandovery area

Cefn Cerrig Quarry part of the Fron road section: There are separate exposures along the (1km) Fron road section with vegetated slopes between them, but, rocks occur close to the surface in many of these gaps. The two quarries, which show the stage boundary occur at the SW end of the road section.

Before clearance: The exposures in both quarries were discontinuous and obscured by scree. It was thus difficult

Figure 1. The SW portion of the Coed Glyn Môch section showing Rhydings Formation muddy sandstones.

to relate the sequences from one quarry to the another and to the road, and even to relate the faces to each other, in the lower, SE level and a higher, NW part of the larger quarry, where the base of the Telychian Stage had been defined.

After clearance: The clearance exposed a complete section linking all the faces in the upper and lower parts of the quarry. The SW face of the lower part of the quarry, which defines the strata-type base of the Telychian Stage was cleaned, and by following the beds around the quarry faces, the location of the base of the Telychian Stage was exposed and identified in the E face. The quarry faces, show (15m) well-bedded, very fossiliferous, sandstones and sandy-mudstones, dipping at 65ºSE, in this quarry: less than the 85ºSE the smaller quarry and the inverted, and deformed sequence in the road section.

Cilgwyn track section: This 40m long forestry road cutting is on the W side of a small tributary of the Ydw, in the forestry of south Cilgwyn. The GCR site exposed the Scrâch Formation of the southern Llandovery area, but did not show any of the stratigraphical boundaries.

Before clearance: Very little outcrop was visible, the whole section being covered with moss, brambles, small conifers and the rotting brash resulting from adjacent conifer clear-felling.

After excavation: During the clearance work no solid geology was revealed in the E part of the GCR site, which was occupied by a great thickness of glacial till. However, to the SW, the section exposed a (63.19m) sequence of the Scrâch Formation in the GCR site and extending further SW down the forestry road.

SUMMARY

The period of site clearance from 1994 to 1999 enabled CCW, with the help of Forest Enterprise, and the contractor, to expose the classical forestry road sections (GCR sites) in the northern and southern Llandovery areas, and the Cefn Cerrig quarry. The Scrâch and Cilgwyn provided the greatest contribution to the stratigraphy, greatly enhancing our knowledge of the uppermost Ashgill sequences in the area and providing a detailed lithostratigraphy to enhance biological control. The sections at Trefawr and Cefn Cerrig were cleaned, but the particular faces which exposed the strata-types were left unmodified.

Figure 2. The Cilgwyn section (2002) a number of years after the clearance; showing the course slumped sandstone and siltstone units at the top of the exposed section.

At Trefawr, the sequence into the Rhyding's Formation has been continuously exposed, and at Cefn Cerrig, the sequence around the quarry faces has been linked together. Cwm-coed-aeron section has been made more accessible, and available for stratigraphical comparison with Trefawr.

Clearing the Coed Glyn Môch section has confirmed the impression of O.T.Jones, that this side of the hill is a fault zone, and that the fault is a thrust, with an associated imbricate zone. This section is thus of great significance as a Caledonian structure: being one of the few locations showing a SE-facing Caledonian thrust in the Welsh Basin. It should thus be considered as a GCR site in the Caledonian Structures block, rather than a stratigraphical site due to the tectonic disturbance of the stratigraphical sequence.

CONCLUSIONS

These sections have never before been so clearly exposed, thus enabling them to be studied in much greater detail. New fossiliferous horizons have been located, and many parts of the stratigraphical sequence, which were not known in detail, have been exposed for the first time. Some sequences which were thought to have been transitional from one formation to another, are now seen to change much more abruptly. At both Scrâch and Cilgwyn track sections, the Ordovician-Silurian boundary sequence was exposed and the character of the sequence was different to what was expected. It was also found to be much more fossiliferous than previously thought. As a result of this clearance work, Wales, which has for many years been considered the home of lower Palaeozoic stratigraphy, now has cleanly exposed standard sections, which will enable renewed international study.

REFERENCES

Cocks, L.R.M. 1971. in Bassett and Bassett, Geological Excursions in South Wales and The Forest of Dean

Cocks, L.R.M., Holland, C.H. and Rickards, R.B. 1992. *A revised correlation of Silurian rocks in the British Isles.* Geological Society of London, Spec. Report. 21. 32pp.

Cocks, L.R.M., Holland, C.H., Rickards, R.B. and Strachan, I. 1971. A correlation of the Silurian rocks of the British Isles. *Journal of the Geological Society of London* 127, 103-136

Cocks, L.R.M, Toghill, P. and Zeigler, A.M. 1970. Stage Names within the Llandovery Series. *Geological Magazine* 107, 79-87.

Cocks, L.R.M., Woodcock, N.H., Rickards, R.B., Temple, J.T. and Lane, P.D. 1984. The Llandovery Series of the Type Area. *Bulletin of the British Museum of Natural History (Geology)* 38, 131-182.

Jones, O.T. 1925. The Geology of the Llandovery District: Part I The Southern Area. *Journal of the Geological Society of London* 81, 1-36.

Jones, O.T. 1949. The Geology of the Llandovery District: Part II, The Northern Llandovery Area. *Journal of the Geological Society of London* 103, 43-64.

Siveter, D.J., Owens, R.M., Thomas, A.T. and., Bassett, M.G. 1989. Silurian Field Excursions. A geotraverse across Wales and the Welsh Borderland. Cardiff, 133p.

In: Parkes, M.A. (Ed.) 2004 *Natural and Cultural Landscapes - The Geological Foundation*, 215-218, Royal Irish Academy, Dublin

STONE DECAY IN THE BUILDING OF THE FACULTY OF SCIENCES IN OPORTO

Arlindo Begonha*, Ângela Almeida, Rosa Oliveira*****
*** - Assistant Professor, Faculdade de Engenharia, Rua Dr. Roberto Frias, s/n, 4200-465 Porto, Portugal, Email: abegonha@fe.up.pt**
**** - Assistant Professor, Faculdade de Ciências, Praça Gomes Teixeira, 4099-002 Porto, Portugal, Email: aalmeida@fc.up.pt**
***** - Researcher, Faculdade de Ciências, Praça Gomes Teixeira, 4099-002 Porto, Portugal**

Abstract: Results obtained from a detailed study concerning the stone deterioration in the building of the Faculty of Sciences of the Oporto University are presented in this paper. The monument was built with the Oporto granite, a two-mica, medium to coarse grained leucogranite, composed of quartz, microcline, plagioclase, muscovite and biotite. In the building, stones were used with different weathering degrees and, consequently, different susceptibilities to deterioration. Granular disintegration, plates, flakes, black crusts, thin black layers and biological colonization are the main deteriorations affecting the granite. Gypsum has been identified in all stone deteriorations, halite in samples of stones affected by granular disintegration and flakes, niter and glauberite in samples of granular disintegration. The origin of the soluble salts was established as well as the mechanisms of degradation of the stone.

INTRODUCTION

The Faculty of Sciences was built in the 19th century in the English Neoclassic style. It was first designed by José da Costa e Silva in 1803 and modified by Carlos Luís Ferreira da Cruz Amarante in 1807 (Oliveira 2001). This building is located in the downtown of Oporto, characterized by high levels of air pollution. At the present time the building is very deteriorated.

The material used in the construction was the Oporto granite, a two-mica, medium to coarse grained leucogranite (Begonha 2001), composed of quartz, microcline, plagioclase, muscovite and biotite. Muscovite is the dominant mica and microcline is frequently perthitic. Apatite, zircon and rutile are the accessory minerals. The Oporto granite is considered as syntectonic with the third phase (D3) of the Hercynian orogeny (Ferreira et al. 1987), and its intra-Westphalian age of 318 million years was determined by the U-Pb zircon and monazite geochronology (Almeida 2001).

The fresh Oporto granite was submitted to a late to post-magmatic alteration process, characterized by the muscovitization of the plagioclase crystals, by a second generation of plagioclase and by the presence of several generations of dioctahedral micas, chlorite, rare chlorite/smectite mixed-layer minerals and a pure smectitic phase (Begonha 2001).

The stones used in the monument had different weathering degrees inherited from the quarry (kaolinite was identified by X-ray in the bulk rock) and, consequently, different susceptibilities concerning their deterioration. Actually, gibbsite and kaolinite are the final secondary minerals of the weathering process of the granite in Northwestern Portugal (Begonha and Sequeira Braga 2002; Sequeira Braga et al. 2002) and these two minerals have been systematically identified by X-ray diffraction in the bulk rock and in the < 2 μm fraction in stones of granitic monuments in this region (Begonha and Sequeira Braga 1993, 2000; Begonha and Teles 2000; Begonha 2001, 2002). In accordance with Begonha et al. (1994) and Begonha (2001), the porous network of the weathered granitic stones is very homogeneous, unimodal and essentially composed of very well interconnected cracks. This allows a fast transfer by capillarity and fast evaporation kinetics of the salt solutions from groundwater and rain. Water evaporates mainly inside the stones, therefore most of the soluble salts crystallizes in the porous network of the granite leading to strong stone decay.

This study is part of a project concerning the protection and conservation of the cultural heritage. It aims to draw the attention to the deterioration processes of the granitic monuments and to promote the discussion of the different threats to stone buildings in present day anthropogenic polluted environments.

MATERIALS AND METHODS

Fifteen samples of different types of stone deteriorations (granular disintegration, plates, flakes, black crusts and thin black layers) were obtained in the Faculty of Sciences.

To identify the minerals of soluble salts responsible for the genesis of stone decay, samples of different types of stone deteriorations were observed and analysed by scanning electron microscopy (SEM) and by X-ray diffraction (XRD). This work used a Jeol JSM-6301F microscope equipped with the Noran Voyager spectrometer, an energy dispersive X-ray system (EDS) and a wavelength dispersive X-ray system (WDS) and a diffractometer apparatus Phillips PW 1710 (40 kVx20 mA) using the Kα Cu radiation.

LOCATION OF THE DETERIORATIONS IN THE MONUMENT

Granular disintegration and thin black layers are the dominant stone deteriorations in the Faculty of Sciences. Plates, flakes and black crusts were also observed. Biological colonization was not studied in this work.

Granular disintegration is associated with outside and inside low areas of the building and submitted to the capillary rise of salt solutions from the ground. The outside areas are wet but not subjected to run off. The loss of material is higher in the areas between 1.5m and 3.0m high. The evaporation of the solutions leads to the deposition of salts in a sequence according to their solubility. Arnold and Zehnder (1989) refer that this process fractionates the solutes to form a sequence in which the salts rise higher the more soluble they are. This explains the concentration of more soluble salts, such as chlorides and nitrates, in higher areas, giving rise to higher degrees of deterioration and the concentration of lesser soluble salts, such as, sulphates in the lower areas of the façades. Granular disintegration is also observed in the rain-exposed areas of the South and Southwest façades.

Plates and flakes are also related to inside and outside low areas of the building submitted to the capillary rise of salt solutions from the ground. Flakes appear above or by the side of stones strongly affected by granular disintegration.

Black crusts are associated with wet sheltered areas not submitted to run off, such as cornices, niches and balconies. Black crusts are formations exterior to the stone being composed by gypsum and dry deposited materials such as fly ashes.

Thin black layers are the second most important stone deterioration in the Faculty of Sciences. They are observed in wet areas, more or less exposed to rainfall and are also exterior to the stone, chemically heterogeneous, composed by a great number of particles cemented by a matrix. The matrix is formed by several organic compounds rich in carbon and by iron rich compounds.

MINERALOGICAL STUDY OF THE DETERIORATIONS

Several minerals of soluble salts were identified in samples of stone deterioration of the Faculty of Sciences.

Gypsum has been identified in all types of stone deteriorations, halite in samples of stones affected by granular disintegration and flakes, niter and glauberite in samples of granular disintegration.

Granular disintegration is normally associated with the presence of small amounts of highly soluble salts such as halite with or without niter. Even when its content is low, halite is the most damaging mineral of soluble salts due to its high solubility and its strong crystallization forces. Owing to its high solubility, halite easily crystallizes, solubilizes or deliquesces in the presence of small quantities of water (ground water or rainwater transferred by capillarity) or simply by the variation of the relative humidity of the air. During the evaporation process, salt solutions become more and more concentrated in highly soluble salts. Halite crystallizes only at the final stages of the evaporation process in small pores and from highly concentrated salt solutions. Besides this, the crystallization forces of halite are the strongest among the minerals of soluble salts since its molar volume is the smallest ($27.06\ cm^3 x mol^{-1}$) (Begonha 2001). Niter is also a highly soluble salt with strong crystallization forces. Gypsum is always present but its content does not seem to influence the degree of intensity of granular disintegration. Begonha (2001) refers that significant amounts of gypsum are found in samples of granular disintegration with low intensity of stone decay while samples with low amounts of gypsum exhibit high intensity of stone decay. Needle shape crystals of glauberite develop also significant crystallization forces.

Black crusts are composed of gypsum crystals, fly ashes (carbon and sulphur rich porous fly ashes and silicon and aluminium rich smooth surface fly ashes) and silicate rich particles.

The following particles have been identified in samples of thin black layers: silicate rich street dust, iron rich particles and smooth surface fly ashes rich in silicon and aluminum or rich in iron. Small amounts of gypsum were identified. The composition of these thin black layers is similar to those referred by Nord and Ericsson (1993) and Begonha and Sequeira Braga (1996).

Figure 1 shows some of the morphologies exhibited by the minerals of soluble salts and particles responsible for the several types of stone deterioration in the building of the Faculty of Sciences of Oporto. Halite crystallizes as perfect cubic forms (Fig.1a) or frequently exhibiting dissolution features (Fig. 1b). Gypsum can show crystals with tabular habit, pseudo-hexagonal shape (Fig. 1c) or no regular shape due to partial dissolution (Fig. 1d). Other frequent aspects comprise fly ashes in the black crusts (Fig. 1e) and thin black layers (Fig. 1f).

CONCLUSIONS

The minerals of soluble salts and air pollution are the main causes of the deterioration of the Oporto granite in the Faculty of Sciences. The main source of the soluble salts is rainwater. The main sources of air pollution are the industries and traffic.

The crystallization of the minerals of soluble salts in the areas of strong evaporation causes the development of stresses responsible for stone decay. This mechanism is

Figure 1. Next page. SEM observations - Different aspects of deterioration in the Faculty of Sciences.

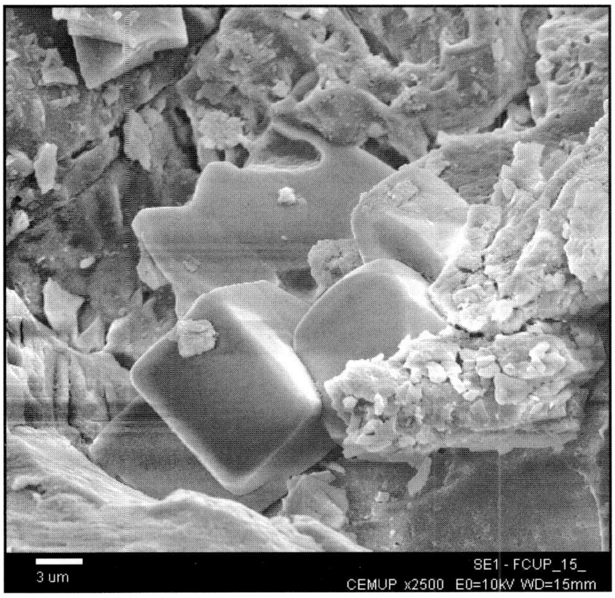

1a. Granular disintegration - cubic forms in halite

1b. Granular disintegration - dissolution features of halite

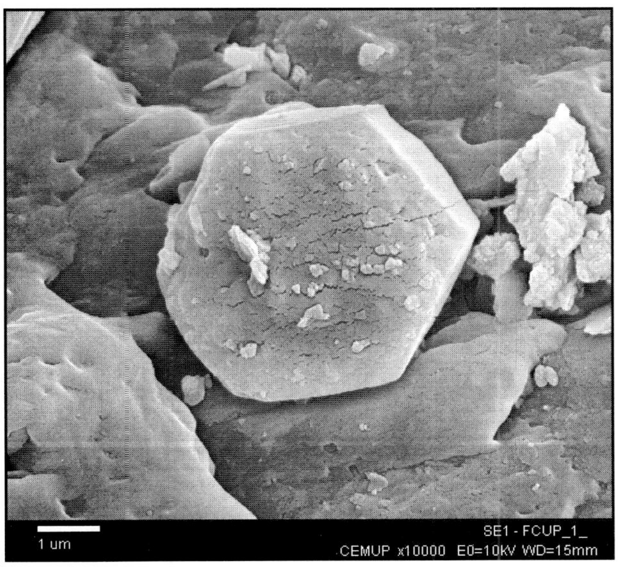

1c. Flakes - Pseudo-hexagonal crystal of gypsum

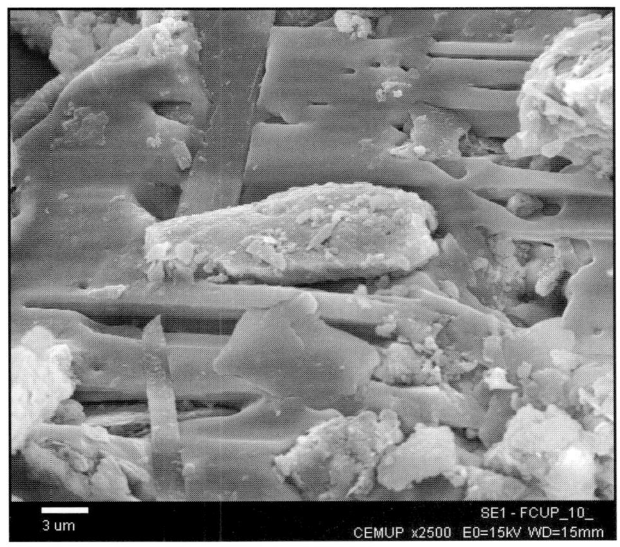

1d. Black crust - gypsum with dissolution features

1e. Black crust - fly ash

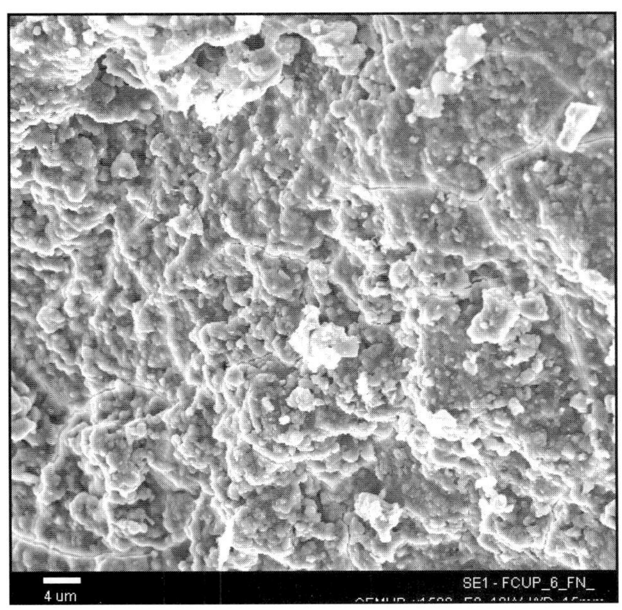

1f. Thin black layer - general aspect

responsible for the genesis of granular disintegration, plates and flakes.

Black crusts are formed when the minerals of soluble salts crystallize on the external surface of the stones. They usually do not contribute to significant stone deterioration but indeed to the unaesthetic aspect of the building.

Dry deposition of air pollutants is responsible for the genesis of thin black layers and black crusts, respectively formed in exposed and sheltered areas of the façades. Thin black layers are free of soluble salts or present small amounts of gypsum, while black crusts are basically composed by gypsum crystals and airborne particles.

ACKNOWLEDGMENTS

This study is being developed under the scope of the Centro de Geologia da Universidade do Porto and has the financial support of the "Programa de Financiamento Plurianual de I&D (FCT)".

REFERENCES

Almeida, A. 2001. Geochemical and geochronological characterization of the syntectonic two-mica granite of Porto (NW Portugal). In *Proceedings IIIer Congreso Ibérico de Geoquímica and VIII° Congreso de Geoquímica de España*, Zaragoza (eds. Lago, Arranz and Galé), pp. 311-315.

Arnold, A. and Zehnder, K. 1989. Salt weathering on monuments. In *Proceedings 1st Symposium La Conservazione dei Monumenti nel Bacino del Mediterraneo* - Bari 1989, (ed. Fulvio Zezza), Grafo, Brescia, pp. 31-58.

Begonha, A. 2001. *Meteorização do granito e deterioração da pedra em monumentos e edifícios da cidade do Porto*. FEUP-Edições, Colecção Monografias, 2, Porto, 445 pp.

Begonha, A. 2002. Estudo da Patologia da Pedra da Igreja de S. Pedro de Rates. FEUP, Porto, 85 pp.

Begonha, A., Jeannette, D., Hammecker, C. and Sequeira Braga, M.A. 1994. Physical characteristics of the Oporto granite related to stone decay in monuments. In *Proceedings 3° Simposio Internazionale La Conservazione dei Monumenti nel Bacino del Mediterraneo* - Venezia 1994, (eds. V. Fassina, H. Ott and F. Zezza), Venezia, pp. 541-546.

Begonha, A. and Sequeira Braga, M.A. 1993. Stone Weathering and Deterioration in a Granitic Monument. In *Proceedings ISRM International Symposium on Safety and Environmental Issues in Rock Engineering EUROCK'93* - Lisboa 1993, (eds. Ribeiro e Sousa and Grossmann), Balkema, Rotterdam, pp. 267-274.

Begonha, A. and Sequeira Braga, M.A. 1996. Characterization of black crusts and thin black layers in granitic monuments: the role of air pollution. In *Proceedings European Commission Workshop Degradation and Conservation of Granitic Rocks in Monuments* - Santiago de Compostela 1994 (eds. Vicente, Delgado Rodrigues and Acevedo). Protection and Conservation of European Cultural Heritage, Research Report nº. 5, pp.121-126.

Begonha, A. and Sequeira Braga, M.A. 2000. Study of stone deterioration in the cloister of the Mosteiro de Grijó, Portugal. In *Proceedings 9th International Congress on Deterioration and Conservation of Stone* - Venice 2000, (ed. Vasco Fassina), Venezia, vol. 2, pp. 689-696.

Begonha, A. and Sequeira Braga, M.A. 2002. Weathering of the Oporto granite: geotechnical and physical properties. *Catena* 49 (1-2), pp. 57-76.

Begonha, A. and Teles, M. 2000. Study of stone deterioration in the Palácio do Freixo in Oporto. In *Proceedings 9th International Congress on Deterioration and Conservation of Stone* - Venice 2000, (ed. Vasco Fassina), Venezia, vol. 2, pp. 593-601.

Ferreira, N., Iglesias, M., Noronha, F., Pereira, E., Ribeiro, A. and Ribeiro, M.L. 1987. Granitóides da Zona Centro Ibérica e seu enquadramento geodinâmico. In *Libro Homenaje a L.C. Garcia de Figuerola*, (ed. Rueda), Madrid, pp. 37-51.

Oliveira, R.M.C. 2001. *Deterioração de monumentos graníticos: o caso da Faculdade de Ciências da Universidade do Porto - abordagem didáctica*. MSc Thesis, Universidade do Porto, 206 pp.

Nord, A.G. and Ericsson, T. 1993. Chemical analysis of the black layers on building stone. *Studies in Conservation* 38, nº. 1, pp. 25-35.

Sequeira Braga, M.A., Paquet, H. and Begonha, A. 2002. Weathering of granites in a temperate climate (NW Portugal): granitic saprolites and arenization. *Catena* 49 (1-2), pp. 41-56.

In: Parkes, M.A. (Ed.) 2004 *Natural and Cultural Landscapes - The Geological Foundation*, 219-222, Royal Irish Academy, Dublin

ADAPTATION OF ECOLOGICALLY BASED LAND USE PLANNING PRINCIPLES IN THE CENTRAL HUNGARIAN REGION

Erika Bodnár MSc, geographer, Geological Institute of Hungary, H-1143 Budapest, Stefánia út 14. Hungary. Email: bodnar@mafi.hu

WHY ECOLOGICALLY BASED LAND USE PLANNING?

"It is about protecting our home and standing watch over the air, water, vegetation, and soil on the Earth's crust that makes life possible on what would otherwise be a planet as barren and inhospitable as our moon" (Honachefsky 2000).

The area investigated is part of the Central Hungarian Region near the capital, Budapest. The task of the work is investigation of the environmental impact of gravel mining. In this work field investigations were made as well as study of literature about the area. During the work it was realised, that economical viewpoints are missing from the now applied land use planning methods. In this article I try to show some different viewpoints of land-use planning, first of all the ecologically based planning, which is the best for preservation of natural value.

The investigated area has 22 working and 27 abandoned gravel mines. The mining is going on under the groundwater table following from geological conditions and high groundwater level (Figures 1, 2). It means that every new mine's opening equals the birth of a new lake. The gravel mining is from the surface, and it has great environmental impact. This impact influences all environment elements. These are partly temporary like noise, dust or disturbing the living world or partly permanent like changing the original natural character. The formation of lakes is a definite change (Figure 3).

In Hungary, there is a compulsory regulation for businesses that they have to prepare: design of landuse planning not later than three years after the opening of a mine. This design should be the answer to several questions. The most exciting questions are the impacts, which have influences on territories away from the mine: direct landscape impacts, which have importance in depositing, back-filling, or transport of mine-waste, impact on water-balance: sinking of ground-water level, changing

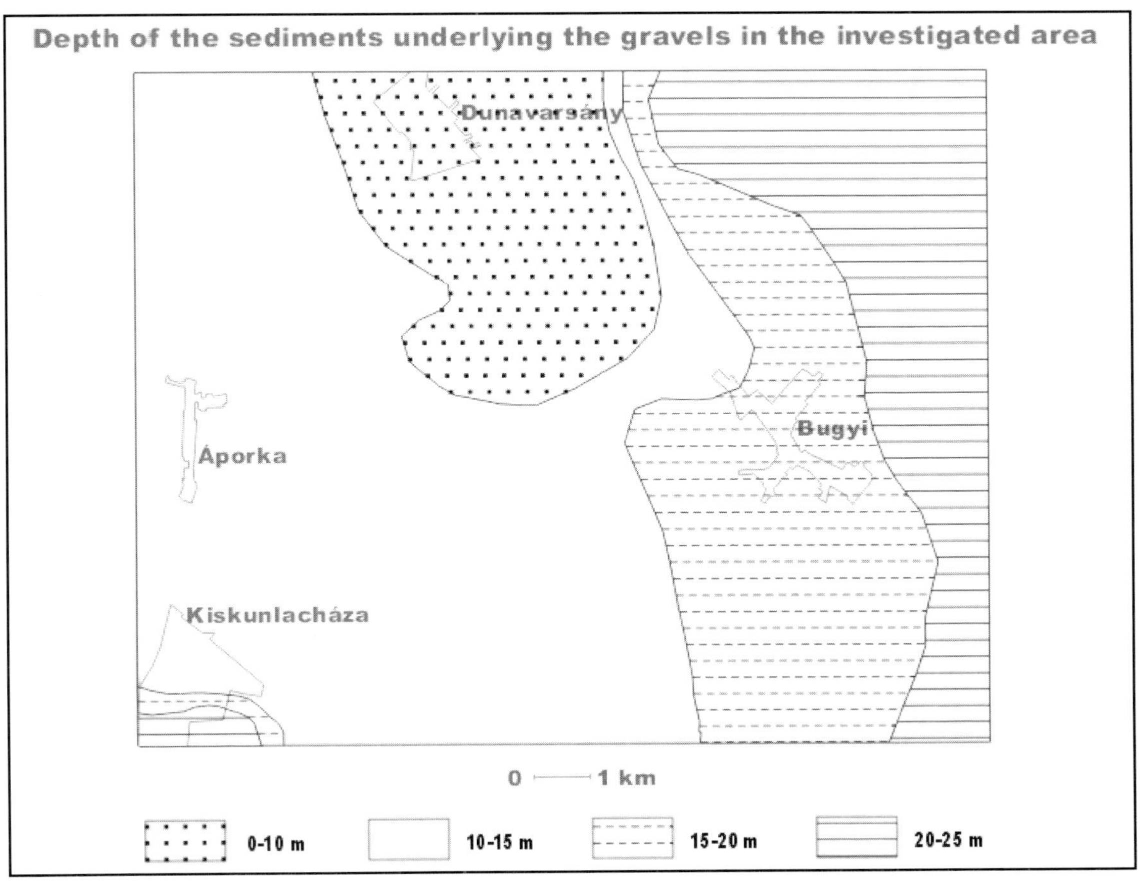

Figure 1. Depth of the sediments underlying the gravels in the investigated area.

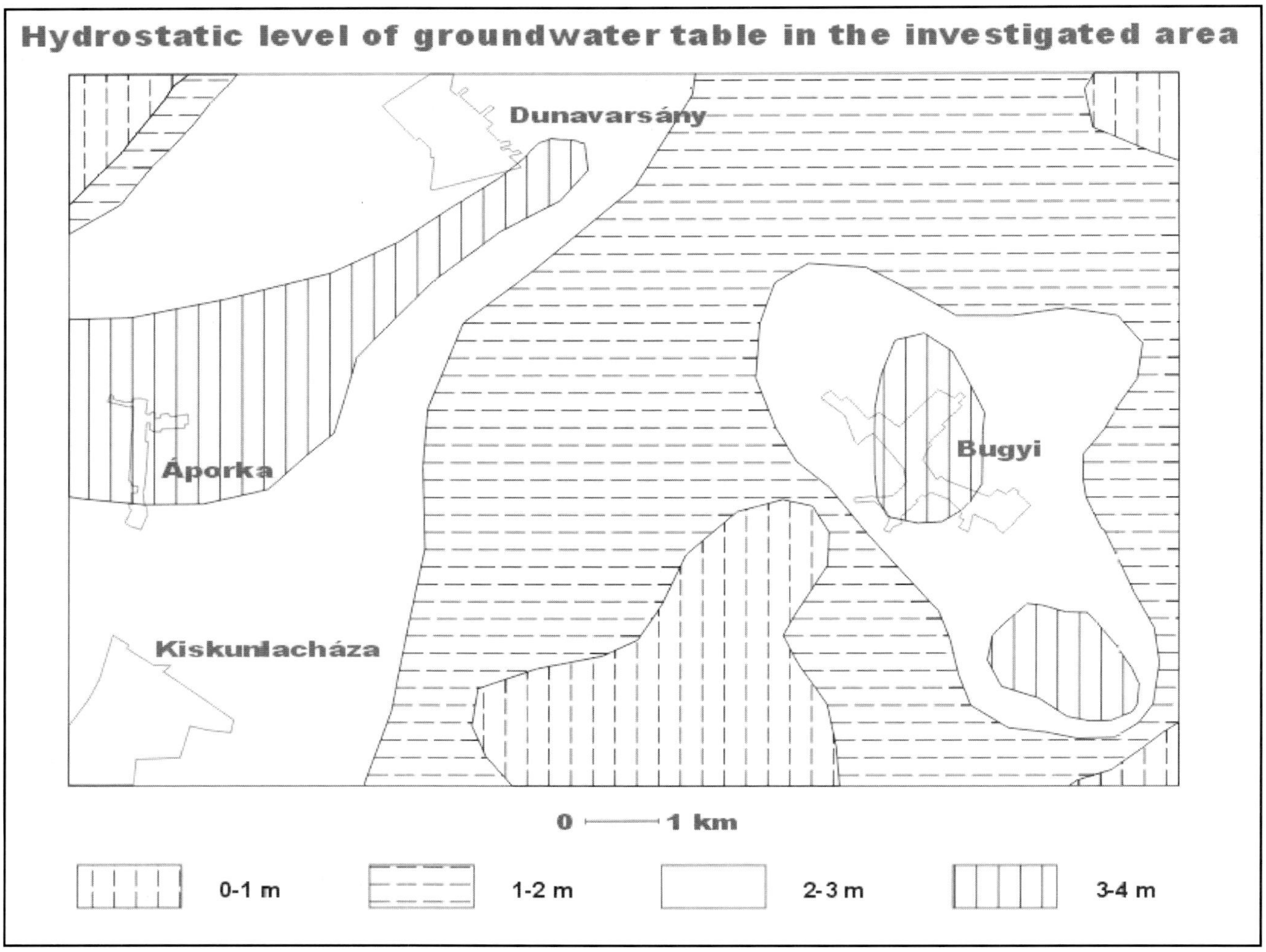

Figure 2. Hydrostatic level of groundwater table in the investigated area.

of floristic circumstances; impacts of subsequent utilisation of lakes which could be more dangerous than original mining impacts.

In the landuse planning we can use two principles. One of them is suited for the social requirements: it becomes angling, fishing, recreation area where the people can find a relaxing amenity environment. The second of them is rehabilitation: the injuries of landscape - if it is possible - need to be repaired so that the site is similar to the original situation. It is easy to realise that for conservation of environment, balance of biodiversity, and long-term investment for people it demands the application of the latter method.

What is the basic principle of ecologically based land use planning?

The most important thing is that after-use and original situation should be maximally close. This is plausible - in this case - mining from below groundwater table is impossible. The original situation basically changes because after the mining there remains a lake. It means that we can't return the sites to their original situation. Therefore we have to choose between two possibilities: formation of a 'natural' habitat or development of a human environment. On the investigated area both of them are present because there is no sharp boundary between the human and ecological utilisation. If we choose natural-close rehabilitation the habitat should be fit to the surrounding environment: after analysis of the involved environmental elements they should be of a form related to the original nature.

What does it mean in practice?

Environmental impact appraisal should be carried out at the beginning of mining. This appraisal includes survey of original conditions of environmental elements, analysis of changes caused by mining. The appraisal contains goal of reutilization and character of planned end-situation too.

The investigated area is bordered by a National Park. Before the beginning of agriculture there was characteristic forestry steppe vegetation. The landscape became cultural "puszta" because of human activity (Figure 3). Significant agricultural activity is going on this territory but we can find the original grassland in some places too. Because of mining newer habitats appeared in this land-utilisation mosaic. Partly lakes, partly wetlands were formed and on many places are unused "cultural footprints" too: abandoned and working waste depositories.

For success of ecological view-points, the methods and

Figure 3. See text for explanation.

type of steepness, directions of lakeshores, bottom morphology of lakes, planting lakeshores with plants should be planned simultaneously with mine-working. Lakes typical of "sodic puszta" - a few metres deep, water quantity hanging on distribution of precipitation, drying up bottom - couldn't form because of significant deepness and the connection with groundwater table. The friction angle of excavated material determines steepness of lakeshores. This angle gives a natural shore form which will not slide as well as become muddy. The handling of ponds is very important for forming their similar development to natural lakes.

At the forming of lake-shore plant-associations, there should be formed not only areas with sedge and bulrush patches, but continuous reed bands too for suiting the requirements of nutritions, multiplication and lair of different aquatic animals/fauna. Small islands could provide for rests and breeding-place of wandering birds. These islands become "stepping stones" for available biological balance. Willow, poplar, alder woods on the shore are good for aquatic birds and birds of prey too, and similar to the original steppe vegetation. It is optimistic to form smaller lakes in a cassette system where they can communicate with each other through dams as a natural filter system (Figure 4). Among the lakes should be created mosaic type habitats with joining paths. With this method undisturbed moving of amphibians (frogs), reptiles (tortoise, snakes), birds (aquatic-, steppe-birds, migratory-, permanent birds) and mammals (deer, fox, rodent and insectivore) could be provided.

Figure 4. See text.

In general cases these corridors can provide for the movement of inferior animals/fauna too.
An important problem on the investigated area is the handling of mine dumps and communal waste dumps. Lakes and the dumps are similar in forming a new living habitat, but they are different in difficult rehabilitation of any original vegetation/plant-association: acacia, black-elder, and other weeds are hardly suppressible.

The mining companies working on this area recognised their own responsibility for shaping of the situation after gravel mining: some of them have made landscape and landuse plans for reclamation of abandoned mining areas. Recultivation of the old mining places demands rigorous environmental regulation for satisfactory result.

REFERENCES

Albert, K. 2001. *Explanatory of the sensitiveness maps.* (in Hungarian: Magyarázó az érzékenységi térképekhez) Environmental Ministry, pp. 3-48. Budapest.

Csorba, P. 1999. *Landscape Ecology.* (in Hungarian: Tájökológia) Kossuth Lajos University Applied Landscape Geographical Division, pp. 4-113. Debrecen.

J. Erdélyi-Szalóki and M. Kajcsa 1993. *Environmental features of Pest Region II.* (Pest megye környezeti jellemzoi II.) Manuscript. Environmental Supervision of Centre-Danube-Valley. pp. 99-156. Budapest.

Honachefsky, W. B. 2000. *Ecologically Based Municipal Land Use Planning.* Lewis Publishers. pp. 1-185. London, New York, Washington.

Kákonyi, Á. 1995. *Mining in terms of landscape and nature conservation.* (in Hungarian: A bányamuvelési tevékenység táj-, és természetvédelmi vonatkozásai) Professional opinion. Manuscript. pp. 1-9. Kecskemét.

Kolucsay 2000. *Land use planning for Bugyi-V. "Gravel" mine-site.* (in Hungarian: Tájrendezési terv. Bugyi-V. "Kavics" védnevu bányatelek területére) Manuscript. pp. 3-24. Budapest.

Savanyú, K. 1995. *Review of mining granting process.* (in Hungarian: A bányászati tevékenységek engedélyezési folyamatának ismertetése) Manuscript. Hungarian Mining Office, pp. 2-16. Budapest.

VÁTI Hungarian Regional Developing and City Planning Company 1999. *National Land Use Planning Design.* (in Hungarian: Országos Területrendezési Terv) Preplan stage. Negotiational Material. Ministry of Agriculture and Countryside Development. Manuscript. pp.7-63. Budapest.

In: Parkes, M.A. (Ed.) 2004 *Natural and Cultural Landscapes - The Geological Foundation*, 223-226, Royal Irish Academy, Dublin

AN EXAMPLE OF CONSERVATION OF CULTURAL HERITAGE IN THE ANTIQUE CITY STRATONIKEIA AND RECLAMATION PROCEDURES IN ESKIHISAR LIGNITE DEPOSIT, IN SOUTHWESTERN TURKEY

Hulya Inaner, Dokuz Eylul University, Faculty of Engineering, Department of Geology, 35100, Bornova, Izmir, Turkey. Email: hulya.inaner@deu.edu.tr
Yuksel Akin, Turkish Coal Enterprises, GELI, Yatagan, Mugla, Turkey
Eran Nakoman, Dokuz Eylul University, Faculty of Engineering, Department of Geology, 35100, Bornova,Izmir, Turkey, Email: eran.nakoman@deu.edu.tr

Abstract: The Eskihisar lignite deposit is located to the west of Yatagan county of Mugla Province in southwestern Turkey. Lignite is produced by open-pit mining methods using excavator-dragline-trucks. The total reserve is approximately 97.900.000 tonnes. The Eskihisar lignite deposit is able to meet the daily need of the power station having a 3X210 MW capacity located close to the mine. The total area of Eskihisar lignite deposit is 7.691.000 m^2. 226.000.000 m^3 of overburden was hauled and approximately 55.500.000 tons of lignite were produced from 1979 until today. Overburden ratio is 4 m^3/tonnes. A further 185.000.000 m^3 of overburden is expected to be hauled in the coming decades. There are 4.358.879 tons of lignite reserves under the antique city of Stratonikeia. This reserve has not been mined to conserve the antique city in its present state. 42.400.000 tonnes of lignite will be produced in the next decades.

The antique city that lies within the boundaries of Eskihisar village in the town of Yatagan in the Mugla Province is situated in an area that was known as "Caria" in antiquity. Principal constructions within this antique city are the gymnasium, the bouleuterion, the theatre, the temple and the propylon. The city preserved its importance even after the Romans, as it housed the diocese of the area. Archaeological excavations started in 1967 and the work to include the city in the country's cultural heritage repository still continues.

Present inside and outside dumping areas total 2.156.000 m^2, and the dumped soil is 50.000.000 m^3. The importance of reclamation and recultivation in Turkey has been recognized by Turkish Coal Enterprises (TKI). Reclamation and recultivation studies were started in 1990 together with Regional Forestry Authority in the Eskihisar deposit. Totally 213.9 hectares of outside and inside dumping areas were reclaimed by planting 575.080 trees of different kinds. In accordance with the established plans, more than 200 hectares of outside and inside dump areas will be reclaimed in consideration of the natural and cultural factors in the area. The preferred post-mining land use for the area is forestation. TKI has been continuing its collaboration with the universities and other companies about the management and improvement of the environment after mining activities.

INTRODUCTION

Lignite coal deposits have been explored and evaluated by the General Directorate of Mineral Research and Exploration (MTA), and exploited using various mining methods by the Turkish Coal Enterprises (TKI). Generally large deposits are mined by TKI and low reserve deposits are mined by private enterprises. Estimates of total lignite reserves in Turkey from studies to date are about 8.4 Gt, out of which 3.9 Gt are the exploitable reserves. Besides the Elbistan lignite deposit having total reserves of 3.759 million tons, the Southwestern Turkey coal formations make up the potentially important reserves after the Soma coal in Western Turkey. Most of the known lignite deposits in Turkey are of low calorific value and have high contents of ash, moisture and total sulphur. Lignite is used mainly to generate electricity in thermal power stations, for domestic heating, industrial factories and internal consumption.

There are low quality lignite deposits of reasonably high reserves in the deposits of, Turgut, Eskihisar, Bagyaka, Tinaz, and Bayir, which are located in the northwest of Yatagan county, and Karacahisar, Ekizköy, Sekköy, Çakiralan, Hüsamlar and Alatepe deposits, which lie in the south of Milas county in the Mugla province (Fig. 1).

Work was first started in the Yatagan Eskihisar lignite deposit in the year 1956 by MTA institute in collaboration with a German firm. A workable potential of coal deposits in the Eskihisar area was determined in the drilling carried out in 1973, and production started in the year 1979. 253 reserve and operational boreholes have been drilled upto now.

GELI mines were established in 1979 as a part of Aegean Lignite Establishment (ELI) which was located in Izmir. After 1984 GELI mines have been a part of the new estab-

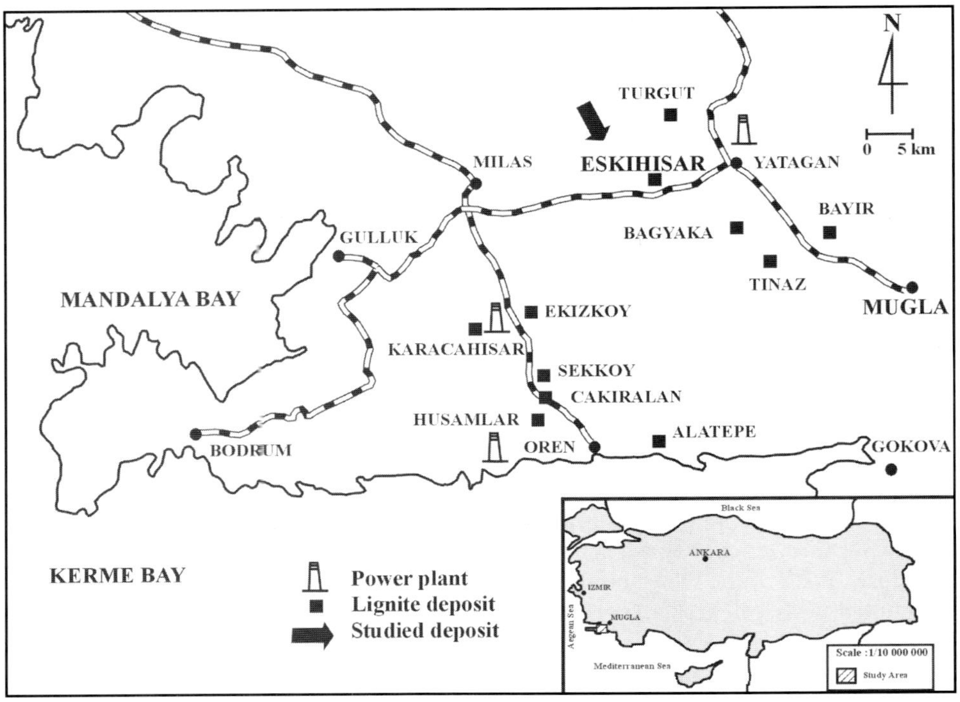

Figure 1. Left. The Eskihisar lignite deposit is located to the northwest of Yatagan in the Mugla Province in southwestern Turkey.

lishment of Turkish Coal Enterprises and were called South Aegean Lignite Establishment (GELI).

The Eskihisar lignite deposit is located to the northwest of Yatagan in the Mugla Province in southwestern Turkey (Fig. 1). Coal have been produced by open-pit mining methods. Previous studies carried out in the Mugla-Yatagan-Eskihisar region were concerned with the general geology and investigation of the coal. In addition to these, some hydrogeological and geophysical investigations have been made (Inaner 1994).

GEOLOGY

The lignite-bearing Neogene sediments cover large areas in and around the districts of Yatagan and Milas, situated in the province of Mugla in southwestern Turkey. These deposits are Turgut, Eskihisar, Bayir, Bagyaka, Tinaz, Karacahisar, Sekköy, Ekizköy, Çakiralan, Hüsamlar and Alatepe (Fig. 1). Stratigraphically these deposits are very similar (Fig. 2).

In the Eskihisar deposit, the Pre-Neogene basement comprises Palaeozoic metamorphic gneisses and schists. Triassic-Upper Cretaceous marbles reach up to 2000m thick. The Neogene Formations areTurgut, Sekköy and Yatagan. These lignite-bearing formations lie unconformably over the basement rocks of the Palaeozoic and the Mesozoic. The Turgut Formation lies unconformably on the basement of

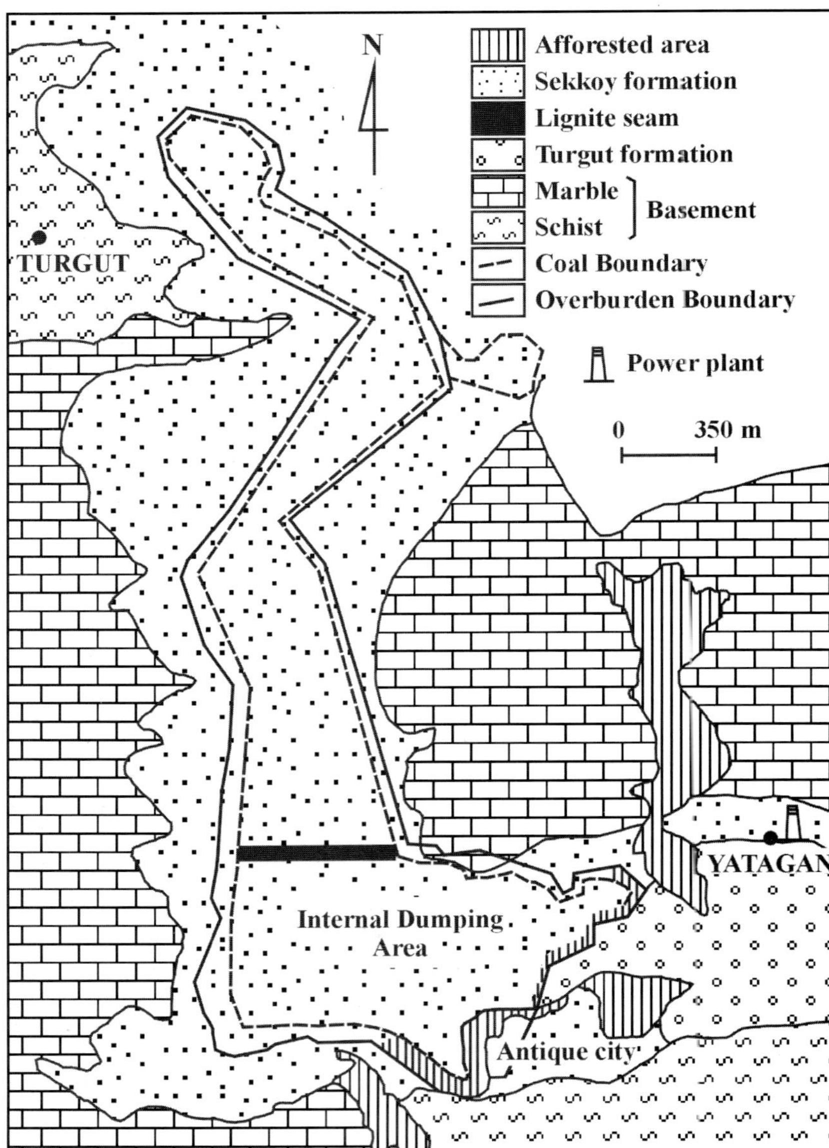

Figure 2. Below. The lignite-bearing Neogene sediments cover large areas in and around the districts of Yatagan and Milas, situated in the province of Mugla in southwestern Turkey.

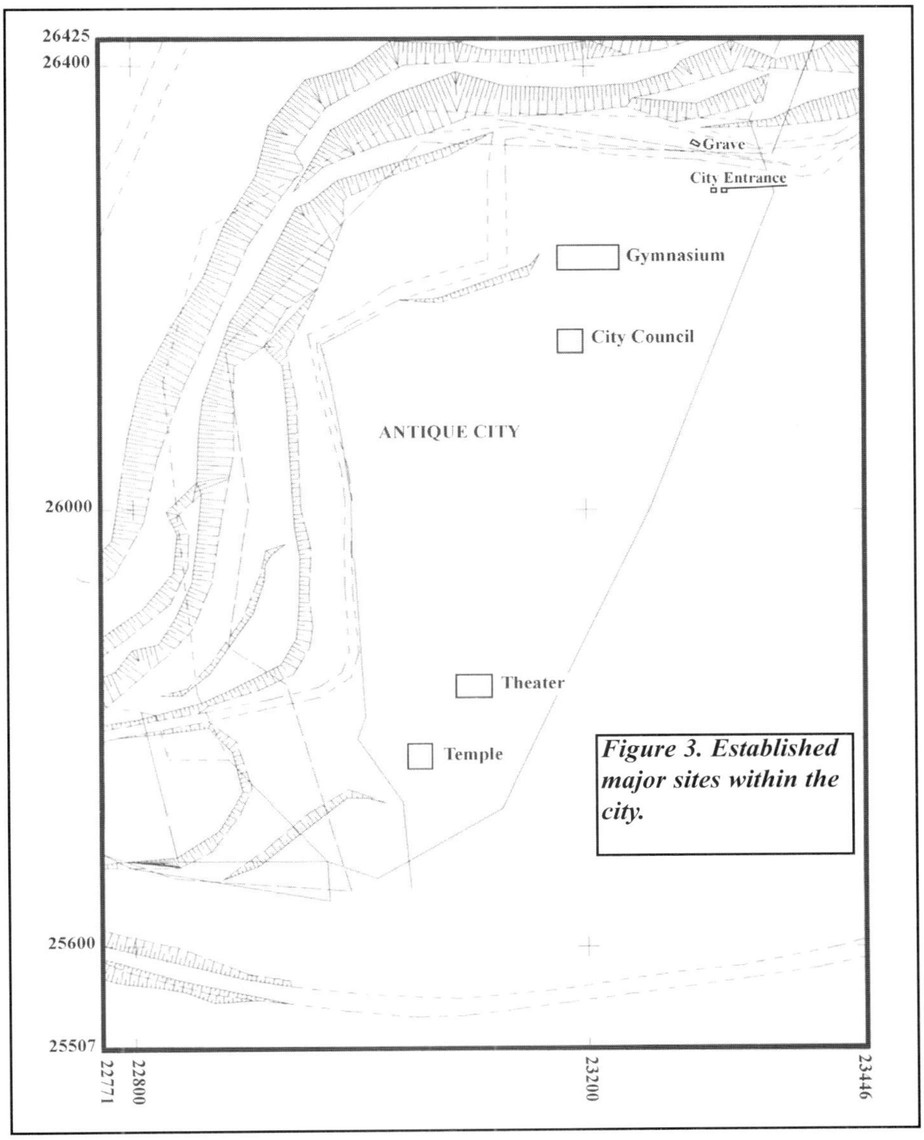

Figure 3. Established major sites within the city.

Palaeozoic gneisses and schists, and Mesozoic marbles. The Turgut Formation is made up of conglomerates, claystones, siltstones, sandstones and gravelstones. A lignite seam formed between the top of the Turgut Formation and the overlying Sekköy Formation. The Sekköy Formation is made up of marl and limestones reaching 150m in thickness. The Sekköy Formation succeeds the Turgut Formation. The Sekköy Formation is succeeded by the Yatagan Formation which consists of gravels, clays, sands, tuffs, tuffites, marls and limestones having a thickness of 400m. Post-Neogene sediments are Alluvium and Slope Debris, which overlie the Neogene sediments.

The Miocene lignites are generally located in fault bounded basins in western Turkey. Extensional tectonics started in the Middle Miocene and still continues in this area. Tectonism has effected the Eskihisar deposit with two kinds of fault, namely: Pre-Neogene faults and Post-Neogene faults. Pre- Neogene faults formed the basin in which coal and coal-bearing sediments accumulated.

ANTIQUE CITY OF *STRATONICEIA*

The ancient town in the Eskihisar villiage of Yatagan, Mugla is a city situated within the area in Southeastern Anatolia which was known as "CARIA" in antiquity. From the written documents obtained in excavations and the studies carried out in the area, it was determined that the ruins within the boundaries of Eskihisar village belonged to the ancient city of Stratoniceia. Archaeological excavations started in 1967 still continue.

Eskihisar, within whose boundaries lies this ancient city, was hit twice in 1948 by two powerful earthquakes. Because of these earthquakes and the lignite deposits closeby, the village was moved twice to two different nearby locations.

Established major sites within the city are the propylon, the gymnasium, the bouleuterion, the theatre and the temple (Fig.3).

The propylon: The greater part of this gate with double entrances located in the North of the city still stands erect. A semi circular pool is situated between these two entrances. The gate leads the way to the sacred temple Lagina dedicated to Hecate.

The gymnasium: The mosque, café and the century old planes which constitute the village centre lie within the area of gymnasium.

The bouleterion: The greater part of this construction situated in the centre of the city is reasonably well-preserved. The inscriptions on the northern wall of the construction are of late Greek era.

The Theatre: The theatre is built into a slope in the south of the city in Hellenistic times. Although the cavea still preserves its original form, the stage was changed later.

The Temple: The temple is situated to the south of the theatre.

In the excavations carried out in the Necropolis coins, ceramic pieces and terra cotta figures of late Hellenistic or Roman times have been discovered and are preserved in the museum in Mugla. Provisions for the museum and the excavation staff are made by GELI.

Every year approximately 3,000,000 tons of coal is produced from the Yatagan Eskihisar Lignite open pit mine by Turkish Coal Enterprises. As of year 2001 total reserves amount to 42,400,000 tons. 4,358,879 tons of lignite reserves which lie under the ancient city Stratonicea is not going to be excavated to preserve the remains of this ancient city.

Geophysical studies of the area were carried out in the years 1996, 1999 and 2000 by the Department of Projects and Constructions of Direcorate of Turkish Coal Enterprises to determine the existence of any ancient tombs. These studies have determined the locations and boundaries of the tombs in the area (Erguder et al. 2000, Kocak et al. 1996, 1999), which lie buried in the tuffs within the junction of Yatagan and Sekköy formations. The excavations can be easily carried out as the tuffs tend to be quite soft. No tombs have been determined within other geological formations.

Inside overburden areas around the ancient city are expected to be turned over by the General Directorate of Yatagan-Eskihisar Open Pit Lignite Works to the institution to be established as the Centre for Carian Archeological Investigations . GELI's greatest contribution to this ancient city is going to be in this respect, and the centre to be established in this large area will provide contributions to the interested researchers. This ancient city, which is going to be reactivated by the research centre, is expected to contribute positively to the tourism in the area.

RECLAMATION AND RECULTIVATION STUDIES

Turkish Coal Enterprises (TKI) and Turkish Electricity Generation and Transmission Corporation (TEAS) have been involved in reclamation and recultivation and studies in the inside and outside overburden dumping areas, and ash dumping areas.

Lignite fired Yatagan thermal power plant with 3X210 MWe installed capacity is located in the Yatagan district of Mugla province and occupies 1.163.000 m^2. Ash have been carried by ash band as moistured. There is no dust problem in the area. Different kind of trees have been planted in the ash dumping area by TEAS.

Thanks to recent environmentally conscious approaches, reclamation of areas which had been damaged by mining activities started in 1990 by reforestation and reorganization of the area. First of all, outside overburden areas that are not going to be used have been made suitable for replanting of trees, and in the year 1991 work was started in collaboration with Directorate for Forestry. An area of 170.6 hectares of outside and 43,3 hectares of inside overburden areas has been planted with 575,080 trees in total. Kind of trees planted are Acacia, Cypress, Aylantlus, Oleaster, Ash tree, Maple, Pine, Ligistrium, Olive and Fig. The most suitable tree in this area is Acacia. Planting efficiency is 95%. Wild life has started in the afforested area.

RESULTS

When lignite coal is mined totally, the mining area will be afforested. The coal mine area and ancient city will be turned into a tourism centre. The Eskihisar lignite coal mine will represent a nice example for conservation of cultural heritage and environmentally conscious mining.

REFERENCES

Akin, Y., 1993. TKI Kurumu Güneyege Linyit Isletmeleri Müessesesi Müdürlügü Eskihisar Açik Linyit Isletmesinin, Mugla ili Yatagan ilçesi Eskihisar Köyünde bulunan Antik Kent'e (Stratonikeia) olan olumlu ve olumsuz tesirleri. TKI Raporu, p 4. Mugla.

Ergüder, I., kizildag, I., güntel, S. Ve Akin, Y., 2000 GELI Bölge Müdürlügü Eskihisar Ocagi Stratonikeia-Lagina antik kentleri arasi yol güzergahi Jeofizik Etüdü., TKI Rap., p. 7, Ankara.

Inanaer, H., 1994. Güneybati Anadolu Linyit Havzalarinin Degerlendirilmesi, Doktora Tezi, p. 318, Izmir.

Koçak, Ç., ergüder, I. ve kizildag, I., 1996. GELI Yatagan Eskihisar Ocagi Antik Mezar Alani Jeofizik-Rezistivite Etüdü Raporu., TKI Rap., p. 7, Ankara.

Koçak, Ç., ergüder, I., kizildag, I. Ve Akin, Y., 1999. GELI Yatagan Eskihisar Ocagi Antik Mezar Alani Jeofizik-Rezistivite Etüdü., TKI Rap., p. 7, Ankara.

TKI, 2001. TKI, GELI BÖLGE MÜDÜRLÜGÜ FAALIYET RAPORU. MUGLA

SESSION 7

HOW CAN WE HELP PEOPLE TO KNOW AND ENJOY THEIR LANDSCAPES?

Geologists' Assocation visit to Killiney Beach, Co. Dublin.
Photograph: Matthew Parkes

In: Parkes, M.A. (Ed.) 2004 *Natural and Cultural Landscapes - The Geological Foundation*, 229-234, Royal Irish Academy, Dublin

READING THE LANDSCAPES OF SCOTLAND: RAISING EARTH HERITAGE AWARENESS AND ENJOYMENT

John E. Gordon, Vanessa Brazier and Colin C.J. MacFadyen, Scottish Natural Heritage, 2 Anderson Place, Edinburgh EH6 5NP, Scotland. Email: john.gordon@snh.gov.uk

KEYNOTE ADDRESS

Abstract: Geology, landforms and soils (collectively our earth heritage) are integral components of all landscapes, including built landscapes, and they underpin many aspects of cultural and economic activity. Raising wider awareness and appreciation of earth heritage and involvement in its conservation is a significant challenge. This applies at all levels - with schools, local communities, industry, public bodies, land managers, planners and policy makers and their advisors. To be effective, raising awareness must be relevant and stimulating, as well as informative and reliable. 'Reading the landscape' is a common theme in several approaches used by Scottish Natural Heritage to promote wider awareness and enjoyment of Scotland's earth heritage. These include publications, visitor attractions, 'Sharing Good Practice' workshops and events such as the Scottish Geology Festival.

INTRODUCTION

Scotland has a remarkable diversity of rocks, fossils, landforms and soils, spanning much of geological time. They record key periods and events in geological history, crucial stages in the evolution of life and more recent earth surface processes associated with the Quaternary glaciations and their aftermath.

Geology, landforms and soils are an integral part of Scotland's landscapes, and they provide the basis of many aspects of cultural and economic activity. Collectively, they comprise our earth heritage. However, to most people, earth heritage remains invisible and apparently irrelevant, despite the dependence of modern society on the exploitation of the Earth's resources. Similarly, understanding of earth surface processes is poor, although awareness of effects is more immediate when people are faced with geomorphological hazards such as landslides, flooding and coastal erosion.

The Scottish landscape itself is varied, ranging from spectacular mountain and coastal scenery to the farmlands, cities and settlements of the lowlands. Landscape, or rather scenery, is a major resource exploited by the tourism industry which contributes significantly to Scotland's economy (McKirdy et al. 2001). Visitors to Scotland ranked scenery as the main attraction of Scotland as a tourist destination, ahead of history and culture (VisitScotland and Scottish Natural Heritage 2002). Rocks and landforms are the basis for most of Scotland's famous landscapes and scenery and they also provide a resource for a variety of recreation and leisure activities. Scotland's hidden earth story could add value to the appreciation of landscapes by both tourists and local people alike. There are themes to catch the imagination (what might be termed the 'wow!' factor) - drifting continents and Scotland's journey from south of the equator to its present location; continental collisions and mountain building; volcanic landscapes 'born of fire'; the 'big freeze' of the ice age; living landscapes of dynamic rivers and coasts; and the story of life recorded in fossils. As acknowledged by McKirdy et al. (2001), it is not the message that is dull, but rather the way in which it has been expressed in the past, through use of over-technical language and without reference to people's everyday experience.

The last decade in Scotland has seen development of significant interest in promoting earth heritage awareness using a variety of media. Much has been achieved through collaboration and partnerships among a variety of interested organisations and wider recognition of the need to adopt better interpretive methods. In this paper we provide specific examples, targeted at different audiences, that use 'reading the landscape' as a common theme.

WHY RAISE AWARENESS?

Raising wider awareness and appreciation of earth heritage and involvement in its conservation are key challenges if we are to protect fully our natural heritage and manage our landscapes in a sustainable way. Modern society depends heavily on earth resources for industrial production, all aspects of modern technology, food production, transport, transport infrastructure, building technology and materials, and waste disposal. At the same time, human activities have significantly modified the landscape, re-shaping the surface of the Earth through movement of rock and soil, building cities, motorways and dams, fixing the coast through concrete barriers, deforestation and soil erosion, and causing extinctions of species (Hooke 1994, 1999). Natural processes have been altered, often with adverse effects on floodplain and coastal processes, which in turn have increased local hazard risk. At a broader scale, global changes in climate and sea level are increasingly likely to affect human activities. Wider appreciation and understanding of earth processes by the public will help lead to more informed debate of the issues and implications for society (Boulton 2001). There is also a further argument that better awareness of the contribution of earth science to society should lead to better support for earth science education, foster an interest in earth science as a career pathway for the next generation and also provide support for funding for academic research.

At another level, Scotland's earth heritage is part of our scientific and cultural heritage The science of geology was born in Scotland 200 years ago when James Hutton investigated the rocks around Edinburgh and further afield. Many of the fundamental principles of earth science have been developed in Scotland and applied worldwide. Today, Scotland's rocks and landforms are a resource of national and international importance (Gordon and MacFadyen 2001; Trewin 2001). They have provided crucial evidence for interpreting past geological processes of global significance, such as volcanism and ice ages. Equally, some of Scotland's rocks contain a rich variety of fossils which have greatly elucidated the evolution of the plant and animal kingdoms. If people appreciate and have an interest in Scotland's earth heritage, they are more likely to support the protection of vulnerable sites, both at a national and local level. For example at a local level, the potential loss of a valued community woodland would likely be met with protest; the loss of a small quarry, rock exposure or landform, with indifference (McKirdy and Threadgould 1994).

Rocks, landforms and soils are also an integral part of Scotland's natural heritage. They provide the basis for the diversity of habitats and species. Scotland's biodiversity also depends on the continued operation of active geomorphological processes that maintain dynamic coastal, river and mountain habitats and ecosystems (e.g. Hansom and Angus, 2001; Soulsby and Boon 2001; Thompson et al. 2001). It is becoming clear that management of sites for the benefit of habitats and species cannot succeed without reference to their physical setting and an understanding of the role of geomorphological processes (e.g. Leys 2001; Hansom et al. 2001; McEwen and Lewis 2001).

The 1999 Scottish Natural Heritage (SNH) Conference on 'Earth Science and the Natural Heritage' recognised several priorities for action based on the need for wider awareness of earth heritage and how earth system processes work (Brazier 2001). The conference demonstrated the great wealth of existing knowledge and understanding of earth systems, and also how these might be applied to sustainable management of natural resources. However, a fundamental message was that this knowledge and understanding had not been communicated to the general public and decision makers in terms that they could understand and act upon. The principal priorities for action therefore centred on improved communication in three main areas:

- among local communities, tourists, the voluntary sector and ranger services;
- at all levels of formal education;
- at policy, planning and decision-making levels.

'READING THE LANDSCAPE' IN EARTH HERITAGE INTERPRETATION

'Reading the landscape' is a theme that has been developed successfully by SNH in partnership with others to promote better awareness of Scotland's earth heritage. Activities have addressed all three groups of target audience identified above, but in this paper we focus principally on the first group. The approaches used include publications, visitor facilities, 'Sharing Good Practice' workshops, Scottish Geology Festival events and self-guided trails.

Publications: 'Landscape Fashioned by Geology'

The reading the landscape approach is best epitomised in the *Landscape Fashioned by Geology* series of publications produced by SNH and the British Geological Survey (BGS) (Threadgould and McKirdy 1999). Ten titles have so far been published, including an overview of Scotland (McKirdy and Crofts 1999), and others are planned. The series is aimed at the public, teachers and tourists with an interest in looking beyond the surface of the landscape or in finding out more about particular landscapes or landmarks and how they formed. The focus is on explaining geology and geomorphology through landscape and scenery, using appropriate text supported by high quality photographs and other illustrations. Classic landscapes such as Skye, Arran and the Northwest Highlands feature in the series, which also includes less well known areas such as Fife and Tayside and East Lothian and the Borders. Urban landscapes are also covered: Edinburgh was the first in the series and a publication on Glasgow is planned.

Visitor facilities at Knockan Crag

In 2001, SNH opened a new visitor attraction dedicated to earth heritage at Knockan Crag in NW Scotland, one of the classic areas of Scottish geology. Here, Peach and Horne unravelled the complexity of the Moine Thrust and provided the foundation for the concept of large-scale, higher-level crustal deformation. With so many potentially exciting stories to tell, it was important to have a detailed interpretation plan in place to ensure that the project remained focused. Throughout the design and installation phases, the project adhered to good practice (SNH 2002a). The overall aim was to interpret the geology of Knockan Crag in a manner that the intended audiences would find lively and entertaining. Aimed primarily at non-specialists and families, the facilities include a 'rock room' with hands-on models, a touch-screen computer, a comic strip, rock art, sculptures and quotations and poetry inscribed into rock (Figure 1). There are also three waymarked trails leading from the rock room to the Moine Thrust. These reveal stunning views across the ancient geological landscapes of Assynt and Coigach. The overall design, architecture, interpretation and rock art are intended to work together to enhance the visitor experience and to encourage people to experience rock and landscape in different ways. Knockan Crag also acts a local hub for a series of 12 'satellite' sites that reveal further links between geology and landscape.

Figure 1. Rock sculpture at Knockan Crag. (© Lorne Gill/SNH).

'Sharing Good Practice': helping others to read the landscape

For several years SNH has operated a programme of 'Sharing Good Practice' workshops and seminars for external participants. Events concerned with earth heritage issues have addressed two themes: sustainable management and raising public awareness. The former includes workshops on 'Minerals and the Natural Heritage', 'Coastal Defence and Erosion Management', 'River Restoration' and 'Working with Rivers' and have been developed in association with external partners, including the HR Wallingford, local authorities, the Scottish Environment protection Agency (SEPA) and the River Restoration Centre. The target audiences for such events are local authority planners, engineers, fisheries boards, conservation bodies and other land and water managers. Such events demonstrate how 'reading the landscape' can lead to better understanding of natural processes and hence better management. They share the benefits of integrated approaches, understanding and working with natural processes, and the development of solutions sympathetic to the natural heritage. These may involve limited intervention approaches that in the long-term may be more cost effective and environmentally friendly than heavily engineered solutions. Some of the events are also supported by best-practice guidance (HR Wallingford 2000; Scottish Natural Heritage 2000).

The second theme, raising public awareness, has included events on 'Selling Geology to the Public', 'Journeys Through Time - Interpreting a Landscape Fashioned by Geology' and 'Hewn from the Living Rock - Storytelling from Stones'. Such events are aimed at participants and organisations who are involved in meeting the public and providing interpretation, including local authority and NGO rangers and tour guides. The 'Journeys through Time' event focused on the development of self-guided trails and was held in the Cairngorms and delivered in association with the BGS and the Cairngorm Ranger Services. 'Hewn from the Living Rock' events have been held at Stirling Castle and Crathes Castle in association with the BGS, the National Trust for Scotland (NTS), Historic Scotland and the Scottish Tourist Guides Association. Their focus is on exploring the links between geology, landforms, buildings and landscape. Such courses help others to provide added value for tourists and other visitors.

Scottish Geology Festival and other events

A more informal means of involving the public, visitors and others in reading the landscape is through particular events. A good example is the biennial Scottish Geology Festival (formerly Scottish Geology Week), organised by SNH in partnership with the BGS, the National Museums of Scotland (NMS), the Hunterian Museum, Our Dynamic Earth, and several geological societies (Glasgow, Edinburgh and Open University). 'Rock On', the third event in 2001, included 145 events throughout Scotland, engaging schools, the public and amateurs through, for example, guided walks, talks, rock and fossil forays and rock surgeries. Among the highlights was a painting competition for school pupils on the theme of their favourite landscape, which attracted hundreds of high quality entries. The schools' competition has become an annual event. In 2002, the International Year of the Mountains, pupils were invited to produce a poster showing how their local mountain was formed and the geological and geomorphological processes involved. This linked with the Environmental Studies - Society, Science and Technology element of the 5-14 Scottish National Curriculum Guidelines. As part of the competition, pupils were also asked to describe their community's interactions with the mountain, in the form of a short story or poem.

As a second example, the bicentenary in 2002 of the birth of Hugh Miller provided a focus for a programme of events in and around Cromarty and Edinburgh, celebrating the life and times of this remarkable geologist, naturalist, writer and folklorist who was a key figure in the 19th century Scottish cultural landscape and played a hugely influential role in popularising geology (Taylor 2002). The programme was led by the Cromarty Arts Trust in partnership with a range of organisations including NMS and NTS, with grant aid from SNH. Associated events included conferences in Cromarty and Edinburgh (e.g. Borley, 2002). In addition, a national trail leaflet, 'In the Steps of Hugh Miller - Exploring the Geology and Scenery of Scotland', has been produced by the Cromarty Arts Trust. This provides details of 25 localities throughout Scotland, as well as places to visit in the Cromarty area associated with Miller.

Self-guided trails

Self-guided trails offer an informal means of helping people to read the landscape at a variety of levels. Geological trail guides and leaflets have been produced by a variety of organisations, including Lothians and Borders and Fife RIGS Groups and the Edinburgh Geological Society. They are aimed mainly at those with an interest in geology or learning more. Currently SNH is piloting a new series, 'Trails Through Time', within the Cairngorms National Park. This will help individuals, groups and families to gain added enjoyment from their visits through exploring the landscape and the opportunity to see how different processes have shaped it. If successful, the series will be extended to other areas.

DISCUSSION

Traditional earth heritage interpretation was generally based on boards or leaflets that were frequently written by scientists assuming too great a technical knowledge by their audience and a tendency to provide an overload of information rather than interpretation. Fortunately, as the above examples illustrate, the overly scientific approach lies largely in the past and there is now provision of interpretation and public information using a wider range of approaches and media targeted at specific audiences and based on sound interpretive principles.

One significant development has been the interaction between earth scientists and professional interpreters, for example in the Knockan Crag and 'Trails Through Time' projects. This has had several important results. First, it has established the importance of interpretive planning with clearly identified themes, objectives and media (Scottish Natural Heritage 2002b). The value of this process is well illustrated in the case of Knockan Crag (Carter 2001). Second, it has ensured selection of appropriate themes, use of the best media and pitching messages at the right level. It has also meant progressing from presentation of information to interpretation based on key principles and best practice as developed, for example, by Tilden (1977) and Veverka (1994). These practitioners stress the value of encouraging people to do something actively and to experience strong feelings so that they remember. Hence the litany: 'provoke, relate, reveal'. It is crucial that these lessons are disseminated more widely. The 'Sharing Good Practice' workshops have been particularly valuable in this respect, bringing together different specialists - earth scientists to provide the themes and stories, interpreters to explain good practice and methods, and interpretation providers (rangers, tour guides and visitor centre staff) to deliver.

One outcome of such interaction has been the use of new approaches and methods to engage people, using the 'doorways' concept to relate geology to their everyday lives (McKirdy and Threadgould 2001). This may include a more light-hearted approach, such as the presentation of information in the form of cartoons or tabloid-style journalism, as employed in some of the displays at Knockan Crag. This approach may not appeal to some purists, but providing that it is based on fact, then it offers a means of engaging new audiences. Other approaches include exploring the links between earth heritage and the arts (Clifford 1993; Pestrong 1994), recognising a more holistic view of science, literature, the arts and the landscape, expressed, for example, much earlier in the ideas of Miller (1893) and Geikie (1905). This is reflected in the use of rock art, sculpture and literature at Knockan Crag.

Another notable success has been the level of collaboration and partnerships between organisations, for example in the delivery of the Scottish Geology Festival programme. There is now a strong momentum across a range of organisations in Scotland to promote earth heritage and develop new initiatives. At a strategic level, one such initiative would be development of the networking concept (e.g. McKirdy et al. 2001), using national gateways both real (e.g. 'Our Dynamic Earth' and the museums) and virtual (e.g. the scottishgeology.com website) linked to a network of regional hubs (e.g. Knockan Crag, Skye, Arran and Glen Coe) each with their own satellite sites. Where appropriate, such hubs could integrate with existing visitor facilities and the new national parks. They could offer important opportunities for more holistic interpretation linking rocks, landforms, soils, habitats, biodiversity, land use, environmental resources and human activity. Involvement with the tourism industry at a strategic level will be crucial, as indicated by the success of the 'Landscapes from Stone' programme in Ireland (McKeever and Gallagher 2001).

The geology of Scotland's built landscapes should also play a significantly greater role in helping to raise earth heritage awareness (MacFadyen and McMillan, this volume). Rocks and landforms are a distinctive part of many of our urban environments and there are close links between urban landscapes, architectural heritage and geology. In some cases, such as Edinburgh and Stirling, volcanic landforms are a striking feature of the urban landscape; in others, the influence is more subtle but nevertheless apparent, for example in the drumlin landscape of Glasgow. Local geology has also strongly influenced the building and paving stone used in many towns and cities, providing further opportunities for reading the built landscape (McMillan et al. 1999).

CONCLUSION

Recent years have seen significant developments in promoting earth heritage awareness through 'reading the landscape' in Scotland and it is important to continue to build on successes such as the Knockan Crag visitor facilities, Scottish Geology Festival, 'Sharing Good Practice' events and the 'Landscape Fashioned by Geology' publications. Exciting opportunities lie on the horizon following the creation of Scotland's first national parks and the potential for geotourism developments there and elsewhere. However, better linking and marketing with the tourism industry will be required. Given that many organisations are involved or have a potential role to play, there is a need for clear vision, co-ordination of activities and development of existing and new partnerships between different interest groups and sectors. It is equally crucial that earth scientists continue to work closely with interpretation professionals to ensure that clear messages are identified for particular audiences and that these are presented in engaging ways.

REFERENCES

Borley, L. 2002. *Hugh Miller in Context*. The Cromarty Arts Trust.

Boulton, G.S. 2001. The Earth system and the challenge of global change. In: Gordon, J.E. and Leys, K.F. (Eds), *Earth Science and the Natural Heritage: Interactions and Integrated Management*. The Stationery Office, Edinburgh, 26-54.

Brazier, V. 2001. Earth science and the natural heritage: developing an agenda for Scotland. In: Gordon, J.E. and Leys, K.F. (Eds), *Earth Science and the Natural Heritage: Interactions and Integrated Management*. The Stationery Office, Edinburgh, 306-313.

Carter, J. 2001. Knockan Crag Interpretation Plan - Final Draft. www.snh.org.uk/wwo/Interpretation/pdf/knockan_plan.pdf

Clifford S. 1993. Sculpture, community and the land. In: O'Halloran, D., Green, C., Harley, M., Stanley, M. and Knill, J. (Eds), *Geological and Landscape Conservation*. The Geological Society, Bath, 487 - 491.

Geikie, A. 1905. *Landscape in History and Other Essays*. Macmillan and Co., London.

Gordon, J.E. and MacFadyen, C.C.J. 2001. Earth heritage conservation in Scotland: state, pressures and issues. In: Gordon, J.E. and Leys, K.F. (Eds), *Earth Science and the Natural Heritage: Interactions and Integrated Management*. The Stationery Office, Edinburgh, 130-144.

Hansom, J.D. and Angus, S. 2001. Tir a'Mhachair (Land of the Machair): sediment supply and climate change scenarios for the future of the Outer Hebrides machair. In: Gordon, J.E. and Leys, K.F. (Eds), *Earth Science and the Natural Heritage: Interactions and Integrated Management*. The Stationery Office, Edinburgh, 68-81.

Hansom, J.D., Lees, R.G., Maslen, J. and McManus, J. 2001. Coastal dynamics and sustainable management: the potential for managed realignment in the Forth Estuary. In: Gordon, J.E. and Leys, K.F. (Eds), *Earth Science and the Natural Heritage: Interactions and Integrated Management*. The Stationery Office, Edinburgh, 148-160.

Hooke, R. LeB. 1994. On the efficacy of humans as geomorphic agents. *Geological Society of America, GSA Today*, 4, 217-225.

Hooke, R. LeB. 1999. Spatial distribution of human geomorphic activity in the United States: comparison with rivers. *Earth Surface Processes and Landforms* 24, 687-692.

HR Wallingford 2000. *A Guide to Managing Coastal Erosion in Beach/Dune Systems*. Scottish Natural Heritage, Perth.

Leys, K.F. 2001. The sustainable use of freshwater resources: a case study from the River Spey. In: Gordon, J.E. and Leys, K.F. (Eds), *Earth Science and the Natural Heritage: Interactions and Integrated Management*. The Stationery Office, Edinburgh, 169-179.

McEwen, L.J. and Lewis, S.G. 2001. A sediment budget approach to inform sustainable habitat management for lampreys: the Endrick water catchment, Scotland. In: Gordon, J.E. and Leys, K.F. (Eds), *Earth Science and the Natural Heritage: Interactions and Integrated Management*. The Stationery Office, Edinburgh, 185-189.

McKeever, P.J. and Gallacher, E. 2001. Landscapes from Stone. In: Gordon, J.E. and Leys, K.F. (Eds), *Earth Science and the Natural Heritage: Interactions and Integrated Management*. The Stationery Office, Edinburgh, 262-270.

McKirdy, A.P. and Crofts, R. 1999. *Scotland. The Creation of its Natural Landscape*. Scottish Natural Heritage, Perth.

McKirdy, A.P. and Threadgould, R. 1994. Reading the landscape. In: O'Halloran, D., Green, C., Harley, M., Stanley, M. and Knill, J. (Eds), *Geological and Landscape Conservation*. The Geological Society, Bath, 459 - 462.

McKirdy, A.P., Threadgould, R. and Finlay, J. 2001. Geotourism: an emerging rural development opportunity. In: Gordon, J.E. and Leys, K.F. (Eds), *Earth Science and the Natural Heritage: Interactions and Integrated Management*. The Stationery Office, Edinburgh, 255-261.

McMillan, A.A., Gillanders, R.J. and Fairhurst, J.A. 1999. *Building Stones of Edinburgh*. Edinburgh Geological Society, Edinburgh.

Miller, H. Jr. 1893. Landscape geology: a plea for the study of geology by landscape-painters. *Transactions of the Edinburgh Geological Society* 6, 129-154.

Pestrong, R. 1994. Geosciences and the arts. *Journal of Geological Education* 42, 249-257.

Scottish Natural Heritage 2000. *Minerals and the Natural Heritage in Scotland's Midland Valley*. Scottish Natural Heritage, Perth.

Scottish Natural Heritage 2002a. Knockan Crag - an interpretive case study. www.snh.org.uk/wwo/Interpretation/casestud.html

Scottish Natural Heritage 2002b. Introducing interpretation. www.snh.org.uk/wwo/Interpretation/default.html

Soulsby, C. and Boon, P.J. 2001. Freshwater environments: an Earth science perspective on the natural heritage of Scotland's rivers. In: Gordon, J.E. and Leys, K.F. (Eds), *Earth Science and the Natural Heritage: Interactions and Integrated Management*. The Stationery Office, Edinburgh, 82-104.

Taylor, M.A. 2002. Hugh Miller and his fossils - a bicentenary appreciation. *The Edinburgh Geologist* No. 38, 10-19.

Thompson, D.B.A., Gordon J.E. and Horsfield D. 2001. Montane landscapes in Scotland: are these natural, artefacts or complex relics? In: Gordon, J.E. and Leys, K.F. (Eds), *Earth Science and the Natural Heritage: Interactions and Integrated Management*. The Stationery Office, Edinburgh, 105-119.

Tilden, F. 1997. *Interpreting Our Heritage*. 3rd ed. University of North Carolina Press, Chapel Hill, North Carolina.

Threadgould, R. and McKirdy, A. P. 1999. Earth heritage interpretation in Scotland: the role of Scottish Natural Heritage. In: Barettino, D., Vallejo, M. and Gallego, E. (Eds), *Towards the Balanced Management and Conservation of the Geological Heritage in the New Millenium*. Sociedad Geológica de España, Madrid, 330-334.

Trewin, N. H. 2001. Scotland's foundations: our geological inheritance. In: Gordon, J.E. and Leys, K.F. (Eds), *Earth Science and the Natural Heritage: Interactions and Integrated Management*. The Stationery Office, Edinburgh, 59-67.

Veverka, J.A. 1994. *Interpretive Master Planning*. Falcon Press, Helena, Montana.

VisitScotland and Scottish Natural Heritage 2002. Tourism Attitudes Surveys, 1999 and 2001. www.scotexchange.net

In: Parkes, M.A. (Ed.) 2004 *Natural and Cultural Landscapes - The Geological Foundation*, 235-238, Royal Irish Academy, Dublin

RAISING PUBLIC AWARENESS OF GEOLOGICAL HERITAGE: A SET OF INITIATIVES

Graciete Dias and José Brilha
Earth Sciences Department, University of Minho, Campus de Gualtar, 4710-057 Braga, Portugal Email: graciete@dct.uminho.pt, jbrilha@dct.uminho.pt

Abstract: Most people are unfamiliar with geological knowledge, a fact that hinders the implementation of geoconservation measures. Additionally, like in many other countries, the strategies of Portuguese Natural Parks are mainly concerned with bioconservation. The information available at these parks for visitors and students is mostly limited to biological issues, further increasing the geological illiteracy of the general public. Hence, putting people in contact with geological subjects is becoming very important for both geologists and policy makers. A project is being developed in Portugal in order to characterise, catalogue and make public the geological heritage of two Natural Parks from the NE region. This project is providing several products with the purpose of increasing public awareness towards geological subjects. Among these outputs we emphasise interpretative panels, geological guides and booklets on specific geosites or trails, and web pages. Two more initiatives are being developed in the scope of this project: the training of staff from the two Natural Parks who, in the future, will be guiding field trips; and the organisation of summer field trips in the region, addressed to the general public.

1. INTRODUCTION

Any nature conservation policy should integrate all aspects of nature: geological, biological, and social ones. Traditionally, international conservation institutions often forget the integration of geological concerns in their projects (Brilha 2002 and references herein). Geoconservation is still a word that causes great indifference among politicians, conservation professionals and park managers. In spite of the importance of geology in the definition of many protected areas, park managers failed to give proper treatment to these themes. In the USA park system, for example, almost 50% of all units have geological interest (Shaver and Wood 2001). These authors confirm that "the public and most park managers have largely viewed these physical features as scenic backdrops to the plants and animals that exist within them" and that "...the geosciences traditionally have not been well integrated into land management or ecosystem planning". Schelhas (2001) confirms that the USA parks are managed mainly for the conservation of representative ecosystems. Nevertheless, this panorama seems to be slowly changing in some countries. The last report of the Geologic Resources Division of the American National Park Service gives a bright perspective of the growing importance of geology in the management of parks (McCoy 2001). The same can be observed in some other places such as Scotland (Grant 1999), Canada (Cline et al. 1984; Varangu 1997) or England (Weighell 1999).

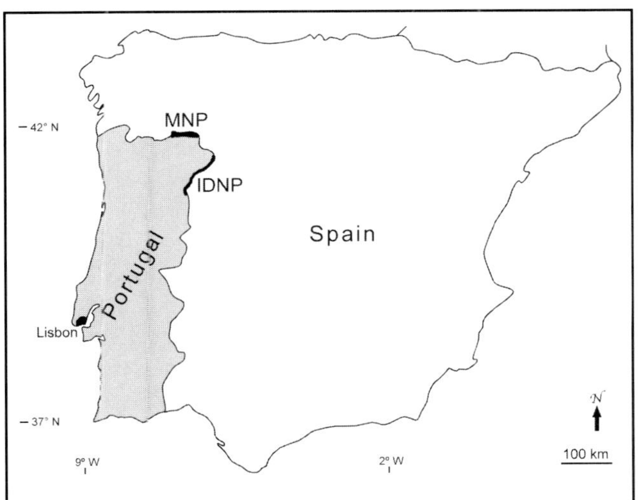

Fig. 1 – Location of the two natural parks under study in NE Portugal. MNP: Montesinho Natural Park; IDNP: International Douro Natural Park.

Figure 1. Location of the two natural parks under study in NE Portugal. MNP: Montesinho Natural Park; IDNP: International Douro Natural Park.

The Portuguese framework is not very different from many other countries. Protected areas were primarily defined regarding biological aspects. This paper shows how geologists are trying to increase the sensibility to geological features inside natural parks from NE Portugal and presents some of the initiatives proposed to raise public awareness of geological heritage.

2. GEOLOGICAL HERITAGE IN PROTECTED AREAS OF NE PORTUGAL: A PROJECT UNDER DEVELOPMENT

Two major protected areas are implanted in NE Portugal: Montesinho Natural Park (MNP) and International Douro Natural Park (IDNP), covering an extensive area of 1601 km^2 (MNP 750 km^2, IDNP 851 km^2) (fig. 1). The first one is characterised by a remarkable geodiversity and by the occurrence of exotic terranes, presenting a complex geology. Due to this complexity, Pereira et al. (2002) highlight the difficulties in the development of an interpretational strategy for general public. The IDNP also presents an interesting geology with notable geomorphological features (Alves et al. 2002).

The geological relevance of these two parks, associated with the absence of geological information in park management plans, leads to the development of a multidisciplinary research project aiming to: (i) improve the geological knowledge in the MNP and IDNP; (ii) create scientific instruments to support a sustainable management of the resources and territory; (iii) proceed with the inventory and characterisation of geosites; (iv) contribute to the increase of the public awareness of Earth heritage.

In order to achieve this last objective, a set of activities is being implemented as described below. Furthermore, some deliverables will be available for each park: (i) geological and geomorphological maps, as well as a geological resources map; (ii) a geosite map; (iii) a geological guidebook; (iv) web pages; (v) geological trails and interpretative panels contents.

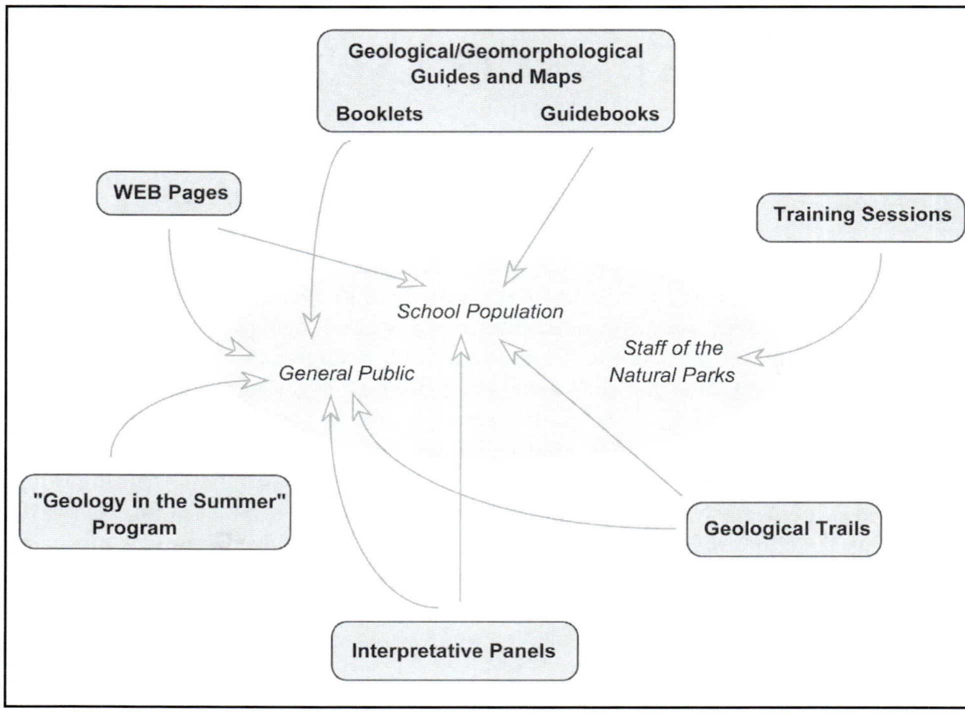

Figure 2. Proposed initiatives in order to increase public awareness of Geological Heritage.

3. RAISING PUBLIC AWARENESS OF GEOLOGICAL HERITAGE

As widely accepted, building awareness and education for various target groups is a big priority for a successful geological heritage conservation. However the challenge is considerable: (i) public understanding of geodiversity, geoconservation and geological heritage issues is poor; (ii) the audience must be well known and (iii) the message needs to be effectively communicated.

3.1. Target groups

In our project three target groups were considered: the general public, the school population and the technical staff of the two Natural Parks.

The general public includes tourists, day-visitors and local population. It must be emphasised that this audience is heterogeneous, including individuals who visit geological sites for intellectual improvement or just for leisure. Furthermore, it is well known that in any conservation project the involvement of local people is essential to the project success (Palmer et al. 1995).

Also heterogeneous is the school population group, of which three levels were taken into account: junior school students, senior school students and teachers.

The technical staff and guides of the Natural Parks frequently support environmental education activities. Therefore, this set of professionals was also considered as a target group with specific needs of training on geological heritage and geoconservation issues.

3.2. A set of initiatives

In the scope of the project various initiatives were proposed, addressed to the above target groups (fig. 2). These initiatives are being carried out in a joint effort with the MNP and the IDNP managers and technical staff.

The adopted strategy considered that, for any audience, rather than just offering *information*, interpretative provision is required. Site-specific geological *interpretation* is a complex process, defined by Hose (2000) as "the revealing of meaning and value of a site by translating geology's technical and scientific language, data and concepts into meaningful commonplace facts, terms and ideas, based upon the experiences, knowledge and understanding of non-specialist persons". It also attempts to develop the sense of *observation* and *discovery*.

Interpretative panels

Panels are a very common interpretative provision. In the last years some authors pointed out basic principles to make geological panels appropriate for users (Wilson 1994; Hose 1998, 2000). Several outdoor and on-site thematic panels will be set up in selected sites, addressed to the general public and school students. Sites were strictly selected among key-outcrops and viewpoints with relevant scientific/pedagogical and aesthetic values. Contents, design and presentation of these panels follow a set of guidelines (fig. 3).

Contents are interpretative rather than informative and based on observed features, with a clear distinction between observation and interpretation. They reflect user-friendly geological themes suggested by site observation, for example: unusual or unique features; internal geodynamics -

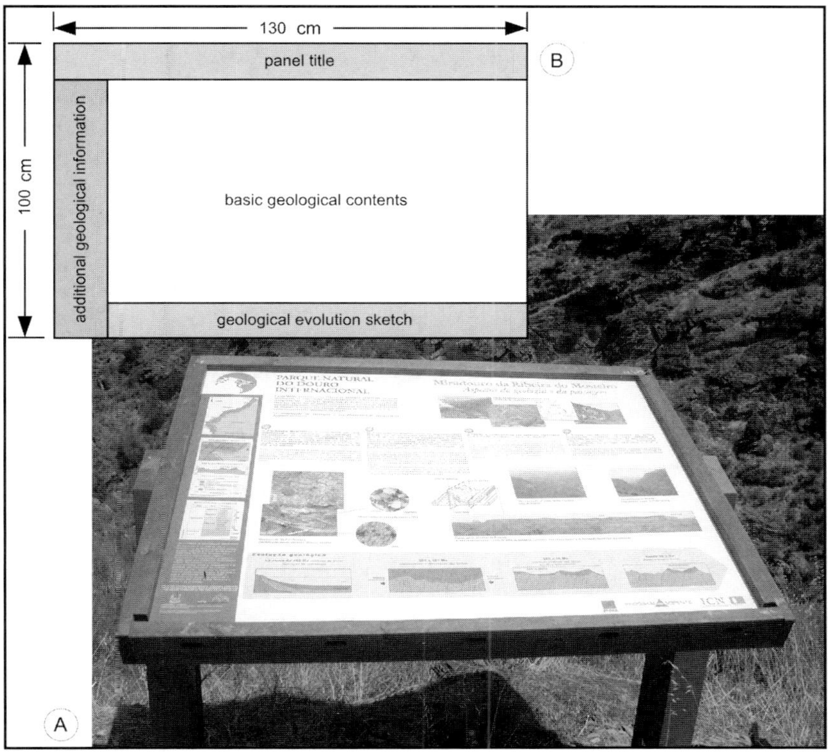

Figure 3. Proposed interpretative panels. (A): design and presentation of a lectern type panel; (B): contents organisation.

faulting, folding; land-shaping processes; environmental, economic, historical features. For some specific sites, traditional themes are also incorporated (lithology, mineralogy, tectonics, stratigraphy, palaeontology).

A lectern-type (130x100 cm) or banner-type (130x110 cm) presentation and a graphics-rich/text-poor layout are used. Basic contents are displayed at the centre of the panel, occupying the majority of the surface (fig. 3), and are addressed to the general public and junior school students. Additional information and/or interpretation is given in a left-side coloured strip and aimed at senior school students and "geo-dedicated" visitors/tourists. At the bottom schematic block-diagrams or cross-sections present the geological evolution of the region.

Trails and viewpoints

Park managers mainly for biological and cultural reasons already selected some trails and viewpoints, for which complementary geological interpretation addressed to the general public is being produced. Under the scope of this project, some new thematic trails will be proposed, mainly for the school population. The geodiversity of these parks constitutes a remarkable classroom that must be used by teachers and students. Several types of trails and bus circuits adapted to different student grades will be proposed.

Geological guides

The geological information will be available as guidebooks and booklets. Guidebooks will integrate geological and geomorphological maps, description of the geological features, suggested stops, trails, interpretative information and a glossary. They are mainly addressed to teachers preparing fieldtrips for students and to the public with some geological background. Booklets are designed to support the interpretation of selected viewpoints and trails. They will present simplified language in order to be understood by the general public.

Web pages

Environmental education must be supported either by traditional media (booklets, interpretative panels, guidebooks, etc.) and by electronic resources distributed on-line on the Internet and off-line on CD-ROMs and DVD-ROMs. Lent (2001) shows the advantages for National Parks in having a good web site, mainly in the preparation of a real visit to the park, and how to use the internet in visitor centres. The use of electronic media in the interpretation of a National Park geological heritage was exemplified in Brilha et al. (1999). Thematic web pages will provide assistance either for those who want to prepare a park visit or for those that cannot access it (travel impossibility, difficulties for disabled people, etc.).

"Geology in the Summer" programme

The Portuguese government (the former Ministry of Science and Technology) promotes, since 1998, a very successful programme called "Geology in the Summer" with the aim of raising public awareness of Geology. During the summer season geologists organise field trips and other activities all over the country, which constitute a good opportunity to talk about geodiversity, geological heritage and geoconservation, and the importance of geology in our society. Our project intends to organize several "Geology in the Summer" activities inside the two natural parks under study. Five distinct activities were developed during the summer of 2002. This experience, as well as previous experiences in the Peneda-Gerês National Park (NW Portugal), shows that the public is very interested in learning about the geological features, probably due to a general insufficiency of available geological information and to the fact that guided walks and personal interaction with geologists is greatly appreciated.

Natural parks staff training

Two-level workshops are designed for graduated (biologists, archaeologists, architects) and non-graduated technical staff. It should be noted that, at present, the technical staff of the two parks does not include geologists. The workshops are essentially based on fieldwork activities centred on the geosites and trails, followed by in-door training. A final workshop is previewed, focused on geoconservation principles and integrating the geology, geomorphology and geological heritage of the Natural Parks.

4. FINAL REMARKS

As stated by Hose (2000), there is still little real knowledge and understanding of the effects of these kinds of initiatives on cognition, informational exchange, emotive response and behavioural modification of the different target groups. This is a great challenge for geoscientists, but also for Educational and Social Sciences researchers.

The overall perspective of the true impact of this project will only be possible after its conclusion by the end of 2003. Nevertheless, some positive feedback already received is very encouraging towards the success of the above initiatives. Park managers, local authorities and cultural associations are very cooperative and they are already asking for help due to the absence of geological experts in their staff. The public reaction to some interpretative panels already placed in-situ is very positive. In addition, the five field trips organised during the summer of 2002 were an excellent opportunity to talk about geological issues to very interested groups of people. In order to evaluate the impact of some initiatives, we are collecting information based on interviews, inquiries addressed to each target group, web pages access logging and e-mail feedback.

The Madrid Declaration, approved in 1999 during the III International Symposium ProGEO on the Conservation of the Geological Heritage, states: "A first step to achieve the integration of geoconservation into nature conservation would be an evaluation of the Geological Heritage in already protected areas, followed by the promotion of interpretative initiatives, in order to increase social awareness of Geoconservation". The project herein described intends to be a contribution to this first step inside natural parks of NE Portugal.

ACKNOWLEDGEMENTS

The PNAT/1999/CTE/15008 project is sponsored by FCT (Portuguese Science and Technology Foundation) and ICN (Portuguese Nature Conservation Institute).

REFERENCES

Alves, M.I.C, Monteiro, A., Ferreira, N., Dias, G., Brilha, J.B., Pereira, D.I. 2004. Landscape as a support for biodiversity: the Arribas do Douro case study (International Douro Natural Park, NE Portugal). This volume.

Brilha, J.B., Dias, G., Mendes, A.C., Henriques, R., Azevedo, I.C., Pereira, R. 1999. The geological heritage of the Peneda-Gerês National Park (NW Portugal) and its electronic divulgation. In: *Towards the Balanced Management and Conservation of the Geological Heritage in the New Millennium*. D. Barettino, M. Vallejo, E. Gallego (eds), Sociedad Geológica de España, Madrid, 315-318.

Brilha, J.B. 2002. Geoconservation and Protected Areas. *Environmental Conservation* 29(3), 273-276.

Cline, D., Erdman, K., Pearce, W. 1984. The dinosaur World Heritage Site: responsible management in Canada. In: *National Parks, Conservation and Development: the role of protected areas in sustaining society*. J.A. McNeely, K.R. Miller (eds), Smithsonian Institution Press, Washington, D.C., 741-743.

Grant, R. 1999. Evaluation of Success or Success of Evaluation? In: *Learning in Protected Areas - How to Assess Quality*. D. Elcome (ed), European Committee for Environmental Education ECEE, Commission on Education and Communication, International Union for the Conservation of Nature and Natural Resources, Gland, 42-51.

Hose, T.A. 1998. Mountains of fire from the present to the past - or effectively communicating the wonder of geology to tourists. *Geologica Balcania* 28(3-4), 77-85.

Hose, T.A. 2000. European 'Geotourism' - geological interpretation and geoconservation promotion for tourists. In: *Geological Heritage: its conservation and management*. D. Barettino, W.A.P. Wimbledon, E. Gallego (eds), Madrid, Spain, 127-146.

Lent, M-J. 2001. What is new media? In: *Challenge for Visitor Centres - Linking Local People, Visitors and Protected Areas*. M. Kyöstilä, A. Leivo, T. Loikkanen (eds), Nature Protection Publication of the Finnish Forest and Park Service, Series A, No 129, 56-59.

McCoy, C. (ed) 2001. *Geologic Resources Division: 1999-2000 Biennial Report*. National Park Service, Denver, CO, USA, 79p.

McNeely, J.A., Miller, K.R. (eds) 1984. *National Parks, Conservation and Development: the role of protected areas in sustaining society*. Smithsonian Institution Press, Washington, D.C., 825p.

Nelson, J.G., Serafin, R. (eds) 1997. *National Parks and Protected Areas: Keystones to Conservation and Sustainable Development*. NATO ASI Series, Ser. G, Ecological Sciences, Springer, Berlin, vol. 40, 292p.

Palmer, J., Golstein, W., Curnow, A. (eds) 1995. *Planning education to care for the Earth*. IUCN, Gland, Switzerland and Cambridge, UK, 169p.

Pereira, D.I, Meireles, C., Alves, M.I.C., Pereira, P., Brilha, J.B., Dias, G. 2004. The geological heritage on the Montesinho Natural Park (NE Portugal) - An interpretation strategy for an area with high geological complexity. This volume.

Schelhas, J. 2001. The USA national parks in international perspective: have we learned the wrong lesson? *Environmental Conservation* 28 (4), 300-304.

Shaver, D., Wood, J. 2001. Geology in the National Park Service. *Geotimes*, April 2001, AGI, 14-19.

Varangu, A. 1997. A clash of values: planning to protect. The Niagara escarpment in Ontario, Canada. In: *National Parks and Protected Areas: Keystones to Conservation and Sustainable Development*. J.G. Nelson, R. Serafin (eds), NATO ASI Series, Ser. G, Ecological Sciences, Springer, Berlin, vol. 40, 65-79.

Weighell, A.J. 1999. Earth Heritage Conservation in the United Kingdom, the World Heritage List and UNESCO Geoparks. In: *Towards the Balanced Management and Conservation of the Geological Heritage in the New Millennium*, D. Barettino, M. Vallejo, E. Gallego (eds), Sociedad Geológica de España, Madrid, 24-27.

Wilson, A. (ed.) 1994. Earth Heritage Conservation. The Open University, Milton Keynes, UK, 272pp.

In: Parkes, M.A. (Ed.) 2004 *Natural and Cultural Landscapes - The Geological Foundation*, 239-240, Royal Irish Academy, Dublin

HOW CAN WE HELP PEOPLE TO KNOW AND ENJOY THEIR LANDSCAPES: AN EXAMPLE FROM DARTMOOR NATIONAL PARK, ENGLAND

Suzanne Goodfellow, Head of Ecology and Wildlife Conservation, Dartmoor National Park Authority. Website: www.dartmoor-npa.gov.uk

Abstract: The paper describes and illustrates a practical example of helping people to appreciate the geological landscape of Dartmoor, Devon, England. The *Dartmoor Rock 370, 000th Millennium Exhibition* and website to accompany it were supplemented by a range of publications, and practical measures such as training guides, and developing new educational sites.

UK NATIONAL PARKS

There are 13 UK National Parks (plus 2 new ones in process of designation in England). They can be classed as IUCN Category 5 national parks. Their statutory purposes are

- to conserve and enhance natural beauty, wildlife and cultural heritage
- to promote opportunities for the piblic understanding and enjoyment of the special qualities (of the Parks)
- to foster the economic and social well being of local communities

In other words the intention is to have sustainable, integrated conservation management in inhabited landscapes.

DARTMOOR NATIONAL PARK

This NP was designated in 1951. It covers 95,311 ha, centred on a granite massif that rises out of the surrounding sedimentary rocks, the National Park boundary following the metamorphic aureole. It is a landscape of high blanket bog, wild moorland, intimate farmland, woodlands and small villages, dominated by rocks - rocky tors and clitter, stone walls and buildings, prehistoric menhirs and stone circles. It is mostly in private ownership with 32,000 inhabitants and 3.5 million visitors annually.

Typical Dartmoor National Park landscape

KEY GEOLOGICAL AND LANDFORM FEATURES

Dartmoor granite and surrounding rocks, associated mines and quarries
Land formation processes
Tors and clitter slopes
Extensive peat deposits
Sticklepath Fault
Caves and vertebrate fossils

There is multiple land use - quarrying (granite, china clay and aggregate), farming, forestry, recreation, water supply, military training and tourism. This diversity presents challenges and opportunities for conservation management of the geological, wildlife and cultural resource!

RELEVANT MANAGEMENT POLICIES AND STRATEGIES

- The Nature of Dartmoor: A Biodiversity Profile (1997) describes geological resource, and sets five key geological objectives.
- Action for Wildlife: The Dartmoor Biodiversity Action Plan (2001) sets targets and actions to achieve these in a Habitat Action Plan for Rocks (including tors, clitter slopes, quarries, caves, mines and buildings). It integrates geological and wildlife management.
- Dartmoor National Park Management Plan (2001) - includes an overall objective to raise public awareness of the value of wildlife and geological conservation at local and global levels

FIVE KEY GEOLOGICAL OBJECTIVES FOR DARTMOOR

- Identify, maintain and enhance rock exposures and natural landforms that are important for understanding the origin and geological development of the Dartmoor landscape
- Encourage appreciation of the Dartmoor landscape and its geology, and an understanding of its links with scenery, tourism and mining through education and interpretation aimed at the public, education establishments, the minerals industry and other bodies
- Protect the processes of natural evolution of the Dartmoor landscape, including drainage patterns, slope development and peat formation
- Encourage surveys of old mine workings and spoil tips, promote appropriate protection measures and responsible use and, where necessary, their positive conservation management
- Ensure there is a Cave Conservation Plan for caves in the Buckfastleigh area which safeguards the remaining cave sediments and fossils

ACTIONS IN 2001/2002

A RIGS survey (Devon RIGS Group funded by Dartmoor National Park Authority (DNPA) was completed in 2001. Forty sites were selected, based on interest and accessibility. Some conservation work was completed on the RIGS sites, and also the implementation of an interpretation programme.

As part of the overall strategy, *Dartmoor Rock: the 370,000th millennium exhibition* was conceived as a major focus for activities and interpretationbased on a partnership of the DNPA, English Nature and the Devon RIGS Group. This exhibition and related activities formed a complete package of geological outreach work - providing not only a dynamic exhibition but also opportunities for self-exploration, guided Rock Walks and teaching material. It was launched in HMVC in Autumn 2001 (in 1 month it was visited by over 12,000 visitors) and moved to Postbridge, starting in the summer of 2002 for two years (with a potential audience of over 100,000 visitors). It was aimed at both local people and visitors of all ages and intended to stimulate 'outdoor' exploration of geological landscape features as well as raise understanding.

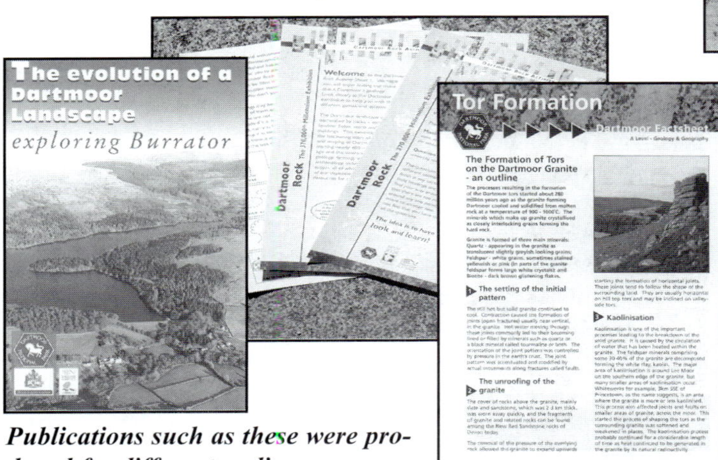

Publications such as these were produced for different audiences.

Associated material was produced for the following audiences:

- Pre-school children - fun sheets with stamps, floor jigsaw
- Primary school - activity sheets (plus above), Dartmoor Rock pencils with web page address
- Secondary school - activity sheet
- Tertiary (A and A/S level) - Tor Formation fact sheet
- Adults - Self guided walks booklet - 'The evolution of a Dartmoor landscape - exploring Burrator'
- All ages - Finding Out More leaflet - accessible places to go (educational sites) plus web sites; Dartmoor Granite video on uses of Dartmoor granite
- All information was available on a web site. Books, maps, and a video were for sale at Information Centres.
- Guided Rock Walks were offered with specially trained guides
- Guides were trained by local geologists so that geology is now incorporated into general guided walks
- Articles in press and Dartmoor Visitor (local twice yearly paper for DNP area produced by DNPA)

Training guides (above); Dartmoor Rock Walk (below)

THE FUTURE

It is intended to continue with the following actions

- Produce and sell publications based on Dartmoor Rock material (sponsored by minerals industry?)
- Continue with guided walks and special events (trips to working quarries etc) especially during the 'shoulder' season
- Prioritise conservation work on RIGS sites with owners support
- Complete the geological data base on GIS allowing integration with wildlife and archaeological interests
- Assess the need for further survey work
- Seek to include payment for protection and enhancement of geological features in future agri-environment schemes

LESSONS LEARNT

- Local partnerships based around shared knowledge, skills and funding are essential and work well when relatively small and project based.
- Great thirst for geological knowledge amongst local people - school teachers, parents, local craftspeople, builders, farmers and other country dwellers, and holidaymakers
- Important to integrate 'straight' geology with human use (past and present) and landscape appreciation and make language accessible!
- It is possible for non-specialists to raise awareness on a relatively broad and shallow front with limited resources (interpretation work above cost DNPA £3000 + staff time)

See Page 151. Figure 1. Traditional boulder style gravestones with minimal dressing.

See Page 162. Figure 1. Dungeness SSSI, Kent, England. Britain's largest shingle foreland reflecting over 5000 years of coastal development today supports a diverse fauna and flora in the shingle beaches, ridges and associated brackish and freshwater pools demonstrating a direct link between geodiversity and biodiversity.

See Page 177. (above) Figure 3. The area of landscape park near Krakow - the gap of the Vistula river bordered by Jurassic limestones with the medieval Benedictine Abbey on the right.

See Page 177. (above right) Figure 4. Babia Gora National Park - MAB Biosphere Reserve - the range elevated above the timberline; on the first plan the northern slope transformed by several landslides.

See Page 177. (right) Figure 5. Bieszczady National Park - International MAB Biosphere Reserve - a typical structural relief formed on resistant sandstones; the highest parts of the range covered with mountain meadows.

See Page 151. Figure 3. Display area of available stones in the Stone Mason's yard. Almost all of these had been cut and polished to give one smooth face.

See Page 177. Figure 2. The castle medieval ruin Ogrodzieniec harmonising with Jurassic limestones as one of the protected landscape component.

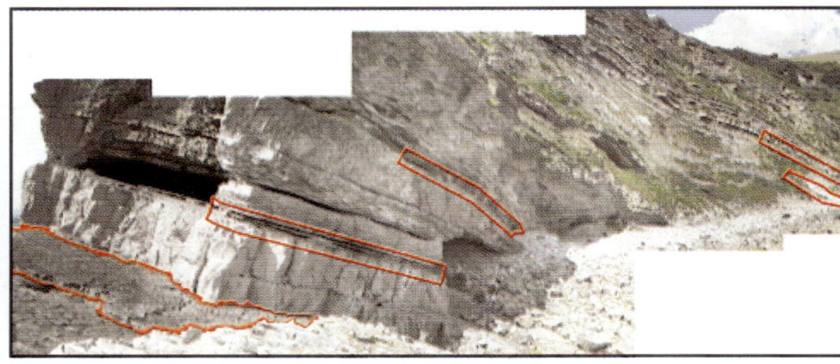

See Page 189. (above) Figure 2. Subfeatures from left to right: 1) exposed bedding plane with ammonites, 2) formation boundary - marine to non-marine, 3) fossil soil, 5) erosion surface on charophyte-bearing lake micrites, 6) marine shell bed, marking transgression. Project: 4) concealed interval to be exposed.

See Page 206. (below) Figure 1. A water diviner (left) using a stopwatch on his daughter's velvet ribbon informing a polite driller (right) that the water supply borehole he has just successfully drilled is wrong.

See Page 206. (left) Figure 2. Trinity Holy Well, Co. Limerick, with a slatted animal house and slurry spreading in the adjacent field. June 2001.

See Page 221. (above left) Figure 4. Rehabilitation of gravel workings in Hungary. Smaller lakes in a cassette system where they can communicate with each other through dams as a natural filter system

See Page 231. (left) Figure 1. Rock sculpture at Knockan Crag.

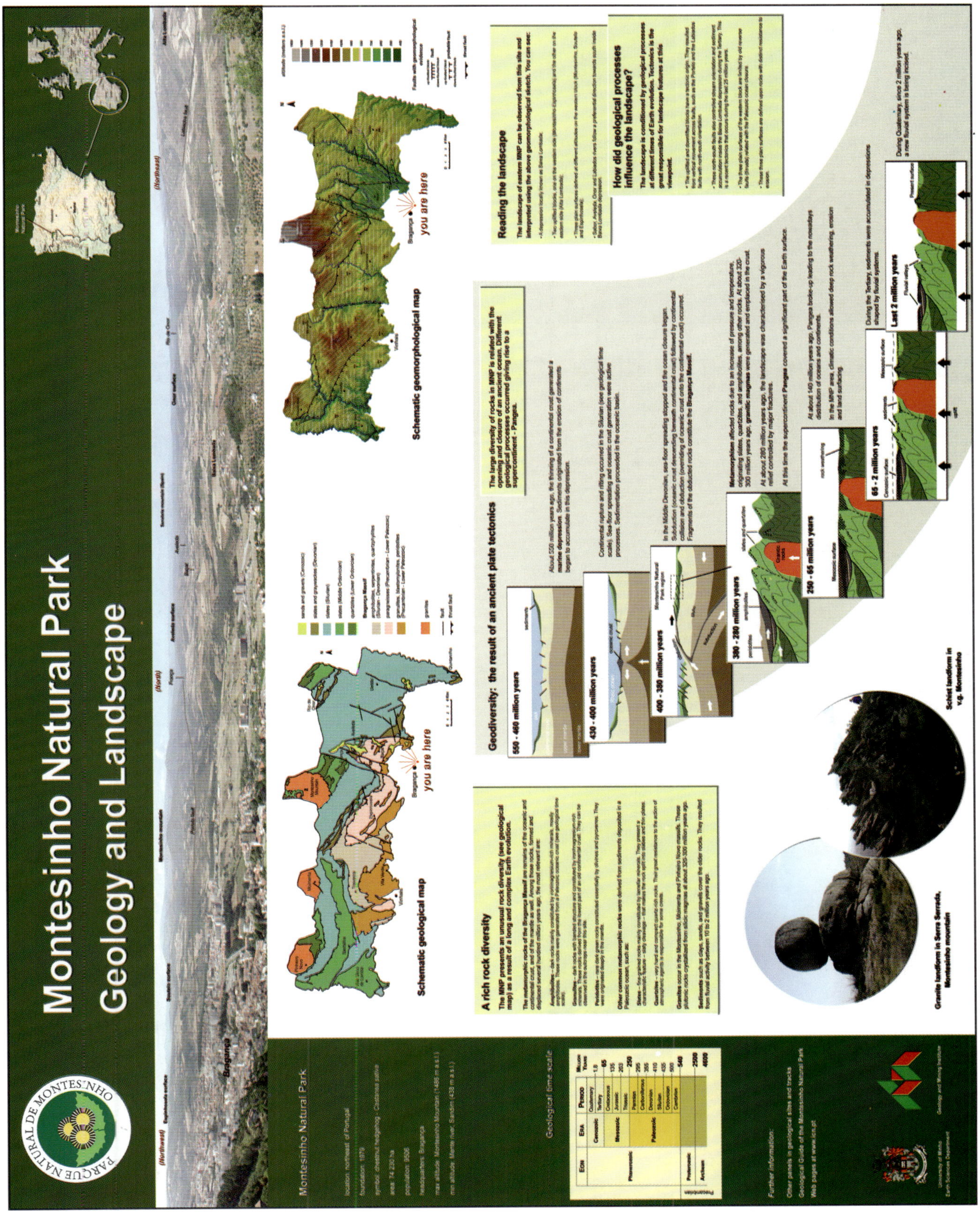

See Page 255. Figure 1. General view of the S.Bartolomeu interpretative panel, Montesinho Natural Park (NE Portugal).

See Page 273. (left) Figure 1. *Our Dynamic Earth*, with Salisbury Crags behind.
See Page 275. (below left) Figure 3. The Restless Earth.
See Page 275. (below right) Figure 4. The Polar Regions.
See Page 274. (below middle left) Figure 2. Siccar Point.
See Page 266. (below middle right) Figure 4a. The Dominican Church in Claddagh, Galway built of locally quarried pink Galway Granite blocks edged with grey Carboniferous Limestone.
See Page 278. (bottom right) Figure 3. Participants from different UK university geology departments on the familiarisation trip, at Horn Head Co. Donegal.
See Page 279. (bottom left) Figure 4. Participants from different UK university geology departments on the familiarisation trip.

See Page 282. (left) Figure 2. The Mourne Mountains. Photo: Environment and Heritage Service.

See Page 281. (right) Figure 1. The Causeway Coast. Photo: Environment and Heritage Service.

See Page 283. (left) Figure 3. Lough Neagh. Photo: Environment and Heritage Service.

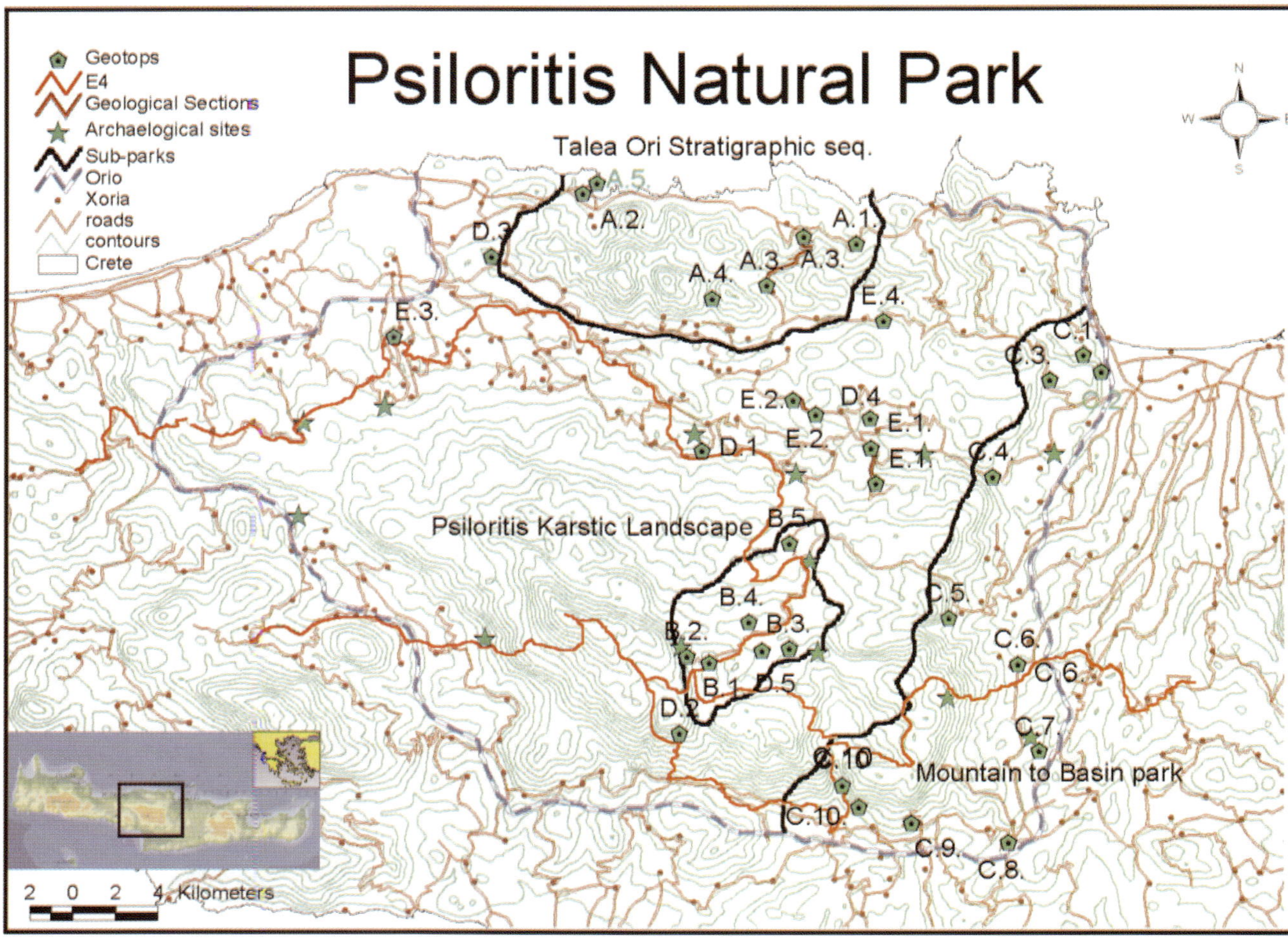

See Page 292. (above) Figure 1. Location and main features of Psiloritis Geopark.
See Page 293 (left) Figure 2. "Sculpture of Nature" in the area of Chonos village.
See Page 293. (below left) Figure 3. The Vossakos folds near the northern coast.
See Page 293. (below right) Figure 4. Information table in the area of the Psiloritis Mountains.

See Page 303. Figure 1. Geotouristic map of Albania.

See Page 305. (above left) Figure 1. Jatulian sandstones (2,2 Ga).
See Page 306. (above right) Figure 3. Karelian moraine.
See Page 306 (left) Figure 2. Polished surfaces of the Archean gneisses.
See Page 308. Figure 5. Kizhi Island.
See Page 314. Figure 2. A puzzle represents, better than any other image, the conceptual action that we must reproduce in an interpretation process. Recomposing single pieces in an image that reveals the real meaning of what we are observing. (La dalle à ammonites de Digne les Bains, by courtesy of the Réserve Géologique de Haute-Provence, FR).

In: Parkes, M.A. (Ed.) 2004 *Natural and Cultural Landscapes - The Geological Foundation*, 241-244, Royal Irish Academy, Dublin

PERCEPTION OF THE GEOLOGICAL LANDSCAPE THROUGH THE PLACE-NAMES OF ROARINGWATER BAY, WEST CORK

Anthony Beese, Geologist, Downeen, Rosscarbery, County Cork

Abstract: More than nine hundred place-names from the archipelago have been interpreted, (a) through interviews, (b) by examining documents and maps, and (c) by visiting the features named. Most of the names are topographical and some are observations on the geological landscape. Bedrock characteristics such as colour, shape and state of weathering are identified by qualifying elements, for example in *An Stocán Buí, Carraig na Mionnán* and The Raggedy Cliff. The resonant *Faill na gCloch Geal*, or 'cliff of the bright stones', refers to clusters of quartz veins. Fishermen memorized sequences of names like these to mark their progress around the hazardous coast. Occasionally, their perception was incorrect: a large offshore rock, actually *in situ*, could be observed as *An Charraig Bhréadach*, or 'the broken-off rock', while the adjacent high coast was *An Fhaill Bhriste*, 'the broken cliff'. On the rocky islands, 'greener' tracts with drift cover were of agricultural value and important enough to be identified by place-names without qualification. The *Fothar* and *An tAonach* refer to glacial deposits on Sherkin and *Oileán Chléire* (Cape Clear) respectively. Many of the islands' natural resources appear in place-names. For example, *Tráigh Leagaigh* described a strand strewn with flat stones, and *Tobar na bPlobaí*, was an imitative name for a spring-well that made a splashing sound. Thus, in the past, the islanders used a rich vocabulary of words for their landscape, with which they had a vital and close relationship. Now, however, owing to the loss of the Irish language and to wide-spread cultural changes, the landscape has become devalued. The islands are being perceived in an increasingly abstract way despite their importance to tourism.

METHODOLOGY

Roaringwater Bay is a picturesque archipelago, located at the south-western end of Ireland. For the purposes of this study, the bay is treated as ten islands or groups of islands (see Figure 1 and Table 1) but there are also numerous islets and rocks. The four larger islands, *Oileán Chléire*, Sherkin, Heir and Long, are inhabited, although with reduced populations. The smaller islands became uninhabited between the 1930s and 1964, Horse Island being the last. *Oileán Chléire*, or Cape Clear Island, is one of the two remaining *Gaeltachtaí* in County Cork but is not primarily Irish speaking (Hindley 1990).

The place-names and folklore of the archipelago have been collected by the author since 1995 and the current paper is an analysis of some of this research. Every field, well and feature of the coast was given a toponym or place-name. These are close to extinction and many have already been lost or corrupted. Detailed maps showing Cape Clear (Ó Síocháin 1940; Lankford 1999) and four of the smaller islands have been published (Beese 1997, 1998 and 2000). The field work of the 'Cork and Kerry Place Names Survey' is currently underway and will form an archive in County Libraries.

The survey method was three-fold: (a) collection of oral names from informants, (b) research into documents and maps, and (c) site visits. About seventy islanders and ex-islanders were interviewed and their assistance is gratefully acknowledged. The name, year of birth and brief biography was obtained for each informant. Irish, English and bilingual names were collected. Many Irish names were difficult to translate; a not unexpected result because the place-name record is complicated by dialect and changes in usage (Matley 1982). Therefore, in order to increase the objectivity of the study, a phonetic record of received names was kept and a short history of the place was obtained. Useful documents containing Anglicisms were the first edition Ordnance Survey (OS) sheets engraved in 1845, the OS name books edited by O'Donovan in about 1841 - the OS letters for County Cork are lost - and the old Admiralty chart, first published in 1852. A visit to each location was considered to be essential.

Dineen's dictionary, published in 1927, is a valuable reference because the form of the toponyms tends to follow the 'old' Munster dialect. Closely allied place-names are found in surveys of other west Cork and Kerry districts (for example, Ó Cíobháin 1978 to 1985). Minor reconstruction of Irish names includes the addition of the definite article to some adjectival and single element names. Inevitably, the distinction between a name which describes geology and one which describes geomorphology is not always clear.

BEDROCK

The stratigraphy of Roaringwater Bay may be summarized as a clastic sequence of Devonian (Old Red Sandstone) to Lower Carboniferous age (Reilly and Graham 1976; Pracht and Sleeman 2002). The two dominant formations are the Sherkin Formation, characterised by well-bedded, greenish-grey sandstones and some mudstones, and the overlying Castlehaven Formation, comprising purple, cleaved siltstones and fine-grained sandstones. These beds are well exposed in coast sections but their dull colours and subtly different lithologies are not distinguished by place-names. Instead, fishermen used distinctive features such as white clusters of quartz veins to mark their progress. The lyrical name, *Faill na gCloch Geal*, cliff of the bright stones, is found at two locations on Cape Clear Island.

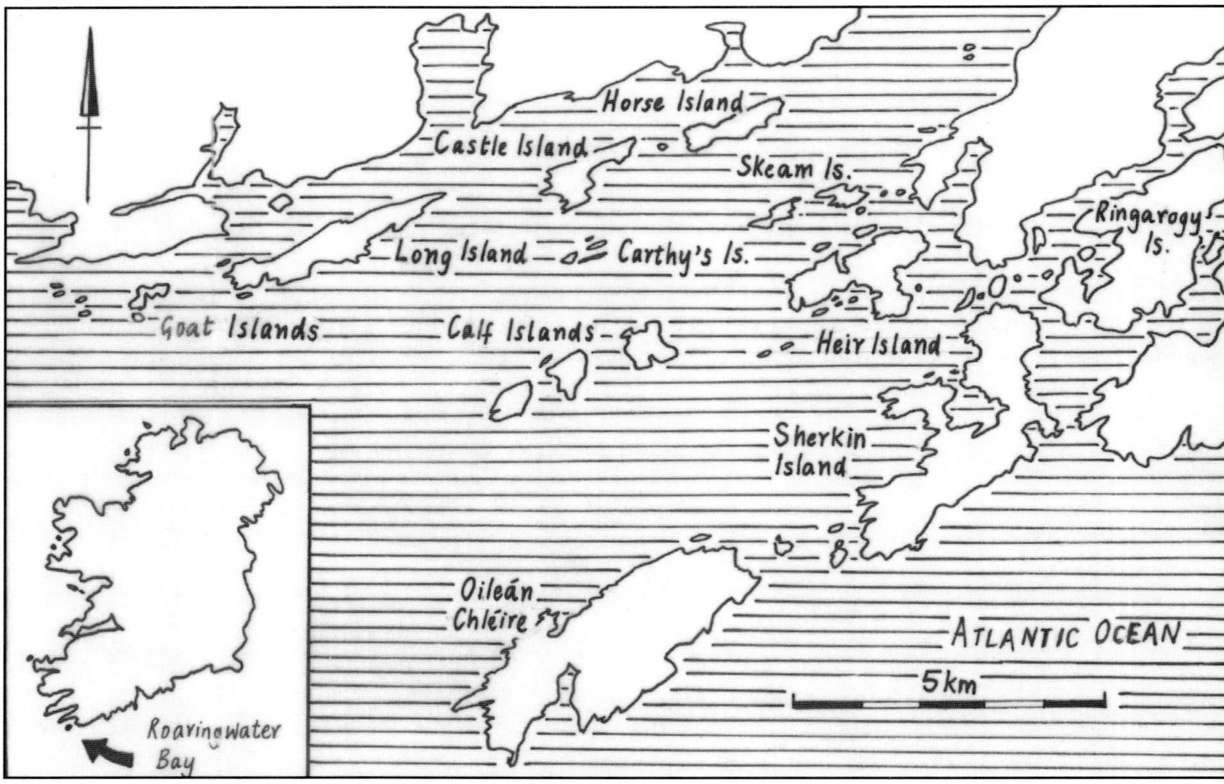

Figure 1. Roaringwater Bay.

The Variscan structure of the sedimentary rocks is expressed in the topography (Figure 1) - Roaringwater Bay is a drowned coast within the 'ridge and valley' province of Munster. Most noticeable is a markedly linear west-south-west grain that follows the fold axial trend. Approximately 1.5% of the place-names describe elongate features parallel to this main trend. The name for Long Island (*Inis Fhada*), which is five times as long as it is wide, may be very old. *An Tráigh Fhada* (Sh, C), *An Fhaill Fhada* (Cl), *An Cuas Fada* (C), *An Charraig Fhada* (C) and *An Pointe Fada* (Sh) are minor names for strand, cliff, inlet, rock and headland respectively. Field names, which are often in English, include the poetic Three Long Ridges (Sk). On Sherkin, there are fields called The Long Field, Long *Coinne*, Long Meadow and Long Lands. Roaringwater Bay itself is known as *An Loch Trasnach*, the crosswise sea. This is a description of the regular layout of the islands, with channels running both along the bay's length as well as across it. The channels running across the bay follow a secondary grain parallel to regular cross faults that have an overall northerly trend.

Occasionally, folds and faults are interpreted in place-names. On the northern mainland, the glaciated dome of Mount Gabriel dominates the bay and is an important landmark. Its Irish name is interpreted here as *An Cnoc Casta*, 'the twisted hill'. This is a neat description of the geological structure of thick beds folded into a syncline. From the bay, the strata look like plaits looped over a giant head. Examples of faults are The Vein, which is coincident with a spring-line on Skeam East, and *Béal na Gáige*, 'mouth of the cleft', where an intermittent waterfall flows over the dramatic cliffs of western Cape Clear.

Bedrock was often viewed according to the degree, shape and colour of its weathering. The Raggedy Cliff (*Faill an Chríonta*) and *An Fhaill Bhriste*, 'the broken cliff', describe strike exposures of the same sequence of deeply weathered sandstones in the southern cliffs of Sherkin. The susceptible beds are near to the top of the Sherkin Formation. An offshore rock below the second cliff is named, *An Charraig Bhréadach*, meaning 'the remnant or broken-off rock', an erroneous perception since the rock is actually *in situ*.

Many names illustrate the detailed shape of coast, which is determined by the main discontinuity sets in the rock mass, namely bedding, cleavage and joints. Three locations on the north coast of Cape Clear include the element, *mionnán*, which is often misinterpreted as meaning a kid goat. On the island, the word clearly describes rocky formations resembling a crown - The *Mionnán* Rock, *Stocán na Mionnán* and *An Mionnán Buí* are all sea stacks and cliffs sculptured from thick beds of sandstone that stand almost vertically. All have tops of half-fallen plates and are within the same stratigraphical interval of the Sherkin Formation. Other elements include *leac*, 'a flat stone', used for a cliff with smooth surfaces, while *binníní*, 'small pinnacles', and *bioráin*, 'points', are used for the irregular weathering of cleaved siltstones. The importance of a site visit is demonstrated by the first example, *leac*, which can also mean a low-lying reef. Other names are occasional. *An Fhaill Fhiaclach*, 'the toothed cliff', is a cliff of interbedded sandstone and siltstone on Cape Clear, while Illaunnagart, off the same island, is interpreted as *Oileán na gCéard*, 'island of the corners'. Jointed sandstone beds have been weathered into sharp, orthogonal shapes.

The vibrant colours of yellow and red are recorded in a few names. *An Mionnán Buí* (Cl), *An Stocán Buí* (Cl, Sh) and *An Leac Bhuí* (Cl) are all coast exposures of thickly bedded

Island(s)	Abbreviation	Total collected to date	Number describing geology	Percentage describing geology
Castle	C	30	8	27
Calf (3)	Cf	28	3	11
Oileán Chléire (Cape Clear)	Cl	210	27	13
Carthy's	Cy	9	1	11
Goat	G	5	0	0
Heir (Hare)	H	112	9	8
Horse	Ho	40	8	20
Long	L	102	12	12
Sherkin	Sh	381	44	12
Skeam (2)	Sk	45	7	16
TOTAL		962	119	12.5

Table 1. Proportion of place-names with an element describing geology

sandstone in the sequence mentioned above. Each location is characterized by a distinctive, pale yellow (*buí*) weathering, which probably mirrors lichen growth and degree of exposure to sea spray as well as geology. The soil of *An Pháirc Dhearg*, 'the bright red pasture', a field on Skeam West, has a pinkish-red colour that is derived from the underlying mudstones of the Kinsale Formation.

UNCONSOLIDATED DEPOSITS

Patches of glacial sediment cover the irregular rock-head, especially at low elevations. These deposits are best exposed in inlets and strands and are characterized by stratified and solifluction fabrics. Small-scale landforms with post-glacial sediments include coast dunes, beaches, peat bogs and an estuary.

Within the context of the rough terrain of the islands, the agricultural value of land was often identified by a place-name without qualification. For example, The *Fothar* (Cl, Sh), The *Aonach* (Sh) and *An tAonach* (Cl) describe 'greener' landforms with glacial drift. These elements appear elsewhere in west Cork and Kerry. *Fothar*, or *fothair*, is defined as 'a grass-grown surface sloping over and down a cliff', and can be an inland as well as a coast feature. In south Kerry, the variant, *fuithear* is sometimes heard. The word seems to have been used to indicate a steep slope that could be used for grazing. The traditional definition of *aonach* is a place of assembly or fair, but Watson (1926), writing about the place-names of Scotland, made the point that the root *aon-* could apply in two senses; firstly, relating to a place of union, and secondly, to a solitary place. In Scotland, *aonach* is used variously for upland features (Matley 1982) and for 'a green plain near the shore on a stony bottom, a green beach' (MacLennan 1970). The latter is close to the usage in west Munster where the element is consistently associated with gently sloping tracts adjacent to the coast and below a rough, moorland landscape. Ó Cíobháin (1978 to 1985) ties the word to 'an isolated expanse'.

Inferior land with extensive rock outcrop and a lack of drift cover is identified by two place-names that are more familiar elsewhere on the Atlantic coast. *An Cladach*, 'the rocky shore', a townland on Sherkin, is better known as a district of County Galway. The Burren, a small ridge of sandstone on the same island, is, of course, the name of the famous karst landscape in County Clare. The Anglicism is probably from *boireann*, equated in Dineen (1927) with 'a rocky hill'.

Other elements of agricultural significance are *réidh*, a level moorland, as in *Faill na Réidhe* (Cl), and mín, 'a smooth green spot amid rough land' (Dineen, 1927), as in *An Mín Bhán*, a townland near Schull, on the mainland. On Castle Island, an east-facing cliff topped with a few metres of glacial sediments, and overgrown with vegetation, is called *An Fhaill Ghlas*, 'the green cliff'. Ringarogy Island, to the east of Roaringwater Bay, has been mapped as a broken landscape, being made up of a patch-work of small cultivated fields and rocky knolls (The Agricultural Institute 1963). Its Irish name is interpreted as *Rinn 'Gearróga*, 'headland of the short furrows'

Natural resources are frequently recorded in toponyms. For example, up to the 1950s, the men of the 'sand boats' came to Sherkin to excavate the dunes at *Doic na Gainidhe* (dialect form of *gainimhe*). The name means 'dock of the sand'. The shelly sand was applied to lime-deficient and heavily textured, acid soils. At *Cuas na Gainidhe*, or 'inlet of the sand', on Cape Clear, the sand was dredged for use on the island itself. *An Coinigéar*, or 'the warren', on Horse Island, identified a small area of sand dunes. High quality building stone was available from storm beaches as well as from small quarries. At the western ends of both Sherkin and Cape Clear, there are boulder-filled inlets called *Tráigh Leagaigh*. The name translates roughly as 'strand strewn with flat stones'. Earthy soil that provided a suitable binding for stone was available at *Gort an Mhoirtéil*, or 'cultivated field of the mortar' on Cape Clear (Lankford 1999). Even small deposits of bog iron ore were labelled. For example, there is a small inlet on Long Island with the name, *Poll an Iarainn*, meaning 'hole of the iron'. Bog iron was known locally as *cac iarainn* or 'iron excrement'.

WATER SOURCES

The hydrogeology of the islands may be summarized as

poor aquifers discharging low yields of groundwater. Potential recharge is limited by the small size of rainfall catchments. Springs and dug wells preceded bored wells and piped supplies from the mainland, and were, in many cases, given names that include the element, *tobar.* On the bigger islands, approximately 5% of the place-names describe sources of drinking water. The quantity of the supply is occasionally recorded, while references to quality appear more often.

An Tobar Mór, 'the big well', is a stone-lined well near Sherkin Friary. There were daily gatherings at *Tobar na mBan*, 'spring-well of the women', on Heir Island. However, on the smaller islands, the supply could not be relied upon and it is significant that there is an increased proportion (11%) of place-names for the low-yielding springs of the Skeam Islands (Beese 2000).

The groundwater sources were highly vulnerable to pollution and it seems likely that important springs were consecrated as Holy Wells in an effort to protect them. *Tobar Chiaráin* (Cl) and *Tobar Mhughaine* (Sh) commemorate the early Christian leaders Ciarán and Mughain respectively. The medieval tradition of locating graveyards close to Holy Wells, as at *Tobar Chill Bhrúin* (Cl), interpreted as 'spring-well of Brún's burial ground', is now recognized as bad practice. A depression spring for cattle, polluted by trampling and dung, was identified separately as *Tobar na mBó* (C), 'spring-well of the cows'.

Flowing, rather than standing, water was considered to be of good quality and worthy of comment. *Cuas an Uisce* (Cl, L, Sh, Sk) is equated with 'inlet of the running water' (Sharrock 1973). It is a name that appears all around the west Cork coast, and was a signal to local and visiting fishermen that a supply of drinking water was available, at least in fine weather (Beese 2000). Related names are *Faill an Uisce* (Cl, Ho) and *Tráighín an Uisce* (H). *Carraig an Fhuaráin* (Sh), or 'rock of the cold spring', and the imitative name, *Tobar na bPlobaí*, 'spring-well of the plashing', also seem to have been indicators of good quality.

Various place-names are associated with small springs on the coast. The *Níochán* (L) and *Lochán a' tSúsa* (Sh), are places where freshwater collects in pools, and were, therefore, suitable for washing clothes. The element, *sruthán*, or rivulet, in the names, *Faill a' tSrutháin* (Sh) and *Cuas na Sruthán* (Cl), seems to have been an indication that a flow of water had never dried (see Ó Croínín 1985). *Cuas a' Ghlas-Lóigh* (Cl), or 'inlet of the light, green weed' was one of a series of landmarks along the south coast of Cape Clear Island. There, growths of the vivid green seaweed, *Enteromorpha*, highlight where freshwater seeps from bedrock fissures into the tide.

DISCUSSION

Analysis of the place-names of the archipelago shows that about one eighth (12.5%) include an element that describes geology (Table 1). The word 'geology' is used here for aspects of the bedrock, glacial and postglacial deposits and groundwater. Percentages of 'geological' names for the individual islands are broadly similar to the overall figure. No doubt, an important factor in the assessment is that Roaringwater Bay is a coast environment, and therefore, has good exposures of bedrock and covering drift.

The interpreted toponyms are an expression of the close relationship between the fisherman-farmer and the geological landscape. They are part of an aesthetic code as well as points of reference for survival. Owing to major language and cultural changes, the modern perception of Roaringwater Bay has become more abstract, and regrettably so, since the non-academic view of the islander and tourist may not be that different.

REFERENCES

The Agricultural Institute (An Foras Talúntais). 1963. *West Cork resource survey.* Dublin.

Beese, A. 1997, 1998 and 2000. Three articles on the place-names and natural environment of Castle, Horse and the Skeam Islands. *Mizen Journal*, 5, 65-74; 6, 36-55; 8, 74-82.

Dineen, Rev. P.S. 1927. *Foclóir Gaedhilge agus Béarla.* An Irish-English Dictionary. Irish Texts Society, Dublin.

Hindley, R. 1990. *The death of the Irish language. A qualified obituary.* Routledge, London.

Hydrographic Office of the Admiralty. 1852. *Chart showing Long Island and Baltimore Bays.*

Lankford, É. 1999. *Cape Clear Island: its people and landscape.* Cape Clear Museum, County Cork.

MacLennan, M. 1970. *A pronouncing and etymological dictionary of the Gaelic language.* Acair and Aberdeen University Press.

Matley, I. 1982. Perception of mountain environments. In MacAodha, B.S. (ed.). *Topothesia. Essays presented to T.S. Ó Máille.* RTCOG, Galway.

Ó Cíobháin, B. 1978 to 1985. *Toponomia Hiberniae,* Volumes I to IV. An Foras Duibhneach, Baile Átha Cliath.

Ó Croínín, S. & D. 1985. *Seanachas ó Chairbre I.* Cómhairle Bhéaloideas Éireann. Interviews with Sean Ó hAodh (page 22).

O'Donovan, J. c. 1841. *Ordnance Survey name books,* County Cork.

Ordnance Survey of Ireland. 1845. *Six inch to one mile scale sheets 148, 149 and 153.*

Ó Síocháin, C. 1940. Revised 1970. *Seanchas Chléire.* Oifig an tSoláithair, Baile Átha Cliath.

Pracht, M. and Sleeman, A.C. 2002. *A geological description of west Cork and adjacent parts of Kerry to accompany the bedrock geology 1:100,000 scale Sheet 24.* Geological Survey of Ireland.

Reilly, T.A. and Graham, J.R. 1976. The stratigraphy of the Roaringwater Bay area of southwest County Cork. *Geol. Survey Ireland Bull.*, 2, 1-13.

Sharrock, J.T.R. (ed.). 1973. *The natural history of Cape Clear Island.* T & A.O. Poyser, Berkhampstead.

Watson, W.J. 1926. *The history of the Celtic place-names of Scotland.* William Blackwood, Edinburgh.

In: Parkes, M.A. (Ed.) 2004 *Natural and Cultural Landscapes - The Geological Foundation*, 245-248, Royal Irish Academy, Dublin

THE LINKS BETWEEN OUR NATURAL AND BUILT HERITAGE: HOW WE CAN USE URBAN LANDSCAPE AND GEOLOGICAL RESOURCES TO AID UNDERSTANDING AND APPRECIATION OF THE WIDER LANDSCAPE

Colin MacFadyen, Scottish Natural Heritage, 2 Anderson Place, Edinburgh, EH6 5NP. Email: colin.macfadyen@snh.gov.uk. Andrew McMillan, British Geological Survey, Murchison House, West Mains Road, Edinburgh, EH9 3LA. Email: aamc@bgs.ac.uk.

Abstract: The majority of Scotland's population lives in an urban environment. This is an important group of people to influence if we are to ensure the conservation and better management of the nation's landscapes as a whole. To do so, we need to utilise fully the urban landscape and its geological resources. Raising awareness, among the urban population, of landscapes and the processes that formed them is made easier if there is a local geological resource, such as Arthur's Seat volcano in Edinburgh. Heavily used by students and amateur geologists, this resource has significant potential for raising wider public awareness of the geological processes that formed our landscape, through self-guided geological trail leaflets and interpretive boards. Elsewhere, in lieu of urban outcrops, buildings and building stones provide excellent links between the natural and built heritage and can be used to stimulate interest in geology and landscapes beyond the cityscape (Bennett et al. 1996). Based on an earlier project in London (Bennett and Doyle 1996) a Caithness flagstone in the Royal Mile in Edinburgh was recently engraved with details of its name and age. It will be seen by thousands of local people, as well as tens of thousands of tourists. Such initiatives are a means of stimulating awareness of the geology and landscape of Scotland. Building stone also displays wonderful examples of sedimentary structures and igneous and metamorphic textures; these characteristics can be successfully used to demonstrate geological origins and processes and properties of stone. Efforts to promote an understanding and appreciation of the wider landscape through the use of urban geological resources, including building stones, will help to close the gaps in the public's perceptions about geology, the built heritage and the wider landscape. It may also lead to change on a generational timescale of the thinking and approach of tomorrow's professional architects and builders. This may foster, in turn, more sympathetic and sustainable design and construction, both in city and wider landscapes, and more sustainable development in general.

SCOTLAND'S EARTH HERITAGE: THE ROLES OF SNH AND BGS

Scottish Natural Heritage (SNH) is the UK government body responsible to the Scottish Executive and to the Scottish Parliament. Charged with working with Scotland's people to care for the natural heritage, our aims are promotion of its care and improvement, its responsible enjoyment, its greater understanding and appreciation and its sustainable use now and for future generations. SNH educational and awareness duties from the Earth science perspective, endeavour to convey the message that the geology, both static and dynamic, underpins all other aspects of the natural heritage, namely landscapes and biodiversity.

The British Geological Survey (BGS) is a UK non-statutory government organisation and a component of the Natural Environment Research Council (NERC). Its role is to advance geoscientific knowledge of the UK landmass by means of systematic surveying, data collection, long-term monitoring and high quality research and to provide comprehensive, objective, impartial and up-to-date geoscientific information. A key area of its activities is to enhance the UK science base by developing knowledge, information, education and training in the geosciences and by promoting the public understanding of the relevance of geoscience to resource and environment issues.

EARTH HERITAGE EDUCATION AND AWARENESS RAISING EFFORTS

Over the last ten years, SNH has produced a variety of educational and awareness vehicles. Some examples of SNH work include:

- interpretive panels and leaflets at designated sites and areas;
- production of 'Earth Heritage' the geological and landscape conservation magazine (in association with English Nature and the Countryside Council for Wales);
- development of the Knockan Crag interpretive centre in Sutherland: and
- input to various other visitor facilities such as the National Trust for Scotland's Visitor Centre at Glen Coe.

SNH has also worked to forge partnerships among those organisations in Scotland involved with raising an awareness of the Earth heritage (National Museums of Scotland, Our Dynamic Earth, the BGS and the Hunterian Museum in Glasgow), leading to joint ventures, such as the organisation of 'Rock On' the biennial Scottish Geology Festival (formerly Scottish Geology Week). The production of the popular *Landscape Fashioned by Geology* books (e.g. McKirdy and Crofts 1999, Browne et al. 2001) represents further SNH and BGS collaboration.

The BGS, following the lead shown by the Geological Survey of Northern Ireland and Geological Survey of

Ireland (this volume) has, in recent years published a range of 'popular publications' including 'Whisky on the Rocks' (Cribb and Cribb 1998), 'Isle of Man: foundations of a landscape' (Pickett 2001; see also Pickett this volume), Holiday Geology Guides (e.g. Gallois 2001), building stones leaflets (e.g. Lott and Barclay 2002), Building Stones Resources of Britain map (BGS 2001), and GeoTourism maps (e.g. BGS 2000). The BGS has also recently provided input to the National Trust for Scotland's new Glen Coe Visitor Centre. Links with the UKRIGS organisation on a local level has led to successful publications including guided geological trail leaflets and interpretive posters. Examples include the Hermitage of Braid and Corstorphine Hill leaflets and posters published by the Lothian and Borders Regionally Important Geological Sites Group (LaBRIGS) and the St Andrews and St Monans leaflets published by geoHeritage Fife.

RAISING AWARENESS OF EARTH HERITAGE IN AN URBAN ENVIRONMENT

Our task in education and interpretation is generally easier in the wider countryside, where the links between geology, landscape and biodiversity can be more easily understood and demonstrated and therefore the need for sustainable management can be appreciated. However, with over 80 per cent of the UK population living in the urban environment in towns/cities of 10,000 or more people (UK DETR 2000), efforts at raising an awareness and understanding of geology have to utilise whatever resources are available. The utilisation of urban geological resources, for geological educational purposes is not new (e.g. Bennett et al. 1996), but it is being applied widely now to Scotland.

Edinburgh is particularly well endowed with urban geological resources. The most obvious example is the Arthur's Seat volcanic complex which is used extensively as an educational resource, with numerous guided walks, some commercially run, taking place. Historic Scotland which manages the site on behalf of H.M. the Queen, has installed interpretive panels at key access points, and a self guide leaflet has been published by the Edinburgh Geological Society (Land and Cheeney 2000). The Castle Rock, an offshoot of the Arthur's Seat complex, has been for many years the subject of an interpretive display in West Princes Street Gardens produced by the same society. All these interpretive products help to raise an awareness and understanding of Edinburgh's 'volcano'.

Edinburgh also has a variety of other urban outcrops, such as Blackford Hill where in the 1840s Louis Agassiz recognised 'the work of ice', and the intertidal exposures of Carboniferous sedimentary rocks on the city's shoreline at Joppa. The latter area has been interpreted by means of a display panel produced by Joppa Community Council and Joppa Amenity Society. The LaBRIGS Group has also published a poster. There is of course the built heritage of the city, with the various building stones illustrating a variety of rock types and structures that tell a geological tale of landscapes beyond the city limits. As with other towns and cities, there are published reference works (e.g. McMillan et al 1999) and also building stones tours (Bunyan in McAdam and Clarkson 1996). Walking tour leaflets linking geology to buildings (e.g. Calton Hill, including Edinburgh's 'Parthenon') have been prepared by LaBRIGS.

USING BUILT HERITAGE FOR THE PURPOSES OF EDUCATION & AWARENESS

Helping people to interpret the Earth stories hidden in the rocks of the built heritage, forms the basis of one of SNH's "Sharing Good Practice" training events. Called 'Hewn from the living rock' the aim of this course, first run in 2001, is to introduce the range of exciting stories that lie hidden in the landscape and buildings that surround us, and which add depth to our understanding of historic and cultural heritage.

Bringing geology alive and drawing the wider landscape and building stone together, was the impetus behind another SNH-led project that took place in Edinburgh in 2001, as part of 'Rock On' - Scottish Geology Week. Based on a similar exercise that took place in London (Bennett and Doyle 1996), Edinburgh had its first named stone, a Caithness flagstone at Hunters Square just off the Royal Mile. The stone was carved with details of its age and where it came from, with the aim of informing passers-by that natural rock floors a large portion of this world famous street. It was hoped that the 'named' stone would help restore the links in people's minds, between the natural and built heritage; the aspiration being to bring meaning to the stone and bring it to 'life'.

THE BUILT HERITAGE AND THE WIDER LANDSCAPE: REBUILDING THE LINKS & CHANGING MINDSETS WITH A VIEW TOWARD SUSTAINABLE BUILDING

In Edinburgh, in the 18th and 19th centuries and the earlier part of the 20th century the links between geology and the built environment must have been very apparent, with the huge stone quarries of Hailes and Craigleith being on the doorstep of the city (McMillan et al 1999). The quarries are long gone. The once massive Craigleith quarry, which supplied much of the stone for much of Edinburgh's New Town, became a landfill site in the 1970s and is now occupied by a shopping centre. Fortunately at Craigleith the uppermost beds of Carboniferous strata of the quarry are still visible and the site is now a RIGS (Browne and Reid 1998) where people may be reminded of the once great quarry which supplied Edinburgh and many other cities with the finest sandstone.

However, with the disappearance of local quarries, many people have lost touch with the geological origins of the built environment of Edinburgh. The same is true of most towns and cities across Scotland and elsewhere. A crucial link between peoples' everyday experience and the wider landscape and environment has therefore been lost. Although natural stone is being used in some new build and for conservation purposes, it commonly arrives shrink-wrapped and on pallets. This has the effect of 'sanitising' the building material, effectively removing the link between it and its source in the wider landscape.

Much new build in Edinburgh and elsewhere involves the use of artificial stone, known as 'processed building material', or in other words concrete. Often this material is fashioned into blocks that are either 'rusticated' or polished and coloured, to mimic indigenous sandstones. Processed building material unfortunately can sit at odds with the natural stone of the cityscape. Not only in the cityscape but in the wider countryside, we have seen the shift from traditional building material to the use of inappropriate processed building materials.

Architects often consider using natural stone for new build projects but reject this option on grounds of cost. This is based on the premise that concrete 'stone' is 'less expensive'. On first analysis, currently from a financial aspect, concrete 'stone' appears to be the cheaper commodity. But what is the true energy life cycle cost of the material? When compared with other building materials (Table 1), stone shows a significant energy saving in terms of process energy requirement (PER). Similar figures could apply in the comparison of brick paving, concrete slabs and natural sawn stone. Riven stone and slate require less energy than sawn stone, but figures for splitting are not available.

Thus it can be shown that concrete 'stone' or other artificial processed building material is more expensive from an energy point of view. Natural stone is therefore from an over-all sustainability perspective, more environmentally friendly. Another consideration is the longevity of concrete 'stone' in terms of weathering; natural stone may last for hundreds of years, but will its artificial replacements?

The issue of recycling also needs to be considered. Since artificially processed building materials are not designed for re-use, the only option is crushing the material for aggregate Building stone although generally valued on its first cost, has a high residual value on demolition, and depending on circumstances, it may be recycled two or three times which reduces the PER in each application.

If we are successful in getting the message across to the population, and if we can close the perception gap in the public's mind between their everyday environment and the wider landscape and environment, then the result may well be that tomorrow's professionals - architects and builders, may design and build using materials that are sustainable, that do not incur huge energy costs and which reflect and are in harmony with their cultural landscape, be it within a city or the wider landscape.

In Edinburgh, fortunately, there are many organisations supporting the use of natural Scottish (and English) stone for the city's conservation and restoration projects. Organisations such as the Edinburgh World Heritage Trust and the Cockburn Association recognise the need not simply to repair and conserve with appropriate natural materials, but also to make best use of natural stone in new build. This theme is being developed more broadly across Scotland by the work of the Scottish Stone Liaison Group (a network of organisations), launched by the Scottish Executive in May 2000, which is promoting the use of indigenous stone and slate (Historic Scotland 2000). The Scottish Stone Liaison Group (SSLG) recognises that the decline in use of stone together with a loss of relevant skills and knowledge during the 20th century poses serious problems for any revival of the indigenous stone industry and the best use of stone for conservation, repair and maintenance and in new build projects. The SSLG has also recently launched the Natural Stone Institute to provide a better understanding of all aspects of natural stone and its use for public benefit (McMillan and McKinney in press). The Institute intends, using various media and seminars/workshops, to encourage good practice through education, training, research, technical innovation and information.

The Natural Stone Institute and the Scottish Lime Centre Trust have recently submitted a bid to the Heritage Lottery Fund for a project entitled "Developing understanding of Scotland's stone-built heritage through effective education". The aim is to develop a package of learning materials that will provide late primary and early secondary school pupils with a real knowledge and understanding of

Material	Specification	Energy requirement (MJ/m^2)	Total energy (MJ/m^2)
Brickwork	60 flettons at 2.5MJ	150	
	0.02 m^3 mortar	60 - -	210
	60 common bricks at 4.5 MJ	270	
	0.02 m^3 mortar	60 - -	330
	60 average non-fletton at 9.1MJ	546	
	0.02 m^3 mortar	60 - -	606
	60 facing bricks or engineering at 9.8 MJ	588	
	0.02 m^3 mortar	60 - -	648
	60 alluvial type bricks at 16.8MJ	1008	
	0.02 m^3 mortar	60 - -	1068
Concrete blockwork	9.5 blocks at 34.4MJ	327	
	0.005 m^3 mortar	15 - -	342
Limestone masonry	9.5 sawn blocks at 11.9 MJ	114	
	0.005 m^3 mortar	15 - -	129
Sandstone masonry	9.5 sawn blocks at 9.1 MJ	87	
	0.005 m^3 mortar	15 - -	102

Table 1. Process energy requirements (PER) of common building materials used in single-leaf load bearing external walls (after Historic Scotland 1997 and references therein).

the stone heritage. Preparation of a high quality CD-Rom product will be an important part of the project.

New build in our towns and cities often incorporates natural stone, albeit in the form of cladding to steel-framed concrete structures. The use of stone in prestigious buildings may well enhance public appreciation of the material and at least opens up the debate about its use. In Edinburgh the use of Clashach sandstone from Elgin as rain-screen cladding in the new Museum of Scotland received much publicity. It remains to be seen whether the new Scottish Parliament, to be clad in part with Scottish stone (including granite from Kemnay, Abderdeenshire) will have the same effect, but the earlier debate over possible use of alternative sources of granite from China evoked much public outcry.

Whether it be for new build or for conservation and repair of our historic buildings it is important that we make best use of available indigenous resources. As transport and energy costs continue to rise, powerful sustainability arguments can be made for the opening or reopening of small local stone quarries to supply vernacular and traditional building material needs. With a better appreciation of the local sources of building materials the public may begin to acknowledge that such quarries, of appropriate scale and sensitively managed, need not be detrimental to the environment. In fact they form a valuable asset and an integral part of the cultural landscape.

CONCLUSION

The recently published document 'A Policy on Architecture in Scotland' (Scottish Executive 2001) sets out a shared approach towards the delivery of a better designed, accessible and sustainable built environment. At the heart of the policy is a commitment to involve Scotland's population in decisions about building design and use of materials in the 21st century. Stone has been one of our fundamental building materials for 5000 years and regionality of stone buildings around Scotland can be linked directly to the nation's geology. Yet the general public is largely unaware of these connections. So the message is that we need to continue to use urban geological resources to raise an awareness of the wider Earth heritage and landscape.

In time it is possible that awareness raising may close the gaps that have opened up in the public perception between people's everyday experience and the wider landscape and environment. This in turn may eventually lead to the use of more traditional building materials and, with the associated energy and sustainability implications, promote a return to local landscape and cityscape distinctiveness.

ACKNOWLEDGMENT

Andrew McMillan publishes with the permission of the Director, British Geological Survey (NERC).

REFERENCES

Bennett, M R, Doyle, P. 1996. The introduction of geology into the urban environment: principles and methods, pp. 239-255 in *Geology on your doorstep: the role of urban geology in Earth Heritage Conservation* Bennett, M R, Doyle, P, Larwood, J G and Prosser, C D (eds). London: The Geological Society.

Bennett, M R, Doyle, P, Larwood, J G and Prosser, C D (editors). 1996. *Geology on your doorstep: the role of urban geology in Earth Heritage Conservation.* London: The Geological Society.

British Geological Survey 2000. *Tourists' rock, fossil and mineral map of Great Britain.* Keyworth, Nottingham: Earthwise Publications (British Geological Survey).

British Geological Survey 2001. *Building Stone Resources Map of Britain* (1: 1 000 000). Lott, G. K. Keyworth, Nottingham: British Geological Survey.

Browne, M. A. E. and Reid, E. 1998. RIGS and geological conservation in Scotland. pp. 79-89 in Proceedings of the First UK RIGS Conference, Worcester 1998 (Worcester: Hertfordshire and Worcestershire RIGS Group)

Browne, M A E, McKirdy, A and McAdam, A D. 2001. *Fife and Tayside - A Landscape fashioned by geology*. Perth: Scottish Natural Heritage.

Bunyan, I. T. 1996. Building Stones of Edinburgh, pp. 56-69 in *Lothian Geology - an Excursion Guide.* McAdam, A D and Clarkson, E. N. K. (eds). Edinburgh: Edinburgh Geological Society.

Cribb, S and Cribb, J. 1998. *Whisky on the rocks - origins of the 'water of life'.* Keyworth, Nottingham: Earthwise Publications.

Gallois, R. 2001. *Holiday Geology Guide: Stonehenge - the secrets in the stones.* Keyworth, Nottingham: Earthwise Publications.

Historic Scotland 1997. *A Future for Stone in Scotland.* Edinburgh: Historic Scotland.

Historic Scotland 2000. Scottish Stone Liaison Group Launch Abstracts. Edinburgh: Historic Scotland.

Land, D H. and Cheeney, R F. 2000. *Discovering Edinburgh's volcano - a geological guide to Holyrood Park.* Edinburgh: Edinburgh Geological Society.

Lott, G H and Barclay, W. 2002. *Geology and building stones in Wales (north) and (south).* Nottingham: British Geological Survey.

McKirdy, A.P. and Crofts, R. 1999. *Scotland: the creation of its natural landscape. A Landscape fashioned by geology.* Perth: Scottish Natural Heritage.

McMillan, A. A. and McKinney, A. in press. A future for stone: a view from the Scottish Stone Liaison Group. Proceedings of conference "Stone in Wales: materials, heritage and conservation". CADW: Welsh Historic Monuments.

McMillan, A. A., Gllanders, R. J. and Fairhurst, J. A. 1999. *Building stones of Edinburgh.* 2nd edition. Edinburgh: Edinburgh Geological Society.

Pickett, E A. 2001. *Isle of Man - Foundations of a Landscape.* Nottingham: British Geological Survey

Scottish Executive 2001. *A Policy on Architecture in Scotland.* Edinburgh: Scottish Executive.

UK Department of the Environment, Transport and the Regions (DETR). 2000. Our towns and cities: the future - delivering an urban renaissance. Government White Paper.

In: Parkes, M.A. (Ed.) 2004 *Natural and Cultural Landscapes - The Geological Foundation*, 249-252, Royal Irish Academy, Dublin

UNDERSTANDING AND COMMUNICATING THE LANDSCAPE HISTORY OF THE CAIRNGORM MOUNTAINS IN SCOTLAND

C. W. Thomas[1], M. R. Gillespie[1], N. R. Golledge[1], C. J. Jordan[2] and V. Kirkbride[3]
[1]British Geological Survey, Murchison House, West Mains Road, Edinburgh EH9 3LA.
Email: cwt@bgs.ac.uk
[2]British Geological Survey, Kingsley Dunham Centre, Keyworth, Nottingham NG12 5GG
[3]Scottish Natural Heritage, Redgorton, Battleby, Perth PH1 3EW

Abstract: The Cairngorm Mountains are famous for their glacial landforms - notably the corries and glacial troughs, for their ancient tors and rolling plateaux that survived the passage of ice, and for a sub-arctic fauna and flora unique in the UK. However, little is known about the influence of the structure of the rocks, particularly within the Cairngorm Granite, on this distinctive and important landscape. The British Geological Survey and Scottish Natural Heritage are jointly investigating the link between geological structure and the Cairngorm Mountain landscape. The project is gathering new data on structure and landscape form in the Cairngorm Mountains to allow us to develop our understanding of the history of landscape formation. We are developing innovative approaches to the interpretation and visualisation of the dramatic earth history of the Cairngorms, so that a wide range of people, including those involved in education, visitors to the Cairngorms and those charged with the management of this unique environment, have the opportunity to appreciate the extraordinary story of the formation of the Cairngorm Mountain landscape.

INTRODUCTION

The Cairngorm mountain massif lies within the Grampian Highlands of Scotland and is almost wholly underlain by the Cairngorm Granite, a late-Caledonian pluton c. 410 million years old (Figure 1). The massif contains the largest area of land above 1000m in the British Isles and is home to a sub-arctic fauna and flora unique in the U.K. The mountain landscape includes spectacular glacial landforms, most notably deeply excavated troughs and corries, and open, rolling upland plateaux of Tertiary age that have largely survived the passage of ice. Abundant and often very large tors are a characteristic feature of the plateaux, particularly in the eastern part of the massif.

The importance and sensitivity of the physical and biological environment of the Cairngorms has long been recognised. The mountains host a National Nature Reserve and numerous Sites of Special Scientific Interest (SSSIs), along with sites of other special landscape and environment designation. However, various and growing human pressures on the sensitive landscape, arising from its amenity value and the concomitant importance to the local economy via tourism, have become matters of increasing concern. The conflict of interest between protection of the environment and the development pressures arising from recreational use has been highlighted recently by the heated debate over the construction of the Cairngorm Funicular Railway to replace an ageing chair lift system in the ski resort area. Such conflicts highlight the importance of the Cairngorms to a wide range of people who often have very different views on how the Cairngorms should best be managed. In recent years, the Cairngorms Partnership has been responsible for developing criteria and consensus for management and appropriate development. The Cairngorms area is now Scotland's second National Park and is included on the 'Tentative' list of World Heritage sites.

Cairngorm ecosystems appear to be well-understood and acknowledged, providing the basis for much nature-orientated tourism. However, relatively little is known about the importance of geology as the foundation not only of the physical landscape, but also of the ecosystems. The special characteristics of the Cairngorm mountains are fundamentally a result of the Cairngorm Granite. It determines their shape, elevation and extent, their soils, vegetation and ecosystems, and the local climate. There is a pressing need for geological information and tailored resource materials appropriate to a wide variety of uses, including both formal and leisure education, leisure pursuits, environmental management and policy development. The jointly-funded British Geological Survey - Scottish Natural Heritage Cairngorms Project aims to address, at least in part, the need for geological information within the context of the development of the Cairngorm National Park.

A core aim of the project is the development of an understanding of the role of geology in the formation of the landscape. In particular, we are studying geological structure and lithological variation in the Cairngorm Granite and the interaction between these and surface processes in determining the landscape. The Cairngorms landscape is commonly presented as 'glacial' in origin. While it is true that glacial processes during the Quaternary have resulted in some of the more spectacular elements of the landscape, glacial and other processes that operated during the Quaternary are only part of the story. To put this in perspective, they have operated only over 0.5% of the time during which the Cairngorm granite has existed. In order to present a more balanced view of the Cairngorms landscape, we need to understand more about the preceding 99.5 % of Cairngorms history.

STUDYING THE LINK BETWEEN GEOLOGICAL STRUCTURE, LITHOLOGY AND LANDSCAPE

Understanding the degree to which there are links between structure and landscape is important for two reasons: firstly, it establishes the influence of structure on landscape; secondly, if such links are established, they have implications for the age of the landscape, both in incipient and developed form.

In areas dominated by sequences of sedimentary rocks, the links between lithology and landscape are commonly clear and may be pronounced. However, such links are less clear in outwardly homogeneous massifs like the Cairngorm Granite. The Cairngorm Granite forms an upland massif surrounded by lower ground underlain by metasedimentary rocks that are generally less resistant to erosion. What is much less clear is the degree to which the Cairngorm mountain landscape is influenced by lithological variations in the granite.

In the Cairngorms project, we are studying the link between structure, lithology and landscape by various means. We have used satellite and stereo air-photograph imagery to measure the orientation of linear landscape and fracture features on the kilometre to tens-of-metre scale. By and large, the satellite imagery provides 'lineament' data reflecting large-scale landscape features such as troughs and corries; these may have a structural control, but it is not generally possible to establish this from the satellite imagery. A conspicuous feature of the Cairngorms in the satellite imagery is a preferred, approximately N-S alignment of major trough features, together with a marked regularity in spacing (Figure 1).

Stereo air-photographs have been used to measure the orientation of medium-scale lineament and fracture data. Some of these medium-scale lineaments are landscape features, such as water courses that may have a structural control. Others are necessarily generalised and of uncertain origin. It is clear from the inspection of the air photographs that many of the data relate directly to structural phenomena, most particularly fractures.

What cannot be readily established from the air photographs is the 3-d geometry of the structures, nor their character (e.g. are the fractures faults or joints?). We have undertaken an extensive field survey across the Cairngorm pluton (summer, 2002) to measure the orientation of, and to characterise, structures at the metre to ten metre scale. The field data will help us determine the distribution and disposition of fracture systems within the Cairngorm Granite and to determine in three dimensions, the relationship between outcrop-scale fractures and the larger-scale fractures and lineaments obtained from air photographs and satellite imagery. Analysis of the data is at an early stage, but preliminary inspection indicates that there is a link between structure at the outcrop scale and the large-scale landscape features so strikingly developed in the Cairngorms massif. The dominant structures within the granite are populations of joints that are dominantly steep and spaced on the metre scale. Analysis of the structural data is expected to provide important information on the origin of the joints.

In addition to collecting structural data during the field survey, we have also recorded the granite lithology, the degree of weathering, and the presence or absence of quartz veins, dykes and sills and alteration zones. These data will help us understand the local and/or regional influence of lithology on the landscape. Already, it is clear that quartz veins and associated haloes of altered granite that formed in the late stages of the pluton's evolution are associated with landscape features, such as cols, on the scale of tens to hundreds of metres. Quartz veins and similar dyke- or sheet-like intrusions are likely themselves to

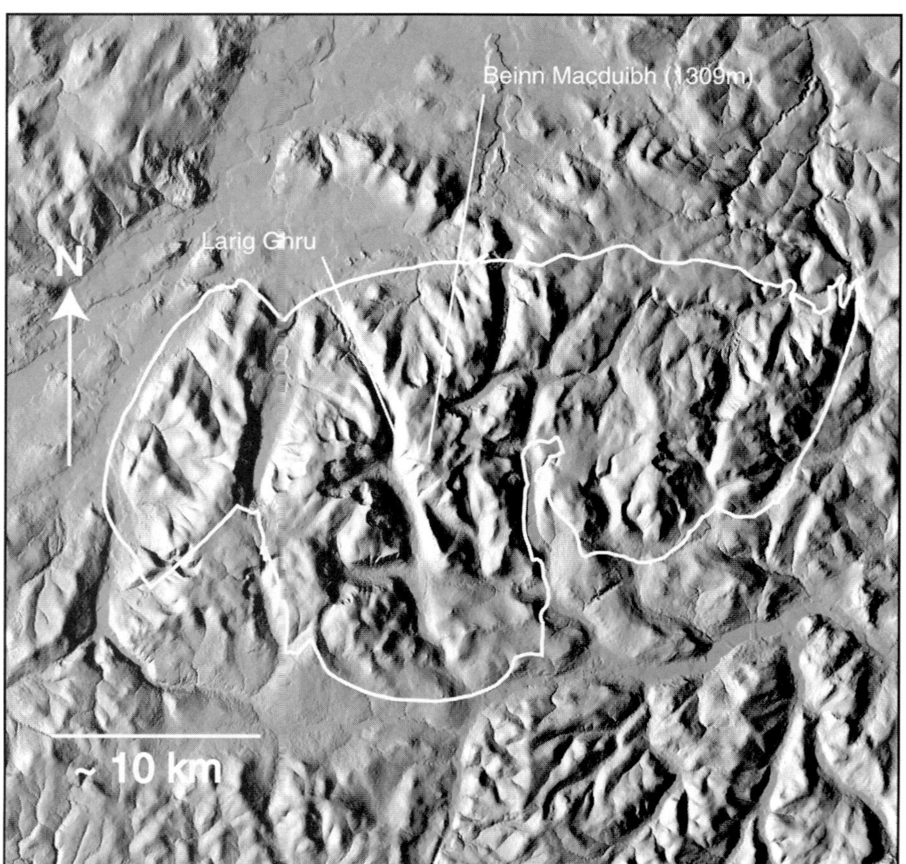

Figure 1. The Cairngorm mountains and the limit of the Cairngorm Granite (white irregular boundary). Shaded relief image of a 10m digital terrain model.

have a structural control; understanding the structure of the granite is thus closely related to understanding the nature of lithological variations.

If the link between structure, lithology and landscape proves to be strong, then it follows that the basic architecture of the landscape (though not necessarily its subsequent development) is an inherent property of the granite itself. The corollary of this is that the landscape is, in essence, pre-determined and that surface processes that would act on the granite as it became exposed only served to develop that landscape, exploiting existing domains of weakness.

STUDYING THE INFLUENCE OF GLACIATION ON CAIRNGORMS LANDSCAPE EVOLUTION

The Cairngorms landscape is often presented publicly as largely 'glacial' in origin. Although glacial action is responsible for many of the spectacular landforms, much of the upland plateaux survived significant erosive glaciation and retain the general form of the Tertiary landscape developed in the last 60 million years. Intensive chemical weathering during the initially warm and humid Tertiary period resulted in formation of a deep regolith over the granite. The spectacular tors, such as those so typical of the Ben Avon massif, developed as solid, massive cores in soft, deeply weathered rock. Subsequent exhumation has created the present-day tors. Our structural mapping of the granite supports a theory that the location, size and form of the tors is controlled by the spacing and disposition of joint sets, i.e. there is a strong structural and lithological control. Layers of Tertiary gruss are preserved in many parts of the plateaux, as are areas dominated by small core-stones that are clearly in situ.

In collaboration with Dr Adrian Hall (Fettes College, Edinburgh), we have attempted to portray the localisation of glacial erosion by 'rebuilding' the pre-glacial, end-Tertiary landscape (Figure 2). The boundary between areas essentially unmodified by glaciation (high plateaux) and areas modified by glaciation (corries and troughs) was first identified using well-established criteria. The unmodified plateau areas have several characteristic landscape features, e.g. dome-shaped hills and summits, wide shallow valleys, and gentle slope gradients. There is a marked absence of steep slopes, abrupt breaks of slope, and cliffs. These characteristics were used to guide extrapolation of contours on glacially unmodified surfaces into glacially modified areas, to recreate an approximation of the pre-glacial surface across the massif. A digital terrain model was built from the resulting contour model. We then digitally 'subtracted' the current landscape from the pre-glacial terrain model to produce a map of relative glacial erosion intensity (Figure 3). The model demonstrates clearly that the western part of the massif is more highly glaciated than the east (which is evident from casual observation of topographic maps). It also reveals a 'core' of intense glacial erosion in the centre-west part of the massif, in Glen Einich, Glen Geusachan, the Lairig Ghru and the Garbh Choire. An important outcome of these models is a clear demonstration that all of the principal drainage features in the Cairngorms (i.e. the valleys) existed prior to the glaciations. Glaciation has therefore served to enhance selected landscape features in the Cairngorms: only the corries can be said to have been created by glaciation alone. The pre-glacial land surface model is only a first-order estimate, but used in conjunction with modern surface modelling software it helps to clarify the role of glaciation in development of the landscape.

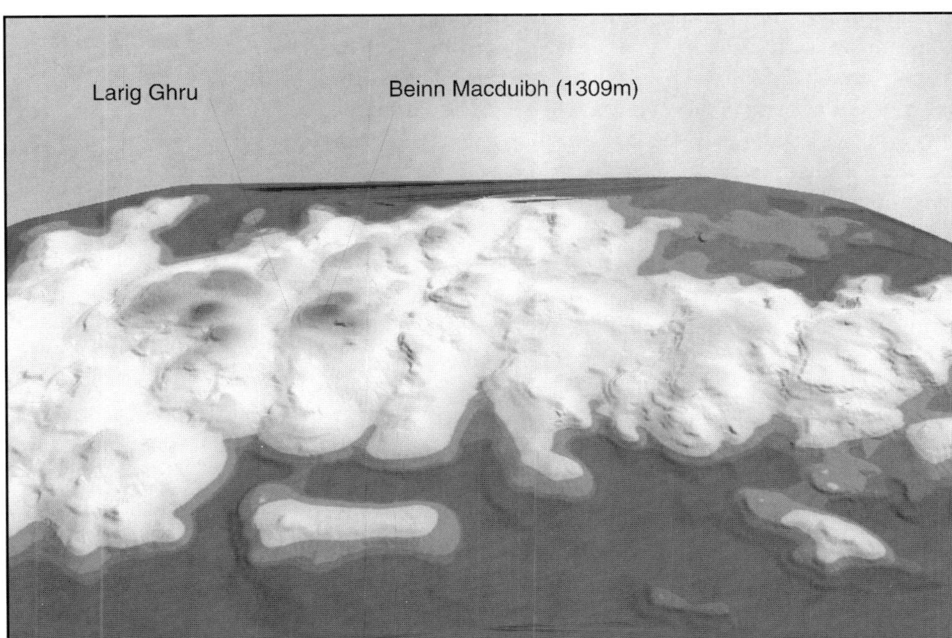

Figure 2. Perspective view of a digital 'pre-Quaternary' landscape model for the Cairngorm mountains. The outer darker greys represent lower-lying ground. The highest ground is represented by medium greys in the core of the image.

COMMUNICATING THE CAIRNGORMS STORY

A key outcome of the Cairngorms project is the conclusion that many Cairngorm landscape features reflect geological events that substantially pre-date the Quaternary glaciations. Large-scale elements of the landscape, such as the shape, height and size of the massif, reflect directly the size and shape of the granite pluton at the time it was emplaced, and its position in the crust. The disposition of medium-scale landscape elements, such as the large valleys and the hills that separate them, probably reflects a combination of

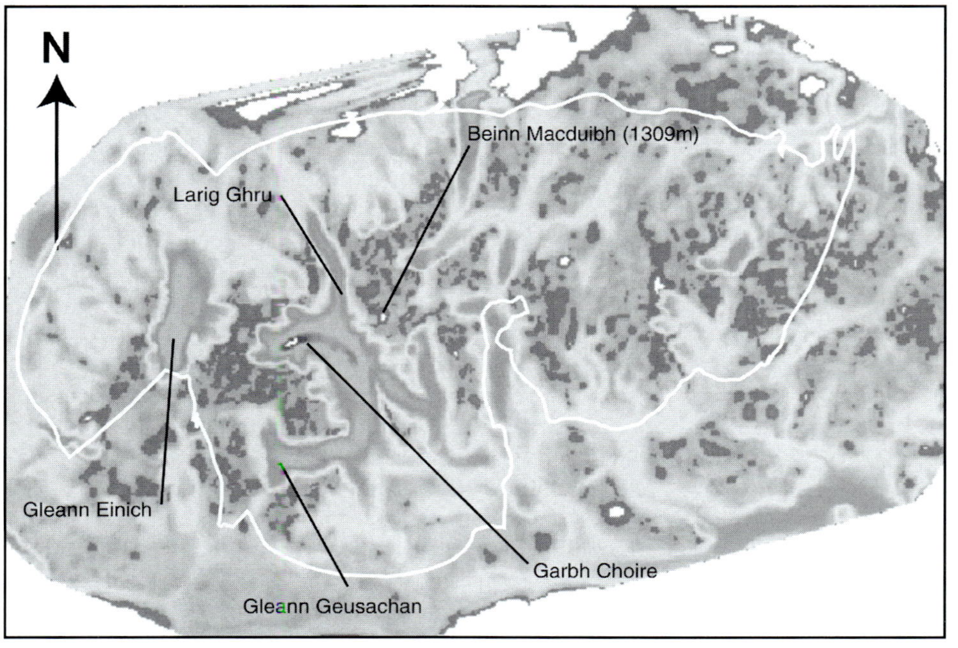

Figure 3. The localisation of glacial erosion, as revealed by the digital 'subtraction' of the current Cairngorm DTM (Figure 1) from the pre-Quaternary DTM (Figure 2). The marked linear areas in mid to dark grey tones are the glacial troughs. See text for discussion.

factors including the shape of the pluton roof, the distribution of granite facies, and the location of structural weaknesses such as fracture zones and alteration zones. All of these factors were 'locked-in' to the granite during and shortly after its emplacement, around 400 million years ago. During times when the granite has been exposed at the earth's surface (first in the Devonian period around 370 million years ago, and then again from around 60 million years ago to the present day), agents of weathering and erosion have exploited these geological features to develop the landscape. Smaller-scale features of the present landscape reflect geologically recent processes, following uplift and exhumation of the granite in the Tertiary period (beginning around 60 million years ago). The tors were formed as a result of differential weathering under warm, humid conditions in the Tertiary. Tor location was controlled by the spacing and orientation of joints in the granite. Rapid deepening of valleys and creation of corries occurred during the Quaternary glaciations of the last 2 million years. It is clear that the Cairngorms landscape is the complex product of an exceptionally long and varied history in which the geology has been fundamentally important.

Communicating the story of landscape evolution in the Cairngorms to a wide public audience is an important goal of the Cairngorms project. Successful communication of geological science to the public is a difficult task, not least because of the immense timescales and the large number of unfamiliar concepts and processes that must be conveyed. We plan to use innovative and state-of-the-art digital media to illustrate key parts of the story, in particular those parts relating to the massif in its present form (i.e. Tertiary and Quaternary periods). Specific themes include pre- and post-glacial DTMs, drapes of various topographical and geological themes, fly-thru models, 3-d anaglyphs, VRML presentations illustrating glacial action, cartoons of pluton emplacement, superimposition of structural features, uplift and exhumation, etc.

CONCLUSIONS

Preliminary results of the Cairngorm project indicate the following:

- much of the 'unique' character of the Cairngorm mountain landscape results from the present altitude of the massif relative to other parts of the UK; this is a direct consequence of geological history
- the landscape is the product of an exceptionally long and varied history, with fundamentally geological foundations, extending back to the time of granite emplacement
- the principal elements of the present landscape, and their disposition, may have existed since the Cairngorm Granite was first exposed at the earth's surface (probably around 370 million years ago)
- irregularities in the pluton roof, variations in granite texture, and linear zones of weakness in the granite created by zones of fracture and alteration have all probably played a role in controlling the size, location and disposition of medium-scale landscape features such as valleys and hills
- comparison of digital terrain models representing the pre-glacial (Tertiary) land surface and the present post-glacial surface reveals that all the main drainage features in the Cairngorms pre-date glaciation
- corries are probably the only (non-depositional) landscape features to have been created entirely by glacial action
- the numerous tors that dot the pre-glacial plateaux in the Cairngorms formed during deep weathering in the Tertiary and survived the Quaternary glaciations either because the enclosing ice sheets were essentially immobile, or because they have been exhumed from weak weathered granite since the last glaciations

ACKNOWLEDGEMENTS

The authors thank Martin Smith for review. This paper is published with the permission of the Director, British Geological Survey (NERC).

In: Parkes, M.A. (Ed.) 2004 *Natural and Cultural Landscapes - The Geological Foundation*, 253-256, Royal Irish Academy, Dublin

THE GEOLOGICAL HERITAGE ON THE MONTESINHO NATURAL PARK (NE PORTUGAL) - AN INTERPRETATION STRATEGY FOR AN AREA WITH HIGH GEOLOGICAL COMPLEXITY

D. I. Pereira*, C. Meireles, M. I. Alves*, P. Pereira*, J. Brilha* & G. Dias***
***Earth Sciences Department, University of Minho, Campus de Gualtar, 4710-057 Braga, Portugal. Email (1st author): insuad@dct.uminho.pt**
****Geological and Mining Institute, Apartado 1089, 4466-956 S. Mamede de Infesta, Portugal**

Abstract: **The geological complexity and the exotic character of several formations of the Montesinho Natural Park (NE Portugal) print peculiar characteristics to the landscape. This part of the Iberia is characterized by the occurrence of a pile of allochthonous units thrusted over autochthonous Late Proterozoic and Palaeozoic metasediments and igneous rocks. The two major exotic terranes comprise a continental allochthonous terrane of Precambrian to Lower Cambrian age and a Silurian-Lower Devonian ophiolite complex thrusted by the allochthonous unit. More recent Cenozoic tectono-sedimentary events left their imprints on the landscape, namely with the presence of three unconformity-bounded sedimentary sequences. This complex geology has been detailed in recent scientific research allowing an accurate reading of the landscape. However, if on one hand that complexity is a valuation factor, on the other hand it renders the knowledge transfer to the general public more difficult. Raising public awareness of the regional geodiversity and its heritage relevance is thus a challenge to undertake. In this paper an interpretative panel is presented where the main features of the landscape and their relationship with the units and the geological processes are pointed out.**

INTRODUCTION

The Portuguese national park system - which includes one national park, twelve natural parks, nine natural reserves and three protected landscapes, among other local areas totalizing about 7% of the Portuguese territory - was created almost exclusively for biological and cultural reasons. As in many other countries, geology is absent from the majority of the portuguese protected areas statutes (Brilha 2002). Some efforts are being developed by Portuguese geologists in order to change this situation. On the other hand, the integration of relevant geological information in the park management plans is now desired by park managers. In this context, a project of inventory and characterization of the geological heritage is under development in two natural parks from north-eastern Portugal (Dias and Brilha 2002).

The Montesinho Natural Park (MNP), one of the areas under study, is limited by Spain in its northern and eastern borders. It was established in 1979 constituting one of the larger national protected areas with about 750 km^2. Its creation was due to the harmony between the resident population and the natural environment. About 9000 inhabitants are distributed among 92 villages where ancient and rich traditions were preserved due to the isolation. A rough relief is the main reason for this isolation. The whole area of the MNP is dominated by mountains, Montesinho on the eastern part and Coroa on the western part.
The importance of the MNP is multidisciplinary:

i) Particular geological setting (rich geodiversity with several exotic formations);
ii) Biodiversity;
iii) Archaeological richness (for instance, evidence of ancient roman gold mining);
iv) Cultural traditions.

The complex and exotic geology and its marks on the landscape are the main geological characteristics of this park. Landscapes and their conservation are very important in protected areas management (Phillips 1997) and in nature conservation as a whole (Gonggrijp 1999). Denoting the high importance of landscapes, the International Union for Conservation of Nature (IUCN) defines Category V (Protected Landscape/Seascape) as one of the six types of protected areas (http://iucn.org/themes/wcpa/pubs/pdfs/WCPAInAction.pdf).

In this paper, we will present the strategy used for the production of an interpretative panel to be located in a selected viewpoint of the MNP. The main difficulty is related with the association of two relevant factors: a complex geology and a general public with low geological familiarity. This hard task is strengthened because 45% of europeans feel that they are neither interested nor informed about science and technology and two thirds consider themselves uninformed, according to a recent survey conducted by the European Commission (http://europa.eu.int/comm/public_opinion/archives/eb/ebs_154_en.pdf). Nowadays, the Portuguese education system integrates geology courses during some of the nine compulsory years. Therefore, the general population is gradually gaining the basis to understand basic geological information, which is a key-factor for the increase of public awareness of geological heritage. This will also contribute for the empathy between local people and MNP management, a key feature to the success of any nature conservation policy (Elcome and Baines 1999).

THE GEOLOGY AND LANDSCAPE OF THE MONTESINHO NATURAL PARK

Autochthonous sequences of the Central Iberian Zone, as well as a pile of thrust units of the "Galiza - Trás-os-Montes" Zone, occur in the MNP area. The mafic and ultramafic rocks of the Bragança Massif, one of the four allochthonous massifs in northwest Iberia, is the main geological feature. The autochthonous sequence is mainly composed of Ordovician quartzites and slates. The nappes comprise from bottom to top (Ribeiro 1974; Iglésias et al. 1983; Ribeiro et al. 1990; Meireles 2000a,b):

1) Parautochthonous thrust complex, including metasediments of Silurian - Devonian age having lithological affinities with the subautochthonous;
2) Lower allochthonous thrust complex of the Lower to the Upper Palaeozoic;
3) A fragmented Palaeozoic ophiolite complex;
4) An upper continental allochthonous terrane of Precambrian to Lower Cambrian age.

These nappes are thrust over a Silurian - Devonian subautochthonous metasediment sequence. Variscan granitic rocks also occur in the MNP area.

Finally, Cenozoic tectono-sedimentary events left their imprints on the landscape, namely with the presence of three unconformity-bounded sedimentary sequences (Pereira 1997, 1999a,b; Meireles et al. 2002).

The late Variscan tectonic episodes caused the rupture of the Iberian Meseta plateau. The Aveleda and Baçal surfaces are inside a graben controlled by the NNE-SSW (Portelo fault) and the N-S (Aveleda fault) fault systems. The Portelo fault belongs to a major lineament of the late-Variscan tectonic episode in NW Iberia (Bragança - Vilariça fault system). This is a left-lateral strike slip fault with uplift of the western block. The Aveleda fault was reactivated as a dextral strike slip fault in late-Variscan episodes (Meireles et al. 2002).

In the MNP, several local surfaces are well defined and clearly identified from different sites, constituting remains of the Iberian Meseta plateau. In general, each surface is defined over distinct lithological units and is tectonically controlled by major faults. This complex geology imprints particular landscape features and plays an important role in the natural heritage of this park. The relation between landscape and geology is the key for a first contact of the general public with the MNP geology.

THE S. BARTOLOMEU PANEL

The use of interpretative panels in geology is common in many places. In the last years, some authors give clues in order to make these panels appropriate for general public (Wilson 1994; Hose 1998; Hose 2000). Interpretation strategies for protected areas must integrate multiple resources. Dias and Brilha (2002) presented different ways to deliver geological information to various target groups in protected areas of NE Portugal.

The S. Bartolomeu panel is mainly addressed to the public with a basic geological background and, in particular, to senior students and teachers. It will be set up on the viewpoint with the same name. This viewpoint was selected because:

i) Allows a complete perspective of the eastern area of the MNP;
ii) Is placed over mafic rocks from the lower continental crust, a good example of the exotic character of the regional geology;
iii) Allows good accessibility;
iv) Presents low vulnerability;
v) Is well known by the local population and mentioned in tourist information.

The S. Bartolomeu panel intends to display the main geological features of this park, based on the following geological features and concepts:

i) Great lithological variety (slates, quartzites, marbles, amphibolites, granulites, peridotites, gneisses, granites, conglomerates, sands, and muds);
ii) Endogenic processes as plate tectonics, magmatism and metamorphism, represented by the Paleozoic lithologies;
iii) Exotic character of various lithologies as a result of Paleozoic dynamic processes;
iv) Tectonics of plate interiors and related stratigraphic units;
v) Geological cycle and orogeny;
vi) Endogenic-exogenic interactions and landforms development;
vii) Geological time and landscape evolution.

This panel (130x100cm) includes different graphic resources and small text boxes (Figure 1). A panoramic photograph illustrates the main geomorphological features - uplifted blocks and surfaces limited by major faults. The caption over this photograph makes these features clear to the public. The simplified geological map represents the main lithologies and tectonics. Another simplified geomorphological map is used to help the identification of several surfaces and fault limits. With this graphical support, the viewer is challenged to recognise the landscape features observed from this viewpoint.

Two photographs illustrate relevant sites suggesting a detailed stop. The text in the panel is presented in three boxes with the following titles:

i) A rich rock diversity;
ii) Reading the landscape;
iii) How did geological processes influence the landscape?

The landscape evolution is presented using a graphical sketch that provides the understanding of the geological processes through the geological time.

Finally, a lateral strip includes complementary information, such as suggestions of trails and sites, indications for further readings and contacts.

Figure 1. General view of the S.Bartolomeu interpretative panel, Montesinho Natural Park (NE Portugal).

FINAL CONSIDERATIONS

The geology interpretation inside Portuguese protected areas is just starting. As recognized by several authors, protected areas are very important for environmental education. The lack of geological information in Portuguese parks is very negative in the effort of raising public awareness of geology. This paper deals with the production of an interpretative panel as an initiative integrated in a wider interpretative strategy for an area with high geological complexity - the Montesinho Natural Park (NE Portugal). This strategy will make available new tools to MNP managers that can be used to promote holistic conservation strategies in this protected area. Additionally, the school population will also benefit with the implementation of these interpretation products, converting the MNP into a true out-door classroom.

ACKNOWLEDGEMENTS

The PNAT/1999/CTE/15008 project is sponsored by FCT (Portuguese Science and Technology Foundation) and ICN (Portuguese Nature Conservation Institute).

REFERENCES

Brilha, J.B. 2002. Geoconservation and Protected Areas. *Environmental Conservation* 29(3), 273-276.

Dias, G. and Brilha, J. 2004. Raising public awareness of geological heritage: a set of initiatives. This volume.

Elcome, D. and Baines, J. 1999. *Steps to success - Working with residents and neighbours to develop and implement plans for protected areas*. IUCN, Commission on Education and Communication, European Committee for EE, Switzerland, 42p.

Gonggrijp, G. 1999. Geodiversity: the key to a holistic approach in renaturation. In *Towards the Balanced Management and Conservation of the Geological Heritage in the New Millenium*. Barettino, D., Vallejo, M., Gallego, E. (eds.), Sociedad Geológica de España, Madrid, 77-80.

Hose, T. 1998. Mountains of fire from the present to the past - or effectively communicating the wonder of geology to tourists. *Geologica Balcanica* 28, 3-4, 77-85.

Hose, T. 2000. European 'Geotourism' - Geological interpretation and geoconservation promotion for tourists. In *Geological Heritage: its Conservation and Management*. Barettino, D., Wimbledon, W.A.P., Gallego, E. (eds.), Instituto Tecnológico Geominero de España, Madrid, 127-146.

Iglésias, M. P. L., Ribeiro, M. L. and Ribeiro, A. 1983. La interpretation aloctonista de la estrutura del Noroeste Peninsular. In *Geologia de España*, Libro Jubilar J.M. Rios, Inst. Geol. Min. España, Madrid, 1, 459-467.

Meireles, C. 2000a. *Carta Geológica de Portugal à escala 1:50.000. Notícia explicativa da Folha 3D (Espinhosela)*. Inst. Geológico e Mineiro, Lisboa, 64 p.

Meireles, C. 2000b. *Carta Geológica de Portugal à escala 1:50.000. Notícia explicativa da Folha 4C (Deilão)*, 2ª edição. Inst. Geológico e Mineiro, Lisboa, 28 p.

Meireles, C., Pereira, D. I., Alves, M. I. C. and Pereira, P. 2002. Interesse patrimonial dos aspectos geológicos e geomorfológicos da região de Aveleda-Baçal (Parque Natural de Montesinho, NE Portugal). *Comun. Inst. Geol. e Mineiro*, Lisboa, 89, 225-238.

Pereira, D. I. 1997. *Sedimentologia e estratigrafia do Cenozóico de Trás-os-Montes Oriental (NE Portugal)*. Tese de Doutoramento. Universidade do Minho, Braga, 341 p.

Pereira, D. I. 1999a. Terciário de Trás-os-Montes oriental: evolução geomorfológica e sedimentar. *Comun. Inst. Geol. e Mineiro,* Lisboa, 86, 213-226.

Pereira, D. I. 1999b. O registo sedimentar em Trás-os-Montes Oriental nas proximidades do limite Neogénico/Quaternário. *Estudos do Quaternário*, APEQ, Lisboa, 2, 27-40.

Phillips, A. 1997. Landscape approaches to national parks and protected areas. In *National Parks and Protected Areas: Keystones to Conservation and Sustainable Development*. Nelson, J.G., Serafin, R. (eds.), NATO ASI Series, Ser. G, Ecological Sciences, Springer, Berlin. vol. 40, 31-42.

Ribeiro, A. 1974. Contribution à l'étude tectonique de Trás-os-Montes Oriental. *Mem. Serv. Geol. Port.*, 24 (Nova Série), Lisboa, 168 p.

Ribeiro, A., Pereira, E. and Dias, R. 1990. Structure of Centro-Iberian Allochthon. In Northern Portugal. In *Pre-Mesozoic Geology of Iberia*, Dallmeyer, R.D., Martinez, E. (eds.), Springer Verlag, Heidelberg, 220-236.

Wilson, C. (ed.) 1994. *Earth Heritage Conservation*. The Open University, Milton Keynes, 272p.

In: Parkes, M.A. (Ed.) 2004 *Natural and Cultural Landscapes - The Geological Foundation*, 257-259, Royal Irish Academy, Dublin

INTERPRETING A UNIQUE LANDSCAPE: THE ASSYNT REGION OF NORTHWEST SCOTLAND

Kathryn Goodenough, Elizabeth Pickett, Maarten Krabbendam and Tom Bradwell.
British Geological Survey, Murchison House, West Mains Road, Edinburgh EH9 3LA
Email: kmgo@bgs.ac.uk

Abstract: The British Geological Survey has recently commenced a new 10-year project with the aim of revising the geological maps of the Northwest Highlands of Scotland, focusing on the Moine Thrust Zone. An integral part of this project will be the development of new products to be used in education and public awareness of earth science. Four main audiences have been identified: schoolchildren; university students; visiting amateur and professional geologists and the wider public. The first material to be developed will focus on the Assynt district, a classic area of geology and among the most spectacular landscapes in the British Isles. Co-funding for this project has been secured from BP, and work is currently in progress on a variety of products. These include a website, being developed jointly with Leeds University, that will provide information on the geology of the Assynt area for all the audiences identified above; and an easy-to-read geological map and accompanying guide to geological walks in the Assynt area, targeted primarily at the general public. We are also collaborating with the National Museums of Scotland and the Edinburgh Geological Society to produce a geological excursion guide aimed at geologists and geology students. Further funding is currently being sought for the development of a larger-scale road map of the Northwest Highlands, which will introduce the geology and other interesting aspects of the landscape to visitors.

INTRODUCTION

The Assynt district has a wide variety of rock-types, found in close proximity, and as a result the link between the landscape and the underlying bedrock is particularly clear. The area has been of the highest interest to geologists for many years, since the Highland Controversy of the nineteenth century (Oldroyd, 1990), which was finally closed by the publication of the classic 'Northwest Highlands Memoir' (Peach et al. 1907) and the Assynt Special Sheet (Geological Survey of Great Britain 1965). Since then, many academics and geology students have worked in the area. Many research papers have been published, including the seminal study of Elliott and Johnson (1980). In the latter decades of the twentieth century the development of better road links to the area has made it a popular destination for tourists and hillwalkers, many of whom are interested in how the stunning landscape around them has developed. This has been recognised by Scottish Natural Heritage, who have developed the excellent visitor centre and walking trail on the Moine Thrust at Knockan Crag, and in collaboration with BGS have published '*Northwest Highlands: A Landscape Fashioned by Geology*' (Mendum et al. 2001).

In April 2001, the British Geological Survey commenced a new 10-year project in the Northwest Highlands with the aim of combining new research and survey work with the geological studies of the last century, in order to create a series of new maps and publications about the area. These products will include not only academic papers and traditional geological maps and memoirs, but also guide books, simplified maps and web material aimed at non-geologists. These latter products are discussed in more detail in this paper.

THE GEOLOGY OF THE ASSYNT DISTRICT

The geology of Assynt is controlled by the Moine Thrust, a major structure in the Earth's crust that runs through northwestern Scotland from Loch Eriboll on the north coast to the Isle of Skye. The Moine Thrust was formed some 430 million years ago, at the time when the continents bearing Scotland, England and Scandinavia moved inexorably together, bulldozing sediments from the sea floor up onto the continents to form a great mountain chain. Deep beneath the mountains, the sediments were squeezed and heated (the process of metamorphism), then forced westwards over older rocks. The major fracture plane along which this movement occurred is the Moine Thrust, and it is one of the most important geological structures in Scotland. Immediately beneath the Moine Thrust, along its entire length, are a number of other smaller-scale thrusts, collectively known as the Moine Thrust Zone. For much of its length, the outcrop of this zone is only a few hundred metres wide, but Assynt is unusual because here the surface outcrop of the Moine Thrust Zone broadens to many kilometres in width, forming a 'window' into the Thrust Zone.

The Assynt region can effectively be divided into three areas of differing geology - the foreland to the west of the Moine Thrust Zone; the Moine Thrust Zone itself; and the Moine schists to the east. Each area has very different landscape characteristics, instantly recognisable by visitors to the area.

The foreland is essentially the area west of the main road between Elphin and Kylesku, and it consists of three main rock types. The oldest rocks, the Lewisian gneisses, are almost 3000 million years old - some of the oldest rocks in Europe. They form the low, rocky hills and the rough coastline around Lochinver, providing many natural harbours for fishing boats. Around 1000 million years ago, rivers flowing across a barren landscape laid down a blanket of red-brown sand and pebbles, many kilometres thick, known as the Torridonian sandstone. Much of this has been eroded away, but remnants can still be seen forming grand isolated peaks, such as Stac Pollaidh, Suilven and Cul Mor, that rise in great layers of sandstone above the surrounding gneisses. Some of the peaks, including Canisp and Quinag, are capped by pale grey layers of Cambrian quartzite, laid down as pure quartz beach sand some 500 million years ago.

Immediately east of the Elphin - Kylesku road lies the 'Assynt window' - the broad outcrop of the Thrust Zone that makes this area unique. The geology of this area is extremely complicated, with the rock layers being highly disrupted. The most important landscape features here are the valleys of the Traligill, Allt nan Uamh and Loyne rivers, stretching east from the road, and the great hills of the Ben More Assynt range above. The valleys are chiefly underlain by limestone, the outcrop of which has been considerably widened by thrusting. This is one of the largest limestone areas in Scotland and the calcareous soil supports a wide variety of lime-loving flowering plants, including mountain avens and saxifrages. The ease of dissolution of the limestone means that these valleys are riddled with caves and underground rivers. Above the heads of the valleys is a great quartzite ridge, culminating in the Munros (peaks over 3000 feet high) of Conival and Ben More Assynt, which attract large numbers of hillwalkers. Beyond this ridge is a broad area of Lewisian gneiss caught up within the thrust zone, which forms wild, remote hills and valleys. To the east lies the Moine Thrust, above which appear the Moine schists that underlie much of the Northern Highlands of Scotland. Moine means bog in Gaelic, and there is no doubt that this is an appropriate name for this wild, peaty expanse.

The landscape of Assynt is strongly controlled by the underlying geology, but the shapes of the peaks and valleys have largely been formed by more recent processes - in particular the action of glaciers, which covered much of this area for a large part of the last two million years. Great sheets of ice streamed across the landscape, scouring and smoothing the rocks, gouging out deep valleys and leaving the peaks standing tall. Multiple episodes of glaciation have not only sculpted the rocks, but also left behind sediments that in places form a thick cover on the rocks below. Since the glaciers retreated, some 10,000 years ago, Assynt has gradually been colonised by plants, animals, and eventually people. Aspects of this later history have been just as important as the geology in shaping this beautiful district of Scotland.

INTERPRETING THE ASSYNT LANDSCAPE

Many visitors come to Assynt to explore the beautiful and varied landscape, which is underpinned by the bedrock geology, and has been sculpted by processes such as glaciation. Scottish Natural Heritage has begun the process of interpreting Assynt for the visitor, targeting the touring motorist with the visitor centre at Knockan, roadside 'satellite sites' and the attractive book 'Northwest Highlands: A Landscape Fashioned by Geology' by Mendum et al. (2001). The visitor centre at Knockan has drawn around 11,000 visitors in its first year (Jan Breckenridge, SNH pers. comm.), indicating widespread interest in the geology. However, there are many other groups of people visiting Assynt, most of whom will spend some time there rather than just passing through: these include hillwalkers; school and university field parties; and geologists (professional and amateur alike), often from abroad. Owing to the attraction of the unique mountains of this area, and the quality of the geological exposures, these groups represent a significant percentage of all visitors to the region; more than 10 university geology groups, each of around 40 people, visit Assynt each year (Chris Rix pers. comm.). A grant from BP has allowed BGS to begin developing a variety of products that will provide information for these groups. These include: a guide to some of the most popular walks in the Assynt region, to be sold with a simplified and clarified geological map; a website, being developed in collaboration with Leeds University, that will be used as a gateway to information about the Moine Thrust Zone; and a new version of the Geological Excursion Guide to Assynt, being developed in collaboration with the Edinburgh Geological Society and the National Museums of Scotland.

Walking guide and simplified geological map

The traditional geological maps produced by BGS are complex, and are geologists' attempts to create 2D representations of a 3D system. They include detailed information that is required by professional geologists and other users, and so are often unsuitable for use by non-specialists. However, simplification of these maps is possible, such that only the more important rock units are shown, and the geology is explained through use of photographs and diagrams - as superbly done by the Geological Survey of Northern Ireland in their Causeway Coast map (Geological Survey of Northern Ireland 1998). The simplified geological map of Assynt will be produced with this in mind although, unlike the Causeway Coast example, a traditional BGS 1:50000 map showing the detailed geology will be produced separately, for use by professional geologists.

The simplified map will be accompanied by a guide to some popular walks in the Assynt region, explaining the geology that can be seen along the walk but also highlighting many other features of interest along the way. The routes to be included in the guide vary from easy coastal walks at Achmelvich and Clachtoll, through visits to the limestone valleys of the Allt nan Uamh and Traligill, up to hill and mountain walks such as Stac Pollaidh, Suilven, Quinag and Conival. These are all walks that are already popular for a range of interests, including the landscape, the natural history and the human history, rather than geological excursions. We anticipate that the map and guide will add significant value to the experience of those who explore Assynt on foot, by increasing their awareness and understanding of the landscape around them. The guide, which is to be published in 2004, incorporates abundant photographs and colourful illustrations (including paintings by Elizabeth Pickett, a professional geologist and talented illustrator) as well as informative, easy-to-read text. It has been 'tested' by a variety of volunteers, including hillwalkers, Duke of Edinburgh Gold Award groups and local rangers.

Website

The Moine Thrust website, being developed in collaboration with Dr Rob Butler at Leeds University, will provide a wide variety of information about the Moine Thrust Zone - including the historical background to mapping in this area, an overview of the geology, 3D models of thrust processes and aerial 'fly-throughs' across the area. This website will provide a gateway through which university and school students from across the world, as well as other potential visitors to Scotland, can learn about the spectacular, varied geology and landscape of the Northwest Highlands.

Excursion Guide to Assynt

The Edinburgh Geological Society publish a Geological Excursion Guide to the Assynt District, which was last revised in 1979 (Johnson and Parsons 1979) but still sells a few hundred copies a year (EGS publications secretary, pers. comm.). The abundance of new research in this area has made it crucial that the excursion guide, now somewhat outdated, is revised, and the BGS team has become involved in the revision. This guide is the primary source of information for professional and amateur geologists, from both Britain and abroad, visiting the Assynt district.

FUTURE PLANS

Funding is currently being sought to extend similar types of interpretive products across the length of the Northwest Highlands. If the walking guide proves to be a success, then there are many other areas along the Moine Thrust Zone that could benefit from similar products, including Ullapool & Loch Broom; Torridon; and Lochalsh. A larger scale road map, with a guide to roadside stops along the Moine Thrust zone, is also planned, as are products for use by schools and ranger services.

CONCLUSIONS

The Assynt region has a spectacular and varied landscape that is underpinned by the bedrock geology, and has been shaped over geologically recent times by the action of glaciers and other natural forces. The soils, vegetation, and hence the human history, are all controlled to some extent by the geology. This area is popular with hillwalkers as well as school and university parties and others with an interest in the earth sciences, and these groups represent a wide audience to whom landscape interpretation can be targeted. BGS is currently developing a range of products in order to inform those who are already in Assynt about the landscape around them. We hope that this new material will encourage more people to visit Assynt to see the fantastic geological heritage of this beautiful area.

ACKNOWLEDGMENTS

Rob Butler, University of Leeds is thanked for advice and work on website development; Joyce Bain and Tom Bain of BGS, Edinburgh are thanked for design work and photographic services on the walking guide. This paper is published with the permission of the Executive Director, British Geological Survey, NERC.

REFERENCES

Elliott, D. and Johnson, M.R.W. 1980. Structural evolution in the northern part of the Moine thrust belt, NW Scotland. *Transactions of the Royal Society of Edinburgh: Earth Sciences* 71, 69-96.

Geological Survey of Great Britain 1965. *Assynt Special Sheet, Solid and Drift Edition.* 1:63360. Ordnance Survey, Chessington.

Geological Survey of Northern Ireland 1998. *Causeway Coast, Sheet 7 Solid.* 1:50,000. British Geological Survey, Keyworth.

Johnson, M.R.W. and Parsons, I. 1979. *Macgregor & Phemister's Geological Excursion Guide to the Assynt District of Sutherland.* Edinburgh Geological Society, Edinburgh.

Mendum, J., Merritt, J., and McKirdy, A. 2001. *Northwest Highlands: A Landscape Fashioned by Geology.* Scottish Natural Heritage, Perth.

Oldroyd, D.R. 1990. *The Highlands Controversy.* University of Chicago Press, Chicago.

Peach, B. N., Horne, J., Gunn, W., Clough, C.T., Hinxman, L.W. and Teall, J.J.H. 1907. *The geological structure of the North-West Highlands of Scotland.* Memoir of the Geological Survey of the United Kingdom.

In: Parkes, M.A. (Ed.) 2004 *Natural and Cultural Landscapes - The Geological Foundation*, 260, Royal Irish Academy, Dublin

LANDSCAPE, COMMUNITY AND UNDERSTANDING

Louise Thornhill, Regional Director, Countryside Agency, UK

Abstract: As part of the LANDMAP process in south Wales the Countryside Council for Wales (CCW) and its partners undertook a series of public perception studies. These studies involved a mixture of focus group work and individual surveys. All the studies provided similar results demonstrating that the 'general public' have no connection with the 'Landscape'. There is little comprehension of who is responsible for the management of the landscape, the role of designations and the contribution of the farming community to the management of the landscape. However the studies did indicate an interest in the some of the individual components of the landscape, a local pride in their own locality and a latent interest in the landscape that could be encouraged. Interestingly there was also clear differences demonstrated in those studies that investigated the differences between across the genders and ages

If we are going to be successful in achieving the engaging all sectors of the community in the long term conservation of the landscape we need to demystify the 'landscape' we need to break down the barriers to its understanding the landscape as entity and raise the understanding of the factors that contribute to its management.

In: Parkes, M.A. (Ed.) 2004 *Natural and Cultural Landscapes - The Geological Foundation*, 261-264, Royal Irish Academy, Dublin

EXPLORING GEOLOGY AND LANDSCAPE: POPULAR PUBLICATION INITIATIVES BY THE INTEGRATED GEOSCIENCE SURVEYS (N BRITAIN) PROGRAMME OF THE BRITISH GEOLOGICAL SURVEY, EDINBURGH, UK

Elizabeth A. Pickett, British Geological Survey, Murchison House, West Mains Road, Edinburgh EH9 3LA, UK. Email: eapi@bgs.ac.uk

Abstract: The Integrated Geoscience Surveys (Northern Britain) programme of the British Geological Survey is establishing a range of popular publication initiatives. Our aim is to raise public awareness and appreciation of our geological and landscape heritage by producing attractive, imaginative publications, websites and displays that will inform and interest visitors and local people in the foundations of the landscapes of northern Britain. These products will complement existing publications such as those produced by Scottish Natural Heritage in collaboration with BGS, and the recent BGS publication, 'Isle of Man: foundations of a landscape'. The latter was the result of a successful collaboration between BGS and a range of contributors, including Manx National Heritage and university researchers, and was sponsored by a consortium of oil companies. Current initiatives include a BP-co-funded project to produce a website, a simplified geological map and an illustrated book describing walks of geological interest in the spectacular Assynt area of NW Scotland. We are also exploring the possibility of producing similar publications for the Cairngorms, Glencoe, Scotland's National Parks and parts of northern England. Popular tourist areas such as the Lake District, Northumberland and the Northern Pennines, with their beautiful natural landscapes and fascinating histories of human impact, represent ideal subjects for future partnerships and publications.

1. INTRODUCTION

From the rugged mountains and glens of the Highlands to the rolling moorlands of the Northern Pennines, northern Britain contains a wealth of varied and fascinating landscapes - regions whose distinctive characters are deeply rooted in the underlying rocks and the processes that have shaped, and continue to shape, the land. Over millions of years, this geological foundation has created the unique and changing landscapes of northern Britain, and, in more recent times, has played a fundamental role in their natural, cultural and industrial heritage. This foundation is evident all around us - in the shape of hills, valleys and coastlines, in soils, rivers and natural habitats, in the building stones of towns and villages, in the mines and quarries that exploit geological resources, and in archaeology, history and culture.

If, as a society, we are to look after and manage our landscapes in a sustainable way, we need to understand the underpinning role of geology and to appreciate our earth science heritage. Raising awareness of earth science and promoting the understanding of its relevance to resource/environmental issues and our wider natural and cultural heritage, are vital aims and together form an important part of the British Geological Survey's mission. In doing this we raise awareness of the geological survey itself, as well as other geological institutions. By informing and stimulating the public we develop geology as part of the broader tourist infrastructure of an area, thus attracting visitors and supporting the local economy (e.g. McKeever and Gallagher 2001; McKirdy et al. 2001). Organisations such as the National Trust for Scotland and English Heritage are increasingly recognising that landscape interpretation is of interest to visitors and should play an important role in displays and visitor centres.

The challenge for earth scientists is to communicate the subject effectively and imaginatively to the public. The time has never been better; tourism and recreation is big business, and public interest in earth science has grown in recent years through visually stimulating television series and the success of attractions such as Edinburgh's *Our Dynamic Earth* and the visitor centre at Knockan Crag in the Northwest Highlands.

2. RAISING AWARENESS OF EARTH SCIENCE THROUGH POPULAR PUBLICATIONS

Popular publications and the integration of geological information and interpretation into visitor centres, websites, interpretation panels and leaflets, all have great potential for promoting an understanding of geology and its role in the natural and cultural heritage of our landscapes and scenery.

Several organisations have been leading the way in the provision of geological material to the interested and enquiring local or tourist in Scotland and Ireland. Scottish Natural Heritage (SNH), in collaboration with geologists from the BGS in Edinburgh, has produced a series of booklets, the *Landscape Fashioned by Geology* series (Scottish Natural Heritage, 1993-2001). Primarily based on stunning photographs, these booklets explore the landscape and geology of a range of areas of Scotland. The BGS Earthwise™ series includes illustrated books such as *Whisky on the Rocks* (Cribb and Cribb 1998), which explains how Scotland's many distinctive whiskies are influenced by the underlying bedrock. In Ireland, the Geological Survey of Northern Ireland and the Geological Survey of Ireland have collaborated in a highly successful initiative entitled

Landscapes from Stone, which includes a book called *A Story through Time* (McKeever 1999), as well as a map, touring cards and walking cards. Following extensive marketing and discussions with the tourist industry and community groups, McKeever and Gallagher (2001) have emphasised the importance of the 'holistic' approach, in which geology is integrated into the existing tourism infrastructure and is also linked with many other aspects of the natural and cultural environment.

2.1 THE IGSN APPROACH: MAKING LINKS AND ILLUSTRATING PROCESSES

The Integrated Geoscience Surveys (Northern Britain) programme (IGSN) of the British Geological Survey (BGS) is based in Edinburgh and provides up-to-date strategic information on the geology of northern Britain through a range of multi-disciplinary projects. The area covered is Scotland, Cumbria and Northeast England, and, for a recent project, included the Isle of Man. The IGSN programme, with its large resource of information, geological expertise and design capability, is becoming increasingly involved in 'popular geology', and is ideally placed to contribute and collaborate in popular publication partnerships and educational ventures. Products for the non-specialist are now an integral part of many projects.

We recognise the importance of linking geology with other aspects of the landscape and advocate a holistic approach in which geology is integrated with flora, fauna, human history and the built heritage (see also Goodenough et al. this volume and MacFadyen and McMillan this volume). We also aim to portray the continually changing nature of the Earth by telling the story of the processes and past environments involved in the formation of rocks and the ways in which the landscape has been shaped and continues to change through a combination of natural processes and human activity.

An important aspect of this story telling is the use of clear, attractive, but accurate illustrations. It is difficult for a non-geologist to visualise how a rock or landform was created from only text or a photograph of how the outcrop looks today. One of the best ways of explaining that the rocks someone can see in a hillside crag or in the wall of a building were erupted from an ancient volcano or created from the shelly ooze on a tropical sea floor, is by painting a picture. Pictures explain, describe and stimulate and can be useful for emphasising geological features in the landscape.

3. A RECENT BGS PUBLICATION: ISLE OF MAN: FOUNDATIONS OF A LANDSCAPE

This integrated approach was adopted in a recent BGS publication entitled *Isle of Man: foundations of a landscape* (Pickett 2001), the result of a successful collaboration between BGS and a range of contributors, including Manx National Heritage and university researchers. The publication was sponsored by a consortium of oil companies (TotalFinaElf Exploration UK PLC in partnership with Enterprise Oil plc and Amerada Hess Limited) and formed part of a major programme of commissioned geological products. This 48-page paperback tells the story of the Island's geology and landscape, from the formation of the oldest rocks to the last Ice Age and subsequent human impact. It is illustrated by photographs and artwork incorporated into a colourful themed design. Themes running through the book include two areas of particular interest to many people: natural habitats and wildlife, and the evidence of human occupation and activity. Both these subjects are underpinned and influenced by geology and landscape, as described below. The everchanging landscape and the processes involved are also explored in the book.

3.1 Archaeology, building stones and mining heritage

The Isle of Man is justly famous for its rich heritage of prehistoric chambered cairns and hill forts, Viking settlements and carved Norse and Celtic crosses. In terms of their locations and the materials from which they were constructed, these all relate directly to the shape of the land and the rocks that lie beneath. The Island's rocky coastlines and hilly spine, formed from deformed Palaeozoic sedimentary rocks, provided excellent defensive sites in the Island's turbulent history. Beautiful crosses were carved from these Palaeozoic rocks by Celtic and, later, Norse sculptors. In the 13th to 15th centuries, red Devonian sandstones from the west coast and grey Carboniferous limestone from the south were quarried to build the imposing castles at Peel and Castletown. Manx granite was used for tools and implements in ancient times and later as a durable building stone. Igneous rocks continue to be extracted today for aggregate, building stone and coastal armour stone. The Quaternary deposits of the Isle of Man have provided flint for prehistoric toolmakers, sea-rounded boulders for building, peat for fuel and, today, sand and gravel. Minerals that formed millions of years ago deep underground in the Island's rocks were mined for thousands of years to produce valuable metals such as zinc, lead and copper. Mining reached its heyday in the second half of the 19th century and, when it declined, left a heritage of ruined mine buildings, old workings and the great wheel at Laxey, which has been restored as a tourist site.

3.2 Natural habitats and the underlying rocks, deposits and landforms

The Isle of Man has a range of rich and varied natural habitats, all of which relate directly to geological processes - past and present - and the underlying rocks, landforms and deposits. The Palaeozoic sedimentary rocks that form the Island's hilly backbone are blanketed in heather moorland, a landscape that may be partly related to prehistoric burning and clearing of the wildwood. Boggy wetlands at the Curraghs on the northern plain are the last vestige of an Ice Age proglacial lake and are home to many birds, insects and plants. Skeletons of the long extinct 'Irish Elk', a huge deer with antlers spanning 3.5 m, have been found in peat and mud-filled depressions left by melting ice some 10 000 years ago. The shingle heathland and dunes on the northern coast have formed as gravel raised beach ridges over the past 7000 years and are an important habitat for many species of insects, plants and sea birds.

3.3 The continually evolving landscape

The Isle of Man as it is known today only came into being about 10 000 years ago with the final disappearance of its land links with Britain. For thousands of years the Island's landscape was shaped only by the action of the sea, rivers and weather. Then, about 8000 years ago, the first settlers arrived and so began a new stage in the evolution of the landscape The landscape continues to change through a combination of natural processes and human activity. Coastal erosion, a major topic of current debate on the Isle of Man, is changing the Island's northern coastline of glacial deposits on the scale of a human lifetime. As the sea erodes in one place, so it adds to the beaches elsewhere; for example, the Ayres on the northern tip of the Island are accumulating material. The Island's rocky coastlines are also changing. At The Chasms, great blocks of Palaeozoic sandstone are breaking along joint planes and slipping into the sea.

3.4 Summary

Aspects of the human and natural heritage of the Isle of Man integrate naturally with the geology, creating a fascinating story that emphasises the fundamental role geology plays in the shaping of the landscape and the ways in which people through the ages have influenced and been influenced by the rocks and landforms around them. The book also highlights the constantly changing nature of the landscape and the fact that the landscape continues to evolve.

4. CURRENT AND FUTURE INITIATIVES

4.1 Glencoe

Geologists from BGS, Edinburgh were recently involved in providing geological information and interpretation for incorporation in the new National Trust for Scotland visitor centre in Glencoe. Recent research in the area by academics is currently being collated by BGS for a new 1:25 000 scale geological map of Glencoe, and for a report aimed at the expertise level of first year undergraduate geology students. We are keen to explore the possibility of producing a simplified popular publication, based on these up-to-date but relatively technical products. This would be aimed at the interested visitor to Glencoe and would focus on its fascinating geological history - from the huge volcanic eruptions over 400 million years ago that created much of the glen's bedrock to the glaciations that shaped the landscape we see today.

4.2 Assynt

As part of a BP-co-funded project to resurvey the geology of the Moine Thrust Zone of the Northwest Highlands of Scotland, BGS is working on a guide to the landscape and geology of popular walks in the spectacular Assynt area, and also an associated website (Goodenough et al. this volume). Using photos and painted illustrations to highlight geological features and to explain processes and past environments, and by linking the geology with the plants, wildlife, archaeology and history of the region, our aim is to produce an attractive and informative guide with an accompanying easy-to-read geological map.

4.3 Northern England

In collaboration with Durham and Northumberland county councils, BGS has been involved in writing and illustrating a book on the lead mining landscapes of the North Pennines Area of Outstanding Natural Beauty (Forbes et al. 2003). This book coincided with the area's designation as a 'European Geopark'. Several other initiatives, for areas such as Hadrian's Wall and the Lake District, are under discussion and development.

4.4 National Parks in Scotland

Scotland's first National Park, Loch Lomond and the Trossachs National Park, was opened in July 2002 and the Cairngorms was designated as a National Park in 2003. BGS is involved in discussions with park authorities and Scottish National Heritage about the ways in which we can contribute to National Parks in Scotland; their aims include conservation and enhancement of the natural and cultural heritage, the promotion of sustainable use of natural resources, and the promotion of understanding and enjoyment of the area.

EARTHSCAPES

5. EARTHSCAPES: OUR NEW BRAND

The Integrated Geoscience Surveys (Northern Britain) programme of the British Geological Survey has created a new brand, Earthscapes, that will allow products to become recognisable as BGS 'popular products' (see below). Featuring a semi-abstract view, the aim of the combined name and logo is to have an inclusive image which is appropriate to any popular publication focused on northern Britain, whether it be about lead mining landscapes in the Pennines or geological walks in Assynt.

6. CONCLUSIONS

The landscapes of northern Britain are both spectacular and varied. They reflect the incredible range of underlying rocks and structures, and the natural processes, such as glaciation, that have shaped the bare bones of the present-day landscape. Soils, natural habitats and human activity, are all underpinned by geology. Public awareness and appreciation of our landscapes is critical for future sustainable management of the land and also promotes geology and the landscape as an important part of our wider natural and cultural heritage. Popular publications are an important way of raising awareness and informing people in a stimulating and accessible way. The Integrated Geoscience Surveys (Northern Britain) programme of BGS is becoming increasingly involved in this area and, in collaboration with a range of organisations and individuals, is looking forward to a wide range of future partnerships and new products.

ACKNOWLEDGEMENTS

E.A. Pickett thanks M. Smith and A.A. McMillan for reviewing the text and publishes with the permission of the Director, British Geological Survey (NERC).

REFERENCES

Cribb, S. and Cribb, J. 1998. *Whisky on the Rocks*. Keyworth: British Geological Survey.

Forbes, I., Young, B., Crossley, C. and Hehir, L. 2003. Lead Mining Landscapes of the North Pennines Area of Outstanding Natural Beauty. Durham County Council.

Goodenough, K., Pickett, E.A., Krabbendam, M. and Bradwell, T. 2004. Interpreting a unique landscape: the Assynt region of northwest Scotland. In Parkes, M.A. (ed.) 2004. *Natural and Cultural Landscapes - the geological foundation. Proceedings of a Conference 9-11 September 2002, Dublin Castle, Ireland.* Royal Irish Academy. Dublin, pp. 257-259.

MacFadyen, C. and McMillan, A.A. 2004. The links between our natural and built heritage: how we can use urban landscape and geological resources to aid understanding and appreciation of the wider landscape. In Parkes, M.A. (ed.) 2004. *Natural and Cultural Landscapes - the geological foundation. Proceedings of a Conference 9-11 September 2002, Dublin Castle, Ireland.* Royal Irish Academy. Dublin, pp. 245-248.

McKeever, P.J. 1999. A Story Through Time. Geological Survey of Northern Ireland - Geological Survey of Ireland, Belfast and Dublin.

McKeever, P.J. and Gallagher, E. 2001. Landscapes from Stone. In Gordon, J.E and Leys, K.E. (eds) *Earth Science and the Natural Heritage*. Scottish Natural Heritage.

McKirdy, A.P., Threadgould, R. and Finlay, J. 2001. Geotourism: an emerging rural development opportunity. In Gordon, J.E and Leys, K.E. (eds) *Earth Science and the Natural Heritage*. Scottish Natural Heritage.

Pickett, E.A. 2001. *Isle of Man: foundations of a landscape*. Edinburgh: British Geological Survey.

Scottish Natural Heritage. 1993-2001. *Landscape Fashioned by Geology* series. Scottish Natural Heritage, Perth.

In: Parkes, M.A. (Ed.) 2004 *Natural and Cultural Landscapes - The Geological Foundation*, 265-266, Royal Irish Academy, Dublin

URBAN GEOLOGY AND EARTH SCIENCE EDUCATION: FACETS OF THE NATURAL BUILDING STONE ELEMENTS IN GALWAY CITY.

Martin Feely, Jon Hunt and Pat O'Connor,
Department of Earth and Ocean Sciences, NUI, Galway.
Matthew Parkes, Geological Survey of Ireland, Haddington Rd., Dublin 4.

Abstract: The natural stone elements of our cities and towns provide a remarkable and tangible resource for geological and architectural heritage education at all levels. Fieldtrips in the urban landscape increase awareness of our architectural heritage in stone. The example of Galway City in the west of Ireland is illustrated.

INTRODUCTION

The natural stone elements of our cities and towns provide a remarkable and tangible resource for geological and architectural heritage education at all levels. Fieldtrips in the urban landscape increase awareness of our architectural heritage in stone. The buildings, sculptures, street furniture and walkways provide us with a fascinating range of local and imported natural stone. It is like visiting an outdoor museum and all the specimens are touchable. The history of local stone (its geological formation and subsequent use by man) in time and space in the urban landscape can also be addressed.

THE NATURAL BUILDING STONE ELEMENTS IN GALWAY CITY

The natural stone elements in Galway highlight links between local bedrock geology and the building stones used in the city. They attest to a long history of marble, limestone and granite quarrying in the area. Imported building stone from China, Finland, Portugal, Italy, South Africa, Norway, Brazil and England (Figure1) together with stone from other parts of Ireland, add exotic elements to the mosaic of Galway's urban landscape. A two-hour walk in Galway's inner city encompasses a rich architectural heritage in stone blended with a global geological dimension. The natural stone fabric of Galway's inner city provides an educational resource which can be used to heighten awareness and respect not only for our heritage in stone but also our environment. On close inspection the stones display textures, structures and fossils which give us clues as to their origin and in turn tell us much about Earth history. The building stones can also be placed into a framework that correlates each building stone type with its geological age (Figure 2). In very few other places is it possible within two hours to inspect such a variety of rock types ranging in age from the Precambrian to the Quaternary except in the urban landscape or in a geological museum.

LINKS BETWEEN GALWAY'S BEDROCK MOSAIC AND ITS BUILDINGS

Galway city stands on a geological mosaic of metamorphic, igneous and sedimentary rocks. The distribution of these rock types reflects a fundamental geological divide between the eastern and western regions of County Galway. To the west the metamorphic and igneous rocks make up the bedrock of Connemara while the sedimentary rocks extend eastwards forming the limestone bedrock of the Irish Midlands. For example, the limestone provides the Galway region with a huge natural resource of building stone. Local quarries provided cut limestone blocks that have been used extensively over the centuries in the building of Galway city. In the past, limestone blocks were exported to England. In the Angliham quarry a particular variety of limestone was called the "London Bed" and was reserved and shipped exclusively for the London market. A succession of presti-

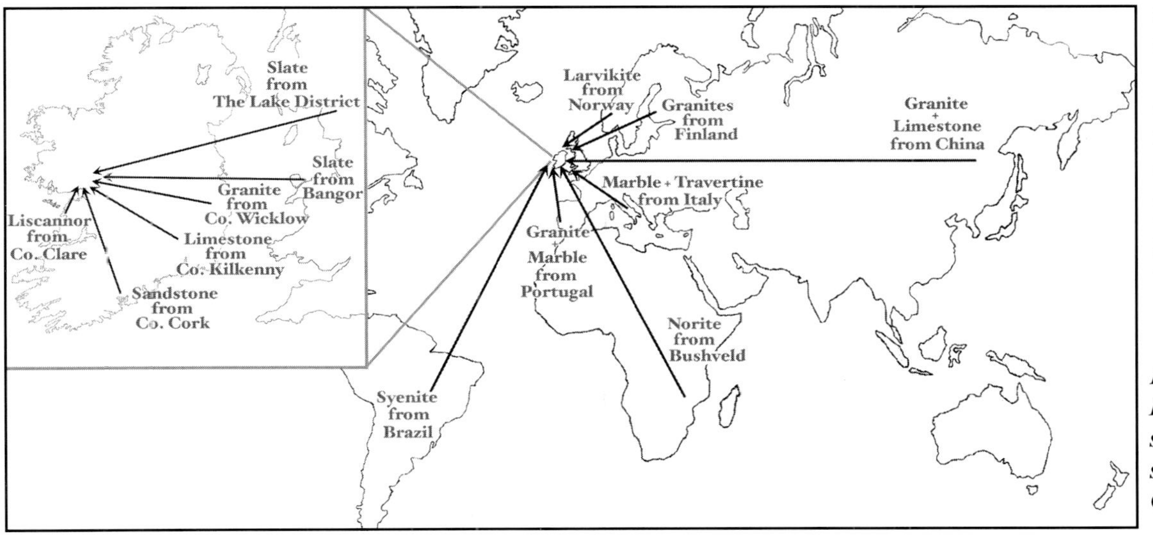

Figure 1. Provenance of some building stones used in Galway City.

Era	Geological Period		Igneous and Metamorphic	Sedimentary
Cenozoic	Quaternary	1		Sands and Gravels (Galway) Travertine (Italy) [28]
	Tertiary	65		
Mesozoic	Cretaceous	144		
	Jurassic	200	Carrara Marble [4]	
	Triassic	250		
Palaeozoic	Permian	300	Larvikite [33] Portuguese Granite [1]	
	Carboniferous	355		Limestone (Galway) [26] Limestone (Kilkenny) [15]
	Devonian	418	Galway Granite [14] Scandinavian Granites [48]	Cork Red Sandstone [4]
	Silurian	440	Burlington Slate [35]	
	Ordovician	490	Amphibolite/Gneiss(Galway)[2] Connemara Marble [4]*	
	Cambrian	544		
Archean	Precambrian Formation of the Earth ~4600		Granite (Rapakivi, Finland)[36] Gabbro/Norite (Bushveld)[35] Oldest Known Rocks on Earth are ~4200 million years old (Acasta Gneisses, Canada)	

Figure 2. Age framework of Galway's building stones

* Connemara Marble is a metamorphosed impure limestone which formed during the Precambrian Period

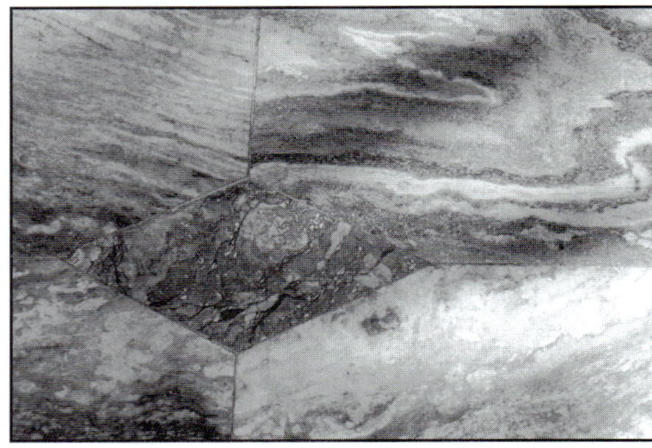

Figure 3. a. Galway Cathedral fossiliferous limestones. b. Connemara Marble tiles used internally.

gious limestone buildings were constructed over the centuries in Galway and include St Nicholas' Church (14th century), Lynch's Castle (late 15th-early 16th century), Spanish Arch (16th century) and the following 19th century buildings: The Quadrangle (NUI, Galway), Bank of Ireland, Great Southern Hotel and The Railway Station (Eyre Square). The Renaissance style Galway Cathedral, built in 1965 is the last great building to be constructed of locally quarried cut limestone blocks while its floor is tiled with Connemara marble (Figure 3a,b). Pink Galway granite and grey limestone blocks were used to construct the Dominican Church in the Claddagh (Figure 4a) while the interior polished stone pillars show off the beauty of the granite's texture especially the salmon pink crystals of feldspar-see Figure 4b.

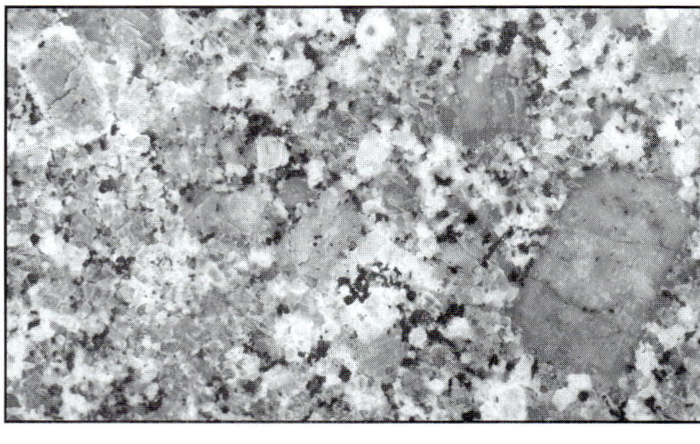

Figure 4. a. The Dominican Church in Claddagh and b. (right) the polished Galway Granite used in internal pillars.

CONCLUDING REMARKS

The urban landscape contains our architectural and geological heritage and provides insights into the processes that created such a rich tapestry of rock types which in turn provides links to the pages of Earth history. Accordingly, a two-hour urban geology walk in Galway's inner city has a global geological dimension and is an educational activity not only for secondary and third level student parties but also for geologists and visitors alike (Feely et al. 2002).

REFERENCES

Feely, M., Hunt, J., Lidwill, G. & Barton, K. 2002. *Galway in Stone-a geological walk in the heart of Galway*. Geoscapes, Dublin. pp48.

In: Parkes, M.A. (Ed.) 2004 *Natural and Cultural Landscapes - The Geological Foundation*, 267-270, Royal Irish Academy, Dublin

THE GRANITE IN THE CITY: A PROPOSAL FOR A WALK THROUGH THE GRANITE OF OPORTO (NW PORTUGAL)

Angela Almeida, Departamento e Centro de Geologia da Universidade do Porto, Praça Gomes Teixeira, 4099-002 Porto, Portugal. Email: aalmeida@fc.up.pt
Arlindo Begonha, Faculdade de Engenharia da Universidade do Porto, Rua Roberto Frias s/n, 4200-465 Porto, Portugal. Email: abegonha@fe.up.pt
Natercia Vieira, Departamento de Geologia da Universidade do Porto, Praça Gomes Teixeira, 4099-002 Porto, Portugal. Email: Natercia.Vieira@netcabo.pt

Abstract: A stroll within the historical centre of the city of Oporto, Heritage of the Humanity, is proposed in order to help people recognise the stone their city, their street and their home are made of, as well as to establish the relationship between architectural and environmental features and the underlying rock. Porto is an old, austere, magnificent city, rising above a granite substratum. Its monumental heritage goes back to Roman times and the profile has been developed by the construction of granite buildings, following history and the social evolution, of an impressive grey architecture according to different styles of granite work that characterize the city in all its aspects such as the old city wall, the medieval and baroque churches, the neoclassic houses and the small humble habitations. The granite can be observed in local outcrops side by side with the urban constructions. It is always affected by weathering processes and the buildings exhibit various aspects of stone decay such as granular disintegration, plates, flakes, black crusts, thin black layers, efflorescences and biological colonization. The aim of this action consists of the description of selected sites within the historical centre of the city where it is possible to recognize the extreme importance of the granite in the character of the city, to call the attention to the deterioration processes that affect houses and monuments and to promote the discussion about the means of conservation of the granite buildings.

INTRODUCTION

The city of Oporto is situated on the right bank of the river Douro, with an E-W elongated shape, occupying an area of about 41 km^2. It rises over a set of granite hills of flattened tops, evidence of a wide erosion platform that spreads towards North descending like a gentle staircase towards the Atlantic Ocean. Southwards, the deep valley of the River Douro abruptly cuts the natural extension of the platform and gives an imposing view of the granite escarpments.

It is an old city, in the origin of the name of Portugal. The oldest existing granite monuments are allegedly of Suevi age. The proposed tour includes six representative stops that put into evidence different aspects of the relation between the granite and the city. Among those aspects attention will be particularly drawn to the granite outcrops within the city, to the topographic irregularities of the city and the urban development, to the instability of the escarpments due to the fracturing of granite, to the great diversity of types of stone deterioration in houses and monuments as a consequence of environmental processes and pollution agents, to the evolution of the use of granite since old ages until present, well patented in the City Wall and medieval churches, with a special emphasis on the flourishing Baroque monuments among which the tower of the Clerigos is the most impressive example of granite carving in Portugal.

TECTONIC AND GEOLOGICAL SETTING

Taking into account that this work deals with the relation between the granite and the city, being the granite present both as small outcrops, still observed locally, and transformed in the imposing granite architecture, a brief description of the main geological characteristics of the city is presented.

Oporto lies essentially in the Central Iberian Zone, close to

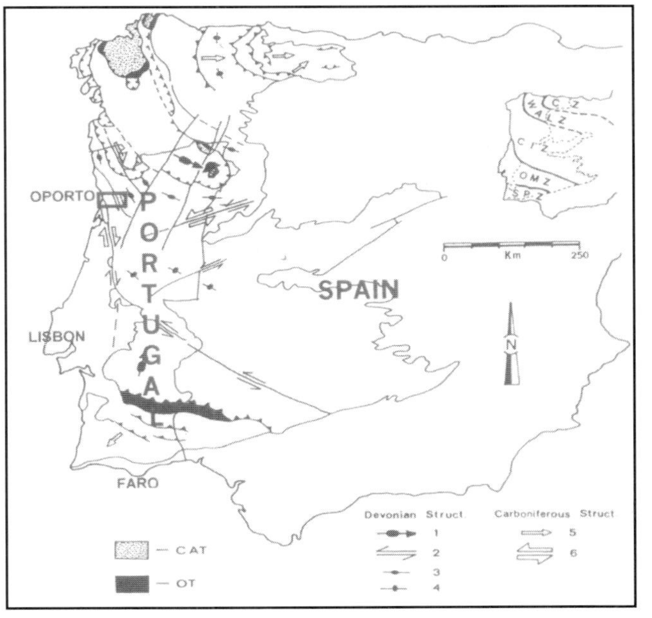

Figure 1. Major Variscan structures in the Iberian Peninsula: CZ-Cantabrian Zone, WALZ- West Asturian-Leonese Zone, CIZ- Central Iberian Zone, OMZ- Ossa Morena Zone, SPZ- South Portuguese Zone, CAT - Continental Alloctonous Terrane, OT- Northern and Southern Ophiolitic Terranes, 1- Devonian shear sense, 2- Devonian shear zone, 3- Stretching in b, 4- Stretching in a, 5- Carboniferous shear sense, 6- Carboniferous shear zone.

the limit with the Ossa Morena Zone (Julivert et al. 1974; Ribeiro et al. 1979) (Figure 1). According to the geological map under the scope of the Geotechnical Map of Oporto (1994) three main geological formations can be observed: surface sediments, metamorphic formations and igneous rocks. The most expressive group is undoubtedly composed by the igneous rocks, particularly the granites of Hercynian (Variscan) age.

Using the classification of Ferreira et al. (1987) two groups of granites can be distinguished in the area of the city: biotite granites with calcic plagioclase and two-mica granites. Based on the time relation between these granites and the third tectonic deformation phase of the Hercynian orogeny, D3, it is possible to distinguish two series within the biotite granite group: an earlier series composed of early- to syn-tectonic deformed granites and a late- to post-tectonic series. The two-mica granites are classified as syn-tectonic relatively to D3. These latter granites are particularly important since they are the most extensive granite group underlying the city and they constitute the essential material the historical buildings are made of.

PETROGRAPHY, GEOCHEMISTRY AND GEOCHRONOLOGY

The two-mica granite the historical centre of the city of Oporto is dominantly settled on is classified as a medium to coarse-grained leucogranite, syn-tectonic in relation with the third Hercynian deformation phase, D3. It is currently known as the Granite of Oporto. Petrographical studies show that it consists of a dominantly non-porphyritic granite, with hypidiomorphic granular texture and a mineral association including quartz, orthoclase, microcline, muscovite, biotite and apatite. Scarce opaque minerals, tourmaline and garnet are occasionally observed. This granite occurs always affected, to various extents, by late- to post-magmatic alteration processes. The potassic feldspar is intensely kaolinized by a process that was responsible for an important kaolin exploitation years ago, being no longer possible to have access to the quarry, presently submerged by the urban development.

Figure 2. Faculty of Sciences of Oporto University (South façade).

Selected geochemistry parameters obtained in bulk rock analyses indicate high values of SiO2 and total alkalis, respectively in the range of 71-74% and around 8%, low ferromagnesian contents and a strong peraluminous character (A/KCN molar ratio ranging between 1.3 and 1.5) probably related with the presence of primary muscovite (Almeida, 2001).

A U-Pb geochronological study on zircon and monazite yielded the minimum emplacement age of 318 Ma (Almeida 2001), being in good agreement with the geological and tectonic setting and with the petrographic and the geochemical data.

ITINERARY

The proposed walk aims to call the attention of the citizen not only to some of the most representative examples of granite architecture in the historical centre of the city, but also to the granite outcrops, pavements, escarpments and to aspects of conservation and cleaning methods, so often controversial.

Figure 3. Hospital of Saint António (main frontal).

The walk starts from the building of the Faculty of Sciences of Oporto University (Figure 2), where a very succinct geological description of the city, accompanied by an illustrated brochure pointing out the objectives of the visit, the selected itinerary and the main aspects to be observed, is presented to the participants. The Faculty has recently been the object of a detailed study on various deterioration processes, both in the exterior and in the interior, caused by pollution and other environmental agents (Oliveira 2001).

Then the visit addresses the Hospital of Saint António, an imposing granite construction of the XIX century (Figure 3). The huge project, of neopalladian style, was settled on a local river valley, being its foundations subjected to many vicissitudes. It is said to be the Portuguese building exhibiting the largest granite façade. The stone, big blocks of two-mica granite, was collected in the nearby quarries. Begonha (2001) presents an exhaustive study of this building whose style has influenced other architectural projects in the city. An example of stone decay and a tentative attempt of conservation is shown in Figure 4.

As the descent towards the historical centre of the city is

Figure 4. Flakes detaching from the stones in the Hospital of Saint António. Note also a bad use of cement to heal granite block joints and fractures.

Figure 5. The Tower of Clérigos.

initiated the view of the Tower of Clérigos dominates the urban landscape (Figure 5). The tower of the church of Clérigos is one of the most interesting Iberian baroque granite monuments of the XVIII century and is considered the ex-libris of Oporto. It has seventy five metres of height and is directly implanted on the granite. The astonishing stonework was possible due the softer nature of the two-mica granite of Oporto. Like other granite monuments in the city the tower exhibits several aspects of stone decay and it has been submitted to periodical cleaning.

One of outcrops of the Oporto granite has been cut by the opening of an avenue leading to the Cathedral. A stop has been selected in this area so that the participants may have the opportunity to compare the bedrock with the nature of the buildings and the monuments (Figure 6).

The Cathedral with its surroundings is a paradigmatic site of the use of the granite of Oporto under various aspects: the original medieval construction, the overlapping of styles and the present reconstruction of parts of the balustrade with the same type of granite (Figure 7).

This tour ends with one more example of a monument built with the two-mica granite of Oporto, the Church of Saint Francisco. This important church exhibits almost every type of stone decay that can be found scattered all over the city monu-

Figure 6. Two-mica granite outcrop within the historical centre of the city.

Figure 7. New granite balustrade in the square of the Oporto Cathedral: an example of reconstruction following the original project.

ments: granular disintegration, plates, flakes, black crusts, thin black layers, efflorescences and biological colonization (Figure 8).

A reflection is proposed in order to be aware of the environmental problems that affect our cities and of the menace to the conservation of an invaluable heritage.

CONCLUSIONS

In the city of Oporto the granite has always been present: in the primitive defensive system, in the austere and solemn medieval monuments, in the rich baroque buildings, in the public departments, in the modest homes. "Oporto is one of those places where memory and senses are renewed and the whispering granite tells us about the past, a history that persists in the present time, carved in stone and in the eyes of its people...".
The six stops that have just been suggested are included in a short pedestrian itinerary within the historical centre of the city of Oporto, along which the granite and the city can be identified. Between each stop the attention is also drawn to the surrounding landscape, with a particular interest in the traces of the medieval city wall and the escarpments that slope towards the river Douro which are presently being the object of detention measures.

Every granite monument that has been selected in this tour exhibits different aspects of stone decay, as a consequence of pollution and other environmental problems. A better knowledge of the phenomena that cause these deterioration aspects impels the citizens to adopt an optimistic attitude and look for the more adequate solutions both to prevent them and to conserve the monuments. After having recognized the causes of damage it is necessary to avoid the continuous action of the deterioration agents and develop multidisciplinary methods of conservation based on the compatibility with the stone nature, reversibility and durability.

Figure 8. Biological colonization on the granite of the Church of Saint Francisco.

ACKNOWLEDGMENTS

This study is being developed under the scope of the Centro de Geologia da Universidade do Porto and has the financial support of the "Programa de Financiamento Plurianual de I&D (FCT)".

REFERENCES

Almeida, A., 2001. Geochemical and geochronological characterization of the syntectonic two-mica granite of Porto (NW Portugal) In *Proceedings IIIer Congreso Ibérico de Geoquímica and VIII° Congreso de Geoquímica de España*, Zaragoza (eds. Lago, Arranz and Galé), pp. 311-315.

Begonha, A., 2001. *Meteorização do granito e deterioração da pedra em monumentos e edifícios da cidade do Porto*. FEUP-Edições, Colecção Monografias, 2, Porto, 445 pp.

Carta Geotécnica do Porto, 1994. Notícia Explicativa. Câmara Municipal do Porto: 36-49.

Ferreira, N., Iglesias, M., Noronha, F., Pereira, E., Ribeiro, A. e Ribeiro, M.L., 1987. Granitoides da Zona Centro Ibérica e seu enquadramento geodinâmico. In *Geologia de los Granitoides y Rocas Asociadas del Macizo Hesperico* (Bea, F., Carnicero, A., Gonzalo, J., Lopez Plaza, M. y Alonso, M. eds). Editorial Rueda, Libro Homenaje a L.C. Garcia de Figuerola, Madrid, 37-51.

Julivert. M., Fontbote, J., Ribeiro, A. & Conde, L., 1974. Mapa Tectonico de la Peninsula Iberica y Baleares. Escala 1:1000000. Memoria Explicativa. Instituto Geologico y Minero de España. Madrid.

Oliveira, R.M.C., 2001. Deterioração de monumentos graníticos: o caso da Faculdade de Ciências da Universidade do Porto - abordagem didáctica. MSc Thesis, Universidade do Porto, 206 pp.

Ribeiro, A., Antunes, M.T., Ferreira, M.P., Rocha, R.B., Soares, A.F., Zbyszewski, G., Almeida, F.M., Carvalho, D. & Monteiro, J.H., 1979. Introduction a la geologie generale du Portugal. Serviços Geologicos de Portugal. Lisboa. 114 pp.

SESSION 8

TOURISM AND RECREATION

Uluru (formerly known as Ayer's Rock) in central Australia, exemplifies many issues relating to geology, landscape and tourism. Despite Aboriginal requests to respect their spiritual values attached to the gigantic rock outcrop, thousands of people continue to climb it, although total visitor numbers to the area are effectively managed. As is common, any scientific or geological explanations for the origins of a world famous landform are part of a mix of cultural values and priorities attached to the place itself. Photograph: Matthew Parkes

In: Parkes, M.A. (Ed.) 2004 *Natural and Cultural Landscapes - The Geological Foundation*, 273-276, Royal Irish Academy, Dublin

LANDSCAPES, TOURISM AND THE ECONOMY

Dr Stuart K Monro, Scientific Director, Our Dynamic Earth, Holyrood Road, Edinburgh

KEYNOTE ADDRESS

Abstract: In Scotland, tourism is an important part of both the urban and the rural economy. Visitors are attracted to Scotland by its culture and heritage and not necessarily by an interest in science. They are, however, attracted by the landscape and this gives an opportunity to communicate this facet of our heritage to our visitors. Earth scientists have an important role to play but geologists are renown for being very good at communicating with each other, but less good at communicating with the public. That is changing and there is an increasing desire among earth scientists everywhere to share their understanding of the evolution of the landscape with an eager and sometimes not so eager public.

In the post-foot and mouth era, there has been a focus on the role tourism can play in the rural economy. By providing a reason for tourists to stay longer in an area, earth scientists are playing a small but significant role in this strategy. It also gives a strong economic reason for helping people understand more about the landscape and a way of enriching the visitor experience, giving added value to the visit.

There is also a role for visitor attractions like Our Dynamic Earth which inspire and excite the imagination of tourists and demonstrate how our landscapes were formed. But that must be followed up by information about the landscapes themselves, described and illustrated in a way that makes understanding accessible. There are many exciting developments in this are giving hope for the future.

INTRODUCTION

Edinburgh, as a capital City, is usually crammed with visitors at all times of the year. The policy has been to make Edinburgh an all-year-round destination and the Edinburgh International Festival has now been supplemented by the Hogmany celebrations, which brings visitors to the City even in the winter. It is a developing City with a new parliament, which will itself attract visitors. Part of the re-development of Edinburgh has been the desire to inject new life into the Holyrood area in the Old Town. Ten years ago this area was the site of a brewery, the Holyrood Brewery, derelict and abandoned when Scottish and Newcastle Brewers Ltd consolidated their activities in the Fountain Brewery to the west of the City. The site was donated to the Dynamic Earth Charitable Trust on condition that they built a landmark building to house an exhibition "in the public good". From these beginnings and with the help of the Millennium Commission, *Our Dynamic Earth* emerged, a visitor attraction telling the story of planet Earth.

The situation of *Our Dynamic Earth* is at the foot of Arthur's Seat, a Carboniferous volcano, 350 million years old and at the end of the crag and tail feature which forms the Royal Mile. It is an appropriate place to bridge the gulf between the tourists who come to Edinburgh for culture and heritage and the scientists who have another exciting story to tell, one which relates to Scotland's greatest asset, its landscape.

The justification for setting up *Our Dynamic Earth* was to communicate the story of the evolution of planet Earth to a wider audience, highlighting the importance of these processes to many important issues facing Society. However it also contributes to government policy on education, particularly life-long learning and it provides a different sort of tourist attraction away from the culture of the festival and the heritage of the castle. In many respects the story of the evolution of the Scottish landscape is part of our culture and heritage, only it is necessary to communicate it as such.

Figure 1. Our Dynamic Earth, with Salisbury Crags behind.

TOURISM IN SCOTLAND

Tourism is particularly important to the Scottish economy. In 2000, UK and overseas visitors spent £4.5 billion in Scotland. Although culture and heritage underpin the industry, over 80% of visitors say that it is the environment that they like the most about Scotland and it is undoubtedly one of the most important aspects in selling Scotland as a tourist destination. The geological community is ideally placed to give added value to visitors and to contribute to this industry. Scotland boasts rocks which date from the oldest to the youngest in the UK. Each represents part of the history of these islands and each has characteristic landforms which reflect that history.

The interaction which a visitor has with the landscape may take many different forms. Some come to take part in sport and may find themselves skiing on the Cairngorm massif or golfing on the raised beaches of East Lothian. Either way, an appreciation of the landscape in which they fine themselves lends added value to the activity.

The recent outbreak of foot and mouth disease had a devastating effect on the rural economy. The agricultural community suffered but so, also, did the tourist industry. It led directly to a 6% decrease in overseas visits and a 14% decrease in spend, taking £110 million out of the Scottish economy. The fallout from foot and mouth has led to a greater interest in diversification of business within rural Scotland of which tourism is a significant part. Historically there has been a focus on small business within the rural community but figures show that new businesses are associated with the principal cities and the main conurbation between Edinburgh and Glasgow. The need to stimulate the development of new small business in rural Scotland has been identified.

STIMULATING THE RURAL ECONOMY

The quest to find new things to do in Scotland has stimulated geologists to re-assess what contribution they can make. Scotland offers some of the finest and most diverse geology in the world which is attractive to an international niche market. Scotland is the home to the most important geological locality in the world, Siccar Point, the place where Hutton, Playfair and Sir James Hall "stared into the abyss of time". It is the locality where the modern science of geology was born, a locality which revolutionised the concept of time. It is a place of pilgrimage for geologists from all over the world. Yet it is Scotland's best kept secret! Sites like Siccar Point could be made more accessible and their significance, in global terms, made more widely known by appropriate interpretation. Siccar Point is as much a part of Scotland's heritage as Edinburgh Castle or Braveheart!

Figure 2. Siccar Point.

Some new enterprises have arisen in association with the classical geological sites of Scotland. Columba 1400 is a £2 million Community and International Leadership Centre at Staffin on Skye. It has the island's first Internet café and IT centre, conference and meeting rooms, accommodation for 24 people and acts as a venue for local community groups. It is also being used by the international geological community as a base for the study of the Tertiary igneous complex of the Cuillins and the other geological localities of the Island.

SCIENCE THROUGH THE LANDSCAPE

While Scotland's landscape is used by geologists to understand the Earth's internal structure and processes, it is also a way of giving added value to the tourist who is attracted by the landscape itself. The geological community has a role in helping visitors from both home and abroad to understand the processes which shaped the landscape. Stimulating and catering for a demanding, all year round, tourist industry often involves development, sometimes in particularly sensitive areas. One such area was the Cairngorm Mountains where skiing development required the building of a funicular railway. Inevitably developments like this, designed to stimulate the local economy, bring conflict. On the one hand there is the need to conserve the landscape and the habitats therein. On the other, is the need to sustain a local community which lives largely off the proceeds of tourism. In this case a balance was struck but not without a great deal of discussion and ill feeling on both sides. Managing the tensions between development and conservation is no easy task. In this case, peace has broken out, tourism was stimulated in both the summer and winter seasons and safeguards were put in place to protect the Cairngorm Plateau.

The far north-west of Scotland is an area of great scenic value that owes its character to the underlying geological structure. Suilven is a beautiful mountain in its own right. It is made up of Torridonian Sandstone resting, unconformably on Lewisian Gneiss. Those admiring this landscape may be even more impressed when they realize that the unconformity represents a period of up to 1000 million years. Or may look at the Lewisian surface in a new light when they recognize that this landscape is 2000 million years old. Earth scientists may not fully realise the extent to which these are completely new revelations to the non-geological tourist. There has been a saying that "geologists are good company, especially for other geologists", in other words, good at communicating with each other but less good at speaking to a lay audience. That is changing and geologists are now more aware that they have exciting stories to tell related to the landscape which can capture the imagination of visitors.

Figure 3. The Restless Earth.

A good example of geologists talking to the public in a way which is engaging and accessible is also in the north-west of Scotland, at Scottish Natural Heritage's Knockan Crag Visitor Centre. It is sited adjacent to the Moine Thrust, an area of complex but important structural geology. It is also where Peach and Horne recognised the significance of large-scale, low-angle, thrust faulting. This, in turn, became part of the evidence to support a plate tectonic model for the evolution of northern Britain. The displays at Knockan make this geology accessible, but it does more than that in that it successfully engages with new audiences. The use of the historical figures of Peach and Horne set the rocks in the context of history. Widened participation is encouraged by the use of art and sculpture to demonstrate parts of the story. In this way there is the opportunity to engage with those more concerned about art than science - but some science will rub off. The cartoon style of some of the presentations addresses the issue of social inclusion and is successful in making these rocks come alive, revealing a story which is as exciting as any children's adventure tale.

The first challenge though, is to get visitors to the far north-west of Scotland. Here there is a role for the "gateway" cities to begin to stimulate an interest in a greater understanding of the landscape. *Our Dynamic Earth* has a role to play here.

OUR DYNAMIC EARTH

The aim of the exhibition is to communicate the dynamic nature of the planet. It begins by showing the ways the Earth is being monitored at the present day set against a backdrop of geological time. This is followed by presentations on the main processes - geophysical, atmospheric, hydrological and biological, culminating in an examination of anthropogenic influences on the planet. The integrated role of these processes is demonstrated in the Earth's major environments.

Equally significant is the second objective of the *Our Dynamic Earth*, to create a national and international educational resource. The subject matter integrates well with the national curricula of both Scotland and England and Wales and this is supported by an educational team and

Figure 4. The Polar Regions.

dedicated "discovery rooms". The rapid evolution of information technology provides exciting opportunities not only to explore related areas but also to disseminate the activities of *Our Dynamic Earth* through the web pages, a method that reaches world wide.

"The State of the Earth" is the first section of the exhibition and looks at the ways in which the Earth is being continually monitored to build up a picture of a dynamic planet. "The Time Machine" conveys the concept of the immensity of geological time, taking the visitor from their present familiar surroundings to the beginnings of the universe in the Big Bang where the story begins!

"How it all started" takes the visitor from the Big Bang and the start of the physical universe to a planet capable of sustaining life. "The Restless Earth" is concerned with the internal processes on the planet: what causes volcanoes and earthquakes, how mountains are built up and why continents move. "Shaping the Surface" examines the surface processes sculpting the landscape, particularly glaciation which has been an important agent in moulding the scenery of Scotland. "Casualties and Survivors" looks at the biological processes that make the planet as it is today. The evolution of the skeleton left fossilised evidence of a wide range of life-forms. Some of these still have living relatives: these are the survivors. Others are only preserved as fossils and are the casualties. Mass extinctions have taken place in the past and will undoubtedly be a feature of the future. In this context will *Homo sapiens*, "The Human Animal" be a casualty or a survivor?

The exploration of "The Oceans" using satellite technology has greatly increased the understanding of the linkages between the oceans and the atmosphere and the transfer of momentum, energy and matter at the ocean-atmosphere interface and their importance in changing climate. "The Polar Regions" represents the extremes of cold climate yet, while the Arctic and the Antarctic are superficially very similar, one is oceanic and the other continental. Here too is the laboratory for exploring changing climates in the past and testing models for the future. The "Journey of Contrasts" takes the visitor from the tundra to the savanna grasslands, through zones where human need has significantly modified the landscape. In "The Tropical Regions" the visitor will experience the hot, wet environment of the tropical rain forest where the Sun's energy encourages the greatest diversity of plant and animal life. "The Showdome" is a summary of the exhibition highlighting the dramatic processes that have shaped the planet Earth many of which may be regarded as "hazards" but, nevertheless, many of these are also the processes of renewal which rejuvenate the planet.

Understanding how the Earth works underpins an appreciation of any landscape. In *Our Dynamic Earth* volcanic processes are illustrated in the Scottish landscape by a helicopter fly-over of Trotternish on Skye and of the islands of Staffa and Eigg. Glacial processes are illustrated in the corries, pinnacles and arretes of the Cuillins of Skye. In this way experiential learning can greatly enhance an understanding of the landscape.

CONCLUSIONS

When tourism and science come together, new opportunities emerge. New opportunities for science to reach a fresh audience and new opportunities to enhance the visitor experience by giving an exciting new insight into the landscape. In doing so the earth science community is stimulating both the rural and urban economies, giving new reasons for exploring the landscape, holding the visitors in an area for longer, increasing spend and stimulating the local economy. The provision of visitor centres, trails and guides is a win - win situation. It is a winner in terms of communicating science and a winner in terms of giving added value to the casual visitor. It is about opening new windows on the landscape… Or, to quote James Hutton; "In this manner, there is opened to our view a subject interesting to man who thinks…. and one which may afford the human mind both information and entertainment." If we are to communicate successfully and engage with the tourist we must seek to both inform and entertain. This is a good way forward for the future of landscape tourism.

In: Parkes, M.A. (Ed.) 2004 *Natural and Cultural Landscapes - The Geological Foundation*, 277-280, Royal Irish Academy, Dublin

LANDSCAPES FROM STONE

Patrick J. O'Connor, Geological Survey of Ireland, Beggars Bush, Haddington Road, Dublin 4, Ireland. Email: patrick.oconnor@gsi.ie

Patrick J. McKeever, Geological Survey of Northern Ireland, 20 College Gardens, Belfast BT9 6BS, Northern Ireland. Email: pmck@bgs.ac.uk

Enda Gallagher, Geological Survey of Ireland, Beggars Bush, Haddington Road, Dublin 4, Ireland. Email: enda.gallagher@gsi.ie

Abstract: Recent years have seen changing attitudes about the relationship between scientists and the general public. There has been a growing awareness and interest within the wider community about issues ranging across all scientific disciplines. Similarly, there is a recognition within the scientific community of the value in effectively communicating our science to the public. These changes have coincided with the growth of the leisure industry in Ireland that today sees tourism as second only to agriculture in importance to the Irish economy.

During 1997, the Geological Survey of Northern Ireland (GSNI) and the Geological Survey of Ireland (GSI) embarked on a programme, under the brand name *Landscapes from Stone*, to develop and popularise the landscapes and rocks of the 12 northern counties of Ireland. Literature aimed at mainstream tourism market is being produced in a concerted effort not only to raise public awareness and appreciation of our geological heritage but to bring added benefit to the tourism industry. By integrating aspects of our geological heritage into the wider natural resource tourism sector, we are ensuring that geology reaches a wide audience.

Additionally, the Geological Surveys embarked on an active marketing programme to promote the geology of the north of Ireland to the international geological community and encourage residential study visits to the region. The success of *Landscapes from Stone* not only demonstrates the appetite of the wider public for science but also illustrates how geology can bring added economic benefit to the region through ecotourism.

INTRODUCTION

The initial concept for a geological public-outreach project managed jointly by the Geological Survey of Ireland and the Geological Survey of Northern Ireland was first given substance in late 1994 through the provision of funding from the International Fund for Ireland for a pilot project in the Cavan-Fermanagh border area (Figure 1) aimed at an evaluation of the minerals development potential and "geotourism" opportunities of the area (O'Connor and Mitchell 1997).

The success of this project encouraged both GSI and GSNI to embark on a far more ambitious institution-building project based on the landscape resources of northern Ireland - the region which embraces counties Antrim, Armagh, Cavan, Derry, Donegal, Down, Fermanagh, Leitrim, Louth, Monaghan, Sligo and Tyrone. Funding was provided from the EU Special Support Programme for Peace and Reconciliation, from the EU INTERREG II Programme and from 27 of the 32 local authorities across the region. The involvement of the vast

Figure 1. The* Landscapes from Stone *Project area. Six counties of Northern Ireland and the six counties bordering them from the Republic of Ireland.

Figure 2. The **Landscapes from Stone** *logo.*

majority of local Councils acted as a great confidence boost to the project team and served to confirm the view that local communities were intensely interested in the documentation, conservation and promotion of their natural heritage.

LANDSCAPES FROM STONE CONCEPT AND MARKETING

At the outset of the new project it was realized from customer feedback and discussions with both Tourist Boards, that a specific branding and marketing initiative would be needed to raise awareness among the public of the value of landscape resources as a basis for economic regeneration of a region previously riven by civil conflict over many decades. In essence, that is how the "Landscapes from Stone" brand concept was developed in an early brainstorming session by the project team. A brand logo (Figure 2) was created and a marketing strategy developed to promote the region to discerning visitors. A niche segment of the educational market represented by those geological institutes wishing to hold residential field courses in the region was specifically targeted (McKeever and Gallagher 2001) and a familiarisation trip (Figures 3, 4) was arranged in October 1998 for representatives of 14 UK-based University geoscience departments. This initiative and further promotion at international geoscience conferences in the U.K., France, Canada and the U.S. proved highly successful in attracting University geology groups to the region in subsequent years.

PROJECT OBJECTIVES

- To assess and develop the natural resource potential of the 12-county region, especially in relation to the development of landscape-based resources in tourism and to inform decision-making in long term land use planning by local authorities.
- To widen the portfolio of environmentally sustainable activities in the region in a manner that also brings people together, from local development groups to local council level, to look at and appreciate not only their own immediate surroundings but also the landscape of the wider region.
- To develop a suite of popular inter-regional guides/booklets to the natural resources of the 12-county region which integrate geology in an holistic view of landscape in its broadest sense and thereby encourage educational visits by schools, universities and other interested parties both from within the region and from further afield.
- To encourage groups from one part of the region to explore and appreciate the environment of another part of the region and thereby promote a degree of reconciliation in developing a shared appreciation of natural heritage.
- To develop a closer working relationship between GSI and GSNI in the interests of the customers they serve and, more generally, to communicate the contribution which geoscience in Ireland makes to our quality of life and economy.

Figure 3. Participants from different UK university geology departments on the Familiarisation trip, at Horn Head Co. Donegal.

Figure 4. Participants from different UK university geology departments on the Familiarisation trip.

POPULAR PUBLICATIONS

Through the period 1997-2002 a series of popular publications were produced to support the tourism product of the region (Appendix 1). In essence, four product types were developed:

- *Landscapes from Stone* map - 14 distinct landscape areas are depicted on a satellite image of the region, each originating through a specific geological event or series of events. In addition to the scenic attractions of the region, established walking trails, forest parks and blue flag beaches are also featured as well as sites of archaeological and cultural interest.
- *A Story Through Time* book - a well illustrated 100-page guide to the geological evolution of the region was written as an accompaniment to the Landscapes map.
- *Explore* Series - a series of 11 touring guides for car or bicycle covering the diverse landscapes of the region.
- *Walk* Series - a series of 7 walks packs, each containing 10 circular walks featuring the geology and landscapes of particular sub-regions in more detail.

Copies of each of these publications have been distributed to retail outlets and tourism offices throughout the region and the uptake by visitors and locals has been very encouraging. Primary schools across the region have also been circulated with information on the availability of these accessible products as a curriculum support. They may also be ordered online through the Survey websites (www.gsi.ie and www.bgs.ac.uk).

CONCLUSION

The successful completion of the *Landscapes from Stone* project was celebrated in November 2001 at an event at Parliament Buildings, Stormont, Northern Ireland hosted by Sir Reg Empey MLA, Minister of Enterprise, Trade and Investment, and Mr. Joe Jacob TD, Minister of State at the former Department of Public Enterprise.

Both Ministers recognized that the project represented a major step forward in cross-border collaboration between two public sector bodies and a sharing of their resources to attain a common goal. Staff in both surveys had worked together in a highly effective and professional manner to produce a range of informative and worthwhile products for use by local communities and visitors to the region.

Further sustained cross-border co-operation between both Surveys in the field of Landscape Resource Tourism will continue with their involvement in the Breifne Mountain Park initiative led by Cavan County Enterprise Board which is seeking financial support under the EU Peace II Programme and which involves five local Councils (Cavan, Fermanagh, Leitrim, Sligo and Roscommon) and the Academy of Irish Cultural Heritages (University of Ulster) as partners.

ACKNOWLEDGEMENTS

The Geological Surveys wish to acknowledge the significant financial support to the project provided by the EU Special Support Programme for Peace and Reconciliation and the EU INTERREG II Programme, together with contributions from local authorities across the region. The support and advice of Bord Failte, the Northern Ireland Tourist Board, North West Regional Tourism Authority and many colleagues in both Surveys is also gratefully acknowledged not least our Directors. Landscapes Project staff who have contributed over the years include Dr. Ian Mitchell, Dr. Michael Flowerdew, Ms. Janis Smyth, Mr. Niall Kearns-Mills, and for the Minerals sub-project Mr. Diarmait O'Reilly, Dr. Wayne Cox, Mr. Alan Bell and Dr. Eibhlin Doyle.

REFERENCES

McKeever, P.J. *and Gallagher, E. 2001. Landscapes from Stone. In Earth Science and Natural Heritage: Interactions and Integrated Management* (edited by John E. Gordon and Katherine F. Leys), Scottish Natural Heritage, 262-270.

O'Connor, P. and Mitchell, I. 1997. Cross-border co-operation in Geosciences. *The Irish Scientist Yearbook*, 5, 126.

APPENDIX 1

McKeever, P., Kearns-Mills, N. and Gallagher, E. 1998. *Landscapes from Stone: a guide to the scenic landscapes and rocks of Ireland (north)*. Geological Survey of Ireland - Geological Survey of Northern Ireland.

McKeever, P. and Smyth, J. 1999. *A Story through Time.* Geological Survey of Northern Ireland - Geological Survey of Ireland. Pp100.

Explore Series

Flowerdew, M., McKeever, P.J. and Smyth, J.D. Explore Donegal. Geological Survey of Ireland - Geological Survey of Northern Ireland.

Flowerdew, M., McKeever, P.J. and Smyth, J.D. Explore West Sligo. Geological Survey of Ireland - Geological Survey of Northern Ireland.

Flowerdew, M., McKeever, P.J. and Smyth, J.D. Explore West Breifne. Geological Survey of Ireland - Geological Survey of Northern Ireland.

McKeever, P.J. and Smyth,J.D. Explore Cuilcagh. Geological Survey of Northern Ireland - Geological Survey of Ireland.

McKeever, P.J. and Smyth, J.D. Explore Oriel. Geological Survey of Northern Ireland - Geological Survey of Ireland.

McKeever, P.J. and Smyth, J.D. Explore Mourne, Gullion and Cooley. Geological Survey of Northern Ireland - Geological Survey of Ireland.

McKeever, P.J. and Smyth, J.D. Explore the Bann and Roe Valleys. Geological Survey of Northern Ireland - Geological Survey of Ireland.

McKeever, P.J. and Smyth, J.D. Explore the Sperrins. Geological Survey of Northern Ireland - Geological Survey of Ireland.

McKeever, P.J. and Smyth, J.D. Explore Lough Neagh. Geological Survey of Northern Ireland - Geological Survey of Ireland.

McKeever, P.J. and Smyth, J.D. Explore Upper Erne. Geological Survey of Northern Ireland - Geological Survey of Ireland.

McKeever, P.J. and Smyth, J.D. Explore South Antrim. Geological Survey of Northern Ireland - Geological Survey of Ireland.

Walk Series

Flowerdew, M. and McKeever, P.J. and Smyth, J.D. 2000. Walk Sligo - Leitrim. Geological Survey of Ireland - Geological Survey of Northern Ireland.

Flowerdew, M., McKeever, P.J. and Smyth, J.D. 2001. Walk Donegal. Geological Survey of Ireland - Geological Survey of Northern Ireland.

Flowerdew, M., McKeever, P.J. and Smyth, J.D. 2001. Walk Mourne, Gullion & Cooley. Geological Survey of Ireland - Geological Survey of Northern Ireland.

Flowerdew, M. and McKeever, P.J. 2002. Walk South Ulster. Geological Survey of Ireland - Geological Survey of Northern Ireland.

McKeever, P.J. and Smyth, J.D. 2000. Walk Cuilcagh. Geological Survey of Northern Ireland - Geological Survey of Ireland.

McKeever, P.J. and Smyth, J.D. 2000. Walk the Sperrins. Geological Survey of Northern Ireland - Geological Survey of Ireland.

McKeever, P.J. and Smyth, J.D. 2001. Walk the Bann and Roe Valleys. Geological Survey of Northern Ireland - Geological Survey of Ireland.

In: Parkes, M.A. (Ed.) 2004 *Natural and Cultural Landscapes - The Geological Foundation*, 281-284, Royal Irish Academy, Dublin

LANDSCAPES: MAXIMISING THE BENEFITS

John Faulkner, Director of Natural Heritage, Environment and Heritage Service, Northern Ireland

Abstract: This paper outlines the arrangements in place to manage three contrasting landscapes in Northern Ireland - the Causeway Coast, the Mourne Mountains, and Lough Neagh. The first is a tourist honey pot, the second an important area for outdoor recreation, and the third a resource which is relatively little exploited for either tourism or recreation. Three conclusions are drawn:

- **Geological history has been important in shaping each of them.**
- **People who live in the area tend to understand its heritage value differently from those who visit.**
- **In managing these landscapes, it is essential for government agencies such as the Environment and Heritage Service to work with andthrough local people.**

I have always enjoyed speaking at events associated with the Royal Irish Academy. The reasons are probably that wonderful epithet "Academy" together with the fact that I am outside my home territory. These provide an environment in which I feel more at liberty to say what I want rather that what I ought as an official of a government agency responsible for landscape conservation.

Conventional wisdom has it that no landscapes in Ireland - or almost anywhere - are truly natural, but are fashioned by the activities of people over the centuries. No one can really doubt the truth of this statement. However I am sure Oscar Wilde would have had something to say about conventional wisdom being a signpost to the errors of the past.

Figure 1. The Causeway Coast. Photo: Environment and Heritage Service.

It does not follow that we should make the rather arrogant leap of supposing that people have had an equal share with nature in determining the character of our landscapes, though it may sometimes be politically expedient to do so.

We and our ancestors have quarried, built, damned, drained, ploughed and planted, but for the most part our role has been relatively superficial. The primary work has been done by nature. Nature has generated the rocks and other materials which underlie the landscape. Nature has given rise to the geomorphological processes which have created its topography. And nature has produced the diversity of life which clothes its surface.

It seems not unreasonable therefore to speak of the more magnificent areas of our countryside as having "natural" beauty. I would like to look at a few of these areas of natural beauty and the issues that surround them, and as this conference is concerned with geology and landscapes it would be remiss of me not to mention the Giant's Causeway.

The Causeway itself is a small area on the shore at the foot of a massive cliff. It is made up of mostly hexagonal horizontal blocks which are sections across the columns into which the basalt cooled some 60 million years ago. Nearby there are other interesting rock features such as the Giant's Organ. The Causeway is near the centre of an outstanding stretch of coast some 15 miles in length from Portrush to Ballycastle, which comprises the Causeway Coast Area of Outstanding Natural Beauty. The Causeway and its immediate surroundings belong to the National Trust and are a World Heritage Site, National Nature Reserve and Special Area of Conservation. There is a visitor's centre on the cliff top at the main access point set back half a kilometre from the stones. Landwards of the cliffs, there is an open farmed landscape, historically occupied by clachan-style settlements.

I have already hinted at a central issue for management of the area by referring to the Causeway rather than, say, the

North Coast or the Antrim Coast. The Causeway itself is only an acre or two in extent, but is very widely used as a marketing image, not only for tourism purposes, but for the area as a whole and sometimes for Northern Ireland as a whole. It gives its name to everything from manufactured goods to service companies and hospitals. Even the world's first crop variety bred for herbicide resistance was named after it.

Not surprisingly, nearly every visitor to Northern Ireland wants to go and see it. They may have no inherent interest in Tertiary geology, or in the debate over vulcanism and neptunism. It's just that it is well known and therefore they want to go and see it, and enjoy themselves in the process.

There is absolutely nothing wrong with this provided that in doing so the site and its surroundings are not spoiled for others, and provided that they benefit rather than damage the local economy. Herein lie some crucial issues.

One is the threat of a gradual erosion of the scientific and scenic value of the site itself and the accretion of paraphernalia to accommodate visitors. Another is the extent to which the economy of the wider area benefits. Many visitors go to the Causeway and the Visitors' Centre and then depart. The Causeway and its immediate surroundings are under pressure from traffic congestion and especially from development proposals while the hinterland is under-visited and some of the local towns are in need of investment and reinvigoration. Inevitably there is controversy over developments, with local interests divided, and wider conservation interests largely against but open to charges of external interference.

Draft proposals for the management plan for the Causeway Coast AONB were published in June 2002 for consultation and a final plan will be published early in 2003. It will also feed into the statutory Northern Area development plan which is also currently under revision. Draft proposals for the management plan were published in June for consultation and a final plan will be published early next year. The draft proposes many measures which could be put in place covering conservation, access, information and liaison and other areas. If it is to succeed, however, it will need active support from a large number of different bodies, in the statutory, voluntary and private sectors. Apart from the steering group for the plan, and a developing regional tourism body, management structures to bring about a common vision and co-operation within the area are not particularly strong. This is probably one area that will need particular attention if the plan is to be followed by meaningful action.

Turning to the south east of Northern Ireland, I would now like to look briefly at the Mournes. These are mountains of hard acidic granite. The granite formed some 50 million years ago when a huge subterranean block of shales sank deep into the earth's crust.

Figure 2. The Mourne Mountains. Photo: Environment and Heritage Service.

The Mournes today are an impressive upland landscape, rising suddenly from the sea to the east and plains to the south and north-west. They are clothed either with acid grasslands or with heath vegetation over shallow peat. The soils are thin and poor, though ecologically the heath is valued as one of the best examples of this type of vegetation.

Around the margins of the hills is a multitude of small farms, few of which would generate enough income to support a family. Walls of large rounded stones are a very characteristic local feature. Newcastle and other small towns are spread around the perimeter.

The Mournes comprises a well defined area of mountains and foothills and associated lowlands. It is by far the most important area in Northern Ireland for hill walking and other outdoor recreational pursuits.

The Mournes has long been regarded as a landscape of great scenic value and was the first area to be given AONB status under the 1985 NCALO. The history of bodies interested and involved in the management of the Mournes is lengthy and complicated, taking in both voluntary bodies and statutory authorities. The current position gives a central role to a voluntary trust - the Mournes Heritage Trust. The Trust is funded primarily by the Environment and Heritage Service, Department of Agriculture and Rural Development, Northern Ireland Tourist Board and the 3 local district councils, and has a management structure which gives ample provision for community level input. The Trust carries out projects of a varied nature - not exclusively environmental ones by any means.

In parallel with the MHT is a liaison group of officials of all the statutory bodies with a local interest. It includes government departments, agencies and district councils.

Over the past 10-15 years, much has been achieved in the Mournes: improved and repaired footpaths; better interpretation and information; restoration of abandoned vernacular houses as tourist accommodation; restoration of many kilometres of stone wall; strategies developed for visitor and environmental management; design guides produced for new buildings; and much else. There remains perhaps a need for a stronger sense of common direction, and for an agreed agenda for sustainable development. The MHT itself advocates national park status for the Mournes, albeit that the details should not merely copy a model from elsewhere. Our current legislation provides for national parks in name, but does not specify anything about the management structure and resources that should be put in place. At present there is no national park in Northern Ireland, but in EHS we fully recognise that national parks do not flourish without planning, management and funding. At the request of our Minister, EHS has recently commissioned a study of the potential benefits of national parks in Northern Ireland. The outcome of this study, and of any bids for funding that might arise from it, will no doubt have a bearing on the specific issue of a national park in the Mournes.

Figure 3. Lough Neagh. Photo: Environment and Heritage Service.

And now for a contrasting site - where the density of tourist visitors per hectare is in the order one million times smaller than at the Causeway. I think we might call it a potential rather than an actual tourist destination. I am referring of course to the largest lake in Britain and Ireland, Lough Neagh. It occupies a roughly rectangular area of nearly 400km in the very centre of Northern Ireland. It is well known in the intellectual sense but not so in the sense of being familiar. Despite its size and position, many people in Northern Ireland have hardly ever seen it (except perhaps from the air) let alone visited its shores. Its reputation seems to revolve round the clouds of midges which appear periodically through the summer, but which bite no one and are a keystone of the ecosystem.

Lough Neagh occupies a shallow basin formed where an area of basalt became downfaulted. The basin has been largely filled with recent sediments of clay, lignite and sand up to 500m in thickness, such that most of the Lough is less than 50m deep.

Scenically its attractions are esoteric. There are no steeply rising shores, or nearby mountains. It is set among lowlands, and around the southern half of the Lough there were formerly large tracts of lowland raised bog. Now mostly cut away, some are now under conservation management as at Peatlands Park, and Brackagh Moss. Traditional dwellings of the area were mud walled thatched cottages set at right angles to the winds off the Lough.

Lough Neagh is a wetland of international importance - a "Ramsar" site - largely because of its vast numbers of overwintering wildfowl. It is home to many species of plant and invertebrate rare elsewhere. The pollan (a fish) was till recently an important seasonal source of food. There is still an eel fishing industry worth several million pounds per annum. Sand extraction from the bed of the Lough is important for the construction industry. Recreationally the Lough is used for wildfowling and there is some limited use for water sports.

There has long been a sense of unrealised potential about Lough Neagh. In the 1980s there grew up a voluntary association of Lough Neagh users - "ALNUS" - to promote its "wise use".

In the 1990s this was translated into a dual structure, set up jointly by the Departments of Environment and Agriculture.

The Lough Neagh Advisory Committee (LNAC) has representatives of all the main users and other interests, mainly local but some province-wide. It includes members of all six of the District Councils bordering the Lough.

The Lough Neagh Co-ordinating Committee has representatives of all the relevant statutory executive bodies - government departments, agencies and officials of the Councils.

Supporting the structure is a Lough Neagh Liaison Officer, funded jointly by the Departments and Councils.

The LNAC has developed a management strategy which was published earlier this year. This is good progress, but there is still one underlying obstacle to overcome - common identity.

Communities around the Lough do not identify with each other so much as with neighbouring inland towns. There are 6 District Councils bordering the Lough. Pride in the Lough is not always evident.

For example the peatland areas around the south are known as Montiaghs. The area has historically been regarded as poor and remote. The idea that it is somewhere with a unique history, a special environment and a heritage of vernacular houses now rare elsewhere is novel to those that live there. It takes time to see what were symbols of poverty in the landscape as something to be valued. Nonetheless, considerable progress is being made.

To conclude, I would like to consider what common strands can be drawn together from these three areas.

First there is the point (that underlies this conference) that much of the heritage value derives from the underlying rocks and the processes that have shaped the local topography. This applies not only to the earth heritage itself, but also to the local biodiversity and even to the heritage of local vernacular structures.

But we should not delude ourselves. The appeal of rocks, stones and earth is esoteric. The associations are not always positive. Consider:

"sink like a stone"

"stonewall defence"

"hit rock bottom"

"fossilised" (for incapable of change).

Overt enthusiasm for earth heritage is likely to remain the preserve of a small minority.

Secondly, there is the point that heritage tends to be perceived differently by the people who live there. They may not instinctively understand the value of what to them is commonplace and/or outmoded, but they may be receptive to encouragement to value it.

Lastly, and most importantly, there is the conclusion that it is not only desirable but essential to work with and through local people. It is their environment. If they do not have the will to manage their environment and the means to do it, it will not happen. Our role as an outside statutory body is to advise, encourage and support. We can also demonstrate good practice. But above all we must understand, not just the scientific origins of the landscape, but also the people that work with and in it. Though nature has largely fashioned it in the past, we are working in human timescales and in these terms it is the people who now have the capacity to manage it in future.

In: Parkes, M.A. (Ed.) 2004 *Natural and Cultural Landscapes - The Geological Foundation*, 285, Royal Irish Academy, Dublin

PROMOTING GEODIVERSITY. THE IMPORTANCE OF SITES AND FEATURES OF COMMON TOURISTIC APPEAL

Lars Erikstad, NINA, BOX 736 Sentrum, N-0105 Oslo

Abstract: Criteria for natural value are important in selecting areas for protection, assessing impact significance in Environmental Impact Assessments and in general planning. Most commonly scientific criteria are given most emphasis, but criteria linked towards education and common experience are also used regularly. Diversity is a criteria that is significant both for scientifically reasons as it is for the non-scientific community. In tourism we will find many attractions that are regularly visited by thousands of people every year and that can be classified as the extremes in the geodiversity spectrum.

These sites are visited because they are reckoned as special, dramatic, odd or fantasy-appealing. They really are interesting because they are showcases of the diversity of natural processes and give people an adventure experience of the power, diversity and wonder of nature. These sites are a big resources in promoting geology, geo-conservation and geodiversity for the public, and to establish strategic alliances between local communities, trade interests and geo-conservation.

This resource have up to now been underused perhaps because the sites have not been regarded as scientifically important or that propagation of geology and geo-conservation has not been given priority. This may well be a big mistake as these sites have the potential of creating direct, in-situ and automatic feelings of the force of geological processes and the value of geological sites and our geodiversity. They are in other words links between the direct experience and attitudes and support from the public as we as geoconservationists depend on.

In: Parkes, M.A. (Ed.) 2004 *Natural and Cultural Landscapes - The Geological Foundation*, 286, Royal Irish Academy, Dublin

PROMOTION OF GEOLOGICAL HERITAGE THROUGH UNESCO

**F. Wolfgang Eder, UNESCO, Director, Division of Earth Sciences,
1, rue Miollis, 75732 Paris Cedex 15, France. Email: w.eder@unesco.org**

Abstract: Through the creation of a worldwide network of national natural parks with significant geological features, labelled 'Geoparks', UNESCO's Division of Earth Sciences is intending to promote the three goals of conserving a healthy environment, educating in Earth Science at large and fostering sustainable economic local development. An individual Geopark, recognized as part of the network, shall be designed to become a tool for a better understanding of the Earth's geological heritage and for the wise use of the Earth, by sensitizing the broad public to a balanced relationship between humankind and the Earth. One central principle of an individual Geopark should be that the area must be capable of acting as a focus for economic activities, particularly through geotourism.

Such territories shall normally represent landscape elements rather than small geological outcrops of limited extent. Therefore, a 'landscape approach' of integrating biological and Earth heritage conservation shall be also widely adopted for the individual Geoparks. Landscape elements which share common geological and biotic characteristics should be managed holistically to protect and enhance their natural characteristics. The relationship of humankind to the landscape should also be recognized through such schemes. The trend is therefore adopted to take a holistic approach (through 'Natural Areas', 'Natural Heritage Zones', 'Cultural Parks' etc.) which combine scientific, cultural and social programmes to encourage sustainable development within the landscape context.

Consequently, the now envisaged UNESCO Network of Geoparks has been modeled on the existing 'Man and the Biosphere' (MAB) Programme and its World Network of Biosphere Reserves which emphasizes a landscape approach to conservation, recognizing the significance of pattern and process at this scale. A feasibility study on 'Developing a UNESCO Geoparks Programme' (T. Weighell, JNCC, UK) concluded in the year 2000 to cooperate with the MAB Programme and to use it as as a vehicle to integrate the Geoparks concept into UNESCO's activities as the best option given the similarity in objectives. This proposal was not accepted by the Council of the MAB Programme .

UNESCO's Executive Board regretted in June 2001 that the Geoparks initiative has not found an appropriate 'home structure' within the existing international programmes of UNESCO (IGCP, MAB, World Heritage Centre), but the majority of delegates felt that countries would benefit from UNESCO's provision of support to, and cooperation with, national geological heritage initiatives. UNESCO's Executive Board therefore decided to request the Director-General 'not to pursue the development of a UNESCO geosites/geoparks programme (within the MAB Biosphere Network), but instead to support ad hoc efforts with Member States as appropriate.

Through the inclusion in a Network of Geoparks UNESCO's Science Sector, and notably its Division of Earth Sciences might be helpful to enhance the international recognition of individual Geoparks, if so requested by national and local communities of Member States.

In: Parkes, M.A. (Ed.) 2004 *Natural and Cultural Landscapes - The Geological Foundation*, 287-290, Royal Irish Academy, Dublin

THE HUMAN FACTOR IN NATURAL AND CULTURAL LANDSCAPES: THE GEOMINING PARK OF SARDINIA

Annalisa Arisci, Jo De Waele, Felice Di Gregorio, Ivan Ferrucci, Roberto Follesa, Giuseppe Piras, University of Cagliari, Geology Department, Via Trentino, 51- 09127 Cagliari, Italy. Email: geoam@unica.it

Abstract: In many landscapes the human imprint is very relevant and can determine its valorisation or deterioration. In most cases human interventions disqualify geological and natural landscapes, but in some they can enrich the already existent values and create new related sites of interest. Among the most explicative areas where human impact on natural and cultural landscapes of great geological interest has both been negative and positive, the Geomining Historical and Natural Park of Sardinia can be mentioned. The mining activities in Sardinia have manifested in many ancient mines throughout the Island and characterised by different minerogenetic, economic, productive, scientific and cultural values, altogether essential for the comprehension of the history of these industries. In the past 8,000 years, these industries have left indelible human traces in the landscape that enrich in an appreciable way the already extraordinary value of the geological, natural and cultural heritage. In these last decades, with the abandoning of these mining activities, the consciousness of these important values has grown rapidly, leading to the development of scientific and cultural initiatives in order to rehabilitate and valorise this heritage from a social, cultural and economic point of view. For all these reasons at Paris in June 1997 the General Assembly of UNESCO has sponsored vigorously the demand presented by the Sardinian Regional Government for the constitution of the Geomining Historical and Natural Park of Sardinia, becoming finally reality in the year 2001. The Authors will emphasise on the differences in cultural and natural landscapes of the eight areas of this important GeoPark.

INTRODUCTION

Sardinia is one of the most interesting geological regions of Europe and represents a "geological school" for many earth scientists of the Italian mainland and Europe. On the Island, in fact, the whole sedimentary sequence beginning from Lower Cambrian and ending up with Holocene is represented with an infinite variety of facies and rocks. Even in small areas granites, volcanic, sedimentary and metamorphic rocks can be observed side by side, showing evidences of three orogenetic events, Caledonian, Hercynian and Alpine. The great geological richness of Sardinia was known since early historical times, and in many areas the "treasures" of the subsoil have been exploited many years before the Roman conqueror entered the Island. For centuries the Island's main economy has been related to the exploitation of its mineral deposits, and these activities have deeply influenced the landscape, history, economy and the culture of the resident populations.

Towards the end of the 80's, with the closure of the most important mine centres, the idea of the constitution of the Geomining Park of Sardinia started to develop: After the recognition of this Geopark by UNESCO the 30th of July 1998 and the Agreement "Carta di Cagliari" signed the 30th September 1998, finally the 16th October 2001 the Italian Ministry of Environment instituted officially the Geomining Historical and Natural Park of Sardinia. The Geopark, divided in eight areas which contain from one to more than hundred mine geosites, covers a total area of 3771 km^2 involving 81 municipalities and representing almost 16% of the entire surface of Sardinia (Figure 1).

THE AREAS OF THE GEOMINING PARK

The institution of eight different areas constituting the Geomining Historical and Natural Park of Sardinia is related to the diversity of the exploitation activities that have been taking place on the Island since 6,000 B.C. These areas are, respectively, Monte Arci (Obsidian), Orani (Talc), Funtana Raminosa (Copper), Gallura (Granite), Argentiera-Nurra (Pb-Zn), Guzzurra-Sos Enattos (Pb-Zn), Sarrabus-Gerrei (Pb, Ag, Zn, As and Sb) and Iglesiente-Sulcis (Pb, Zn, Ag, Ba and C).

Monte Arci

Along the Eastern border of the Campidano Graben in Central-West Sardinia the Pleistocene (5-2.5 Ma) volcano of Monte Arci rises up to a height of 812m a.s.l. offering a wide panorama over the Gulf of Oristano and the Campidano plains. This impressive volcanic complex is constituted, from bottom to top, of basal rhyolites and rhyodacites followed by trachites and alkaline trachites, andesites, dacites and basalts. Underneath this extrusive complex earlier Oligo-Miocene volcanic products (pillow lava's) can be found, intercalated in marly and sandstone beds (Assorgia et al. 1976). In the lower acid volcanic sequences the rapid cooling of the lava flows has enabled the formation of perlite and obsidian that has been exploited during Ancient Neolithic (6,000 B.C.) in many small quarries. In the territory more than 100 centres of cutting have been discovered where obsidian fragments can be easily found. More recently large quarries of perlite have been opened, sometimes destroying ancient archaeological exploitation sites, and represent the most important Italian perlite production centres. This area has been included in the Geomining Park for its historical importance, being the best example of ancient extraction and commerce in the Western Mediterranean, exporting Sardinian obsidian and artefacts already 8,000 years ago.

Orani

Figure 1. The areas of the geomining Park and the main mines.

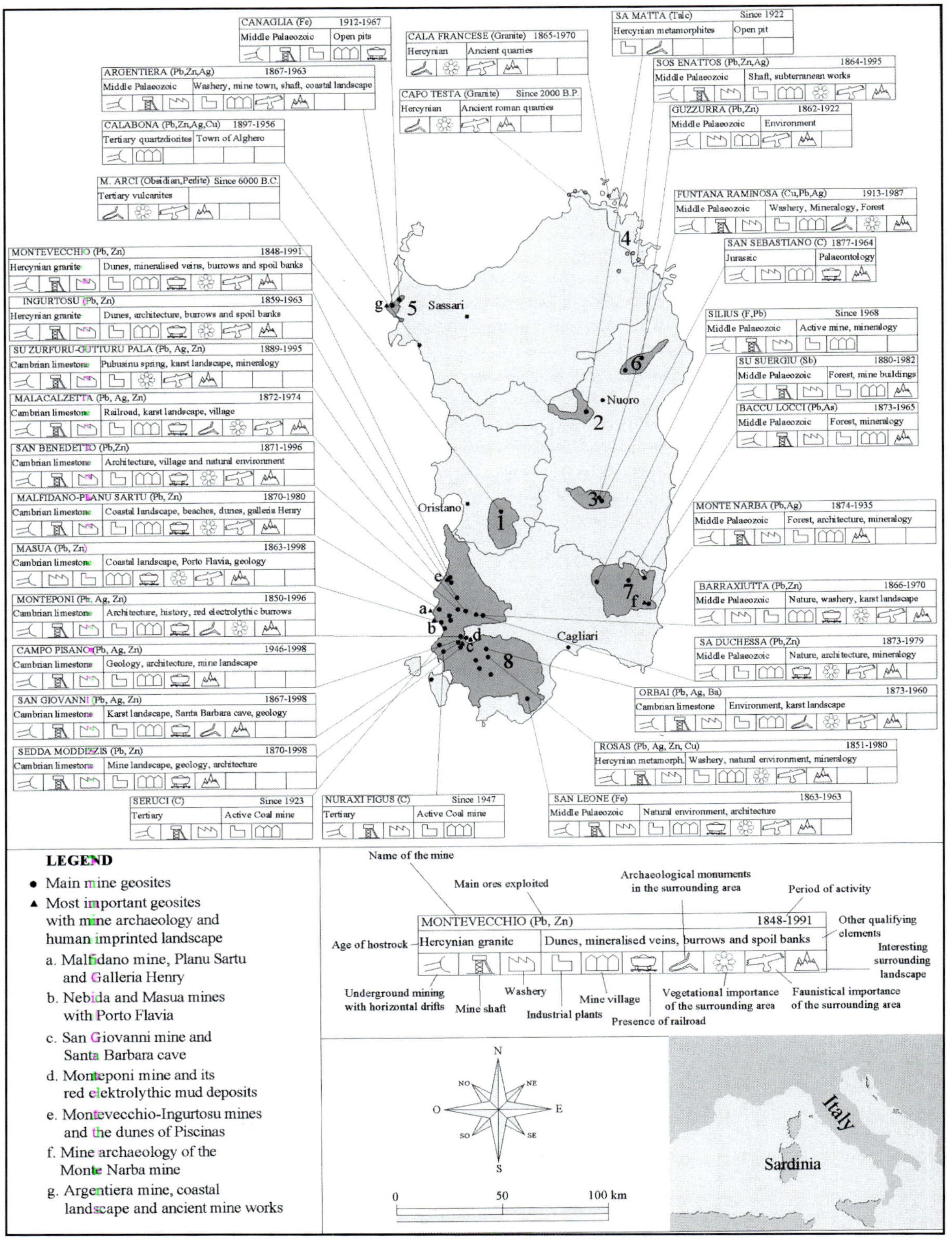

Another prehistoric mine activity in Sardinia is related to the excavation and use of Steatite and Talc, diffusely present near Orani in the centre of Sardinia. In many ancient villages in North Sardinia (3,300-2,400 B.C.) steatite medallions and sculptures, aside matrixes for fusion of bronze statues, have been found. The Orani talc deposit is related to the Sardinian-Corsican batholith that has thermo-metamorphosed the Palaeozoic carbonatic sequences essentially forming marbles and talc. The modern exploitation of these deposits started only in the early 20's of the 20th century in several places near Orani and is still going on only in the Sa Matta open pit mine. Aside the talc deposits, recently, many feldspar deposits are cultivated in the territories of Orani and Ottana (Massoli-Novelli 1971; Marini and Mannini E. 2000).

Funtana Raminosa

On the Southern slopes of the Gennargentu massif (Central-Sardinia) other two historical exploitation centres are situated: the Funtana Raminosa copper mine near Gadoni and the Anthracite mine of Corongiu near Seui. The first mine has great historical importance being exploited since Neolithic times (since 2,500 B.C.) for the copper which was used for the making of the famous bronze sculptures typical of the nuragic culture. Old excavations are still clearly visible in the area, and along the natural pathway that follows the Saraxinus torrent and the Flumendosa valley are known several archaeological sites that demonstrate the ancient commercial track. Only during the 19th century the deposit was exploited by modern techniques and the mine continued working until the late 80's. The mixed sulphur deposit of Funtana Raminosa, containing minerals of Pb, Zn, Cu and Fe, is hosted in a skarn related to the thermometametamorphic evolution of Palaeozoic carbonatic sequences (Dessau 1937). The Corongiu anthracite deposit, instead, was discovered by Alberto Della Marmora in 1826, and was initially known for its fossil flora of 33 species. From 1876-1964 the mine worked a bed of Jurassic anthracite with thickness of 2-3 meters (Mezzolani and Simoncini 1993; Fadda 1997).

Gallura

In NE Sardinia the main economic activities are tourism, the cork-industry and, particularly, the exploitation of granite for ornamental use. While the first two industrial sectors are relatively recent, the granite industry goes back to Roman times. Ancient roman granite quarries are known at Capo Testa (Santa Terera di Gallura) where, on vertical walls close to the sea, the granite was quarried and from there exported to Rome to enrich some of the most important temples and palaces. In the 19th century at Cala Francese (Island of La Maddalena) the massive granite was quarried and exported for some of the major constructions of the century, such as the Suez channel, and for many roads in North Italy (Fanni and Mezzolani 2001). Other minor historical granite quarries are located near Olbia.

Argentiera-Nurra

On the northwestern coast of Sardinia another important mine district is situated; at Argentiera since Roman times, and surely in early Medieval, silver-rich galena was exploited in short excavations and shallow pits. The modern explcitation of silver first, and lead-zinc later, started in 1865 and ended after almost 100 years of activity in 1963 (Ruju 1996; Ottelli 1997). Some km to the North another important mineral deposit was exploited in the period 1912-1967: at La Canaglia-Monte Rosso, in fact, several economical iron deposits, containing magnetite, have been quarried creating a series of open pits which have now become artificial lakes. At Calabona, close to Alghero, another small silver-lead-copper mine has been exploited since the end of the 19th century: this mine was closely related to the Monteponi mine (Iglesias, SW Sardinia), giving the pyrite necessary for its electrolyte plant (Mezzolani and Simoncini 1993).

Guzzurra-Sos Enattos

North of the carbonatic Mesozoic massif of Monte Albo (Lula, NE Sardinia), one of the most interesting geo-ecosystems of this part of the Island, two abandoned lead-zinc mines have been included in the Geomining Park. The Guzzurra-Argentaria and the Sos Enattos mines, considering the easily distinguishable mineralised veins and the high percentages in silver, were probably discovered by the Romans 2,000 years ago; modern cultivation started respectively in 1862 and 1864 and lasted until the early 90s at Sos Enattos. The mineralisation is hosted in metamorphic Palaeozoic rocks intruded by granites and with extensive hydrothermal veins (Mezzolani and Simoncini 1993).

Sarrabus-Gerrei

In South-East Sardinia the second most important mine area of the Island is situated, comprising large parts of the geographical regions of Sarrabus and Gerrei. In this district different mines cultivated lead, zinc, copper, silver, tin, antimony and iron ores. Two main productions are to be mentioned, related to the silver and to the antimony ores, while a third and more recent one opened in 1968 and still working (Silius), is characterised by fluorite with galena. The silver mines all closed down before the end of World War II, while the antimony mines were abandoned much later. The Baccu Locci mine (1873-1965), instead, situated some km to the North, cultivated a rich deposit of arsenopyrite (Stara et al. 1993). The presence of silver (also in its native form) and the importance of some of the mineralised veins have attracted Romans 2,000 years ago, although no important and convincing archaeological evidences have been found yet. The archaeological-industrial importance of this mine district is enriched by an almost uncontaminated natural environment and a geological landscape with many sites of geomorphological interest (Massoli-Novelli 1996; Barca and Di Gregorio 1999).

Iglesiente-Sulcis

By far the most important and interesting mine district in Sardinia is the Sulcis-Iglesiente area, in SW Sardinia, characterised by almost a hundred abandoned mines which cultivated deposits of lead, zinc, copper, silver, tin, barium and iron (Boi et al. 1996; Stara et al. 1996). Mine activities started around 600 B.C. with nuragic people first, Phoenicians and Romans later. These exploitations, discontinuously, went on in the Medieval and developed very rapidly in the second

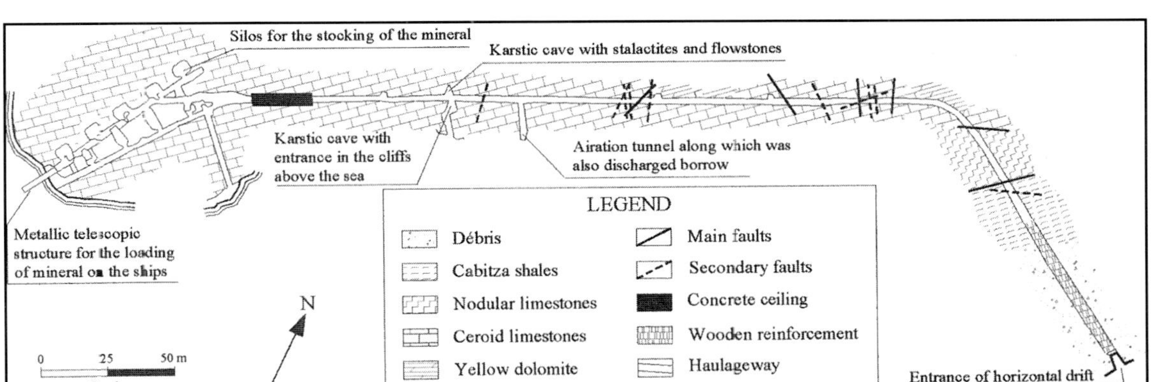

Figure 2. Geological scheme of the industrial plant of Porto Flavia.

half of the 19th century. The most important of these mines have closed down in the past decades and nowadays only the two coal mines of Seruci and Nuraxi Figus near Carbonia are still active (Fanni and Mezzolani 2001).

The main Pb-Zn ores are hosted in the carbonatic lithologies of the Cambrian sequence that characterises this part of Sardinia. This Cambrian stratigraphic succession, constituting the oldest outcropping rocks of Italy, starts with metasandstones of the Nebida Group, followed by the dolomites and limestones of the Gonnesa Group and ending with the Nodular limestones and slates of the Iglesias Group. An angular unconformity, clearly visible near Nebida, separates these rocks from the overlying Ordovician and Silurian clastic sediments (Bechstadt and Boni 1996; Pillola 1989). The whole area is rich in sites of geological and geomorphological interest (De Waele et al. 1998; Barca and Di Gregorio 1999) and in many cases the mines have enriched an already beautiful natural landscape. In fact, many mines have left impressive open pit excavations and subterranean works, burrows and spoil banks, mineral industrial plants and mine villages, some of which have recently been rehabilitated and restored (e.g. Santa Barbara cave-Iglesias, mine tunnel of Galleria Henry, Buggerru, Industrial plant with subterranean works of Porto Flavia, Iglesias - see Fig.2 -, Montevecchio industrial plants and subterranean works).

CONCLUSIONS

The end of most mine activities in Sardinia towards the end of the 20th century introduced the beginning of a new economical development mainly related to cultural and environmental tourism. The industrial archaeological remains have become a cultural heritage and the recognition of the Geomining Historical and Natural Park of Sardinia by UNESCO first (on the 30th of July 1998) and its institution by the Italian Ministry of Environment the 16th October 2001 recently have given form to this new project of cultural and economical sustainable and integrated development of the Island. The first results of this re-conversion are already visible, with the opening to tourism of some of the most important geosites and the enormous success of these initiatives evidences the feasibility of such valorisation projects, enriching the tourist offer of these regions already very much appreciated for their natural landscapes. The natural landscape, enriched by the historical signs of the passed geomining activities, make Sardinia among the most important and interesting mine regions of the world. These preliminary results encourage the continuation of this rehabilitation, covering all eight areas included in the Geomining Historical and Natural Park of Sardinia, but leads also to the conviction that specialised personnel has to be created in order to be able to give an adequate service to the many tourists that will come and visit these historically and naturally important landscapes.

ACKNOWLEDGEMENTS

This research has been in the framework of the National Project COFIN 2001-2003 entitled "Geosites in the Italian Landscape", regional co-ordinator Prof. Antonio Ulzega.

REFERENCES

Assorgia A., Beccaluva L., Di Paola G.M., Maccioni L., Macciotta G., Puxeddu M., Santa Croce R. and Venturelli G. 1976. Il complesso vulcanico del Monte Arci (Sardegna centro-occidentale), nota illustrativa alla carta geopetrografica 1:50.000. *Boll. Soc. Geol. It.* 95, pp. 371-401

Barca S. and Di Gregorio F. 1999. *Paesaggi e Monumenti geologici della provincia di Cagliari*, Saredit srl, Cagliari, 391p.

Bechstadt T. and Boni M. 1996. Sedimentological, stratigraphical and ore deposits field guide of the autochthonous Cambro-Ordovician of Southwestern Sardinia. *Memorie descrittive della Carta Geologica d'Italia* 48: 390 pp. Roma.

Boi G., Alba L., Angelelli F., Castelli P., Del Bono G.L., Di Gregorio F., Melis R., Napoleone I., Naseddu A., Pintus S., Saba F., Sanna C. and Todde F. 1996. Il parco geominerario ambientale e storico dell'Iglesiente-Sulcis - Guspinese. Atti del Congresso Internazionale per il centenario dell'Associazione Mineraria Sarda, 1896 - 1996, Sessione IV, Iglesias, 12-13 ottobre 1996, pp. 75-94.

Dessau G. 1937. Studi sulla miniera di Fontana Raminosa (Sardegna). *Periodico di Mineralogia* 8(1), 177-215.

De Waele J., Di Gregorio F. and Piras G. 1998. Geosites inventory in the Paleozoic karst region of Sulcis-Iglesiente (South-West Sardinia, Italy). International Congress PROGE0 '98, 7-13 June 1998, Belogradchik, Bulgaria. In *Geologica Balcanica* 28(3-4), pp. 173-179.

Fadda A.F. 1997. *Siti minerari in Sardegna ambiente e riutilizzo dopo l'abbandono*. CoEdiSar Cagliari, 208 p.

Fanni S. and Mezzolani S. 2001. *Guida al Parco Nazionale Geominerario Storico e Ambientale della Sardegna.* SiMe Cagliari, Carta geoturistica.

Manconi F. (editor) 1986. *Le miniere e i minatori della Sardegna.* Silvana Editoriale Milano, 240 p.

Marini C. and Mannini, E. 2000. Contributo alla conoscenza del giacimento di talco e clorite di Sa Matta, Orani (Sardegna Centrale). *Rend. Sem. Fac. Sci. Univ. Cagliari* 70, 333-351.

Massoli-Novelli R. 1971. Contributo allo studio genetico del talco di Orani (Nuoro). Nota 1°-Il giacimento di "Sa Matta". *Resoconti dell'Associazione Mineraria Sarda* 76(2), 1-31.

Massoli Novelli R. (1996), Geotopi e siti minerari, *Geologia dell'Ambiente*, anno 4, n. 2, 6-9.

Mezzolani S. and Simoncini A. 1993. *Paesaggi e architetture delle miniere*. Archivio Fotografico Sardo. 364 p.

Ottelli L. 1997. *L'Argentiera. Edizioni Gallizzi Sassari*, 236 p.

Pillola G.L. 1989. Trilobites du Cambrien inférieur du SW de la Sardaigne, Italie. *Paleontographica Italica* 78: 1-174.

Ruju S. 1996. *L'Argentiera.* Franco Angeli Editore Milano, 416p.

Stara P., Rizzo R. and Brizzi G. 1993. *Sarrabus, miniere e minerali.* Edizioni Arti Grafiche, Milano, 208 p.

Stara P., Rizzo R. and Tanca A.G. 1996. *Iglesiente, Arburese, Miniere e minerali.* Edizioni Arti Grafiche, Siena, 2 vol., 192 + 238 p.

In: Parkes, M.A. (Ed.) 2004 *Natural and Cultural Landscapes - The Geological Foundation*, 291-294, Royal Irish Academy, Dublin

PSILORITIS GEOPARK: PROTECTION OF GEOLOGICAL HERITAGE THROUGH DEVELOPMENT

Fassoulas Charalampos, Natural History Museum, University of Crete, Knossou Av., Heraklion 71409, Greece

Abstract: Psiloritis is not only the highest mountain of Crete but also the only area where all rock units of Crete outcrop. It is also an area with many geomorphological and tectonic features, such as gorges, caves, plateaus, fossils, folds, fault planes, valleys and high mountain tops. Culture and civilization follow their own paths for thousands of years adapted and inspired by a unique natural environment, which is enriched by a rich endemic flora and fauna.

The area is also one of the less developed in Crete, even though big tourist attractions (Knossos palace, Heraklion and Rethimno towns) are close. Agro-tourism and eco-tourism are the only tools for development. Recently, the geological heritage of an area extending almost in the whole of central Crete was combined with history, culture and environment giving rise to the Psiloritis Natural Park. As part of the European Geopark Network, the park aims to develop geo-tourism by attracting in the area all-season tourists who are also interested in the culture, history and tradition of Crete. A special kind of educational tourism seems to be realistic as many universities and school classes are visiting the area every year. Thematic courses related to speleology or to target groups (tourist guides, teachers etc.) can also be established.

Such kind of development seems to be the only tool for protection and maintenance of geological heritage in areas where legislation is not so strong, but it is also an action consistent with the tradition and culture of Crete.

INTRODUCTION

It is maybe only geologists who know how long it takes soft sediment to become a rock, a fold to be created, or buried wood to petrify. However, all of us know how quick a mountain can be totally excavated, a new road can be constructed or a coalmine can be increased. The development of our civilization is based on natural resources and on speed.

Unfortunately, rocks host most of the natural resources and most human constructions are founded on rocks. Therefore, rocks are facing a great amount of human pressure, having also the disadvantage that they are not alive. If the natural environment is regarded in our days as one of the greatest values on earth, this is mainly because animals and plants are alive. It is however obvious that if man hadn't gained a tolerable living level he wouldn't even think that animals are living creatures. After centuries of efforts we are pleased to say (at least in places) that we have become developed and civilized. We can think of the value of any living creature and we are willing to offer the money needed for its protection, but can we think the same about rocks? In Crete once, someone was accused of cutting a stalagmite in a cave. The court asked for scientific assistance and the only question from the judge was: "How much does the broken-piece cost?" This indicates how the value of geological heritage is underestimated or even ignored in Greece, as well as how the geotopes are treated.

The value of geological heritage had no sense over a few decades and was the major problem in thinking of its protection. It was mainly a problem for geologists that they couldn't present the value of a rock to the community and in some cases they couldn't even estimate it. In our days the value of geological heritage can be estimated in local, national or international level; it is well considered as an important parameter of an area and thus can be protected and preserved (Page 1999, Reimold 1999, Zouros 2001). Isolation from nature, urbanism, easy access to information and educational purposes have assisted the great to the establishment of a value for rocks.
The value of geological heritage in Psiloritis Mountains, occurring in southern Greece is still a matter of consideration. Located in one of the less developed countries in EU, and bearing a fantastic geological heritage, the mountainous Psiloritis area is facing serious development problems. Protection of this heritage is of high priority and a synthetic way has been found to strike a balance between development and protection of geological heritage.

STATE OF LEGISLATION

In Greece ecological awareness was growing fast over the last decades but still is not comparable to other north European countries (Modinos 2001). Country or local legislation however, don't follow at the same rates. Even though protection of natural environment (living environment) is at a good level the same is not the case for the geological heritage (Fassoulas 2001).

Geotopes can only be protected through the law for "The protection of Natural Environment" (1650/86), characterized either as National Park, Natural Monument, Aesthetic landscape or else. The process is however so complicated, mainly because it is based on the ideas for protection of the living environment, that it is actually impossible to go on. A second possibility is through the law for "Sustainable Development" (2742/99). This gives the right to the Secretary General of a Region to nominate an area as

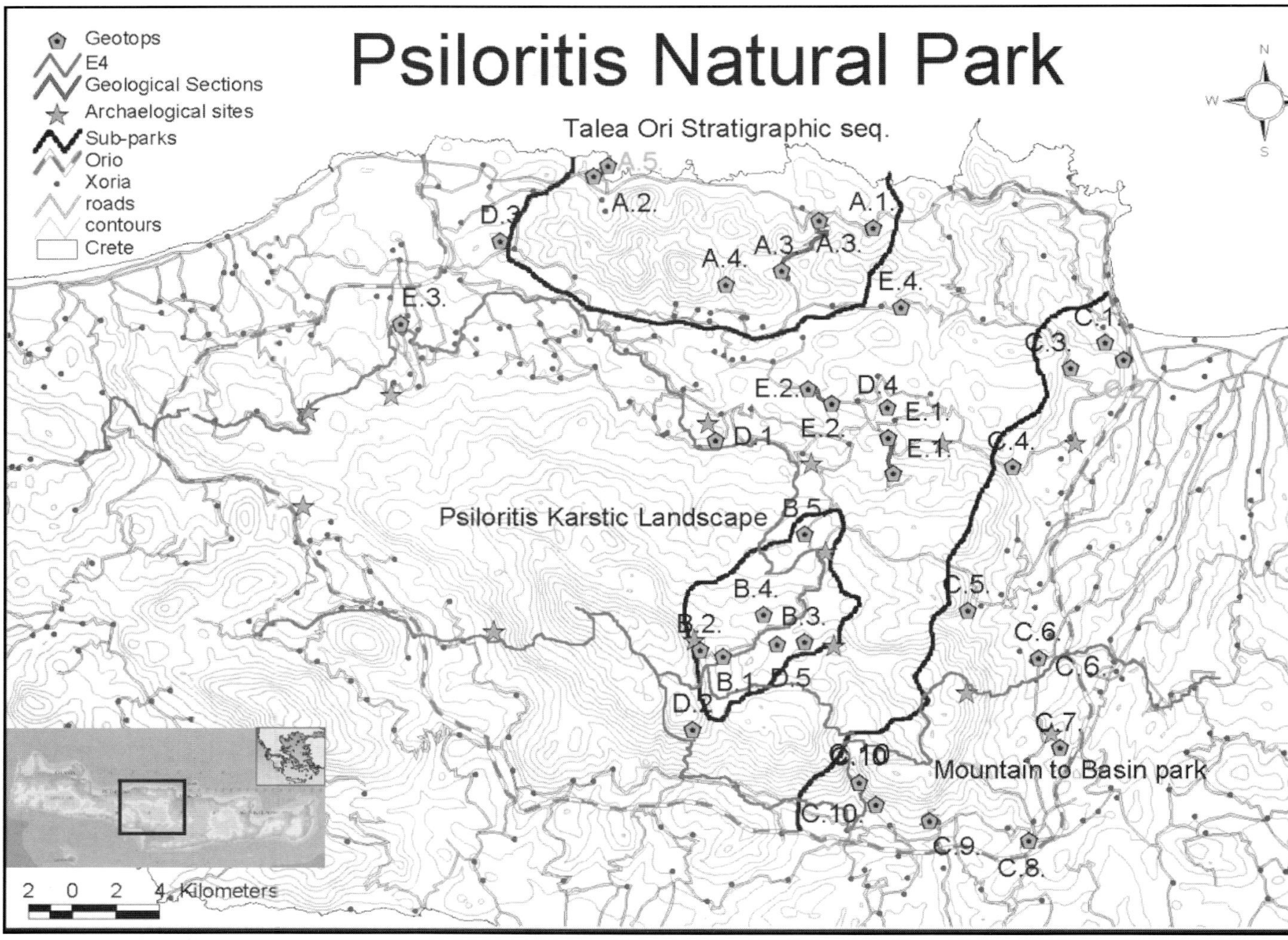

Figure 1. Location and main features of Psiloritis Geopark.

Special Protected Area due to Environmental reasons. In cases where geotopes are related with archaeological sites or are located within forestry lands, these can be protected through the Archaeological or the Forestry Services laws.
Practically, no geotope has been protected yet in Crete except the Samaria gorge, which has been nominated as National Park due to its individual endemism. Olympos Mountains in northern Greece and the Petrified Forest of Lesvos Island are the only protected geotopes in Greece (Katsadorakis 1999).

THE PSILORITIS GEOPARK

Psiloritis Natural Park is located in Central Crete, Greece, covering the Psiloritis Mountain chain and its northern hills (Figure 1). It is a mountainous area with many specialities and problems. Inhabitants are mainly engaged with agriculture and cattle raising. Even though the area is in the neighbourhood of the main archaeological sites of Crete (Knossos, Phaistos, Eleytherna), the main towns (Heraklio and Rethimno) and the main tourist centres of the island, it faces serious development problems. Since 2000 the park is member of the European Geopark Network. AKOMM-Psiloritis S.A., a local development association, is responsible for the management of the Geopark and Natural History Museum of the University of Crete is scientifically supporting the park.

Within the territory of the Geopark, the whole nappe pile of Crete, which consists of seven individual nappes, and the majority of the petrologic types of the island are presented in excellent outcrops and sections. All kind of geological structures are visible, from microscope to regional scale. Big faults with excellent and imposing fault surfaces, caves, impressive gorges hosting many endemic species of the island, unique fold associations and geomorphologic structures are feeding the culture, tradition and customs of the inhabitants for thousands of years.

As the Geopark covers a huge area geotopes are subdivided into sub-parks each one presenting a major geological feature of the area. Psiloritis Karstic Landscape is located at the Psiloritis tops comprising structures and features related to the uplift of the mountains and the karstic weathering. Furthermore, the main tectonic structures formed during nappe emplacement also outcrop. Talea Ori Stratigraphic sequence located at the northern coast offers a journey to the early geological history of the island. The whole statigraphic sequence of the Plattenkalk series is exposed only in this area in an inverted form. Rock types, folds and fossils comprise the geotopes of this sub-park. The Basin to Range Park located at the eastern slopes of Psiloritis Mountains offers a unique opportunity to study the development of the Neogene basins of Crete in respect to the high mountains. SpeleoPark is planned to be organized at the Psiloritis Mountains as an individual park in which the main aspects of speleology will be properly demonstrated in order to be experienced by the visitors (cave development and speleothems, archaeology, history and folklore, cave biology, adventure of exploration, etc).

Many other geotopes located within the geopark were assigned as individual sites presenting main features of the geological history of the island: the Gonies section presents along a short road-cut of 2 km almost all nappes of Crete, the "Sculptures of Nature" near Chonos (Figure 2) is an

Figure 2. "Sculpture of Nature" in the area of Chonos village.

area where karstic weathering and erosion of carbonate rocks has formed unique-shaped rock formations and landscape, and the Almyros spring is a unique example of the complex ways the underground water follows to reach the sea, having also a huge capacity.

All these areas are suitable for every educational activity (higher or lower level) that is related to weathering processes, tectonic activity, petrology and landscape development. This is the reason that many universities from all over the world have chosen Crete and Psiloritis mountains for their field trips. On the other hand, several geotopes can be characterized as unique and rare because of their appearance and preservation. The Messinian evaporites preserved in our days at an altitude of 500 metres give one of the

Figure 3. The Vossakos folds near the northern coast.

fastest uplift rates in the world; the Vossakos folded layers (Figure 3), which have been formed deep in crust under high pressure conditions, appear today in every scale; the "high pressure fossils" were living 300 million years ago and now are excellently preserved within the high pressure metamorphic rocks in Fodele area, and; the Ideon Andron cave is one of the most important archaeological sites of Crete, as Zeus the king of the ancient Greek gods is believed to have grown up there.

DEVELOPMENT OF THE GEOPARK

Geological heritage is not such an important feature of the island as is archaeology and culture. A model for the promotion and protection of the geological heritage in Crete has to encompass all the aspects that are related with it (Fassoulas 2001). Tradition, customs and songs are inspired by the landscape and environment; history and civilization were based on the natural wealth.

Figure 4. Information table in the area of the Psiloritis Mountains.

The main idea for the development of Psiloritis Geopark is to combine all the important and characteristic features of the area with the geotopes and the Geological heritage, presenting the natural and human environment as a whole. The establishment of the Geopark as an important value for the area has to be the first and most important target that will enforce local people's awareness and respect for the geological heritage.

Sustainable development of the Geopark is based on two main activities: education and eco-tourism (or geotourism). Education can be achieved by vertical and horizontal actions. The vertical actions can be related with the development of special courses and field trips addressed to universities and groups with special interests (speleologists, botanists, tourist guides etc.). Horizontal actions include general geology field trips, school projects, entertainment events and other activities for the public.

Eco- and geo-tourism are development activities that fit to the main targets of the Geopark and to the features of the area. The Geopark can be an attractive reason for visitors who want to combine culture, traditions and the unique environment of mountainous Crete. Establishment of such tourism will support many of the Leader initiatives that have been undertaken and will improve many of the living parameters of the area.

Basic infrastructure for the development of the Geopark (Tables, pathways, signs, information centres etc.) and information products (books, leaflets etc.) have already been studied and proposed for funding through EU programs (Leader+, Intereg etc.), whereas signs and information tables have already been placed in several geotopes (Figure 4).

In addition, a serious effort has been taken for the implementation of special management plans in some areas of the Geopark that participate in the Natura 2000 network.

CONCLUSION

During the last decades, development and protection of natural environment are two concepts conflicting with each other. However it is true that both are necessary for our community. Development offers jobs and the right of the people to live at their birthplace. Protection of natural environment is obvious as being the result of the intense degradation of the environment, but also because it offers a qualitative way of living and is a duty of the humans. It is thus necessary to find the best way to combine these concepts in such a way that the one can benefit from the other.

Development of the Psiloritis Geopark is based on the promotion of geological heritage and its relations with culture and tradition of the area. Success of this effort will assist and improve the local economy and knowledge of the geological heritage. Therefore, the value and respect for the geotopes will be increased, resulting finally in a higher possibility for preservation and protection and probably can lead to the improvement of existing legislation.

REFERENCES

Fassoulas C. 2001. Protection and promotion of geological monuments in Crete. In: *3rd Inter. Congr. For Management of Protected areas and Natural Monuments* (proc.). Lesvos, 309-316.

Katsadorakis G. 1999. *The Natural heritage of Greece.* WWF. Athens, 225p.

Modinos M. 2001. *Greece: State of Environment.* N.C.E.S.D., Athens.

Page K.N. 1999. Geoconservation in Devon - The developing infrastructure. *Read Ann. Conf. Ussber Soc.*, 352-357.

Reimold W.U. 1999. Geoconservation - a southern African and African perspective. *J. African Earth Sci.*, 29, 469-483.

Zouros N. 2001. Natural monuments and alternative tourism: Geo-tourism paths in the Aegean and Thrace. In: Protection and of geological monuments in Crete. In: *3rd Inter. Congr. For Management of Protected areas and Natural Monuments* (proc.). Lesvos, 127-142.

In: Parkes, M.A. (Ed.) 2004 *Natural and Cultural Landscapes - The Geological Foundation*, 295-296, Royal Irish Academy, Dublin

EUROPEAN GEOPARKS: THE IRISH EXPERIENCE - CUILCAGH MARBLE ARCH

Richard Watson, Fermanagh District Council

Abstract: Cuilcagh Mountain Park is of international importance for a variety of nature conservation reasons, including a rich diversity of earth science interests. The establishment of the Cuilcagh Mountain European Geopark is described.

The distinctive sandstone summit ridge of Cuilcagh Mountain dominates the Fermanagh countryside. Cuilcagh straddles the Irish border and is rich in globally and nationally rare habitats. The middle slopes are known as Marlbank and display one of the best examples of active blanket bog in Europe, as well as Northern Ireland's finest upland karst landscape. The European Commission has declared active blanket bog as a priority habitat for conservation within their Habitats Directive (*Council Directive 92/43/EEC on the conservation of wild fauna and flora*) because blanket bog is a rare European habitat and is being destroyed around the World by damaging human activities.

Cuilcagh has been heavily glaciated and has a wealth of geomorphological interest. The region has impressive archaeology and is internationally important for some of its species. For example, birds such as the Green White Fronted Goose which is a threatened species throughout most of its European range. The mountain has panoramic landscapes that are popular for outdoor pursuits such as hillwalking, caving, rock climbing and orienteering. Fermanagh District Council developed the Cuilcagh Mountain Park in 1999 to conserve an internationally important area of active blanket bog and to increase public awareness of related Earth Science and environmental issues. The Cuilcagh Mountain Park has facilitated environmentally sustainable land management over an area of 1500 hectares.

A hillwalker looks out over the sweeping expanse of blanket bog from the sandstone cliffs of Cuilcagh Mountain.

The Boat Journey in Marble Arch Caves brings visitors along the underground course of the Cladagh River.

The lower, limestone slopes of Cuilcagh Mountain contain extensive cave systems including Marble Arch Caves. These caves were first explored in 1895 by the famous French speleologist, or cave scientist, Edouard Martel. Fermanagh District Council developed Marble Arch Caves as a showcave in 1985 and the Caves are famous as one of Northern Ireland's leading tourist attractions, with almost one million visitors from all over the World to date.

Cuilcagh Mountain Park is also managed by Fermanagh District Council and opened on 18th June 1999 with support from the European Community's LIFE programme for nature protection. *LIFE Contract No: B4-3200/94/770 Conservation of Active Blanket Bogs in Scotland and Northern Ireland.* Additional funding for the Cuilcagh Mountain Park was provided by the UK Heritage Lottery Fund and the Environment & Heritage Service in Northern Ireland. The University of Huddersfield funded ongoing research into the hydrology and Earth Science interest of Cuilcagh Mountain.

Fermanagh District Council works in close co-operation with the Environment & Heritage Service, the Geological Survey of Northern Ireland, the Geological Survey of Ireland, the Royal Society for the Protection of Birds and others to protect the landscape of Cuilcagh Mountain. A steering group consisting of Environment & Heritage Service, Department of Agriculture & Rural Development, Forest Service, the Geological Survey of Northern Ireland and Fermanagh District Council co-ordinates the conservation of Cuilcagh.

The Cuilcagh Mountain Park interpretative centre is included within the visitor centre for Marble Arch Caves. The visitor centre facilities include an audio-visual theatre, exhibition area, classroom, restaurant, shop and adjacent woodland nature reserve. The centre is also the administrative base for both projects and contains offices, garages and workshops. Tucked away amongst limestone hills the visitor centre has minimal visual impact in the landscape. The building successfully combines vernacular and contemporary architectural styles and is partly built from local limestone. Good examples of local fossils have been built into the walls at various points to provide interesting teaching aids.

Right. Scientists from the Limestone Research Group of the University of Huddersfield downloading an automatic rain gauge in the Cuilcagh Mountain Park.
Left. A cave tour party in the Skreen Hill Passage of Marble Arch Caves.
Below left. Primary School children engaged in a Field Study project with one of Fermanagh District Council's education staff.

Council staff have developed an extensive environmental education service that offers field study and nature walk opportunities for schools, universities and adults in subjects such as geology, caves, peatlands, woodlands, invertebrates and plant identification. Individual educational packages have been tailored to meet the National Curriculum's for both Northern Ireland and the republic of Ireland. A range of educational publications, video packs and lectures supplements the Field Studies programme.

In the summer of 2000 the Geological Survey of Ireland advised Fermanagh District Council of a new international development in geotourism; the European Geoparks Network. Members of both the Irish Geological Surveys were going to attend the inaugural European Geoparks conference in the Maestrazgo Cultural Park in Spain in November 2000. They suggested that both Marble Arch Caves and the Cuilcagh Mountain Park could jointly make a convincing case for Geopark status. Acting on this suggestion Fermanagh District Council decided to send a representative to the conference.

Following the Maestrazgo conference Fermanagh District Council prepared a joint application dossier for Marble Arch Caves and the Cuilcagh Mountain Park. The application addressed the various qualifying criteria set out by the European Geoparks Network. For example the local geology is of national interest and can be interpreted at a number of good sites. Both the showcave and the Mountain Park had been developed with the assistance of European Commission funding and both have made a significant contribution to the local economy. Both the Caves and the Park are environmentally sustainable and have proper management procedures in place.

The application was scrutinised and approved by the Expert Committee of the European Geoparks Network. Subsequently both Marble Arch Caves and the Cuilcagh Mountain Park were awarded joint European Geoparks status at an awards ceremony during the 2nd European Geoparks conference at the Museum of the Petrified Forest on the Greek Island of Lesvos.

Fermanagh District Council's strong commitment to both promoting sustainable development and to protecting the environment of County Fermanagh has been recognised by their notable success in gaining joint European Geopark status for Marble Arch Caves and the Cuilcagh Mountain Park. This is a major achievement by the Council as, so far, there are only 12 European Geoparks (increased to 15 in October 2002) in the whole of Europe and, notably, this is the first European Geopark in the United Kingdom.

At present Fermanagh District Council is working with neighbouring local authorities in the Republic of Ireland, tourism bodies and both Geological Surveys to develop a regional sustainable tourism strategy for the Breifne Mountains that run along the Irish border in the counties of Cavan, Sligo, Roscommon, Leitrim and Fermanagh. A funding package of some 2 million Euros has been secured and the project is scheduled for completion by 2005. Much of the planning and proposed development work for the Breifne project has been influenced by experience gained within the European Geoparks Network. It is hoped that European Geopark status will be extended on a cross-border basis to a larger area of the Breifne region in the future.

As the European Geoparks Network continues both to grow in numbers and to expand its activities, it seems certain to become an increasingly important player in geotourism on the World stage. Individual Geoparks who are members of the Network will benefit from corporate marketing, increased public awareness, extra visitors, networking with other Geoparks, joint funding opportunities, staff exchanges and by being part of a strong, influential lobby group.

In: Parkes, M.A. (Ed.) 2004 *Natural and Cultural Landscapes - The Geological Foundation*, 297. Royal Irish Academy, Dublin

EUROPEAN GEOPARKS: THE IRISH EXPERIENCE - THE COPPER COAST/EUROPEAN GEOPARK, COUNTY WATERFORD

Des Cowman, "Knockane", Annestown, Co. Waterford

Abstract: The background and the establishment of the Copper Coast European Geopark is described.

Five adjoining community groups along the Waterford coast had been in existence since the early to mid-1990s. Though small, each was innovative and developmental in its own way but came together in 1997 to work together towards developing the tourist potential of the area in a structured manner. We decided that we wanted niche-tourism rather than the type of holiday 'trippers' that our nearest town, Tramore, encouraged.

The five community groups therefore decided to base themselves around the mining heritage of the area and hence "Copper Coast". Collectively they developed a series of walks with pamphlets interpreting the ecology and culture of the area. In collaboration with the Geological Survey of Ireland a package of pamphlets was developed explaining in popular terms the dominant cliff-scapes of the area. A common strategy of signposting and of interpretative plaques plus notice boards was also put in place. This was done with the help of Waterford Leader Partnership.

Other aspects of the area are independently developed by the individual communities, Therefore working from the east (Fenor) to west (Stradbally) the individual initiatives comprise the following:

FENOR/KILFARRASY

The major development has been the purchase by the local community of an area of fen-bog and the creation of an elevated walkway though it. Its main use will be for research and education.

DUNHILL

The community here have bought the medieval castle and church, the former of which is dramatically visible inland from the coast at Annestown. They have also developed an enterprise park, conference and communications centre as well as building community housing.

BOATSTRAND/ANNESTOWN

The main concentration here has been on preserving and presenting attractions of the village of Annestown and the small harbour at Boatstrand. However, along the river Anne a pioneering wetlands project has been put in place and Boatstrand is in the process of building a rescue and community resource centre.

BUNMAHON

A mining trail has been created as well as a small interpretative museum. A geological park has been constructed interpreting the local rocks and featuring a solar alignment using two pre-Christian inscribed stones. Architectural and geological surveys have been conducted of the nearby engine house complex.

STRADBALLY

Amongst other awards, this village represented Ireland in he *Entente Floriale* in 2002 and was European winner in its category. It also got an award for the best village green in Ireland because of its large "geological" sundial. An archaeological survey has also been conducted on the medieval ruined parish church.

Each of the communities has been responsible for its own green areas, flower arrangements and other decorative features.

In 2002 three feasibility studies of the area was conducted, their results collated and a development plan drawn up. Relationship with Waterford County Council and the Geological Survey of Ireland were formalised as public authorities that would help us avail of funding through the Interreg programmes in collaboration with other members of the European Geoparks Network. The next twelve months, therefore, should see substantial developments in terms of signposting, information points and marketing.

In: Parkes, M.A. (Ed.) 2004 *Natural and Cultural Landscapes - The Geological Foundation*, 298, Royal Irish Academy, Dublin

GEOLOGY IN AN INTEGRATED CULTURAL LAND-SCAPE: TRAINING AND RESEARCH ACTIVITIES

Mario PANIZZA and Sandra PIACENTE
(Dipartimento di Scienze della Terra, Università di Modena e Reggio E.)

Abstract: Within the framework of Geology, a stimulating perspective of training and research activities is given by the relationship between this discipline and cultural heritage or, more in general, between these assets and the whole landscape in which they are located. Training and research activities may then be carried out according to three approaches: environmental, historical and cultural-philosophical.

The environmental approach is based on the relationships between Geology and cultural heritage, in conformity to a scheme in which the environment may be considered according to two aspects: on one hand the geological assets, and, on the other hand, the geological hazards. On the contrary, cultural heritage may be considered as both vulnerability and places of fruition. These relationships may produce situations of risk or impact (Panizza, 1992).

The historical approach is based on history considered as a means of interpretation, i.e. on the concepts of continuity and integration between the present geomorphological context and - proceeding backwards in time - its history, prehistory up to its geohistory. The aim is to recognise the integration and relationships which have been established in the landscape through time, by interpreting all its "historical" components (lato sensu) in an interdisciplinary manner (Panizza & Piacente, 2000).

The cultural-philosophical approach concerns the cultural dialogue and integration between humanistic and scientific disciplines. More in general, this means to provide an answer to the ever-growing need for a "neo-humanistic" culture, that is for the unity of culture (Panizza, 1989).

Some examples of specific investigations, carried out according to the conceptual principles previously exposed, will be illustrated below, together with a new degree course in "Sciences for Cultural Assets", which has recently been activated by the Faculty of Sciences at the University of Modena and Reggio Emilia.

In: Parkes, M.A. (Ed.) 2004 *Natural and Cultural Landscapes - The Geological Foundation*, 299-300, Royal Irish Academy, Dublin

GEOECOLOGICAL TOURISM IN EAST KAZAKHSTAN

Yulja Kazakova, Institute of Geophysical Researches, Kurchatov, Kazakhstan

Abstract: East Kazakhstan region has a large territory with exceptional variety of landscapes and great possibilities of geoecological tourism development. Nowadays not the whole territory of region is involved in the tourism industry because of the presence of the former Semipalatinsk Test Site. Inclusion of Geological Heritage objects of whole region in tourist routes will allow the enriching of the ecological tourism programs.

INTRODUCTION

Ecological tourism is one of the most efficient ways of realising the conception of sustainable development. The main object of ecotourism is a natural diversity, which consists of biodiversity and geodiversity (geoecological tourism).

There are great variety of landscapes and geological structures at the large (283.300 sq.km) territory of East Kazakhstan region - from the Mountain Altai on the east to former Semipalatinsk Test Site steppes on the west, from taiga forests on the north to salt lakes on the south. There are great opportunities here for a visual demonstration of the linkage between geodiversity and biodiversity. The mountain forests of the Altai are being replaced by the steppe and semi-desert territories. The attractive dissected relief was created by the recent tectonic movements. This resulted in high mountains, Alpine meadows, mountain tundra and the deep depressions of rift origin with badlands and full sections of the Mesozoic-Cenozoic sediments. The biologists have a chance here to study life of rare animals, geologists - to investigate numerous geosites. The area draws mountaineers, rafting, walkers and numerous technogenic objects are very interesting too.

Investigation of the Palaeozoic sections helps to understand the history of Central Asian paleoocean - its origination, development and movements of the island arcs, marginal seas and microcontinents. The visual manifestations of the recent tectonic processes permit to trace development of geological structures in the centre of the Eurasian continent. The list of the geological heritage of East Kazakhstan region far from completion consists nearly of 100 geosites. Among them there are objects worth including in the World Heritage List.

OUTLOOK OF GEOECOLOGICAL TOURISM DEVELOPMENT AT THE SEMIPALATINSK TEST SITE AND ADJACENT TERRITORIES

The majority of geosites have been found in the eastern part of the territory, the rest were insufficiently researched because of the nearby location of the Semipalatinsk Nuclear Test Site. The area was inaccessible for a long period and are now in the rehabilitation stage but have a large tourist potential. The technogenic objects are of great interest: sites of nuclear tests of Experimental field, Balapan and Degelen. The experimental field was the site for ground and air tests, numerous craters are results of those. And also there were buildings and constructions where the equipment to record the explosion parameters were installed. The most interesting object on the Balapan area is the Shagan Lake, which was formed as the result of nuclear explosion. Shagan Lake is not only technogenic, but also a palaeontological object because fragments of Carboniferous sandstone with excellent plant replicas were thrown on the earth's surface by the explosion. The Degelen site is not only a place of underground nuclear explosion testing, but also a picturesque landscape. There are tunnels composed of granites with mattress parting and picturesque relics in the slopes of mountains, now sealed, where the underground nuclear tests were conducted. As the result of explosions in the slopes and tops of mountains the manifestations of latest tectonics of man-caused origin were formed: craters and collapses above explosion chambers, also seismovibration

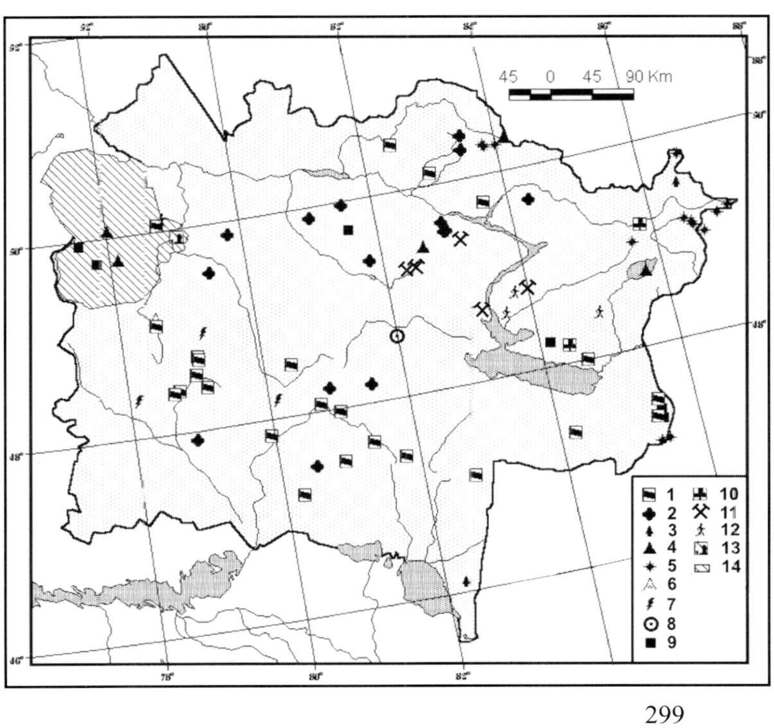

Figure 1. Map of Geosites of East Kazakhstan Region. Types of the geosites: 1 - Stratigraphical: Reference sections and stratotypes; 2 - Mineralogical and petrographical: Gem material; 3 - Hydrogeological: Mineral water; 4-6 - Geomorphologic: 4 - Landscapes, 5 - Glaciers, 6 - Caves; 7-9 - Structural: 7 - Seismotectonic, 8 - Impact, 9 - Others; 10 - Paleontological: Plant fossils; 11-13 - Anthropogenic: 11 - Ancient mine workings, 12 - Petroglyphs, 13 - Modern man-made constructions; 14 - Semipalatinsk Test Site territory.

and seismogravitational structures identical to the structures of catastrophic earthquakes epicentres.

Natural geosites of the Semipalatinsk Test Site and adjacent territories also varied. They are reference sections, tectonic processes manifestations, palaeontological fossils, semi-precious stone deposits, picturesque landscapes, ancient mine workings and petroglyphs. The stratigraphic sections serve as a key for decryption of the geologic history in the broad range of time (>half-billion years). The relevant example is the unbroken section of coal-bearing sediments from Upper Triassic to Middle Jurassic, which can be seen in the quarry of the Karazhyra coal deposit.

The ophiolite sections with the tectonic ultrabasic plates, which are usually substituted by the auriferous carbonate metasomatites, are extremely interesting. The typical example is the Chingiz ridge listvenite - carbonate gold-bearing substitution product of the ultrabasic rocks, which is one of components of Lower Cambrian ophiolite triad. Its outcrops are clearly seen on aerial- and space photos. Its observed thickness is up to 500 m, length is up to 100 km. The volcanic structures of the ancient island arcs and the olistostrome complexes determined the geotectonic development, and demonstrate the lateral zoning of the processes, Paleozoic volcanism and accompanying tectonic movements. Evidences of neotectonics - the recent and modern relief-forming processes which have the composite interference relations with the mechanical consequences of the nuclear explosions, carry the great interest.

There are numerous deposits and localities of the gem material: agate, chrysolite, blue microquartzite, wollastonite, chrysotile-asbestos, decorative jasper, rauchtopaz on the Test Site and adjacent territory. Thus the tourist potential of this territory may be greatly extended because of the inclusion of not only man-made, but also nature objects, interesting for both specialists and popular tourism. The region can become the object of geoecological tourism development. It is necessary for it to conduct deep research of the local geosites and their popularization with production of clear, geologically exact, widely available tourist information and publications.

GEOECOLOGICAL TOURISM IN THE EAST PART OF REGION

The experience on ecological tourism development in the east part of the region is exclusively interesting. This territory always was opened for revelation and account of geological heritage objects, which are traditionally used in tourism here. Markakol and West-Altay reserves, Katon-Karagay national park situated in this region occupy the area of nearly 300 000 sq. km. Besides there are numerous monuments of Nature here, including a big diversity of geosites. There are the reference sections from the Cambrian to Cenozoic age, fossils, gemstones, the picturesque cliff landscapes, the powerful waterfalls up to 80 m high, the medicinal mineral springs, numerous ancient mines, caves and one meteoritic crater among them.

I would like to enumerate the most interesting geological monuments of Nature.

- Markakol is a tectonic mountain lake (1485m asl).
- The Belukha Mountain is a symbol of Altai and the highest peak of Altai and Siberia.
- Geological section Tarkhansky is a stereotype of Tarkhan suite of Carboniferous Tournaisian stage. It is characterized by abundance of plant fossils and represents a standard for making a boundary between Devonian and Carboniferous sediments.
- The combined geological, geomorphologic and landscape monument of Nature Kiin-Kerish consists of the most complete cross-section of the Mesozoic sediments. This section contains five floristic horizons with numerous imprints of the ancient flora from the subtropical (palm-tree, magnolia, araucaria, ginkgo) to the thermophilic (lime-tree, alder, platan, elm, oak) plants and four horizons with bone remains of rhinoceroses, pigs, ruminants, crocodiles, turtles, salamanders and other subtropical forest animals. "Badland" relief is very peculiar.
- Shekelmes Mountain is a geological/ landscape nature monument. Numerous erosion cuts uncover the variegated sand-clayey sediments with many Neogene fossils.
- Delbegetei granite massif is a natural geological museum of minerals and geological processes. This is the only emerald extraction place in Kazakhstan. Pegmatite granite and enclosing rock contains the topaz, rock crystal, morion, schorl, heliodor, fluorite, aquamarine, beryl, muscovite, microcline and others.
- Ashutas Mountain is a reference geological section of the East Kazakhstan of Paleogene - Neogene sediments with unique fossils of the Turgai broad-leaved vegetation.
- Nature monument "Stone Sity" (West Altai Reserve) is a unique landscape complex: granite monadnocks, Alpine lakes and swamp of the tundra type. The picturesque cliffs are composed of biotite, muscovite, binary and leucocratic granite of the Permian Beloubinsk complex.

CONCLUSIONS

Geosites of the East Kazakhstan region are interesting both for specialists and wide circle of tourism lovers. Programs of the test sites and adjacent lands rehabilitation have to provide revelation, protection and use of geosites. These unusually earlier closed territories can successfully use in ecological tourism, which development is very perspective in Kazakhstan. Organization of uniform system of tourist routs at whole territory of the East Kazakhstan region allows to extend the ecological tourism programs.

REFERENCES

Fishman I. About Development of Ecological Tourism in Kazakhstan. Informational Ecological Bulletin of Ministry of Natural Resources and Environment Protection of Kazakhstan, 3 quarter 1999, p. 112-117.

Fishman I.L., Kazakova Yu.I. Geological Heritage of the Test Sites: Geoecological Monitoring and Geoecological Tourism. Monitoring of Nuclear Tests and Their Consequences. Abstracts of International Conference, Borovoye, Kazakhstan, 2002, p. 71-72.

Kazakova Y. Geological Heritage of the Semipalatinsk Test Site. In *Towards the Balanced Management and Conservation of the Geological Heritage in the New Millennium*, Madrid, 1999, p. 365-366.

In: Parkes, M.A. (Ed.) 2004 *Natural and Cultural Landscapes - The Geological Foundation*, 301-304, Royal Irish Academy, Dublin

AESTHETIC LANDSCAPE AND GEOTOURS IN ALBANIA

Afat Serjani, Adil Neziraj, Halil Hallaçi, Tonin Deda, Geological Research Institute, Blloku "Vasil Shanto", Tirana, Albania. Email: aserjani@yahoo.com

Abstract: Albania is a sunny Mediterranean country with a lot of natural beauty. The wild and untouched landscapes, the diversity of glacial, erosion and karst phenomena, the richness of water, the springs and the river landscapes of Albania form a great potential for development of tourism. The tourist potential is enormous due to the heterogeneity of the landscape and the immense wealth of cultural heritage. Geological evolution, various types of rocks, climate changes and atmospheric activity have caused formation of varied landscapes that astonish with their natural beauty. They occur in mountainous units, along with the Adriatic coast and the Ionian Riviera. The most part of these landscapes represent tourist spots of education, didactic, geoecotourist and climate-curative-recreational values. The longest geotour passes through ancient towns of Berati and Gjirokastra, to Butrinti archeological town and then along the beautiful Ionian Riviera. The Northern geotour runs through historical towns to the Albanian Alps. To the east a recommended geotour crosses Albanian ophiolites up to the Peshkopi salt domes and on to the Korabi Highland. Separate geotours are recommended to Mali me Gropa Mountain full of karst forms and to Lura Geopark with some beautiful glacial lakes. Short geotours from Korphy to Saranda-Gjirokastra region, from Monte-Negro to Western Albanian Alps, and from Kosovo to Northeastern Albanian Alps can be undertaken as well.

INTRODUCTION

Albania, in spite of criticism during the last 10-12 years after the transition period, remains one of the Mediterranean countries with great natural beauty. Our country it remains as beautiful as it has been during the times when written about by our great writers and philosophers: Naimi, Samiu, Shiroka, Çajupi, Mjeda. Our landscape it is so aesthetic as to have been written about during the last centuries by Byron and Edith Durham from United Kingdom, Pouqueville from France etc. Due to its geological-geographical position in the Eastern Alpine Mediterranean Chain and its accidental relief, Albania is a rich country in special geological-geomorphologic phenomena and full of beautiful landscapes which in most cases represent geological sites.

In mountainous units of Albania there are landscapes of predominantly glacial origin especially in higher levels, with erosional and karst forms on the surface of carbonate and evaporite rocks. Natural river landscapes with canyons and narrow gorges along with the upper streams of Drini, Vjosa, Shkumbini, Semani and Mati rivers are formed, while to the western lowland to the Adriatic coast they form wide fluvial beds and delta mouth landscapes to the sea. There are formed also sandy accumulative landscapes along the Adriatic coast and rocky erosion beaches in the Ionian Riviera. Geotourist values of Albania are often linked with geomorphological-hydrological spots associating outstanding natural landscapes with cold water springs.

The absolute tourist leaders are the Ionian Riviera, the Adriatic coast and Ohrid, Shkodra and Prespa lakes. Some geoparks such as: Lura geopark with many glacial lakes 1800-2000m above the sea level, Valbona Mountainous Amphitheater with high erosion peaks and glacial moraines, Llogara is full of pine forest and cold water springs; Dajti national park east of Tirana and Thethi, Boga, Razma, Voskopoja, Qafe Shtama, Dardha, Çajupi aesthetic landscapes possess climate-curative-recreational values. For protection of landscape and for development of tourist industry the central and local administration of Albania must contribute, especially the Ministry of Tourism established recently, but also institutions, organizations, societies, private firms and companies. We are of the opinion that it is necessary to give rights to foreign tourist companies for management and use of some geoparks and tourist spots in Albania.

LANDSCAPES OF ALBANIA

In Albanian territory, there are structural-erosional, karstic, river-erosional, erosion-denuded, glacial and seacoast erosion or accumulative types of geomorphologic landscapes. The main factors which influenced the landscape formation are as following: geological-tectonically evolution of Dinarides-Albanides-Hellenides chai; ophiolite magmatic rocks formed as result of Mirdita Oceanic spreading, metamorphic rocks in the Korabi zone and sedimentary rocks: platforms and basins in external tectonic zones; climate changes; continuous activity of the atmospheric agents; accidental relief.

The geological-geomorphologic evolution belongs to neotectonical and landscape formation processes in Albania as part of Dinarides-Albaniades-Hellenides Chain. The main processes of the evolution there are:

1.Relations and mutual action of tectonic plates and especially Adria Microplate (Apulian zone) with Euroasian Plate (Orogen).

2.Compression stage after Mastrichtian period which caused the folding and formation of main geological structures. The Albanides were established during the Mastrichtian and changed later, while during Pliocene-Quaternary compressive uplifting phase mountain formation happened and the Scutari-Pec transverse deep fault was reactivated.

3. Clockwise rotation of Albanides domain (by a total of about 45o to 50o degrees) opposite to Scutari-Pec (Shkoder-Peja) transform fault.

Referring to the concept of " Geodiversity" (Erikstad et al.

2001) and according to the geological construction and to geomorphologic-landscape features in Albania, we can determine some areas of geodiversity which make up the solid ground for separate ecosystems:

- North of Shkoder-Peja (Scutari-Pec) transform fault there are found the Albanian Alps - high mountainous area constructed mainly of carbonate rocks full of glacial, karst and erosion forms caused by the same geoprocesses.
- Mirdita zone represents specific geodiversity because of Mirdita Oceanic Spreading and variety of magmatic and metamorphic rocks with predominance of ophiolites. This zone is situated from the Shkoder-Peja transform fault north to Leskoviku town in the southeast.
- To the northeastern Albania there is risen the Korabi Highland where outcrop Palaeozoic rocks influenced by some old denudation and orogenic stages.
- External zones where are included Krasta-Cukali and Ionian pelagic basins and Kruja and Sazani (Apulian) shallow carbonate platforms represent common geodiversity and biodiversity features as well.
- Western hilly and lowland zone from Vlora and Vjosa river south to Shkodra and Buna river north of smoothly denuded relief constructed by Pre-Adriatic Depression (PDA) molasses and flysch formations.
- Adriatic coast from Velipoja north to Vlora bay with Velipoja, Lezha, Divjaka-Karavasta and Narta lagoons and with some sandy beaches.
- Ionian Riviera from Vlora Cold Water to Saranda (Ksamili islands) rocky seaside with many native beaches.

Glacial landscapes are formed mainly at 1600-1800m above the sea level. In the Albanian Alps, in Korabi Highland and in Lura, Bulqiza, Shebeniku, Vallamara ultrabasic massifs there occur nice glacial lakes, while in Munella, Gjallica, Gramozi, Nemeçka, Griba and Çika mountains there are many moraines, cirques and glacial traces. Erosion and river-erosion landscapes are widespread mainly in Northern (Albanian Alps), Central-Eastern and South-Western mountainous geographical units of Albania.

Karst landscapes covers :

- Carbonate rocks in Albanian Alps, mainly in western part in Malesia e Madhe subzone, in thick shallow water carbonate sequences on the ophiolites of Mirdita zone, and in external zones with predomination on the surface of Kruja and Sazani (Apulian) platforms. In Ionian zone it is Kurveleshi Plateau which is considered as a natural karst museum with high narrow canyons, a lot of caves, erosional towers etc.
- Evaporite rocks occur especially in Dumrea, Kardhiqi, Delvina, Mursia salt diapirs.

The most part of aesthetic landscapes in Albania are linked with geoparks, with climate-curative mountain spots, with cold water springs or with nice lakes and lagoons along the Adriatic coast. From North to south there are placed the following national geoparks (Serjani et al. 1999): Valbona in the most upper stream of Valbona river at the foots of Jezerca mountain, Thethi in Albanian Alps, Lura east of Mirdita region in Lura ultrabasic massif, Dajti east of Tirana-capital, Drenova east of Korça and Llogara south of Vlora. In Valbona Mountainous Amphitheater, placed at 1100m above the sea level, surrounded by high erosional pyramids there are some glacial cirques and moraines and on the slopes there is grown forest of *Fagus sylvatica*, *Pines negra* and *Picea exselsa*. Thethi national park and forest is at 700- 2500m above sea level. Lura national park (Serjani 2002) found in eastern Albania is beautiful with its 12 glacial lakes. Dajti national park with its forest, wild animals, different erosional and karst forms and with fresh climate it's used by Tirana's inhabitants. Drenova geopark placed to the east of Korça city is famous for its pines and abies forest. In Llogara geopark in high forest of Pines and Abies at cold water springs 800-1200m above the sea level you feel fresh mountainous air and at the same time you can smell sea air from Orikum and from Dhermi Riviera at the foots of Çika mountain.

The best climate-curative spots of Albania there are: Boga, Razma and Vermoshi in Albanian Alps, Qafe Shtama east of Kruja old town, Gjinar in Elbasani region, Prespa, Dardha and Voskopoja in Korça region, Çajupi in Gjirokastra region. In some specific tourist spots there are landscapes situated next to cold water springs where commonly there are grown high old trees. Such cold water springs are those of Vlora, Tepelena, Kelcyra, Borshi, and Bogova etc.

Very important for tourist values are Shkodra, Pogradeci (Ohrid) and Butrinti lakes. Shkodra and Ohrid lakes represent international ecological areas. Shkodra lake represents a neotectonic landscape with Shegan, Koshenj and Rjoll underwater springs, with a specific lake green flora and many rare kind of birds. The region around the Ohrid lake it is considered as absolute tourist leader of the region (Baumann 2000; Jozja et al. 1998) which can be used as climate health resorts and for other leisure activities. The lagoon and above all Butrinti lake around the archeologically important place, represent a geocomplex of great interest because of their high biodiversity. For tourists coming from Corfu island, Butrinti could reduce the threshold of a visit to Albania and lead to seeing more of the country. They can see the beautiful Ksamili islands on the most southern part of Albania's beautiful Riviera.

MAIN GEOTOURS THROUGH ALBANIA

The main geoecotours which we recommend in this paper presented in the Geotourist Map of Albania (Serjani et al. 2002) there are as following:

1. Tirana - Durres - Divjaka - Berati - Tepelena - Permeti - Gjirokastra - Saranda - Vlora - Tirana.
2. Tirana - Kruja - Shkopet - Lezha - Shengjini - Shkodra - Komani - Bajram Curri - Valbona and back.
3. Tirana - Elbasani - Librazhdi - Pogradeci - Korça - Prespa lake and back.
4. Tirana - Kruja - Shkopet - Burreli - Bulqiza - Peshkopi - Bellova and back.
5. Tirana - Kruja - Rubiku - Rresheni - Lura geopark and back.

Figure 1. Geotouristic Map of Albania.

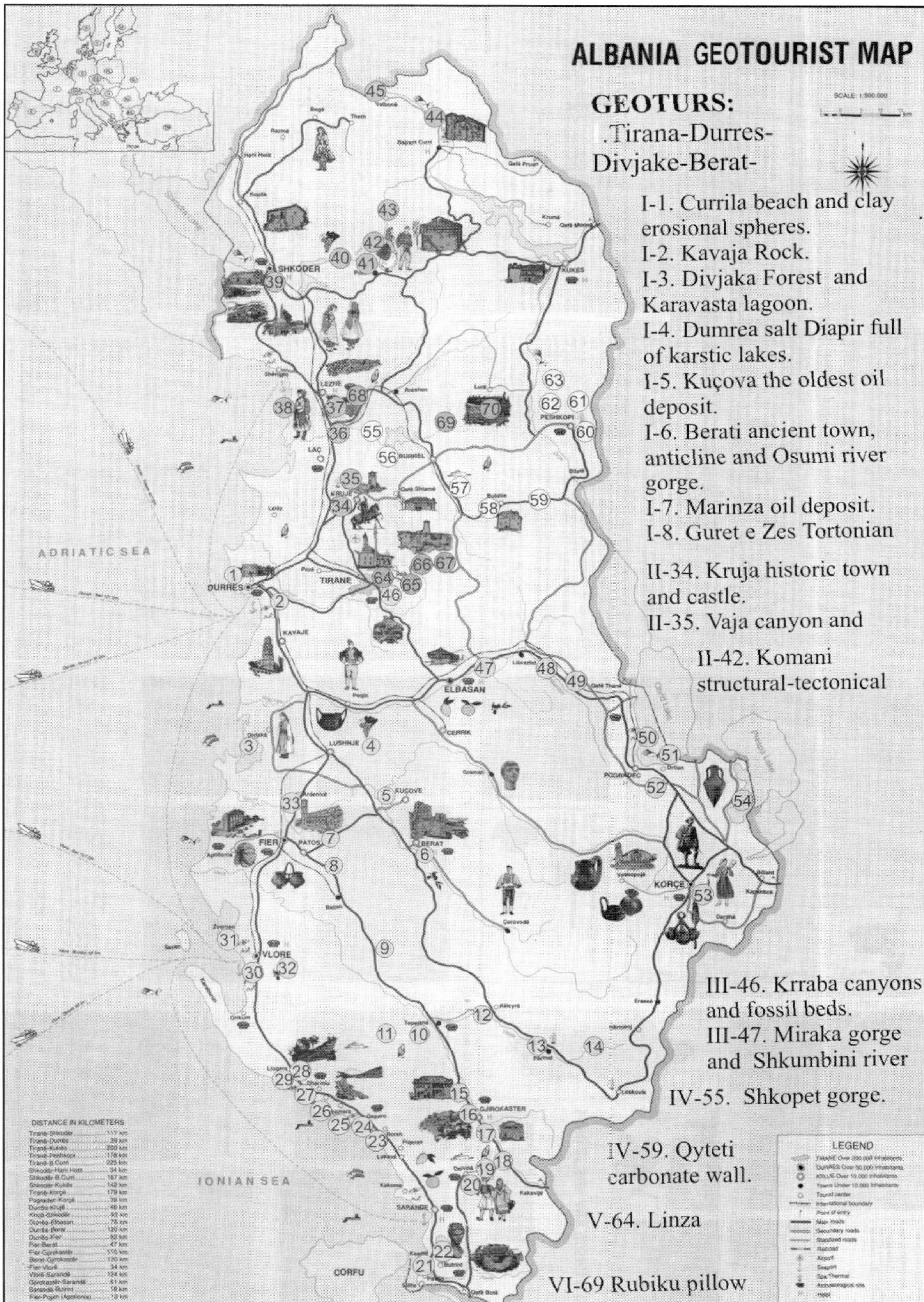

6. Tirana - Dajti national park - Selita - Mali me Gropa (Honey Comb Mountain) and back.

The above mentioned itineraries are presented in the Geotourist Map of Albania (Figure 1). In the first geotour to the south of Albania a visitor can see Berati and Gjirokastra ancient characteristic towns, Butrinti archeological spot, Divjaka-Karavasta lagoon, complex of Gurra canyons, erosion towers and waterfalls, Kelcyra and "Blue Eyes" cold water tourist spots, beautiful Ionian Riviera from Ksamili islands to Vlora cold water spring and Logara geopark.

In second geotour to the north of Albania a visitor can see

Kruja, Lezha and Shkodra historic towns and their castles (with views), Shkopeti gorge, Komani structural-tectonic site full of folding of carbonate rocks of different sizes and colors, Valbona valley and Valbona Mountainous Amphitheatre. The third tour goes to the east of Albania to Korça cultural city. Pogradeci (Ohrid) and Prespa lakes there, are part of the best tourist complexes in this region. Along with this itinerary Librazhdi red series, Skanderbeg Table, Guri i Kamjes and Guri i Capit erosional sites can be visited as well.

The geotour to the northeast of Albania from Tirana to Peshkopi visits ophiolite sequences of Bulqiza ultrabasic massif and Peshkopia salt domes. Lura geopark may be the best place for glacial and ecological research and exploration. The last geotour is dedicated to special karst phenomena, on the large surface of carbonate rocks in Mali me Gropa (Mountain with holes) full of different karst forms.

CONCLUSIONS & RECOMMENDATIONS

- In Albania due to geological-geomorphological evolution there are some genetic kinds of landscapes of predominantly erosional, river-erosional, glacial and karst types.
- Albania is a country of many natural monuments, geoparks and forests of geoscientific, pedagogical, educative, climate-curative, tourist-geotourist values.
- Albania is a Mediterranean country with many sunny days during the year, with good hot climate and many virgin sandy and rocky beaches, with a lot of natural geological and geomorphological landscapes of aesthetic view, with some geoparks and high old forests, many cold water springs, glacial and karstic lakes, with some aesthetic climate mountainous spots, with some archaeological, historical and architectural towns or " towns of stone".
- Albania is a country with a good perspective for development of tourist and geotourist industry. Here we can remember the evaluation of Baumann Ch. (2000): "*I imagine* (developed) *tourism in Albania to be comparable to Swiss tourism. Albania attracts tourists who want to experience nature as well as they want to visit and like romantic and idyllic places*".
- Preservation, conservation and management of landscapes and their use for tourist and geotourist purposes must be included into environment policy of our country and State.
- Natural landscapes, monuments, geoparks and forests must be used for education of young generations and for rest of common people, for scientific research, for tourism, geotourism, sport and medical treatment.

For intensive development of tourism we recommend:

- To establish necessary legislation in accordance to The Resolution of The Tenth Assembly of the United Nations Organization for the Protection of the Nature;
- To establish new infrastructure and buildings and especially to pave the roads which lead to landscapes and tourist spots;
- To establish rules for central and local management of tourist places.
- To compile and to publish geotourist guide books for separate geotours and regions of Albania in some foreign versions.
- We recommend foundation of regional centres in some parts of Albania for knowledge, inventory, description, protection and management of landscapes, sites and geoparks.
- Stimulation and financial support of geoecological and tourist business by Albanian State is necessary. At the same time, we recommend management and use of some geoparks by foreign tourist companies.

REFERENCES

Gjeografia e Shqiperise, 1990. (Geography of Albania).Vols. I., II, Tirane.

Harta turistike e Shqiperise. 1990. (Tourist Map of Albania). Tirane.

Baumann Ch., 2000. Some outstanding Geotopes of Albania and their Geotourist Values. Presentation in the 8-th Congress of Albanian Geosciences. Tirana, Nov., 6-8,2000.

Cialvola, P., Arthurton, R., Brew, R., Lewis, P. M. 1996. Karavasta Lagoon and Surrounding Area. Stud. Gjeog. Nr. 9, Tirane.

Costamagna, Alberto 2002. The inventory of Geomorphologic Geosites in the framework of landscape management and tourist improvement. Proceedings of Workshop: Geomorphological Sites: research, assessment and improvement. Modena (Italy), 19-22 June.

Erikstad, L., Johansson, C. E. and Suominen, V. 2001. *Geodiversity in Nordic Nature conservation.*

Gonggrijp, G. P. 1981. Earth Sciences Conservation: no science without conservation. *Geol.* 60, No 3, Wiggers issue (433-445).

Jozja, N., Neziraj, A. 1998. Geological Heritage Conservation. of Ohrid Lake. *Geol. Balc.* 28, 3-4.

Larwood, J. and Prosser, C. 1998. Geotourism, Conservation and Society. *Geologica Balcanica* 28, 3-4.

Palmentola, G. 1994. Glacial Features and snow-line during the last glacial age on Albanian and Greek mountains. Proceedings of the 7-th Congress of the G.S.G. Thessaloniki.

Serjani, A., Heba, G. 1996. Gjeotrashegimia ne Shqiperi (Geological Heritage Conservation in Albania). Studime Gjeografike, No 9, Tirane.

Serjani, A., Neziraj, A., Jozja, N. 1998. Geomorphological Sites of Albania. ProGEO '98 Meeting, Belogradchik, Bulgaria. *Geologica Balcanica* 28, 3-4. (129-136). Sofia.

Serjani, A., Neziraj, A., Jozja, N. and Niklekaj, V., 1999. Geoscience Significance and tourist values of some "UNESCO/IUGS Geosites and Geoparks" in Albania. Proceedings of ProGEO'99 Meeting: Towards the Balanced Management and Conservation of the Geological Heritage in the New Millennium. (94-101), Madrid, Nov. 1999.

Serjani, A. 2002. Lura Geopark, Albania. ProGEO news, No. 1. (1-3). Stockholm, Sweden.

Serjani, A., Hamdi Beshku and Ndoc Vukzaj, 2002. Geomorphological Sites in Albania and their geotourist values. Proceedings of the Workshop: Geomorphological Sites: research, assessment and improvement. (84). Modena (Italy), 19-22 June.

In: Parkes, M.A. (Ed.) 2004 *Natural and Cultural Landscapes - The Geological Foundation*, 305-308, Royal Irish Academy, Dublin

UNIQUE LANDSCAPES OF KARELIA (NW RUSSIA) FOR TOURISM

V.V. Makarikhin, Institute of Geology, Karelian Research Centre RAS, Pushkinskaya 11, 185610 Petrozavodsk, Russia. Email vmakar@krc.karelia.ru
Y.J. Systra, Mining Institute, Tallinn Technical University, Kopli 82, 10412 Tallinn, Estonia. Email systra@staff.ttu.ee

Abstract: The Republic of Karelia (NW of Russian Federation) is located in the southeastern part of the Precambrian Fennoscandian Shield and occupies its oldest and the most stable nucleus. Hard crystalline rocks are divided into blocks by numerous deep and narrow elongated valleys of the fault zones. Rocky fault scarps, 60 thousand lakes, hills covered by coniferous forest with woodless mires form a big variety of picturesque landscapes. Well-exposed Precambrian bedrock, polished by ice, shows the evolution history of the shield from Early Archean (3.5 Ga) to Late Proterozoic (1.1 Ga).

Processes extending back to early Precambrian time, when the oldest Archean rocks of the crystalline basement of the Fennoscandian Shield were formed, have shaped the natural landscape of Karelia. Nearly two billion years of geological evolution has passed since these strongly metamorphosed and repeatedly deformed volcanic and sedimentary rocks and intrusive magmatic bodies were first exposed at the surface (Figure 1). In some areas we can see how rocks have been exhumed from depths of 15-20 km during continental collisions, supplying vast amounts of sediments to rivers. In other places we can see remnants of the old erosion surfaces, upon which these sediments were deposited, and trace the record of inundation by the sea several times over nearly 1.5 billion years.

Figure 1. Jatulian sandstones (2,2 Ga).

Some rock types, such as quartzite, gabbro, mafic volcanics are more resistant to erosion to others, so that even where they are folded into narrow synclinal folds, they are expressed as topographic ridges. Intervening areas of lower relief tend to be underlain by less resistant, highly fractured and strongly jointed rock of fault zones, and are often associated with deep narrow valleys. Successive advance and retreat of Quaternary glaciers scoured and removed the deeply weathered regolith and rejuvenated the landscape, exposing fresh and often highly polished rock surfaces (Figure 2). Accordingly, the eastern part of the Fennoscandian Shield shows rather subdued relief. The thickness of the glacial and glaciofluvial deposits is generally only a few metres, although in some places, notably where there were temporary hiatuses in glacial retreat, glaciofluvial accumulations of tens of metres are found (Figure 3), and old valleys may be buried to depths of 100-150m. The vegetation cover is dominated by coniferous forest. There are spectacular fault scarps, many wetlands and 60 thousand lakes, including the largest bodies of freshwater in Europe, Lake Ladoga and Lake Onega. Shoreline gneiss outcrops on the White Sea, together with the 'white nights' during summer at high northern latitude, combine to give extremely picturesque landscapes.

Figure 2. Polished surfaces of the Archean gneisses.

Several hundred sites of geological interest have been documented and some tens of unique geological monuments have been designated (Figure 4, Lapo 1996; Makarikhin et al. 2001; Systra 1998). Because of the natural assets, and the relatively low population density (only 4 people per square kilometre), Karelia is becoming a popular destination for outdoor-based tourist activities, including wilderness camping and hiking, fishing and picking. In the last eleven years three national parks and one nature reserve have been declared and several more are being gazetted. There are also negotiations underway for improving access for European tourists to the Paanajärvi National Park, in the northern part of Karelia, by establishing new border-crossing stations near Pyaozero.

Although cultural landscapes occupy only some percent of territory, they have great historical interest. They are living places for indigenous European peoples such as the Vepsian and Karelian nations, who still retain their own fishing-hunting culture. Old wooden settlements located along the lakes and seashore may have interest for tourists.

Figure 3. Karelian moraine.

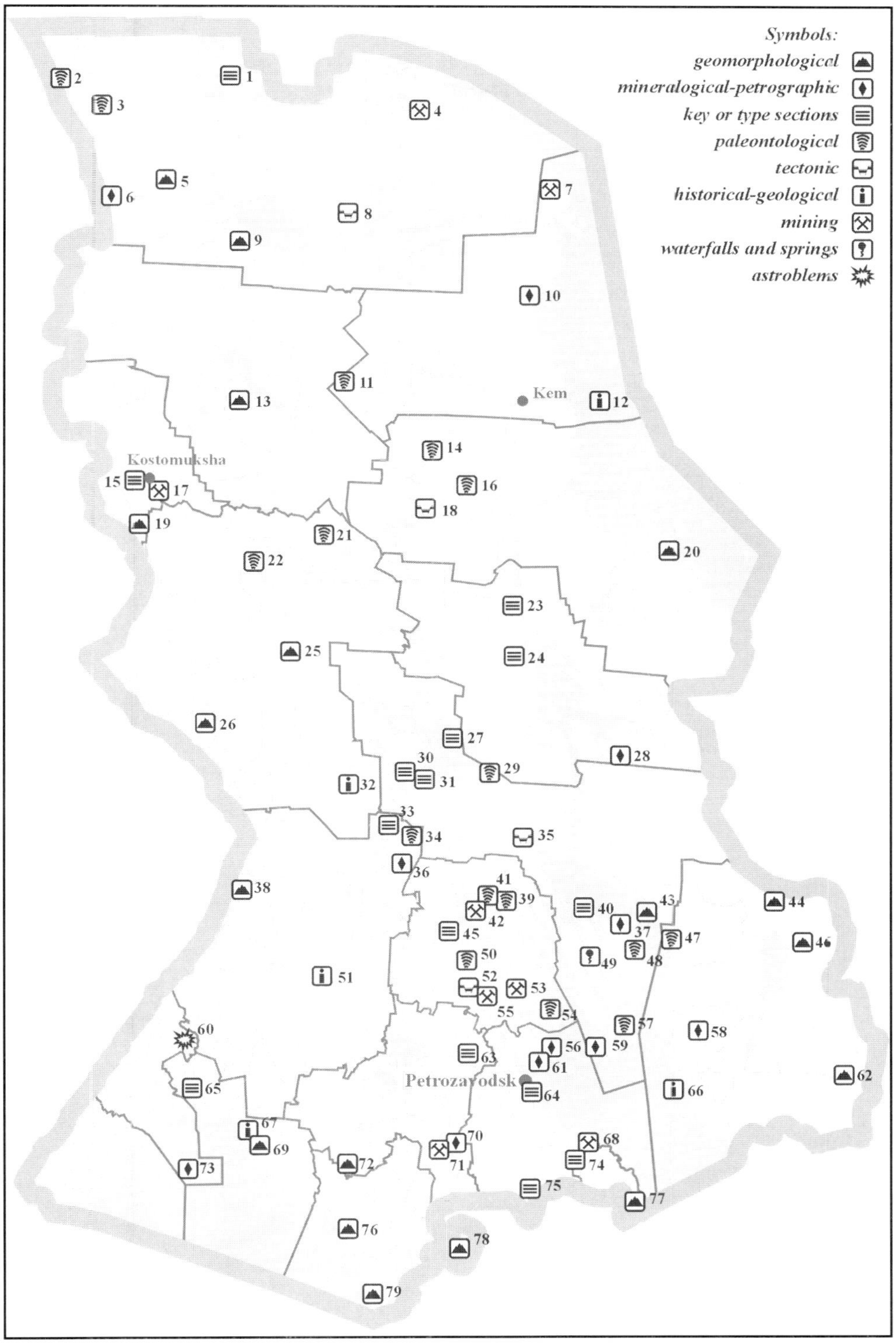

Figure 4. 1-Kukasozero, 2-Sovajarvi, 3-Kapokallio, 4-Chkalovski, 5-Paozero, 6-Poncha, 7-Gridino, 8-Keret, 9-Topozero, 10-Mramorny, 11-Ozhijarvi, 12-Kuzova, 13-Nizhnee Kujto, 14-Ylaostrov, 15-Kostomus II, 16-Hangaslambina, 17-Kostomus I, 18-Kozledjoki, 19-Druzhba SR, 20-Sorokski Park, 21-Chirka-Kem, 22-Bolshozero, 23-Voitsa Mine, 24-Segezha, 25-Sonozero, 26-Lendery, 27-Lehtassari, 28-Vyg, 29-Dulmek, 30-Luzhma, 31-Gormozerka, 32-Vottovaara, 33-Gumarino, 34-Kumsa, 35-Medvezhegorsk, 36-Sovdozero, 37-Tolvua, 38-Kojtajoki, 39-Kappeselkka, 40-Shunga, 41-Lizhmozero, 42-Belaya Gora, 43-Cholmuzhi, 44-Ileksa, 45-Janis, 46-Sukh, Vodla, 47-Palma, 48-Kuzaranda, 49-Tri Ivana, 50-Rajguba, 51-Suojarvi, 52-Spasskaya Guba, 53-Nigozero, 54-Bol.Gangozero, 55-Nadezhda, 56-Sujsar, 57-Yuzhny Oleny, 58-Burakovo, 60-Janisjarvi, 61-Jalguba, 62-Koloda, 63-Besovets, 64-Kamenny Bor, 65-Laskela, 66-Golets, 67-Hopunvaara, 68-Shoksha, 69-Uksinski oz, 70-Vazhinka, 71-Beloe, 72-Kinelahta, 73-Valaam, 74-Anashkino, 75-Pai, 76-Tuloksa, 77-Niva, 78-Mikhailovskoe, 79-Nizhnesvirski SR.

Figure 5. Kizhi Island.

Monasteries and convents dating from the Middle Ages are preserved as well, the biggest on Valaam Island in Ladoga Lake and on Solovets Islands in the White Sea. The old wooden church complex on Kizhi (Figure 5) and Solovets monastery are included into World Historical Heritage List of UNESCO. One of the ways for sustainable development that the government of Karelia determined, is nature tourism, to use its landscape richness.

REFERENCES

Lapo A., 1996. Russian contribution to the global indicative list of geological sites. II Intern. Symp. On Conservation of our Geol Heritage. Rome, p. 62-63.

Makarikhin V., Medvedev P. and Rychanchick D., 2001. Geological natural monuments in Karelia as tourist objects. Man and environment in the Barents Region at the threshold of the XXI century. Petrozavodsk, p. 135-137 (in Russian).

Systra Y., 1998. Main principles of geosite selection in Karelia, NW Russia, ProGEO'97, Estonia Proceedings Tallinn, 1998, p.68-72.

Systra Y.J. 2000. Geological heritage of Northwestern Russia: experience in conservation. *Mem. Descr. Carta Geol. d'It. LIV* (1999), p.361-366.

Systra Y., 2003. *Geologic characteristics. Biotic diversity of Karelia: conditions of formation, communities and species*. Gromtsev A.N., Kitaev S.P., Krutov V.I., Kuznetsov O.L., Lindholm T., Yakovlev E.B. (eds.), Petrozavodsk: Karelian Research Centre of RAS, p.7-12.

In: Parkes, M.A. (Ed.) 2004 *Natural and Cultural Landscapes - The Geological Foundation*, 309-312, Royal Irish Academy, Dublin

GEOLOGY IN AN INTEGRATED CULTURAL LANDSCAPE: EXAMPLES OF PROMOTING SUSTAINABLE AND ECONOMIC DEVELOPMENT IN EMILIA-ROMAGNA REGION (ITALY)

Milena Bertacchini, Paola Coratza, Mario Panizza, Sandra Piacente
Earth Science Department, University of Modena and Reggio Emilia,
Largo S. Eufemia 19, 41100, Modena, Italy

Abstract: Research previously carried out for the Landscape, Park and Natural Heritage Service of the Emilia-Romagna Region, gave us the idea of selecting those aspects of the geological landscape which, more than others, should be appreciated both for their intrinsic and contextual meaning and also for the "symbolic" value which they have gained through time. Thus, a project concerning two particular itineraries was developed. The first itinerary is more strictly scientific and is linked to the features of the physical environment and the anthropogenetic aspects characterising ophiolitic ridges. The second itinerary can be defined as being geological-literary, since it was developed by considering how the morphological features of the landscape in Emilia-Romagna have inspired poetry and culture.

INTRODUCTION

In modern society, environmental problems of a most pressing and urgent nature have arisen, but Earth Sciences still occupies only a marginal role. This is difficult to justify, considering the importance these sciences have always had and will have in the history and development of life and civilization. It is therefore necessary to find new paths and communication strategies which may fill this gap. In Italy, as in most countries, the geological component of the landscape has not yet been given its real value as a cultural asset. It is therefore necessary to find new ways and strategies to increase the awareness and sensitivity not only of the scientific world and institutions but also of society in general.

One new strategy could be *Cultural Geology* (Panizza and Piacente 2003), which is based on the concepts of *Geodiversity* and *Geohistory*. The former refers to the substratum of unique landscapes and basis for the variety of life on Earth; the latter deals with the relationships with other cultural components (history, archaeology, architecture, art etc.) and is a conditioning factor in the history of the various civilizations.

Geology, therefore, is to be understood, interpreted and appraised as a cultural heritage resulting from each specific territory's diverse characteristics, which are properly expressed by Geodiversity and geo-historical itineraries.

From this viewpoint, a project has been activated with the collaboration of the Department of Earth Sciences of Modena and Reggio Emilia University and the Emilia-Romagna Region - Landscape, Parks and Natural Heritage Service. The aims of this project are to census and rate the most important sites of geological interest and develop cultural-tourist itineraries and educational routes. Such geological itineraries have a double aim: to promote the knowledge of Earth Sciences, so little widespread among media or public opinion; to offer new professional perspectives to young un-or unemployed geologists (Piacente and Poli 2003).

In this article two particular itineraries are presented. The first itinerary is more strictly scientific and is linked to the features of the physical environment and the anthropogenetic aspects characterising ophiolitic ridges (Bertacchini et al. 2003). The second itinerary can be defined as being geological-literary, since it was developed by considering how the morphological features of the landscape in Emilia-Romagna have inspired poetry and culture (Bertacchini et al. 2002).

Particular attention was given in translating the complex environmental system into opportunities expressed in a language accessible to the widest possible public, without going outside the local context and social and cultural framework, in order that an environmental awareness might become an integral part of daily life. In this sense the material produced aims at creating new ideas and stimuli providing better knowledge and tourist promotion for all, including the elderly and the disabled. This can be achieved by using local resources giving priority to the young and the elderly. The former in the educational and professional phase, the latter in the conservation phase, by transmitting values and experiences, utilising the past to pave the way for the future (Piacente et al. 2003).

1ST ITINERARY: "THE OPHIOLITES, THE DEVIL'S STONES AMIDST HAMLETS AND CASTLES OF THE EMILIA APENNINES"

The network of minor routes is differentiated by the way an area is expanded; this provides a thematic itinerary with particular targets (Figure 1).

Figure 1. The main ophiolitic outcrops in the Emilia Romagna region.

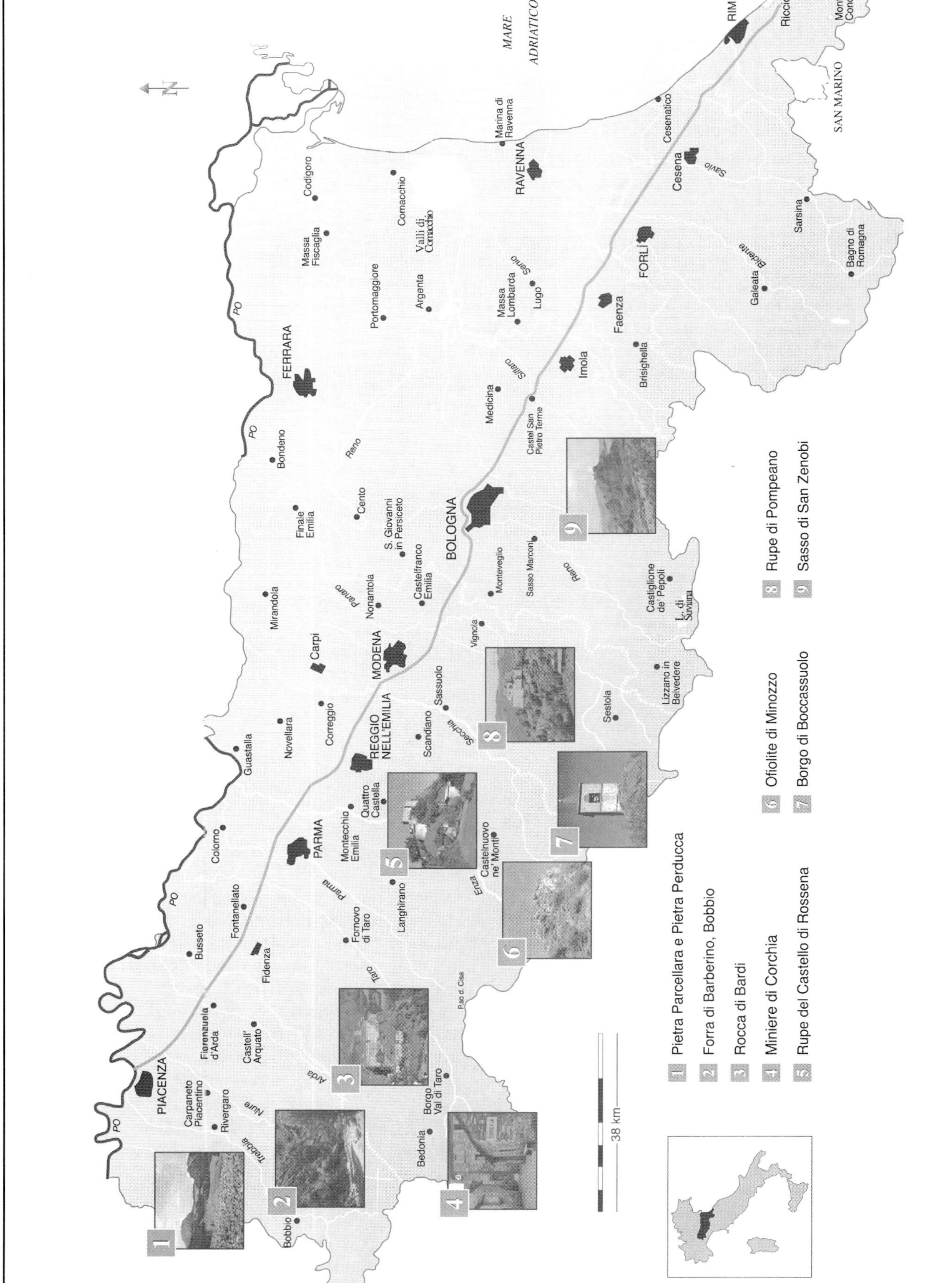

Pietra di Bismantova

La bancata arenacea della Pietra di Bismantova, con le sue pareti strapiombanti sulle circostanti vallate argillose, rappresenta un elemento paesaggistico ben visibile già dalla pianura. Le calcareniti della Pietra, appartenenti alla Formazione di Bismantova, giacciono sulla Formazione delle Marne di Antognola. L'inconfondibile forma di questo spettacolare geosito deriva dall'effetto combinato dell'erosione selettiva sulle argille e sulle calcareniti e della particolare struttura tettonica. L'origine marina della Pietra è testimoniata dall'abbondante presenza di frammenti di organismi marini fossili, come gli Echinidi, i Briozoi, le Alghe coralline e i Lamellibranchi; destano inoltre grande interesse i denti di squali appartenenti ai generi *Isurus* e *Carcharodon*. I primi documenti storici su Bismantova risalgono al VII sec. allorquando a valle della Pietra passava un'importante strada transappenninica che collegava la Toscana con la Lombardia, attraverso il Passo Pradarena. Anche Dante ricorda la celebre montagna da lui paragonata al monte del Purgatorio.

Raffaele Crovi
(Paderno Dugnano, Milano, 1934)

Bismantova aveva una cristallina bellezza orrida. Si stagliava verso il cielo con un impeto improvviso, una irruente forza che traeva a elevarsi lassù, dove nemmeno una nuvola corrompeva il chiarore del cielo. Lo strapiombo sembrava incantarsi in una carezzevole magia che si sprigionava dai crepacci muschiosi, dai piccoli arbusti che crescevano, su su, fino al culmine della roccia.

Opera: *I luoghi della vita*

Figure 2. A page of the volume "Paesaggi Culturali. Geologia e letteratura del '900 in Emilia Romagna": together with the photos of the geosite, its description, the literary passage of reference and a short biography of the author.

The ophiolites, from the Greek *ofis*, snake and *lithos*, rock, "rocks similar to snakes", or also "*green stones*", are an association of intrusive and effusive igneous rocks (peridotites, serpentinites, gabbros, basalts and jaspers) which are found as fragments within sedimentary rocks. Typical of the Northern Apennines, they represent remnants of the oceanic crust of the Tethys basin which in the Mesozoic separated the African plate from the European one.

There are numerous specific features which have contributed in transforming the ophiolites into objects of fantasy and superstition: their dark colour, sharp outline, heavy specific weight, higher resistance to erosion than the surrounding rocks, scanty or lacking vegetation cover and their impossibility to be exploited as farming land. Therefore, the "devil's stones" have been used also as amulets against the hazard of lightening and snakes or for protecting new born babies. The "Savignano Venus" - a famous archaic female statuette, ascribed to the last Palaeolithic phase, made of a serpentine rock common in the Modena district - seems to be connected to ancient magical-religious functions.

Use. The economic exploitation of the ophiolites goes back to the mid-Palaeolithic, when jasper was used for making weapons and tools and is further developed in Roman times through the opening of iron mines in the upper Nure valley and the important centre of Ferriere (Province of Piacenza), where the extraction of ore continued until the First World War. The chromatic nuances of some of them such as the red of basalts, the green of serpentines and the varicoloured breccias have justified their use as ornamental elements.

Human settlements. The stability and difficult accessibility of the ophiolitic ridges favoured the presence of human settlements and strongholds, as they could be easily defended. This can be deduced from some local place names: *petrum* (Pietramogolana) or *saxum* (Sassomorello); *rauca* or *roca* (Roccaprebalza, Roccamurata); or, finally, *castrum* after castle or fort (Sasso di Castro). Similarly, also the geological features and the absence of vegetation are found in various place names: Monte Nero, Pietranera, Sassomorello, Sasso Tignoso, Monti Pelati.

The itinerary unfolds across the main valleys of the Emilia Apennines, in the provinces of Piacenza, Parma, Reggio Emilia, Modena and Bologna. Most of the ophiolitic outcrops quoted in the itinerary are easily accessible to most tourists. In any case, also minor paths, accessible to everybody, are proposed.

2ND ITINERARY: "CULTURAL LANDSCAPES LINKING GEOLOGY AND LITERATURE IN THE 20TH CENTURY IN EMILIA-ROMAGNA"

In the second itinerary "Cultural landscapes linking Geology and Literature in the 20th century in Emilia-Romagna" particular aspects of the geological-geomorphological landscape have been selected which deserve, more than others, to be known not only for their intrinsic significance but above all for the "symbolic" value they have assumed in the past century.

A geological-literary route was therefore defined considering the specific poetic and cultural heritage of the Emilia-Romagna Region, which was influenced by the typical morphological and geological features of the territory.
The poetical-geological itinerary was defined as a mini-guide for the traveller - wandering both in space and time - eager to find Nature in its most poetical form. It is our deepest conviction that Nature can acquire a value for Man when it finds a way to touch the complete range of his sensitivity and feelings, including the most spiritual ones.

Our itinerary through the Region is also a journey through the human mind and spirit. It runs through literary sceneries and images where physical places and poetic narratives are combined in permanent display. Images are intertwined with natural and mental spaces with their identity as landscapes and sensations. In this way, one can go beyond a frontier, which is increasingly sensed intuitively but hardly ever consciously crossed. This frontier links places and culture and the only passport needed is a different kind of perception and sensitivity. But what eye or sensitivity could we hope to use if not what others before us have left behind, as a testimony of their inspiration, expressed in the language of the imagination and impressed indelibly on our consciousness as the signs left in the landscape by Nature itself? The place so much sought after is the place of the soul.

No place is far or unreachable, no land is stranger if one can find that common thread which carries us along and emotionally involves even the most distant and absent-minded visitor to the vast and widespread heritage which is the landscape of our Country.

As an example, a page of the volume "Paesaggi Culturali. Geologia e letteratura del '900 in Emilia Romagna" is shown in figure 2, together with the photos of the geosite, its description, the literary passage of reference and a short biography of the author. The geomorphological sites of the Emilia-Romagna region selected by our research in the Emilia landscape represent the cultural identity and literary trend of the 20th century. They can be appreciated on different levels, among which the notion of landscape shines forth in all its beauty, as if by magic (Emiliani 2000).

CONCLUSION

The itineraries presented in this article are an example of how geology can be a starting point and provide continuity for a journey offering close encounters with our region's science, art, history and literature. This research aimed to transform geomorphological sites into a culturally accessible and shared heritage and therefore into a resource for social and economic development in their own territory. Indeed geomorphological sites are considered as new elements that can catalyse the potentialities of a territory which are often neglected or pass unnoticed. In this way the two proposals become an instrument for territorial upgrading, by means of qualified measures on abandoned or secondary areas, thus promoting them to the rank of development models. Knowing and preserving the traces of a territory's history - that is the expression of its specificity - means to enhance its uniqueness, which is the fundamental basis for social and economic interventions.

REFERENCES

Bertacchini M., Coratza P. and Piacente S. 2002. *Paesaggi Culturali - Geologia e Letteratura nel Novecento in Emilia Romagna*. Università degli Studi di Modena e Reggio Emilia - Regione Emilia-Romagna, Edizioni L'inchiostroblu, Bologna, 137 pp.

Bertacchini M., Coratza P. and Piacente S. 2003. *Le "Pietre del Diavolo" tra borghi e castelli nell'Appennino emiliano. Un percorso culturale nel paesaggio ofiolitico.* In: S. Piacente and G. Poli (eds.), *La memoria della Terra, la terra della memoria.* Università degli Studi di Modena e Reggio Emilia - Regione Emilia-Romagna, Edizioni L'inchiostroblu, Bologna, 62-76.

Emiliani A. 2000. *Il paesaggio, la poesia e la pittura.* IBC, Bologna.

Piacente S., Bertacchini M., Coratza. P. 2003. *Geomorphological tourism amidst nature and culture: examples from Emilia-Romagna (Italy).* In: E. Reynard, C. Holzmann, D. Guex and N. Summermatter (eds.), *Géomorphologie et Tourisme*, Actes de la Réunion annuelle de la Société Suisse de Géomorphologie (SSGm), Finhaut, 21-23 septembre 2001, Institut de Géographie de l'Université de Lausanne (IGUL), 145-155.

Panizza, M. and Piacente, S. 2003. *Geomorfologia Culturale*, Pitagora Editrice, Bologna, 350 pp.

Piacente S. and Poli G. (eds.) 2003. *La memoria della Terra, la terra della memoria.* Università degli Studi di Modena e Reggio Emilia - Regione Emilia-Romagna - Edizioni L'inchiostroblu, Bologna, 159 pp.

In: Parkes, M.A. (Ed.) 2004 *Natural and Cultural Landscapes - The Geological Foundation*, 313-315, Royal Irish Academy, Dublin

GEOSITES: BEYOND CONSERVATION

M. Bini*, G. Poli**
*** PhD School on Earth Science- University of Bologna, and C.N.R., Istituto Geodinamica e Georisorse, Pisa ;**Regione Emilia-Romagna - Servizio Valorizzazione e tutela del paesaggio.**

Abstract: The landscape is the most important key to fully understanding an area. It is a direct and friendly interface which reveals the knowledge of geosites to a public with no specific preparation. For this reason it seems necessary to have a strategy of landscape interpretation, which goes beyond the simple information. In order to develop our process of interpretation our choice of the most appropriate technique depends on the user target, on our established aims and obviously on the basis of the options offered by the site or sites we have available.

In recent years there has been a great deal of interest regarding geosites. However, very little has been done in concrete terms regarding efficient communicative initiatives concerning the general public. This growing interest needs to be focused on an affirmation of the scientific, historical, and cultural importance and value of geosites. In effect, the numerous initiatives started in recent years regarding sites of geological interest reflect clearly aspects of the Italian situation. Firstly, there is essentially a catalogue-based heritage. Secondly, the limits and criticalities of the Italian conservation system consist of knowledge without a practical aim and circumscribed to each single situation studied. Conservation has been predominantly inspired by a restrictive model. Furthermore, there has always been the lack of a systematic and integrated vision of territorial resources. There has been a lack of financial support for the recovery and sustainable exploitation of sites. These factors have frequently resulted in projects regarding intervention or conservation arriving "from above", with no consultation or involvement on the part of the local population, who are obliged to accept legislative measures whose importance or meaning is obscure to them. The above characteristics all combine to produce a model that must be overcome if we really want geology, and more generally nature conservation, to emerge from its narrow-minded shell of scientific knowledge and make geosites a credible instrument for the communication of science and local development of the territory. This is essential both for the definition of conservation policies in sites of geological interest and also for the sustainable exploitation of the landscapes in which the sites are located or of which they are particularly representative.

As Robert Threadgould (1999) affirms, using an effective metaphor, "it's not enough to cover a rock with fur or stick a few feathers on a stone, to increase the perception and obtain the right assessment of our geological heritage "; a rock mass, a series of outcropping strata, a mountain ..., become "monuments of nature" only at the moment at which we give them a value and we all indistinctly recognise this value. In order to achieve this result it is necessary to start a process of interpretation and communication that must enable anyone to look at and consider the physical world and its phenomena with full awareness, observing from a different perspective what surrounds him and what he usually is not able to perceive and assimilate in all its complexity.

Only an efficient process of interpretation and transmission of knowledge can make the most of single sites, transforming them into an added value for the social and economic development of the area in which they are located.

The positive exploitation of a geosite must necessarily follow an integrated approach with the other natural, economic and cultural resources of the area, so as to link, for example, tourism and leisure activities to nature conservation, fruition to activities of environmental education, exploitation to reclamation and land planning.

Geosites are very often the central element of a landscape. Popularising the information relative to the sites means creating full knowledge of the area, so allowing a wise exploitation of the same.

As asserted in "Declaration of the rights of memory of the earth" at Digne (1991), "Our history and that of the earth are inseparable. Its origins and history are ours, one has been adapted or modified for the needs of the other and vice versa". Just as the History of Man relates its evolution, its conquests, its affirmation over other living beings and its fragility, in a similar way the History of the Earth tells us the story of our planet, of the moving continents, of mountain building, of the disappearance of oceans or mass extinctions of animals and plants, and it does this mainly through the places in which these events have occurred.

The Earth is alive, it changes, it moves, and the signs of this evolution all come together and show themselves at the earth's surface and in the landscape like the disordered pieces of a broken mosaic. It is up to us to identify them, decipher and recompose the mosaic, so creating a fully understandable image by means of a procedure of interpretation that tells the full story. The user should be involved in a sense of discovery. The landscape is the most important key to fully understanding an area. It is a direct and friendly interface which reveals the knowledge of geosites to a public with no specific preparation. It enables them to understand the events that have produced the current landforms and those phenomena that will modify and generate future landscapes. If guided appropriately the observer can recognise signs, forms and contexts. "The assumption of

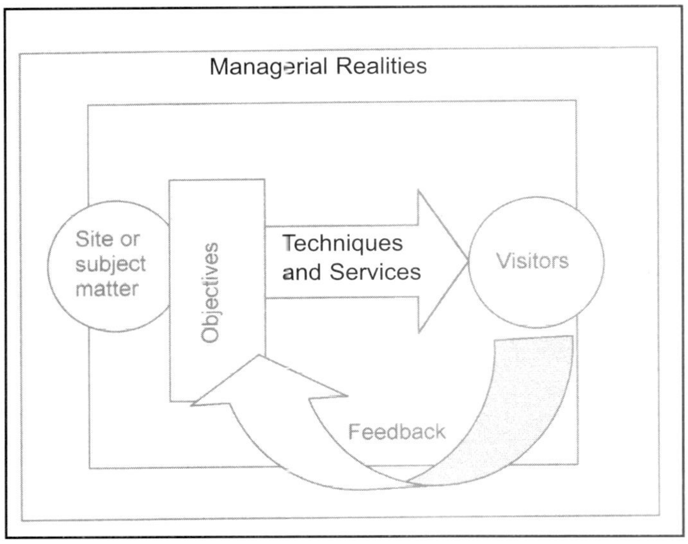

Figure 1. Interpretation model (Cherem 1977).

signs, forms and physical contexts, as points of reference and order in the vast space characterising the landscape, does not mean however having an awareness of what they really are or what they represent in scientific research and in the evolutionary history of the earth and of the landscape in which they are present" (Barca and Di Gregorio 1999).

In effect, the landscape is what is most frequently preserved in the mind of a visitor, but this does not necessarily mean a conscious perception. To move on from an emotive perception to a conscious perception we need to develop a process of interpretation of the landscape which decodifies its meanings and reveals particular complexities of the story that exist in parallel with what we are experiencing in the present. It is a story able to give a precise identity and uniqueness to common places which otherwise would not have drawn our attention.

So which are the ways and tools of communication regarding geosites that we have available? Which are the most efficient and the most suitable? Above all, what does a strategy of interpretation consist of? It has been defined as a "process of communication intended to reveal to the public meanings and relations of our cultural and natural heritage by means of direct involvement with objects, artefacts, landscapes and sites". Interpretation is a specific communication process in which it is important to "clothe" the concepts, putting the attention therefore on the means of communication. The information in itself is not interpretation. The interpretation comes from revelation based on the information. In order to be able to interpret it is essential to understand who the learning process is aimed at. It is not only the site that is important, that is the object of our planning, but also and above all the user, with whom we must establish a relationship that is only possible if we can enter into his everyday experience. The same site, even though we may start with the same aims and scientific information, will be interpreted differently by different users. One of the problems to face regarding relations with visitors is that of translating the scientific language into terms accessible and understandable for everyone. This is, however, not enough. Understanding the information is only the first step in the interpretation process. The information understood must, in fact, be associated with the experience of the visitor so that it does not remain sterile. Any measurements (the length of a valley, the load of a river in full flow, the height of a waterfall, etc.) remain sterile data if they are not correlated to everyday experience (e.g. the equivalent in football pitches, the proportion compared with the height of a house, etc.). Only the link to the everyday enables the visitor to develop a certain process of sensitivity - awareness - understanding - appreciation and, finally, commitment.

Geosites can therefore be narrated to the general public following a strategy of interpretation and using various angles. In an interpretative planning model it is above all necessary to identify the planning elements, that is the sites, resources, establishment of aims, identification of techniques to be adopted and services required in order to be as incisive as possible as regards potential users. Finally,

Figure 2. A puzzle represents, better than any other image, the conceptual action that we must reproduce in an interpretation process. Recomposing single pieces in an image that reveals the real meaning of what we are observing. (La dalle à ammonites de Digne les Bains, by courtesy of the Réserve Géologique de Haute-Provence, FR).

for the planning not to produce a sterile result it is necessary to identify clearly the user target and on the basis of this how we can develop the best relationship and obtain a return (Figure 1).
Among the most relevant and typical ways of presenting geosites are the following:

- by THEMES (geology, geomorphology, palaeontology,...) this enables in depth knowledge of a sector by bringing to life past and future chronological evolution, but from a partial vision of an area and it does not allow us to see the interactions between the different sectors of geology;

- by EPOCHS (Jurassic, Cretaceous, Quaternary ...) - this helps to understand the dynamicity of the physical environment, a not always intuitive concept, and gives indications as to possible future evolutions, but this approach can be dispersive in a visiting area because the chronologically successive sites are close together or in sequence;

- by geographical AREAS or geological CONTEXTS (Alps, the Po Valley, petrographic Province or tectonic-structural units, lagerstatten or distribution areas of fossil fauna ...) - this gives a good idea of the area and present formations so as to have a good general picture, but it does not always give a chronological view or deal with evolutionary problems;

- by EVENTS (extinctions, orogenies, glaciations, salinity crises...) - this goes into depth on aspects relative to a "phase of change" and enables us to read in the various fields and places what there is in relation to that particular event. It also gives an idea of the implications and relations between different fields not only within geological sciences but also with other sectors of science and life. It does not however position the event in a wider history so a 'before' and 'after' is not evident.

- by PROCESSES (forms generated by volcanic or seismic activities, by erosive cycles and deposition ...) - this explains in detail a phenomenon in all its parts, so enabling an understanding of all the factors that influence it and vice versa how the process interacts with external factors. However, there is no temporal or spatial context;

- by TYPES (lakes, waterfalls, dunes, volcanoes ...) - this enables greater depth of knowledge of the reasons for similarities and differences between landform types associated with the same phenomenon and allows comparison of the various developmental stages of a landform. It gives indications of possible future development. But there is no positioning in the chronological development of the area and the landform appears rather isolated from its context.

Referring to what has already been mentioned, in order to develop our process of interpretation our choice of the most appropriate technique depends on the user target, on our established aims and obviously on the basis of the options offered by the site or sites we have available. Equally important in reaching our objective are the means of diffusion of the information to the general public (books, guides, parks, exhibitions, itineraries, audiovisual aids, gadgets, etc.), to be used in various combinations.

The earth is our past, our present and our future. It preserves and watches over many stories that we can read only by interpreting geosites. An intelligent use of geosites is the essential element for an appropriate plan of sustainable exploitation in which conservation is the result, produced spontaneously, in the long path towards increasing perception and awareness in civil society.

REFERENCES

Barca, S. and Di Gregorio, F. 1999. *Paesaggi e Monumenti geologici della provincia di Cagliari.* Saredit srl.
Cherem, G. J. 1977. *The professional interpreter:agent for an awakening giant.* Association of Interpretive Naturalist Journal, Vol. 2.
Poli, G. 1999. *Geositi, Testimoni del tempo. Fondamenti per la conservazione del patrimonio geologico.* Edizioni Pendragon, Bologna.
Threadgould, R. 1999. *Tra l'incudine e il martello.* In Geositi, Testimoni del tempo.
Veverka, J.A. 1998. *Interpretative Master Planning.* Acorn Naturalists, Tustin, California.

SESSION 9

WHERE DO WE GO FROM HERE?

Flooding in the Gort Lowlands of County Clare and County Galway in 1994 and 1995, was due to very high rainfall in those years. Photograph: Conor MacDermot.

In: Parkes, M.A. (Ed.) 2004 *Natural and Cultural Landscapes - The Geological Foundation*, 319-329, Royal Irish Academy, Dublin

WHERE DO WE GO FROM HERE?

GEOLOGICAL HERITAGE DECLARATIONS, 1991 - 2002

John Morris and Matthew Parkes

SESSION 9 BACKGROUND

When the conference organising committee considered the programme for this conference, it was clear that it could not just end with presentations, but that some plenary discussion was essential. It was also felt that in the tradition of previous geological heritage conferences, we would attempt to develop and agree a Dublin Declaration, as a manifesto for future intent. The aspiration was also strong, to find new mechanisms to implement the objectives of the proposed declaration, rather than sitting back with a fine manifesto - and nothing else.

To achieve these ends, the final session was structured as a plenary introduction by John Feehan, a widely respected and well-known figure in the area of the Irish landscape. John, with a background as a geologist, and long experience in presentation and explanation of landscape management, had an excellent pedigree to capture the essence of the previous sessions and steer the culmination of the meeting in a session entitled "*Where do we go from here*?"

After his plenary introduction, the conference broke into three workshop sessions led by Paul Lyle, John Morris and Patrick McKeever. The topics discussed were as follows:

Workshop 1 Documentation and Conservation

Resourcing the agenda. The material heritage of geology and landscape. Identification and designation, protection and development. What strategies and instruments can we employ to protect and conserve the heritage?

Workshop 2 How to deliver

How do we tell the story? Identifying the media, the mechanisms for delivering the geoheritage agenda.

Workshop 3 Resourcing the agenda

Identifying the resources and personnel necessary to fuel, to power that delivery.

The Rapporteur of each workshop then reported back to the conference, with the key points developed in discussion as a preliminary stage to plenary discussion of a draft Dublin Declaration which was supplied earlier for people to have time to consider it adequately.

In order to put in context the ambition of the Dublin Landscapes Conference, the organising committee looked in detail at previous geoheritage declarations and they are collated here with some comments representing the views of the Conference organisers, as they have not previously been juxtaposed in sequence, and the common threads of the developing European geological heritage scene drawn together. The main strand of this scene is the ProGEO organisation - the European Association for the Conservation of Geological Heritage. Whilst familiar to some of the organisers through the professional work in geological heritage for the GSI, it was introduced to many conference participants for the first time through the event.

INTRODUCTION TO GEOLOGICAL HERITAGE DECLARATIONS

Rocks are the physical mass we call planet Earth. They define and provide the physical underpinning of both the biotic and the abiotic realms. They are the foundation upon which the cultural, economic and historical developments of all human societies owe their very existence - now, in the past, and in the future. Without rocks, humans, their societies and cultures, would not exist. We, in time present, are the beneficiaries of a natural endowment generated largely in time past, holding it merely as temporary custodians for the benefit of our successors in time future. But how we act now to cherish and conserve our geological inheritance will determine what they inherit from us. Actions in the present can take many forms, each leading to many different potential futures - but the future will only see in the past, the course of action that we choose to pursue now.

We consider that as much as all biological life has a right to exist, so too has our geological environment. As much as respect for our biological inheritance is now enshrined in national and international BIODIVERSITY declarations and actions, we contend that equal attention now be given to our geological inheritance - the realm of GEODIVERSITY.

GEODIVERSITY

The principle of GEODIVERSITY, and along with it a series of declarations and action plans, has evolved out of a succession of International Geological Heritage Conferences, DIGNE (1991), MALVERN (1993), ROME (1996), and MADRID (1999). The concept has been further refined in the "GEODIVERSITY STRATEGY" statement, and action plan, of the Royal Society for Nature Conservation (UK), as published in the ProGEO Newsletter 2001-1, in 2001. To this we now add the declarations made in DUBLIN - 2002. Copies of the declarations and resolutions, as recorded at each of these Conferences, or publications, follow here. Each of these has contributed to successive developments, although none, to date, have been integrated into a single, comprehensive Geodiversity statement. We now take this opportunity to draw together the fruits of the deliberations by so many, over so many years, to create a single, consolidated declaration to serve the advancement of GEODIVERSITY in Ireland.

We take as our definition of Geodiversity that proposed by the Royal Society for Nature Conservation (UK), which states:-

> **Geodiversity is the variety of geological environments, phenomena and active processes that make landscapes, rocks, minerals, fossils, soils and other superficial deposits which provide the framework for life on Earth. Geodiversity is the link between people, landscapes and their culture through the interaction with biodiversity, soils, rocks, fossils, active processes and the built environment.**

The objectives of this consolidated declaration are as follows.

- To agree and provide a shared vision on GEODIVERSITY
- To use this shared vision as:
- the framework within which complementary geological heritage actions and initiatives are developed and advanced
- a platform for developing a strong and coherent voice for advocating conservation and advancement of geological heritage actions at national and international levels throughout Europe
- to establish a contact mechanism between all groups active in geological heritage activities, nationally and internationally

CONSERVATION OF GEOLOGICAL HERITAGE

Principles

Geoconservation is regarded by society as an essential part of the protection of our natural and cultural heritage. [ROME, 1996]

Integration of geoconservation into nature conservation policies should take place, with full cognisance of the fundamental importance of geodiversity. [MADRID, 1999]

This integration must be achieved applying proven methods founded on good geoscience. Achievements in geoconservation (=geotope conservation), as well as any successes making the public aware of geological heritage, are based on modern Earth-science data and up-to-date descriptions of geological features.
This symposium approves fully and recommends the systematic approach adopted for GEOSITES (including soils and landscapes), working in clear regional, structural and stratigraphic frameworks, avoiding the ad hoc methods of the past. We look for the incorporation of geological sites and terrains in World Heritage sites as soon as possible: however, this should be done with care and balance, and inclusions under World Heritage should be based on the sound GEOSITES method. [ROME, 1996]

GEOSITES offers great, previously unanticipated, gains in terms of improved communications and working. It is proposed to form a Task Force to examine the possibilities for enlarging and building upon the geosite concept, and the possibilities for Lithosphere/Geosphere reserves. [ROME, 1996]

A major step foreseen is the promotion of a national inventory of geoconservation sites/geotopes which have international and/or national value. Concurrently we will promote regional inventories, inviting the Authorities to operate autonomously, both within the technical-scientific context agreed at a national level, and in accordance with the IUGS GEOSITES criteria.

We intend to have a geological/tectonic framework and the network of GEOSITES established for Europe in 1998, meshed with national and regional networks, developing descriptive documentation gradually. [ROME, 1996]

Responsibility

We intend to promote the opinion that full weight should be given by States and Regions to geotope selection, conservation and management, placing this in the wider context of environmental, territorial and cultural planning.

Nature conservation organisations have responsibility for the protection of our geological heritage (including the physical landscape) as a component of the natural heritage. They need to be assisted by Geological Surveys and the Geoscientific Community.

We support the effective application of existing nature/geological conservation legislation.

Advocacy

We intend to exert continuous pressure on the competent authorities and public opinion, so that existing laws are applied. In that way the Conservation of the Geological Heritage may be guaranteed, as well as offering wider employment possibilities for new generations of geologists and other competent specialists. [ROME, 1996]

We are aware that an important factor in the protection of the cultural heritage is knowledge. Therefore we consider of paramount importance the propagation of Earth Sciences at various educational levels, ranging from elementary school to university.

One vital element in any strategy towards more effective

geoconservation is raising awareness of geoscience and of geological heritage. This must be achieved in school curricula, and through training.

Mobilisation of scientific opinion and concerted contributions towards geoconservation make it possible to undertake integrated consideration of natural environment protection, including both biotic and abiotic elements.

A first step to achieve the integration of geoconservation into nature conservation would be an evaluation of the Geological Heritage in already protected areas, followed by the promotion of interpretative initiatives, in order to increase social awareness of Geoconservation.

NATIONAL AND INTERNATIONAL REPRESENTATION

The MALVERN resolution determined the need for an international Earth science conservation convention [MALVERN 1993]

We endorse the parallel, and complementary initiatives advanced by ProGEO and the European Geoparks Network, as mechanisms to advance conservation of geological heritage at local, national and international levels in Europe. [DUBLIN 2002]

ProGEO

ProGEO is the European geoconservation network - we hope that its structure is adopted elsewhere in other regions of the world. We agree to work towards links with other regional networks.

We uphold ProGEO's move towards regionalisation as the ideal functional framework.

ProGEO supports the regional and national contextual criteria and frameworks already adopted for GEOSITES.

We restate our enthusiasm and commitment for the regional working groups of ProGEO and their co-ordination of the work of national ProGEO committees.

European Geoparks

The Geopark concept which originated within the Earth Sciences Division of UNESCO was adapted by a group, established as the European Geoparks Network. The initial members in France, Greece, Spain and Germany have now been joined by many others, including the Copper Coast in Ireland and Cuilcagh Mountain in Northern Ireland. Many other communities are now interested in this established structure, fully exploiting their geodiversity in locally driven, sustainable economic development.

EMBEDDING GEODIVERSITY INTO LOCAL AND NATIONAL ACTIONS AND PROGRAMMES

The scope of local and national actions and programmes aimed at protection and promotion of geological heritage is as diverse as the countries engaged in these actions. The papers in this volume illustrate some examples of these actions.

In Ireland, the Irish Geological Heritage Programme (Parkes and Morris 2001) of the Geological Survey of Ireland is the primary mechanism for statutory protection of geosites, through selection of sites for designation as Natural Heritage Areas. A complementary programme to define County Geological Sites has gained widespread acceptance, with individual local authorities commissioning specific projects to supplement the data available for inclusion of these sites in the County Development Plans, which are the primary strategic policy and control of development.

In Northern Ireland the Earth Science Conservation Review undertaken by the Environment and Heritage Service (Enlander 2001) has identified Areas of Special Scientific Interest, which are in the process of designation.

Across the island, actions undertaken through Earth Science 2000 (ES2k) provide a primary opportunity and mechanism for geologists to raise awareness in the wider community and within the educational system.

EMBEDDING GEODIVERSITY INTO INTERNATIONAL PROGRAMMES

The most encouraging development in this context is the potential recognition of geological heritage at a Council of Europe level, through the *Working Group of the Committee for the activities of the Council of Europe in the field of biological and landscape diversity* (CO-DBP). In September 2003 a draft Recommendation was agreed, with Ireland represented at the meeting by the editor. This Recommendation should see its way through to the Council of Europe for approval in the first half of 2004. Whilst the Recommendation is the least powerful instrument of European government, it is a very significant development to actually have geological heritage on the table at all at this level, and marks an important progression.

REFERENCES

Enlander, I. 2001. The Earth Science Conservation Review: conserving the earth heritage resources of Northern Ireland. *Irish Journal of Earth Sciences* 19, 103-112.

Parkes, M.A. and Morris, J.H.M. 2002. Earth Science Conservation in Ireland: The Irish Geological Heritage Programme. *Irish Journal of Earth Sciences*, 19 (2001), 79-90.

THE DIGNE DECLARATION, 1991

International declaration of the rights of the memory of the Earth

1 - Just as human life is recognized as being unique, the time has come to recognize the uniqueness of the Earth.

2 - Mother Earth supports us. We are each and all linked to her, she is the link between us.

3 - The Earth is 4.5 billion years old and the cradle of life, of renewal and of the metamorphosis of life. Its long evolution, its slow rise to maturity, has shaped the environment in which we live.

4 - Our history and the history of the Earth are closely linked. Its origins are our origins, its history is our history and its future will be our future.

5 - The aspect of the Earth, its very being is our environment. This environment is different, not only from that of the past, but also from that of the future. We are just the Earth's companion with no finality, we only pass by.

6 - Just as an old tree keeps all the records of its growth and life, the Earth retains memories of its past... A record inscribed both in its depth and on the surface, in the rocks and in the landscapes, a record which can be read and translated.

7 - We have always been aware of the need to preserve our memories - i.e. our cultural heritage. Now time has come to protect our natural heritage, the environment. The past of the Earth is no less important than that of human beings. Now it is time for us to learn to protect, and by doing so, to learn about the past of the Earth, to read this book written before our advent: that is our geological heritage.

8 - We and the Earth share our common heritage. We and governments are but the custodians of this heritage. Each and every human being should understand that the slightest depredation mutilates, destroys and leads to irreversible loss. Any form of development should respect the singularity of this heritage.

9 - The participants of the 1st international symposium of the protection of our geological heritage, including over a hundred specialists from over thirty nations, urgently request all national and international authorities to take into consideration and to protect this heritage by means of all necessary legal, financial and organizational measures.

Written on the 13th June 1991, in Digne, France.

THE MALVERN RESOLUTION, 1993

Malvern Conference on Geological and Landscape Conservation 1993: Resolution endorsed by closing session of conference

Resolution (amended version)
The Malvern International Conference 1993

- believes that there is need for an international Earth science conservation convention;
- recognises that the justification for, potential scope of, and objectives of, such a convention should be examined in depth; and
- supports the establishment of an international task force which will pursue, and report back on, these propositions.

Action Plan

(1) The following delegates at the Malvern Conference have agreed to form a task force, for a period of two years in the first instance, with the purpose of:
- progressing the Malvern International Conference 1993 Resolution taking account of the conclusions of the Workshops at the conference;
- facilitating the organisation of a successor conference to the Malvern International Conference 1993;
- expediting the creation of an international organisation for Earth science conservation which will, on formation, take over the functions of the task force.

Dr P. Creaser (Australia)
Prof. P. Jacobs (Belgium)
Mr G. Martini (France)
Ms A. Spiteri (Malta)
Mr C. Stevens (United Kingdom)

(2) The international task force will seek the support of delegates and others in carrying forward its report, and widen its membership as necessary.
(3) The Joint Nature Conservation Committee of the United Kingdom (Monkstone House, City Road, Peterborough PE1 1JY, UK) will provide a secretariat to support this task force for a period of two years after which time it is anticipated that a permanent international organisation will be in place.
(4) The task force will report back annually in writing to the delegates who attended the Malvern International Conference 1993 until either a permanent international organisation is in place or a successor conference is held, whichever is the earlier.

22 July 1993 Malvern

In O'Halloran, D., Green, C. Harley, M., Stanley, M. and Knill, J. (eds), Geological and Landscape Conservation. Proceedings of the Malvern International Conference 1993. Geological Society, 530pp.

THE ROME DECLARATION, 1996

ProGEO is an open association whose primary aim is the promotion of geoconservation. GEOSITES is a project, under the auspices of IUGS and with the support of UNESCO, to devise a global inventory of geological, geomorphological and landscape sites.

Italy has an important place in the geology of Europe. Through this conference it has been shown that a wide range of activities are already making an important contribution to European geoconservation. The following declaration sets the European context. It is followed by an integrated Italian strategy.

1. Geoconservation is regarded by society as an essential part of the protection of our natural and cultural heritage. Achievements in geoconservation (=geotope conservation), as well as any successes making the public aware of geological heritage, are based on modern Earth-science data and up-to-date descriptions of geological features. Thus scientific knowledge underpins all worthwhile activity on GEOSITES: upon this cornerstone ProGEO policies are founded.
2. ProGEO is the European geoconservation network - we hope that its structure is adopted elsewhere in other regions of the world. We agree to work towards links with other regional networks.
3. We uphold ProGEO's move towards regionalisation as the ideal functional framework.
4. ProGEO supports the regional and national contextual criteria and frameworks already adopted for GEOSITES.

This symposium approves fully and recommends the systematic approach adopted for GEOSITES (including soils and landscapes), working in clear regional, structural and stratigraphic frameworks, avoiding the ad hoc methods of the past. We look for the incorporation of geological sites and terrains in World Heritage sites as soon as possible: however, this should be done with care and balance, and inclusions under World Heritage should be based on the sound GEOSITES method.

5. GEOSITES offers great, previously unanticipated, gains in terms of improved communications and working. It is proposed to form a Task Force to examine the possibilities for enlarging and building upon the geosite concept, and the possibilities for Lithosphere/Geosphere reserves.
6. We intend to have a geological/tectonic framework and the network of GEOSITES established for Europe in 1998, meshed with national and regional networks, developing descriptive documentation gradually

A STRATEGY FOR ITALIAN GEOCONSERVATION

7. We intend to exert continuous pressure on the competent authorities and public opinion, so that existing laws are applied. In that way the Conservation of the Geological Heritage may be guaranteed, as well as offering wider employment possibilities for new generations of geologists and other competent specialists.

8. A major step foreseen is the promotion of a national inventory of geoconservation sites/geotopes which have international and/or national value. Concurrently we will promote regional inventories, inviting the Authorities to operate autonomously, both within the technical-scientific context agreed at a national level, and in accordance with the IUGS GEOSITES criteria.

9. We intend to promote the opinion that full weight should be given by States and Regions to geotope selection, conservation and management, placing this in the wider context of environmental, territorial and cultural planning.

10. We are aware that an important factor in the protection of the cultural heritage is knowledge. Therefore we consider of paramount importance the propagation of Earth Sciences at various educational levels, ranging from elementary school to university.

Postscript
This conference looks forward to meeting again next year to advance the cause of European geoconservation, to consider the evolving Geosite list, and in two years in Sofia for ProGEO "98", to discuss a draft European list which ProGEO will submit to IUGS and UNESCO.

The MADRID DECLARATION, 1999

IIIrd INTERNATIONAL SYMPOSIUM ProGEO ON THE CONSERVATION OF THE GEOLOGICAL HERITAGE, MADRID - DECLARATION

The Madrid Declaration was approved unanimously by the plenary session of the III rd International Symposium on the Conservation of the Geological Heritage. Before its reading, it was introduced to the plenary session by the executive secretary of ProGEO. In this section, both words of introduction and the Declaration are included.

Introduction to the Declaration (words addressed to the plenary session by W.A.P. Wimbledon, Executive Secretary of ProGEO)

I have been asked by our chairman, Dr García Cortés, and the organising committee of this excellent meeting, to introduce the final Symposium Declaration. Before I do that, I would like to steal a few moments to set the scene and the context for this meeting, and this declaration. This is the III rd International Symposium on the Conservation of the Geological Heritage, following those in Digne and Roma. The organisers of those important meetings are with us here: very great thanks to them. Digne gave us an ethos and showed us what we wanted to do (and did not want to do) in Europe as an association, and Roma gave us the balance and made the association a whole, with all regions fully engaged.

ProGEO has come a long way since Digne declaration, a colourfully worded exposition of our ethos. In the time since, ProGEO has made many and major practical contributions, the "geo" part of the Council of Europe's European Landscape and Biodiversity Strategy came from ProGEO. Our association played a major part the drafting of that document, and also in getting geology more to the fore in the Operational Guidelines for the selection of World Heritage Sites, when UNESCO (WH centre) revised them. The proposed Geoparks initiative of UNESCO (ES Division) has been guided by a number of leading ProGEO members, determined to give scientific sites a prominent place. Through its determination to identify a European geosite list, based on the best scientific opinion, ProGEO has led the way in the work of the International Union of Geological Sciences, work that is now being replicated in Asia and Africa.

ProGEO continues to work for the strengthening of ties and the healthy integration of Geoconservation within Nature Conservation. This holistic, all-embracing, approach was promulgated long ago by ProGEO. Though we see that there is enormous sense in working with other disciplines for practical conservation, we regard geoconservation as indivisible, like geosciences it supports; it is varied, but its individual parts constitute a continuum, the same principles of geoconservation encompass all types of site. This is important, because it is a key element in our fight to have geoconservation recognised in all countries and at all levels, not to have it seen as some subsidiary to archaeology, biology, or another discipline. It would be the biggest mistake to let geoconservation be weakened by fragmentation so that, for instance, fossil or mineral sites or landforms were arbitrarily protected, and geology in all its richness forgotten. In this, cross-border co-operation helps avoid harmful factionalism and parochialism. There is a growing determination to conserve geodiversity wherever it is significantly represented in the rock and landform record. New and productive links have been formed with the International Union of Geological Sciences through participation in its Geosite programme and Working Group. In 1998, at a ground-breaking meeting in Bulgaria, ProGEO complied the first ever European listing of Geosite frameworks, for consideration by our community of geoscientists and conservationists. Recently has come Grecel, a magnificent first for its dynamic organisers, leading the way in geoconservation training and education. In all these things, ProGEO regional groups should play their part.

I would like to use this opportunity to say a few words of thanks, to all who contributed to this creative meeting: we have all learned much - on tourism, on training, on modern technology, as well as our established preoccupations of site identification, selection and management of statute and the planning process - the things that keep many of us busy in our daily working lives. Anyway, thanks and many thanks to all who have reached out to us to communicate their message.

I have said various words of adulation and genuine words of thanks. Words play an important part in the modern world. Buzz words, catch phrases, words that people readily remember. ProGEO has given the world several new words. You hear them now every day: "geoconservation", "geosites", "geodiversity", geoheritage". This commonplace, easy language is important, if we are to catch people's imaginations, to involve them, to overcome the lingering mystery of geoscience and geoconservation. We are mostly specialists and we love our fancy long words, and we love to confuse all with our jargon. But there is no great mystery. As the Digne Declaration said, the geological record that we try to conserve is a book, it our geological heritage, and it can be read by all with little help. It is our shared heritage. It is also the record of life and the substrate on which all present life now exists.

Of course, the greatest achievement of ProGEO in recent years - in parallel with papers, innumerable national and regional workshops, and three international symposia - is the fact that we have formed a movement, a geoconservation movement!

This has been a very busy, productive and informative meeting. Paul Dingwall kindly came all the way from New Zealand to represent IUCN and attend this conference. It was encouraging to hear him say so clearly in his keynote talk that ProGEO was globally taking the lead of geoconservation, and setting the agenda. ProGEO's easily remembered name has now a certain popularity! I am overjoyed at the way in which our Spanish colleagues have guided us at this meeting, playing their rightful role in our movement. We are mostly privileged foreigners here. I take this opportunity, the only one I will get, as the Secretary of ProGEO, to thank IGTE for its leadership, to thank our Spanish colleagues, whom we esteem very much, for their unfailing courtesy and care. I thank them all, and the organising committee in the person of Dr Daniel Barettino.

Why are we here? We are here to take forward the agenda ProGEO has already published and which all the association has agreed. We have been here to work, to aid our Spanish colleagues and to work with them. We have come to take forward our shared aims, and to give some help in pushing geoconservation in Spain and Iberia along its road. This has been a successful venture.

We, as an association, have delayed coming to Spain for far too long, though all have been well aware of the progress made here by the Spanish Commission, ITGE, and our colleagues in ProGEO. In the years since Dr García Cortés and colleagues attended the Digne meeting, much has happened in Spain, and the proceedings of the meeting show that in full measure. Of course, I talk much of Spain, but we have in the last days been furthering co-operation in Iberia as a whole. Last night I had the honour to attend he first meeting of the Southern European Working Group of ProGEO covering Spain, Portugal, Italy and southernmost France, the latest of the association's regional working groups. The meeting discussed the first steps forwards regional collaboration and joint projects. The commitment and energy of colleagues in Spain, Italy and Portugal are obvious.

The Declaration. The two previous International Symposia, at Digne and Roma, produced declarations, laying down the guidelines, the fundamentals, for geoconservation and a way forward. The principles are therefore set. In introducing this declaration, I only need to say that it builds on those sound foundations, and much else that has been agreed and published, in Europe and in the Spanish Commission. This declaration is particularly addressed to supporting the agenda for geoconservation as it exists today in Spain through our colleagues. It is designed to support the efforts of those who work to make geoconservation a success in Iberia.

Mr Chairman, with your permission, and on behalf of the organising committee, I ask the support of all colleagues for this declaration, and wholeheartedly recommend it to this assembly.

W.A.P. Wimbledon

THE MADRID DECLARATION, 1999

Preamble: Geodiversity is a natural phenomenon. It is the physical underpinning of both the biotic and the abiotic world.

Our cultural, economic and historical development is intimately linked to the complexity and controlling influences exerted by the physical (geological) world.

1- Integration of geoconservation into nature conservation policies should take place, with full cognisance of the fundamental importance of geodiversity.

2- This integration must be achieved applying proven methods founded on good geoscience.

3- Nature conservation organisations have responsibility for the protection of our geological heritage (including the physical landscape) as a component of the natural heritage. They need to be assisted by Geological Surveys and the Geoscientific Community.

4- Mobilisation of scientific opinion and concerted contributions towards geoconservation make it possible to undertake integrated consideration of natural environment protection, including both biotic and abiotic elements.

5- A first step to achieve the integration of geoconservation into nature conservation would be an evaluation of the Geological Heritage in already protected areas, followed by the promotion of interpretative initiatives, in order to increase social awareness of Geoconservation.

6- One vital element in any strategy towards more effective geoconservation is raising awareness of geoscience and of geological heritage. This must be achieved in school curricula, and through training.

7- We endorse the production of a southern European integrated Geosite framework and site/terrain list

8- We restate our enthusiasm and commitment for the regional working groups of ProGEO and their co-ordination of the work of national ProGEO committees.

9- We support the effective application of existing nature/geological conservation legislation.

Written on the 25th November 1999 in Madrid

THE DUBLIN DECLARATION, 2002

DECLARATION ADOPTED BY DELEGATES
at the Natural and Cultural Landscapes Conference held in Dublin Castle, Dublin, Ireland, September 9 - 11, 2002

Sustainability of our environment relies upon informed, integrated and balanced decision-making on all land use issues relating to our natural and cultural landscapes. Therefore, the delegates will work to promote the geological heritage as a central and essential consideration in the management of landscape.

To achieve this, we :

- commit to developing an integrated, holistic approach by establishing a Forum open to all individuals and organisations interested in the geological heritage of Ireland.

- acknowledge that local landscapes can best be managed through the participation of all stakeholders in the decision-making process. Opportunities need to be identified whereby all stakeholders can benefit from a caring approach to this heritage.

- advocate that maintenance and conservation of geological heritage sites in private owner ship be placed upon the same footing as that attaching to archaeological and architectural sites.

- commit ourselves to producing easily understandable geological information, and making it readily available to individuals, communities and organisations.

- advocate that public bodies include earth science considerations in their decision making, and seek appropriate professional advice on such matters.

- recommend that all geological site documentation be prepared to a transparent and common standard for dissemination.

- endorse the initiatives advanced by UNESCO, the IUGS Geosites programme, ProGEO and the European Geoparks Network, as complementary mechanisms to advance conservation of geological heritage in Europe and elsewhere.

Agreed and adopted in Dublin, September 11, 2002.
APPENDIX

We also consider that:

- the European Landscape Convention, ratified by the Council of Europe, provides a frame work to further enhance awareness of the geological foundation of all landscapes.

- the Geodiversity Strategy statement developed by the Royal Society for Nature Conservation, UK, provides a constructive set of principles for consideration and applica tion elsewhere.

THE SITUATION SINCE SEPTEMBER 2002

Following the Conference, a Steering Group (SG) was formed, composed of members of the Organising Committee plus others who had expressed a strong willingness to participate or assist in the deliberations. The initial primary aim of the Steering Group was to establish an "Irish Geodiversity Forum" reflecting both the Dublin Declaration itself, and the wishes of Conference members for a means to continue the momentum of the Conference.

As with many such actions, what initially seemed simple, unravelled with layers of complexity as we examined ways of achieving the objective. Many approaches were considered and the potential membership, functional operations and legal status of the proposed Forum all examined. However, given the already heavy work commitments of Steering Group members in the general area of advancing an Earth Science Agenda, there was an understandable reluctance to establish a formal structure which could not be clearly be sustained by those already committed to the purpose. The National Committee for Geology (NCG) of the Royal Irish Academy was at the same time examining recommendations upon developing public outreach activities which would serve to advance, broaden and deepen awareness and appreciation of geology across a broad cross-section of society: specifically in primary and secondary schools and the general public. The potential for duplication of effort was apparent and so the Steering Group met with NCG working group members (some of whom were on both groups!) to avoid problems.

At some point in 2003 a solution appeared, which although blindingly obvious, had not been thought of in the Steering Group's previous discussions. The solution was not to set up another organisational structure, competing for people's scarce time and resources, but simply to add the efforts and enthusiasm of any "Irish Geodiversity Forum" potential members to the already existing effort of an organisation with a parallel and equivalent objective. The new approach was to pool the limited energies of those involved with an already established and successful group, in an organic and evolutionary way that could develop the strengths of all without excessive bureaucracy and administration. The efforts could go into getting our message across, rather than developing a structure.

Earth Science 2000 (ES2k) was set up in Northern Ireland, as a means of raising awareness of Earth Science generally and specifically in schools and colleges. It was initiated at a time when geological education was under severe threat, with closure of the Geology Department in Queen's University Belfast, and a contraction of A-Level geology teaching opportunities. Although the primary focus was perhaps in Northern Ireland, the clearly defined remit of ES2k was to cover the 12 northern counties of Ireland, reflecting a practical scope of ambition with a new venture.

So by 2003, the group had become well established and respected in geological circles, but more importantly, had become a recognised voice and organisation outside of the geological community. The primary mechanism was the Newsletter, which had rapidly grown into a full colour lively magazine, under the editorship of Tony Bazley, a former Director of the Geological Survey of Northern Ireland. However, a variety of other initiatives had also been developed. The organisation, and especially free distribution of several thousand copies of the magazine to schools and libraries and individuals was funded by sponsorship from a range of organisations including:

Environment and Heritage Service
Geologists' Association (The Curry Fund)
The Geological Society
The Belfast Geological Society

The outcome of discussions held with the Committee of ES2k was that a proposal was put to the membership at the Annual General Meeting of ES2k in September 2003. This proposal allowed a gentle expansion, in an organic manageable way, into an organisation covering all of the island of Ireland, including both political jurisdictions of Northern Ireland and the Republic of Ireland. Although the geology does not respect human political boundaries, the realities of funding sources and educational systems north and south mean that some activities will have to run in a dual way. In most respects, the ambition of ES2k is identical to that of the original vision of the Conference for an "Irish Geodiversity Forum".

In practice, new committee members from Munster, Leinster and Connacht will bring new energy to the expanded outlook. The Magazine will expand its page numbers, its print run and its coverage to all Ireland. New sponsorship will support the free distribution of the Magazine to additional schools, libraries and museums in the Republic of Ireland. The Heritage Council has agreed to support this in 2004, and both the Geological Survey of Ireland and the Geological Survey of Northern Ireland have committed to support ES2k, but additional sponsors are also being sought.

In summary, ES2k is an organisation which can deliver the aspirations of the Dublin Declaration and the Natural and Cultural Landscapes Conference. Through a variety of mechanisms, as both the organiser, and as an 'umbrella' organisation publicising and promoting other activities organised amongst the geological community, it can raise awareness of the importance of geology to many aspects of everybody's lives, and the integral links that geology and geomorphology have as the foundation to the natural and cultural landscapes. We encourage you to become a member and to get involved at whatever level is right for you.

http://www.habitas.org.uk/es2k/